PURINE METABOLISM IN MAN – IV

Part A: Clinical and Therapeutic Aspects;
Regulatory Mechanisms

ADVANCES IN EXPERIMENTAL MEDICINE AND BIOLOGY

PURINE METABOLISM IN MAN—IV

Part A: Clinical and Therapeutic Aspects; Regulatory Mechanisms

Edited by

Chris H. M. M. De Bruyn

University of Nijmegen Faculty of Medicine
Nijmegen, The Netherlands

H. Anne Simmonds

Guy's Hospital Medical School
London, United Kingdom

and

Mathias M. Müller

University of Vienna
Vienna, Austria

Springer Science+Business Media, LLC

Library of Congress Cataloging in Publication Data

International Symposium on Human Purine and Pyrimidine Metabolism (4th:
Maastricht, Netherlands: 1982).
Purine metabolism in man—IV.

(Advances in experimental medicine and biology; v. 165)
Includes bibliographies and index.
Contents: pt. A. Clinical and therapeutic aspects. Regulatory mechanisms—pt. B.
Biochemical, immunological and cancer research.
1. Purines—Metabolism—Congresses. I. Bruyn, Chris H. M. M. De. II. Simmonds,
H. Anne. III. Müller, Mathias M. IV. Series. [DNLM: 1. Purine-
Pyrimidine Metabolism, Inborn Errors—Congresses. 2. Purines—Metabolism—
Congresses. W3 IN918RP 4th 1982p / WD 205.5.P8 I605 1982p]
QP801.P8P864 1983   612′.01579   83-8097
ISBN 978-1-4684-4555-8   ISBN 978-1-4684-4553-4 (eBook)
DOI 10.1007/978-1-4684-4553-4

Proceedings of the Fourth International Symposium on Human Purine and
Pyrimidine Metabolism, held June 13–18, 1982, in Maastrich, The Netherlands

© 1984 Springer Science+Business Media New York
Originally published by Plenum Press, New York in 1984
Softcover reprint of the hardcover 1st edition 1984

Preface

These two volumes, entitled "Purine Metabolism in Man IV" con-
tain the paper presented at the "IV. International Symposium on
Human Purine and Pyrimidine Metabolism," held in Maastricht (The
Netherlands), June 1982. The proceedings of the three previous meet-
ings in Tel Aviv (Israel, 1973), Baden (Austria, 1976) and Madrid
(Spain, 1979) were also published by Plenum Press.

In the past few years interest in purine and pyrimidine metabolism under normal and pathological conditions has been growing rapidly. Apart from the more or less classical topics such as hyperuricaemia, clinical gout and urolithiasis, an increasing number of papers
relating to other fields have been presented at successive meetings.
Knowledge derived from the study of purine metabolism in relation to
lymphocyte function, for instance, has opened up new possibilities
for immunomodulation and leukaemia chemotherapy, with eventual consequences for other types of cancer.

At previous meetings there have been pointers implicating purine
metabolism in relation to normal cardiac and skeletal muscle function.
During the present meeting much new data on both issues have been reported which indicate clear differences in the pathways of ATP metabolism. The widening of the field of interest is also illustrated by
the recent work on infectious disease: exploitation of the differences
in purine metabolic pathways in certain parasites compared with those
in human cells has resulted in new rationales for therapy being developed.

Another extension of the scope is represented by studies on the
molecular genetics of the genes controlling different purine metabolic pathways. For the first time papers from the recombinant DNA
field are included. It was hoped that this meeting would also see
more contributions from the lesser known field of pyrimidine metabolism and the equally important contributions on the overall control
at the cellular level, together with new developments in the field
of the known inherited disorders of purine metabolism, will stimulate further work in these interesting areas.

We are grateful to the members of the Organising and Scientific Committees for the support received during the preparations for the Symposium. We thank the Medical Faculty of the University of Maastricht, especially its Dean, Prof. Dr. J. M. Greep, the Province of Limburg and the City Council of Maastricht for their warm support. We would also like to congratulate the director of Hotel Maastricht, Mr. Benoit Wesly and his staff for their efforts to make the Symposium a success both from the social and scientific point of view.

The meeting would, however, not have been possible without the devoted and cheerful help of the local organising team from the Department of Human Genetics, University Hospital, Medical Faculty, Nijmegen: Ronney De Abreu, Cor van Bennekom, Foppe Brolsma, Winand Dinjens, Sipke Geerts, Joop Jansson, Frank Oerlemans, Gert Spierenburg, and J. Van Laarhoven, and our special thanks are due to them.

C.H.M.M. De Bruyn
H. A. Simmonds
M. M. Müller

Contents of Part A

Adenosine Deaminase Deficiency

Purine Nucleoside Phosphorylase Deficiency

Ecto-5'-Nucleotidase Deficiency

Xanthine Oxidase Deficiency

Myoadenylate Deaminase Deficiency

III. RENAL HANDLING OF URIC ACID

IV.   PURINE AND PYRIMIDINE METABOLISM IN RELATION
TO INFECTIOUS DISEASE

V.   CLINICAL ENZYMOLOGY AND BIOCHEMISTRY

VI.   PARAMETERS OF PURINE AND PYRIMIDINE METABOLISM UNDER
      HEALTHY AND PATHOLOGICAL CONDITIONS

VII.   DIETARY EFFECTS ON PURINE AND PYRIMIDINE METABOLISM

VIII.   PURINE METABOLISM IN ERYTHROCYTES

X.   REGULATORY MECHANISMS

De Novo Purine Synthesis

Purine Interconversions and Catabolism

PROBLEMS IN DIAGNOSIS AND TREATMENT OF ADENINE AND HYPOXANTHINE-GUANINE PHOSPHORIBOSYLTRANSFERASE DEFICIENCY

M.J. Dillon, H.A. Simmonds, T.M. Barratt, L.D. Fairbanks and P.C. Holland

Renal Unit, Hospital for Sick Children, London WC1 and Purine Laboratory, Department of Medicine, Guy's Hospital London SE1, United Kingdom

Adenine (APRT) and hypoxanthine-guanine (HGPRT) phosphoribosyl-transferase deficiency were originally identified in children.[1,2] The spectrum of manifestations in both is broad with similarities, but also important differences. Both are associated with urinary calculi and can cause severe renal damage.[1,2] Severe neurological problems are also found in complete HGPRT deficiency[1] but have not been noted in APRT deficiency. We have investigated 2 children presenting in renal failure that emphasise the problems of diagnosis and treatment of these 2 deficiencies and the particular diffi-culties encountered in the presence of impaired renal function.

PATIENTS

Case 1 (S.B.) A 4-year-old girl was the second child of unrelated healthy parents who lived on vegetarian diets. From 2 years of age she had episodic mild abdominal pain and aged 3 years an isolated occurrence of haematuria. Following a 6 day illness consisting of abdominal pain, vomiting, diarrhoea and increasing drowsiness she became anuric and comatose. On admission she was dehydrated, clinically anaemic, normotensive, deeply comatose but without any focal neurological signs. Serum creatinine was 815 umol/l and urate 0.58 mmol/l. Cerebrospinal fluid was normal and a toxi-cology screen negative. A C.T. brain scan showed diffuse cerebral oedema. A plain abdominal X-ray revealed no radio opaque calculi and there was no excretion of contrast medium by either kidney. There was no vesico-ureteric reflux on a micturating cystogram. Renal ultrasound demonstrated enlarged kidneys, dilated collecting systems with bright echo clusters casting strong acoustic shadows suggestive of bilateral renal calculi confirmed by retrograde pyelography.

Initial treatment consisted of assisted ventilation, peritoneal dialysis, rehydration and blood transfusion. Consciousness was regained and renal function improved. A renal biopsy (at pyelo-lithotomy) examined under polarised light demonstrated numerous bire-fringent crystals suggestive or uric acid. The crystals were in the lumina and epithelial cell cytoplasm of the cortical tubules and also in the interstitium.

The stones and urine were analysed for purine content by methods described previously.[2] The stones were reported to be uric acid by thermogravimetric anlysis but spectral analysis showed 2.8 dihydroxy-adenine. The urine contained 2.8 dihydroxyadenine, 8-hydroxyadenine and adenine in addition to the normal purines. APRT activity in S.B., the mother and sister was measured in lysed erythrocytes. This was initially raised in S.B. because of the recent blood transfusion but 3 months later was nearly undetectable. The mother had heterozygote levels (25% of normal) but the sister was normal. (Table 1)

Treatment was started with allopurinol (5 mg/kg/24h) and a low purine diet. Urinary purines and allopurinol metabolites were measured to monitor treatment. Aluminium hydroxide was administered to control hyperphosphataemia and co-trimoxazole as a prophylactic urinary antiseptic. Renal function remains impaired (plasma creati-nine 160 umol/l) and she has had several episodes of unexplained coma with cerebral oedema since her initial presentation. These have responded to dexamethasone, correction of fluid and electrolyte imbalance and allopurinol. Lack of compliance with allopurinol medi-cation was not confirmed but at least 1 attack of neurological disturbance was linked to excess intake of high purine foods. An adenosine-like substance and other unidentified compounds were found in the urine.

Table 1. APRT Activity In Lysed Erythrocytes (mmol/mg Hb/h)

S.B. (7-10 days after blood transfusion)	11.0
S.B. (3 months later)	<0.01
Mother	5.4
Sister	20.0
Control range	16-32

Case 2 (L.W.) A 5-week-old boy was the first child of healthy unrelated parents. From 3 weeks of age he thrived poorly, had feeding difficulties and was extremely irritable. On admission he was markedly irritable and the thumb and first 2 fingers of the right hand were red, swollen and painful. Neurologically he was slightly hypotonic. Plasma creatinine was 350 umol/l and urate disproportion-ately high at 1.13 mmol/l. Plain abdominal X-ray showed no radio

opaque calculi but renal ultrasound showed that both kidneys were bright suggesting a crystal nephropathy.

Urine uric acid was 0.34 mmol/24h, hypoxanthine 0.004 mmol/24h and xanthine 0.003 mmol/24h. The uric acid/creatinine ratio (mmol/mmol) was 1.17 (ie within the normal range for control children of this age) and the hypoxanthine/xanthine ratio (mmol/mmol) was 1.3. A renal biopsy demonstrated crystal nephropathy in a cryostat section under polarised light. There was much tubular atrophy with extensive tubular epithelial giant cell transformation in the cortex and medulla. All crystals dissolved on fixing in formalin.

HGPRT activity in L.W. was less than 0.01 nmol/mg Hb/h in lysed red cells, less than 0.9 nmol/mg protein/h in skin fibroblasts and less than 0.2% of normal in intact erythrocytes. Table 2. Incorporation of labelled hypoxanthine into purine nucleotides by intact fibroblasts was approximately 6% of control which is low on the criteria of Page et al[5]. These results confirm severe HGPRT deficiency. Heterozygote levels were found in the Mother's fibroblasts but not in the red cells (Table 2). Normal skin fibroblast levels were found in other family members.

Treatment consisted of allopurinol 5-10 mg/kg/24h and sodium bicarbonate. Aluminium hydroxide was given for hyperphosphataemia and Paracetamol for the pain in his fingers. Plasma uric acid on discharge after 7 weeks had fallen to 0.5 mmol/l. Plasma creatinine had dropped from 350 umol/l to 90 umol/l over the same period.

Table 2. HGPRT Activity In Intact Erythrocytes, Lysed Erythrocytes And Skin Fibroblasts

	Intact Red Cells (% Nucleotide formed)	Lysed Red Cells (nmol/mg Hb/h)	Lysed Fibroblasts (nmol/mg protein/h)
L.W.	<0.2%	<0.01	0.9
Mother	-	101	17
Control	>90%	80-130	63-82

DISCUSSION

These 2 patients emphasise some of the diagnostic and therapeutic difficulties encountered when renal failure complicates APRT or HGPRT deficiency, particularly in childhood.

The value of ultrasonography in identifying crystal nephropathy when the diagnosis is obscured by renal failure has been demonstrated in both cases. This is especially relevant when renal function is so

impaired that other radiological techniques would be undesirable
or unhelpful. Surgical biopsy confirmed the crystalline nephropathy in both and the presence of radiolucent stones in the APRT
deficient child. The stones were initially reported as uric acid
when examined routinely. The crystals in both biopsies showed the
characteristic birefringence of uric acid under polarised light.
However, in the APRT deficient child the crystals were spherical
rather than the characteristic needle shape[4] of uric acid and they
were not dissolved out by formalin fixation. In the HGPRT deficient
child's biopsy, by contrast, they were only visible in the cryostat
section. This underlines the importance of the latter in the
identification of uric acid nephropathy. The presence of microtophi
in the interstitium in both patients as well as intraluminal
crystals and extensive tubular and interstitial damage in both is
considered evidence that the crystals reached the interstitium
through tubular disintegration and supports our earlier
observations on the origin of the lesion in the gouty kidney.[4]

The very early onset of clinical gout in L.W. at 5 weeks must
also be unique and as far as we are aware not previously reported at
such a young age. Since infants maintain a low plasma urate through
a high urate clearance the initial insult would appear to be the high
urine urate associated with the enzyme defect and the raised plasma
urate must have been secondary to the subsequent renal damage. The
presence of a high plasma uric acid and so called "uric acid stones"
in the APRT deficient child could also complicate the diagnosis if
sophisticated techniques of stone analysis or facilities for the
assay of purine enzymes had not been available. Furthermore, plasma
urate levels (after the immediate neonatal period) are extremely low
but to those accustomed to adult values the raised levels in S.B. or
L.W. due to the renal failure may not have seemed abnormal. However,
the levels in L.W. were certainly disproportionately high despite
the severe renal damage (see Cameron et al, this symposium).

One of the hallmarks of HGPRT deficiency in infancy is the
characteristically raised ratio of urine uric acid relative to
creatinine and the high hypoxanthine excretion and ratio relative
to xanthine. Again because of the renal failure in L.W. none of
these values were abnormal for age. The necessity of blood transfusion on admission resulted in surprisingly high APRT levels in S.B.
which again would have obscured the diagnosis of the underlying
metabolic abnormality which was only demonstrated by the unusual
presence of adenine and its metabolites in the urine and
2.8 dihydroxy adenine (not uric acid) in the stone. Consequently in
both children detection of the underlying purine metabolic defect
would have been difficult by conventional methodology and criteria.
Whether the high dietary purine (adenine) intake could have
contributed to the nephropathy in S.B. is uncertain but obviously
low purine diets are essential in both children.

Treatment in both children involved the use of allopurinol - but at a lesser dosage (5 mg/kg/24h) because of the retention of the principal metabolite oxipurinol in severe renal failure. Careful monitoring of plasma oxipurinol levels is desirable. For the HGPRT deficient child alkali was also given to enhance uric acid solubility. It was not prescribed in the APRT deficient child since 2.8-dihyroxy adenine solubility is unaffected by alkali and its use may even be contraindicated.

The coma seen initially, as well as subsequently, in S.B. is unexplained and has not been noted in other cases. The appearance of some adenosine-like substance and other unidentified compounds in the urine at the time of the neurological symptoms suggest these may be neurotoxic metabolites occurring because of the combination of the APRT deficiency and the severe renal impairment. Adenine is known to be toxic in vitro. In vivo circulating levels are normally kept at undetectably low levels by the action of APRT and xanthine oxidase plus effective renal clearance in the presence of the enzyme defect. Obviously cases with such reduced renal function must be monitored carefully. There is evidence suggesting that adenine derived compounds have properties which enable them to regulate neuronal function.[3] Whether they might be toxic in the above circumstances is, however, not known. At present the neurological features in L.W. are mild, but it is well know that the long term outcome is impossible to predict! However, the hypotonicity together with the extremely low enzyme activity in intact as well as lysed cells suggests a poor prognosis.[4]

Finally, the difficulty of heterozygote detection[1] in HGPRT deficiency is exemplified in this study. Erythrocyte lysates from the mother and her sisters had normal HGPRT activity, the low fibroblast levels exclusively in the mother suggest a new mutation. By contrast the mother of the APRT deficient child had erythrocyte lysate levels much less than the anticipated 50% of the normal mean (ca 25%). This is characteristic of the defect and will be discussed elsewhere (Wilson et al - this symposium).

In summary these children had features in common which made it difficult to identify the underlying enzyme abnormality and its relationship to their presentation in renal failure. Both had raised plasma uric acids not unusual in severe renal failure but in these children the magnitude was the important factor. The crystals or stones causing the nephropathy could both have been identified as uric acid. Urinary uric acid on a creatinine basis was not raised in either. The severe renal damage at such an early age in both underlines the importance of early and correct identification of these defects, especially APRT deficiency, which can be treated successfully with allopurinol and such severe renal damage need never occur[2].

REFERENCES

1.   W.N. Kelley and J.B. Wyngaarden. The Lesch Nyhan Syndrome. In:
"The Metabolic Basis of Inherited Disease." Ed.   J.B. Stanbury,
J.B. Wyngaarden, D.S. Fredrickson,   P 1011.   McGraw Hill, New York,
4th Edition, 1978
2.   H.A. Simmonds, C.F. Potter, A. Sahota, J.S. Cameron, G.A. Rose,
D.I. Williams, D.G. Arkell and K.J. Van Acker.   Adenine Phosphoribosyltransferase Deficiency Presenting As "Uric Acid" Stones; Pitfalls
Of Diagnosis. J. Roy Soc. Med. 71: 791 (1978)
3.   T. Page, B. Bakay, E. Nissinen and W.L. Nyhan.   Hypoxanthine-
guanine Phosphoribosyltransferase Variants: Correlation of Clinical
Phenotype With Enzyme Activity. J. Inherited Metab. Dis. 4:
177 (1981).
4.   D.A. Farebrother, J.R. Pincott, H.A. Simmonds, D.J. Warren,
M.J. Dillon and J.S. Cameron.   Uric Acid Crystal-Induced Nephropathy:
Evidence For A Specific Renal Lesion In A Gouty Family. J. Pathol.
135: 159 (1981)
5.   T.W. Stone.   Physiological Roles For Adenosine And Adenosine 5'
Triphosphate In The Nervous System. Neuroscience 6: 523 (1981)

PROBLEMS OF DIAGNOSIS IN AN ADOLESCENT WITH HYPOXANTHINE-GUANINE

PHOSPHORIBOSYLTRANSFERASE DEFICIENCY AND ACUTE RENAL FAILURE

J. S. Cameron, H. A. Simmonds, D. R. Webster, V. Wass, and A. Sahota

Renal Unit and Purine Laboratory
Department of Medicine
Guy's Hospital, London, SE1 9RT, U.K.

In hypoxanthine-guanine phosphoribosyltransferase (E.C 2.4.2.8, HGPRT) deficiency[1,2] the clinical manifestations usually parallel the amount of residual enzyme. Thus, patients with the full Lesch-Nyhan syndrome, including mental retardation, athetosis, hypotonia and compulsive self-mutilation, show no detectable enzyme (<0.01 nmol/mg protein/h), whilst those presenting as adult gout usually show low but detectable levels of enzyme.[1,2] However, patients with severe neurological defects and detectable levels of enzyme have been described,[2] as well as at least twelve patients whose red cell lysates lacked enzyme (<0.1 nmol/mg protein/h) but who were neurologically intact.[3-8] We describe a patient who presented initially with acute renal failure, without any neurological manifestations and later developed tophaceous gout. Erythrocyte lysates demonstrated almost undetectable levels of HGPRT activity.

METHODS

All methods used for the estimation of purine levels and enzyme activity in intact as well as lysed cells are referenced in detail in a previous publication.[9] Creatinine values were measured by standard autoanalyser(R) protocol. The lymphoblast line was established using transformation with EB virus, through the kindness of Dr. S. Pereira of the Clinical Research Centre, Northwick Park, Harrow.

Case History

CL (♂, born 9.10.1961) was well until April 1978, when he

7

developed large cervical lymph nodes and was given ampicillin;
shortly afterwards a macular rash appeared and his face swelled;
the drug was stopped, but the rash became worse, affecting the
whole body except hands and feet.   Polyuria appeared, he became
generally unwell and had a plasma creatinine of 830 µmol/l, blood
urea of 95 mmol/l;   at this time his urate was >1.79 mmol/l.   He
was also desalinated, hypotonic and hyporeflexic, and an atrial
septal defect was confirmed at cardiac catheterisation.   After
treatment with intravenous saline he made a rapid recovery, and
by late May creatinine clearance was 105 ml/min (see Table 1), but
plasma urate remained elevated (Table 1).   He was lost to follow-up
for two years, when he presented with acute gout affecting the left
ankle and obvious tophi on both ears.   Further investigation
revealed almost absent HGPRT in erythrocyte lysates (Table 1).
Treatment with allopurinol was started initially 300, then 400mg/24h.
Neurological examination revealed nothing abnornal, except a
history of enuresis until the age of seven.   He is of normal
intelligence but is unemployed.   Until 1981 he continued to have
further attacks of clinical gout, but when last seen in May, 1982,
he was well with a plasma creatinine of 91 µmol/l.

In 1980 enquiry revealed a cousin, AR, with gout and a diagnosis
of Lesch-Nyhan syndrome.   He was born on 2.3.1943, and was
diagnosed by Dr. J. Seegmiller as HGPRT deficient in 1967[3].   He
had an IQ of 77 with mild retardation and epilepsy.   He also had
an episode of megaloblastic anaemia, and spasticity from cervical
cord compression.   His gout is controlled on allopurinol 300mg/24h
and he continues to take anti-epileptic drugs.

RESULTS

Table 1.   Purine concentrations in plasma and urine, and plasma
           creatinine concentrations in CL over a 3-year period:

| Date | State | Plasma | | Urine | | | Oxypurine* |
		Urate mmol/l	Creatinine µmol/l	Uric Acid mmol/24h	H*	X*	mmol/mmol Creatinine
8.5.78	Acute	>1.79	830	–	–	–	–
11.5.78	renal failure.	0.83	230	–	–	–	–
1.6.78		0.53	79	–	–	–	–
12.6.78	Discharge	0.48	75	3.13	0.11	0.07	0.34
22.6.80†Gout		0.57	67	7.92	0.63	0.14	1.10
19.1.81†On allo-purinol		0.28	76	3.16	1.74	1.10	1.03

* Oxypurine = uric acid + hypoxanthine (H) + xanthine (X)

† On a low purine diet.

The initial plasma urate concentration (Table 1) was disproportionately high; levels greater than 0.6-0.9mmol/l in cases
of acute renal failure are rare.   The plasma urate concentration
remained high when renal function had returned to normal (C_{Cr}
115 ml/min).   However, the urinary uric acid, hypoxanthine
and xanthine were normal for age, as was the excretion of total
oxypurines in relation to creatinine excretion on a millimolar
basis.   Only after clinical gout appeared two years later was
purine overproduction demonstrated, with plasma and urinary urate
increased, as were hypoxanthine and xanthine excretion.   The
oxypurine/creatinine ratio was raised.

Table 2. HGPRT and APRT assays in lysed and intact erythrocytes
          in CL and his cousin AR:

	HGPRT*			APRT			
	CL	AR	Control	CL	AR	Control	
Erythrocyte lysate	0.12	<0.1	(180 -130)	40.1	50.0	(16 -32)	nmol/mgHb/h.
Intact RBC	24.3	9.4	(>90.%)				% conversion of base to nucleotide.
Cultured lymphoblasts	298* 28	– –	(212) (158 -361)				nmol/mg protein/h.

* hypoxanthine or guanine as substrate.

Lysates of cells from CL showed negligible conversion (mean 0.12
nmol/mg Hb/h) at protein concentrations from 0.2 to 7.1mg.   AR
showed undetectable levels (< 0.1 nmol/mg Hb/h).   In intact red
cells from CL, in contrast, 24% of either hypoxanthine or guanine
were converted to the corresponding nucleotide, compared with 93%
and 94.5% in control cells, respectively.   The cousin AR showed
almost 10% conversion under similar conditions (5 µmolar substrate,
40' incubation with 18 mM Pi).   APRT activity was increased in
erythrocyte lysates from both, as anticipated[1,2] in HGPRT deficiency.

Leucocyte extracts from CL showed negligible activity (<0.02nmol/
mg protein/h - not shown).   In lymphoblasts, Dr. J. Allsop
(Clinical Research Centre, Northwick Park) found HGPRT activity
to be 298 nmol/mg protein/h, compared with control of 212 nmol/mg
protein/h.   Subsequent cultures from the same line gave mean
values in our laboratory of 28 nmol/mg protein/h;   extracts of a
separate cell line, established by Dr. J. M. Wilson, University of
Michigan, gave values of 1-2% of control.   Studies using intact
lymphoblasts from CL have shown up to 50% incorporation of

hypoxanthine into nucleotides, whilst a control cell line from an
unrelated patient with a full Lesch-Nyhan syndrome failed to show
any incorporation (<1%).

Detailed studies of the lymphoblast enzyme by Dr. J. M. Wilson
(personal communication) showed no increase in the presence of
elevated substrate concentrations, ruling out a Km mutation[10].
The concentration of immunoreactive enzyme protein was less than
5% of normal (Wilson et al, in preparation).

DISCUSSION

This patient illustrates some of the difficulties in the diagnosis
of HGPRT deficiency when the patient presents in acute renal failure,
the clinical syndrome falls short of the full Lesch-Nyhan syndrome,
and especially highlights the problems attendant on the evaluation
of HGPRT activity from erythrocyte lysates alone.

In general, individuals with an HGPRT activity in erythrocyte
lysates exceeding 1% of normal (ca 1 nmol/mg protein/h) have only
gout.[1,2,4]   Several patients have been reported, however, with
activities in excess of this and a full Lesch-Nyhan syndrome.[2]
Of more relevance to the present case are the rare examples of the
opposite situation, in which patients with normal intelligence
and neurology are found to have very low activities of HGPRT in
red cell lysates.   At least twelve patients with HGPRT activities
of 0.1 nmol/mg Hb/h or less but without neurological manifestations
have been described.[3-8]   In three of these patients, enzyme activity
could be demonstrated in whole nucleated cells, as in the present
case, and in vitro instability of the mutant enzyme was demonstrated.[5,8] The concentration of enzyme is highest in young red
cells (which are often discarded with the buffy coat), and
declines with time.   In another case[6] no activity was demonstrated
in intact fibroblasts; in the remainder, studies on intact cells
were not available.   After many earlier studies on red cells and
fibroblasts,[1] Wilson et al,[10] using the larger amounts of enzyme
available from lymphoblastoid cells in culture, have demonstrated
different mutant enzymes of varying physical, immunochemical and
kinetic characteristics in unrelated patients with HGPRT deficiency.
In our patient, a similar unstable enzyme to that described by
Dancis et al[5] seems likely, present only in small amounts (ca 5%
of normal), but above the threshold necessary to protect against
the appearance of neurological symptoms.   In the few patients
whose brains have been examined, with a full Lesch-Nyhan syndrome,
HGPRT activity was undetectable[1,11].

Our data indicate that caution is necessary in the interpretation
of low or absent enzyme activities in erythrocyte hemolysates, and
must be checked in intact cells.   This was emphasised by Sperling
et al[12] more than ten years ago, and subsequently by others[13]

including de Bruyn,[2] but has not been widely applied.   More
recently, Page et al[13] have shown that fibroblast incorporation
into nucleotides correlates well with clinical severity of disease
over the whole range of expression of HGPRT deficiency.

From the clinical point of view, the diagnosis here was initially
obscure, the nature of the presenting illness giving no clue as to
the underlying abnormality.   In the episode of acute renal failure
plasma urate was disproportionately high, and high urinary uric
acid concentration may have contributed to the acute renal failure,
in addition to dehydration from severe glandular fever.   Urate
production may have been exaggerated by the illness; white cell
count rose to above 20,000 /μl,   with 11% atypical lymphoblasts:
a virally-induced "leukemoid" white cell reaction.

The diagnosis was rapidly established when the patient later
developed tophaceous gout, but it is important to note that
immediately after discharge from hospital, the patient was <u>not</u>
overproducing purines.   From 1981 onwards, overproduction was
clearly present, more than a mole of oxypurine being excreted per
mole of creatinine whilst the patient was on a low purine diet.
Hypoxanthine excretion has remained above xanthine excretion
throughout, as would be expected.[1,14]

In summary, the presentation of patients with partial deficiency of
HGPRT may be more varied than is usually described, and before
absence or near absence of HGPRT activity is accepted, intact
erythrocytes and/or nucleated cells must be examined.   We suggest
that intact erythrocyte studies, together with red cell GTP levels
(see Simmonds et al, this volume) may provide a better guide to
prognosis in affected subjects.

REFERENCES

1. W. N. Kelley, J. B. Wyngaarden.   The Lesch-Nyhan Syndrome.
   In: The Metabolic Basis of Inherited Disease.   J. B. Stanbury,
   J. B. Wyngaarden, D. S. Fredrickson, eds., McGraw-Hill Book Co.,
   New York.   4th edition, p.1011 (1978).

2. C. H. M. M. de Bruyn.   Hypoxanthine guanine phosphoribosyltransferase deficiency.   <u>Hum.Genet</u>, 31: 127 (1976).

3. W. N. Kelley, M. L. Greene, F. M. Rosenbloom, J. F. Henderson,
   J. E. Seegmiller.   Hypoxanthine-guanine phosphoribosyltransferase deficiency in gout.   <u>Ann.Intern</u>.Med., 70: 165 (1969).

4. B. T. Emmerson, L. Thompson.   The spectrum of hypoxanthine-
   guanine phosphoribosyl transferase deficiency.   <u>Quart.J.Med.</u>,
   42: 423 (1973).

5. J. Dancis, L. C. Yip, R. P. Cox, S. Piomelli, M. E. Balis. Disparate enzyme activity in erythrocytes and leucocytes; a variant of hypoxanthine-guanine phosphoribosyl transferase deficiency with an unstable enzyme. J.Clin.Invest., 52: 206 (1973).

6. R. A. Geerdink, W. H. M. de Vries, J. Willemse, T. L. Oei, C. H. M. M. de Bruyn. An atypical case of hypoxanthine-guanine phosphoribosyl transferase deficiency (Lesch-Nyhan syndrome) I. Clinical Studies. Clin.Genet., 40: 348 (1973).

7. D. N. Buss, I. K. Moss, A. Nicholls, J. T. Scott, M. L. Smith, M. R. Watson. Clinical and biochemical observations on three cases of hypoxanthine-guanine phosphoribosyl transferase deficiency. Ann.Rheum.Dis., 34: 249 (1975).

8. B. Bakay, E. Nissenen, L. Sweetman, U. Francke, W. L. Nyhan. Utilisation of purines by an HGPRT variant in an intelligent, non-mutilative patient with features of the Lesch-Nyhan syndrome. Pediatr.Res., 113: 1365 (1979).

9. H. A. Simmonds, A. R. Watson, D. R. Webster, A. Sahota, D. Perrett. GTP depletion and other erythrocyte abnormalities in inherited PNP deficiency. Biochem Pharmac. 31: 941 (1982).

10. J. M. Wilson, B. W. Baugher, P. M. Mattes, P. E. Daddona, W. N. Kelley. Human hypoxanthine-guanine phosphoribosyltransferase. Cells derived from patients with a deficiency of the enzyme. J.Clin.Invest., 69: 706 (1982).

11. R. W. E. Watts, E. Spellacy, D. A. Gibbs, J. Allsop, R. O. McKeran, G. E. Slavin. Clinical, post mortem, biochemical and therapeutic observations on the Lesch-Nyhan syndrome with particular reference to the neurological manifestations. Quart.J.Med. 51: 43 (1982).

12. O. Sperling, G. Eilam, R. Schmidt, G. Mundel, A. de Vries. Purine hypoxanthine-guanine phosphoribosyl transferase deficiency. Biochem.Med., 5: 173 (1971).

13. T. Page, B. Bakey, E. Nissinen, W. L. Nyhan. Hypoxanthine-guanine phosphoribosyl transferase variants: correlation of clinical phenotype with enzyme activity. J.Inher.Metab.Dis., 4: 203 (1981).

14. L. B. Sorensen. Mechanism of excessive purine biosynthesis in hypoxanthine-guanine phosphoribosyltransferase deficiency. J.Clin.Invest., 49, 968 (1970).

HYPOXANTHINE-GUANINE PHOSPHORIBOSYL TRANSFERASE (HGPRT) DEFICIENCY IN A GIRL

N. Ogasawara, S. Kashiwamata, H. Oishi, K. Hara*,
K. Watanabe*, S. Miyazaki*, T. Kumagai*, and S. Hakamada*

Institute for Developmental Research and
*Central Hospital, Aichi Prefectural Colony, Kasugai,
Aichi 480-03, Japan

INTRODUCTION

Lesch-Nyhan disease is an X-linked recessive disorder characterized by hyperuricemia, physical and mental retardation, choreoathetosis, and compulsive self-mutilation. The disease is associated with absence of activity of an enzyme involved in purine metabolism, namely hypoxanthine guanine phosphoribosyl transferase (HGPRT), and is believed to affect male only. We present here, however, an unusual case of a girl with the Lesch-Nyhan syndrome, whose mother is not heterozygous for a deficiency of the enzyme.

Case Report

Detail clinical data will be reported elswhere. Briefly, the girl (S. S.) was born on February 2, 1975, weighing 2675 g and with the body length of 48 cm in the 39th gestational week, The 33-year-old mother and the 47-year-old father were both healthy. At the age of 6 months she visited our hospital complaining of unstable head control. Athetoid movement of hand was observed at the age of one year and 7 months. Ballisms became apparent at the age of 2 years and 9 months. At the age of 5 years her lower lip was partially lost, and further, mild dysarthria, athetoid movement, intermittent scissoring legs and hyperactive deep tendon reflexes were noted. Serum uric acid ranged 7.8-9.5 mg/dl and the urinary uric acid/creatinine ratio ranged 2.4-3.5.

METHODS

　　Erythrocytes prepared from heparinized venous blood were washed twice with 0.154 M NaCl. The lysates were prepared by two cycles of freezing and thawing, dialyzed and centrifuged. The supernatant was used for assay of HGPRT and APRT activities.

　　Fibroblasts were harvested by brief incubation at 37° C in 0.05 % trypsin in PBS. Cells were collected by centrifugation and washed with 0.154 M NaCl. Extracts were prepared by 4 cycles of freezing and thawing. Samples of the supernatant after centrifugation were used for HGPRT or APRT activity, and assays of protein content. The enzyme activities were determined in the presence of 100 μM α, β-methylene-adenosine-5'-diphosphate an inhibitor of 5'-nucleotidase as described by Hershfield et al[1].

　　Cell fusion was mediated by UV inactivated Sendai virus supplied by Dr. Y. Okada (Osaka Univ.).

　　For autoradiography, cells were plated onto coverslips in Falcon plastic dishes. After 24 hr of growth, H^3-hypoxanthine (1.5 C/μmole) was added at a concentration of 5 μC/ml. After 16 hr of growth, coverslips were removed, washed with saline, and fixed in methanol for 5 min. The coverslips were then treated with ice cold 5 % trichloracetic acid for 25 min, washed in water, and dried. The coverslips were dipped in Sakura NR-M2 liquid emulsion, and after exposure for 1 week, the autoradiographs were developed.

RESULTS AND DISCUSSION

HGPRT and APRT Activities in Erythrocytes and Fibroblasts

　　HGPRT and APRT activities in erythrocytes from normal individuals, the patient and her parents are shwon in Table 1. The patient had no detectable HGPRT activity, but had a significantly higher APRT activity, characteristic of the Lesch-Nyhan syndrome.

　　Activities of HGPRT and APRT in cultured skin fibroblasts from the normal, a male patient with Lesch-Nyhan syndrome, heterozygous females, the proband and her mother were shown in Table 2. Like the male patient, the girl showed a virtually complete deficiency of HGPRT activity. The activities of two heterozygotes were about one half of the activity of normal individuals. The patient's mother had apparently normal level of HGPRT activity.

The patient is undoubtly the female

　　The patient is the female, since firstly her external genitalia were those of a normal female, Secondly, the chromosomal analysis of lymphocytes by staining with acridine orange after treatment with deoxybromouridine showed a normal female karyotype, 46, XX, including one late replicating X chromosome. Thirdly,　no Y body

Table 1. HGPRT and APRT activities in erythrocytes

	HGPRT	APRT
	nmoles/min/mg Hb	
Normal (9)	0.173 ± 0.26	0.323 ± 0.075
Patient	< 0.003	0.809
Mother	1.19	0.272
Father	1.58	0.227

Table 2. HGPRT and APRT activities in fibroblasts

	HGPRT	APRT
	nmoles/min/mg prot.	
Normal 1	1.13	1.74
2	1.48	2.16
3	1.17	2.31
Known heterozygote 1	0.455	2.29
2	0.550	2.11
Male Lesch-Nyhan	< 0.003	1.65
Patient	< 0.003	1.64
Mother a	1.25	2.14
b	1.27	2.47
c	1.18	2.39

was detected in interphase nuclei or mitotic chromosomes by the quinacrine-fluorecsence staining.

The patient's mother is not heterozygous for a deficiency of HGPRT

The girl's mother was identified as normal, not as hetero-zygote for deficiency of HGPRT on the basis of following results. The enzyme activities of the fibroblasts obtained from three skin biopsy specimens from the mother were normal, while those from known heterozygotes were approximately half of the normal (Table 2). The heterozygosity of the mother was also not supported on the basis

of demonstration of only one population of cells by cell selection;
no 6-thioguanine resistant cell was detected in the mother's
fibroblasts from three different skin specimens[2] and thus, none of
the mother's fibroblast is deficient in HGPRT. Furthermore, the
ratios of HGPRT/APRT in the single hair follicle were almost in the
normal range, and there was no HGPRT negative follicle. These results clearly showed that the mother is not heterozygous for a
deficiency of HGPRT, but apparently normal in HGPRT gene.

Genetic Mechanism of HGPRT Deficiency in the Girl, Whose Mother is

not a Heterozygote and Whose Father is the Normal

The most attractive is that the deficiency of HGPRT activity
in this female patient with Lesch-Nyhan syndrome represents a mutation in a regulatory gene. The existence of a regulatory gene
necessary for the expression of the HGPRT locus is suggested by
Watson et al[3], Bakay et al[4] and Croce et al[5]. If the girl is deficient in a regulatory gene, fusion of her cells with the cells from
the male Lesch-Nyhan syndrome would furnish the regulatory gene and
the fused heterokaryons would produce active enzyme. We have done
this using fibroblasts with Sendai virus. Deficiency of HGPRT activity in the male patients is apparently due to a defect in the
structure gene for HGPRT, since their mothers are evidenced as het-
roztgotes. The resulting cells were grown and examined for HGPRT
activity by autoradiography after 2, 3, 5 and 7 days. In a parallel
experiments, the female HGPRT deficient fibroblasts were fused with
LTK$^-$, a derivative of the mouse L cell lacking thymidine kinase,
and the cells were selected in HAT medium for 7 days, during which
TK$^-$ cells were all killed. The survived cells were then replated,
grown for 24 hr and autoradiographed. These experiments showed
that the cells from the female patient/LTK$^-$ fusion were well labeled
following exposure to H^3-hypoxanthine, whereas none of the cells
from the female patient/the male patient fusion were significantly
labeled. These results suggest that there was no complimentary
nature between the female and the male patients with Lesch-Nyhan
syndrome, and that the deficiency of HGPRT in the female patient is
not due to the mutation of a regulatory gene for the expression of
HGPRT.

Also, mouse/human fusion experiments were carried out to select
the hybrids which retain the patient X chromosome, and to test if
the HGPRT gene of patient is expressed in mouse cells, and both
mouse and human forms of HGPRT are present in the hybrid cells.
The hybrids were derived from patient skin fibroblasts and LTK$^-$,
lacking thymidine kinase but wild type of HGPRT. Selection for
mouse/human hybrids was carried out in HAT medium. After 1 month,
hybrid clones were picked with glass cylinder and transferred to
individual dishes. Of the 21 clones selected on the basis of thymidine kinase activity, only a single clone, C7, had human glucose-6-

phosphate dehydrogenase, indicating the presence of human X chromosome. Analysis of the HGPRT pattern from this hybrid is in progress to test whether there are the mouse, heteropolymer and human isozymes as the result of human HGPRT gene expression, or whether the clone has neither human HGPRT nor heteropolymer.

Another possibility could be that dificiency of HGPRT activity in our patient with Lesch-Nyhan syndrome is due to a structural gene mutation of both maternal and paternal chromosomes. Although Hooft et al[6] reported a girl with Lesch-Nyhan syndrome, her HGPRT activity was not determined. Therefore, our case is probably the first as to a female case with Lesch-Nyhan syndrome evidenced clinically and biochemically. Assuming the incidence of $1/3-5 \times 10^4$ for HGPRT deficiency in the male and one third being the new mutation[7], it is possible to calculate the incidence of the female HGPRT deficiency due to simultaneous mutation of both maternal and paternal genes as 1 for $3.24-9 \times 10^{10}$. Of course, it should be very rare but not an impossible incident.

Another possible cause for deficiency of HGPRT activity in the female patient is the mutation of HGPRT gene on either maternal or paternal X chromosome and then nonrandom-selective, instead of random, inactivation of the X chromosome carrying the normal HGPRT gene, in a very early developmental stage. However, it should be very rare.

It is also possible that the half chromatid mutation[8] in the father's HGPRT gene and the simultaneous structural gene mutation of maternal X chromosome. For this, the masaicism must be determined in the father's cells.

REFERENCES

1. M. S. Hershfield, F. F. Snyder, and J. E. Seegmiller, Adenine and adenosine are toxic to human lymphoblast mutant defective in purine salvage enzyme, Science 197:1284 (1977).
2. B. B. Migeon, X-linked hypoxanthine-guanine phosphoribosyl transferase deficiency: Detection of heterozygotes by selective medium, Biochem. Genet. 4:377 (1970).
3. B. Watson, I. P. Gormley, S. E, Gradiner, H. J. Evans, and H. Harris, Reappearance of murine hypoxanthine phosphoribosyltransferase activity in mouse A9 cells after attempted hybridization with human cell lines, Exptl. Cell Res. 75:401 (1972).
4. B. Bakay, C. M. Croce, H. koprowski, and W. L. Nyhan, Restoration of hypoxanthine phosphoribosyl transferase activity in mouse IR cells after fusion with chick-embryo fibroblasts, Proc. Natl. Acad. Sci. USA 70:1998 (1973).
5. C. M. Croce, B. Bakay, W. L. Nyhan, and H. Koprowski, Reexpression of the rat hypoxanthine phosphoribosyltransferase gene in rat-human hybrids, Proc. Natl. Acad. Sci. USA 70:2590 (1973).

6. C. Hooft, C. Van Nevel, and A. F. De Schaepdryver, Hyperuricos-
    uric encephalopathy without hyperuricaemia, Arch, Dis. Childh.
    43: 734 (1968).
7. J. B. S. Haldane, The rate of spontaneous mutation of a human
    gene, J. Genet. 31: 317 (1935).
8. S. M. Gartler, and U. Francke, Half chromatid mutations: Trans-
    mission in humans?, Am J. Hum. Genet. 27: 218 (1975).

CLINICAL AND BIOCHEMICAL CORRELATES OF A NEW HPRT MUTATION

H. E. Gruber, M. Vuchinich, T. A. Marlow, M. M. Plent,
R. C. Willis and J. E. Seegmiller

Dept. of Medicine, University of California San Diego
La Jolla, California 92093

J. Bartley, J. W. Hanson and H. Zellweger

Division of Medical Genetics, University of Iowa
Iowa City, Iowa 52242

In the biochemical analysis of patients with Lesch-Nyhan disease the activity of HPRT found in cell lysates does not always correlate with HPRT activity in the intact cell (1-3). Neither analysis consistently predicts the degree of neurologic impairment in the patient (4,5).

Evaluation of a patient (D.D.), with many of the neurologic complications of the Lesch-Nyhan syndrome has revealed the most extreme discrepancy we have yet encountered between lysed cell and intact cell HPRT activity. This manuscript will describe the clinical features of the patient and the initial biochemical evaluation of HPRT activity in various cell types.

The patient has a less severe clinical expression of the Lesch-Nyhan disease. Early development was somewhat delayed, as he did not sit until age one year and did not walk until 20 months of age. He had no words until two years of age. At age five he was extensively evaluated for developmental delay. He had slightly below normal intelligence and mild spasticity with a "cerebral palsy" gait. His reflexes were somewhat exaggerated and he could not perform rapid movements of the tongue. He was dysarthic and about 75 to 80% of his conversational speech could be understood. He has continued his education in a regular classroom, however, he receives special tutoring, especially for dyslexia. Apparently the patient's motor skills have improved over the past few years and he can now run. Currently, on physical examination, he displays mild spastic paraplegia and poor coordination. He has not had

uric acid nephrolithiasis but had red blood cells in his urine at
age 9. There is no history of self-mutilatory or aggressive behavior but he bites his fingernails when frightened.

The patient was diagnosed, at age 10, by a program designed
to screen for Lesch-Nyhan syndrome patients in clinics and hospitals
caring for individuals with cerebral palsy or learning disabilities.
A dipstick of his urine revealed an elevated uric acid to creatinine
ratio of 2.57 and 2.43 (normal range at age 10 is 0.22 - 0.94).
Follow-up evaluation revealed a serum uric acid at 9.4 mg/dl and a
24-hour urinary urate of 1410 mg with 590 mg creatinine. HPRT activity in a erythrocyte lysate was less than 0.2 nmole/mg/hr (See
Table I). However, on radioautography, ^{3}H-hypoxanthine incorporation
into the patient's fibroblasts could not be distinguished from incorporation into fibroblasts of normal individuals.

TABLE I

HPRT ACTIVITY (nmole/mg/hr)

	Patient D.D.	Normal Range
Erythrocyte lysate	<0.2 nm/mg/hr	50-100
Lymphoblast lysate	<0.2 nm/mg/hr	86-135
Fibroblast lysate	<0.2 nm/mg/hr	71-113

Fibroblast and lymphoblast lysates have <0.2 nmole/mg/hr.
In contrast, quantitative ^{3}H-hypoxanthine uptake by the patient's
fibroblasts was 44% of normal while his lymphoblasts incorporated
18% of the amount shown by normal lymphoblasts (See Table II).

Prenatal monitoring of a maternal aunt of D.D. was recently
accomplished. The fetus was male and his amniotic cells were normal by radioautography with ^{3}H-hypoxanthine. Amniotic cell lysates
also had normal HPRT activity with 583 nm/mg/hr as compared to the
normal range of 528-583 nmole/mg/hr. The normal prenatal diagnosis
was confirmed by demonstrating a normal HPRT activity of 59 nmole/
mg/hr in an erythrocyte lysate of cord blood obtained at delivery.

The extreme disparity between undetectable HPRT activity in
cell lysates and mild deficiency in intact cells presents several
important problems. On a clinical basis, screening for affected
patients by radioautography or other assays involving intact cells
can yield a falsely normal result. Furthermore, heterozygote detection by conventional radioautography will not be valid in this

family. Also, if conditions for optimizing HPRT enzymatic activity in lysates were determined, perhaps application of these principals could result in amelioration of the clinical symptoms in this patient. Biochemically, it is important to determine whether the discrepancy between assays of lysates and intact cells is due to either an extremely unstable enzyme or to differences in the conditions of the assay such as changes in pH or substrate concentrations.

TABLE II

QUANTITATIVE ^{3}H-HYPOXANTHINE UPTAKE

Origin of Cell	Cell Type	^{3}H-hypoxanthine Uptake* cpm/cell/24 hr	% Normal Uptake
D.D.	Fibroblast	2.75	44%
Normal	Fibroblast	6.26	100%
D.D.	Lymphoblast	0.178	18%
Normal	Lymphoblast	1.00	100%

*Cells were grown for 24 hours in MEM media for fibroblasts and RPMI-1640 for lymphoblasts containing 10 percent dialyzed fetal calf serum to which 4μCi/ml (10 Ci/mM) ^{3}H-hypoxanthine was added. Radioactivity in nucleotides absorbable on Whatman DE-81 from the lysates of washed cells was determined in a liquid scintillation counter.

Future work will be directed at determining the cause of the discrepancy between the lysate and intact cell assays. The biochemical findings will be applied toward devising an accurate method of heterozygote screening and hopefully toward means of enhancing or stabilizing the HPRT activity in the affected patient.

ACKNOWLEDGEMENTS

Research supported in part by PHS grants GM17702, AM13622, HD10847, The Kroc Foundation and the Clayton Foundation. Dr. Gruber is a Fellow of the Arthritis Foundation.

REFERENCES

1. Holland, M. J. C., A. M. DiLorenzo, J. Dancis, M. E. Balis, T. F.
    Yu, and R. P. Cox. Hypoxanthine phosphoribosyltransferase
    activity in intact fibroblasts from patients with X-linked
    hyperuricemia. J. Clin. Invest. 57: 1600-1605, 1976.

2. Dancis, J., L. C. Yip, R. P. Cox, S. Piomelli and M. E. Balis.
    Disparate enzyme activity in erythrocytes and leukocytes.
    J. Clin. Invest. 52: 2068-2074, 1973.

3. Emmerson, B. T., and L. Thompson. The spectrum of hypoxanthine-
    guanine phosphoribosyltransferase deficiency. Quart. J. Med.
    42: 423-440, 1973.

4. Uitendaal, M. P., C.H.M.M., deBruyn, T. L. Oei, and P. Hosli.
    Molecular and tissue-specific heterogeneity in HPRT deficiency.
    Biochem. Genetics 16: 1187-1202, 1978.

5. Page, T., B. Bakay, E. Nissinen and W. Nyhan. Hypoxanthine
    guanine phosphoribosyltransferase variants: Correlation of
    clinical phenotype with enzyme activity. J. Inherit. Metab. Dis.
    4: 203-206, 1981.

ENZYME REPLACEMENT IN THE LESCH-NYHAN SYNDROME WITH LONG-TERM ERYTHROCYTE TRANSFUSIONS

N. Lawrence Edwards, Walter Jeryc, and Irving H. Fox

The Human Purine Research Center, Departments of Internal Medicine and Occupational Therapy, The University of Michigan School of Medicine, Ann Arbor, Michigan, USA

INTRODUCTION

The pathogenesis of the neuro-behavioral abnormalities associated with the Lesch-Nyhan syndrome remains obscure despite recent reports of neurotransmitter abnormalities in these patients[1]. Many attempts to correct the characteristic manifestations of spasticity, mental retardation, choreoathetosis, and compulsive self-mutilation have been reported but none have reported sustained clinical efficacy. Many pathogenic mechanisms have been proposed over the past two decades to explain the relationship between the known aberration in purine metabolism and the observed neurologic dysfunction. One of these proposed mechanisms is that the absence of the purine salvage pathway in the central nervous system (CNS) results in (1) the accumulation of oxypurines in the spinal fluid which then may act as toxic endogenous mediators[2] and (2) the depletion of guanine and adenine nucleotides that are important to normal CNS function[3-5]. Supplementation of purine intermediates with dietary adenine, guanosine, inosine, and GMP have not altered the clinical course of the disease.

Two previous attempts at enzyme replacement therapy in the Lesch-Nyhan syndrome have used transfusions with normal (HGPRT+) erythrocytes[6,7]. In both cases single transfusions were performed and the period of reported follow up was short. In the present study we describe a four-year experience with repeated erythrocyte transfusions in three boys with Lesch-Nyhan syndrome.

METHODS AND RESULTS

Two boys with classic features of Lesch-Nyhan syndrome and no
detectable erythrocyte HGPRT (ages 4 and 8 years at the time of entry)
and one boy with a strong family history of Lesch-Nyhan syndrome,
mild spasticity, uric acid crystaluria and no erythrocyte HGPRT (age
8 months) were treated in the Clinical Research Center with the
protocol outlined in Table 1.

Table 1.   Protocol for HGPRT Replacement Therapy

---

1.   Transfusion Therapy
            - 3 Lesch-Nyhan patients (ages 0.7 to 8 years)
            - 1-2 units (250 to 500 ml) packed RBC's
            - Every 2 months for 2 to 4 years
            - Maintenance drugs (diazepam and allopurinol)

2.   Biochemical Evaluation
            - Erythrocyte (pre- and post-transfusion)
              HGPRT levels
            - Serum, urine, and CSF purines
            - Isotopic evaluation of purine catabolic rate

3.   Neuro/Behavioural Evaluation
     Videotape sessions every 2 months
            - Gessell Developmental score
            - Milleni-Comparetti scale
            - Alpen-Boll developmental test

Erythrocyte HGPRT levels were maintained at 15 to 70% of normal
enzyme activity (72 mmol/hr·mg prot).   The effect of the transfusions
on purine metabolism was monitored by serum urate, urinary purines
and CSF oxypurine values (Table 2).

In general, motor skills, adaptive skills, and social skills
progressed at a rate commensurate with the natural course of the
disease. Two patients showed modest improvement in speech and self-
care while the third subject (A.C.) had an actual decrease in athe-

Table 2.   Effect of Transfusion Therapy on Purine Metabolism

Patient		Serum Urate (mg %)	Urinary Purines (µmol/mg creat)	CSF[a] Oxypurines (µM)
J.H.	pre-treatment	9.1	15.5	33.2
	after 30 months	8.0	14.0	17.0
W.S.	pre-treatment	10.8	15.9	28.4
	after 26 months	9.5	18.9	31.5
A.C.	pre-treatment	8.3	19.6	19.1
	after 16 months	8.3	29.9	13.0

a. Post-treatment values of CSF oxypurines were all made after
    16 months of therapy.

tosis and hypertoxicity.   Transfusion therapy had no substantial results
effect on systemic purine metabolism as measured by the turnover rate
of the adenine nucleotide pool following radiolabeling with [^{14}C]-
adenine.

CONCLUSION

This study shows that long-term enzyme replacement therapy with
partial exchange transfusions in the Lesch-Nyhan syndrome results in
only modest improvement in the characteristic neurologic symptoms.

REFERENCES

1. K.G. Lloyd, O. Hornykiewicz, L. Davidson, et al., Biochemical
    evidence of dysfunction of brain neurotransmitters in the Lesch-
    Nyhan syndrome, N. Engl. J. Med. 300:1106 (1981).
2. E, Boyd, M. Dolman, L. Knight, and E. Sheppard, The chronic oral
    toxicity of caffeine, Canad. J. Physiol. Pharmacol. 43:995 (1965).
3. D. Rosenburg, P. Monnet, J. Mamelle, M. Colombel, B. Salle, and
    M. Bovier-Lapierre. Encephalopathie avec troubles du metabolisme
    des purines, Presse Med. 76:2333 (1968).

4. W.N. Kelley, M.L. Greene, F.M. Rosenbloom, J.F. Henderson, and
    J.E. Seegmiller, Hypoxanthine-guanine phosphoribosyltransferase
    deficiency in gout, Ann. Intern. Med. 70:155 (1969).
5. P.J. Benke and J. Anderson, Use of folic acid, adenine and bicar-
    bonate in newborn twins with the Lesch-Nyhan syndrome, Pediatr.
    Res. 3:356 (1969).
6. R.W. Watts, R.O. McKeran, E. Brown, T.M. Andrews, and M.I.
    Griffiths, Clinical and biochemical studies on treatment of
    Lesch-Nyhan syndrome, Arch. Dis. Child. 49:693 (1974).
7. A.J. Pawlak, J.S. Zaremba, J. Barankiewicz, E. Zdzienicka, and
    B. Czartoryska, Effect of blood transfusion on activities of
    hypoxanthine-guanine phosphoribosyltransferase and adenine
    phosphoribosyltransferase in circulating red blood cells of a
    patient with Lesch-Nyhan syndrome, Acta. Med. Pol. 19:331 (1978).

ADENINEPHOSPHORIBOSYLTRANSFERASE (APRT)-ACTIVITY IN PATIENTS WITH NEPHROLITHIASIS OR RENAL FAILURE

P. Banholzer, W. Gröbner, W. Löffler, S. Reiter, and N. Zöllner

Medizinische Poliklinik der Universität München

Pettenkoferstr. 8a, 8ooo München 2

INTRODUCTION

The homozygous state of APRT-deficiency is associated with high urinary levels of 2, 8-dihydroxyadenine (2, 8-DHA) and may lead to nephrolithiasis and renal failure (Cartier and Hamet 1974; van Acker et al., 1977). Heterozygous patients are usually without symptoms (Fox et al., 1977). Recently Kuroda et al. (1981) reported of heterozygous patients with elevated 2, 8-DHA excretion and 2, 8-DHA stones.
Because of the high incidence of low values of APRT and the difficulties in diagnosing the defect it was suggested that the homozygous state of APRT-deficiency might be a frequently undiagnosed defect in patients with renal failure and/or nephrolithiasis (Simmonds 1979, Gault et al. 1981).

METHOD

We measured APRT-activity from erythrocytes lysates of 524 unselected patients of the Medical Policlinic Munich micro-radiochemically using the method of Kelley (1967). Moreover the activity in 127 patients with renal failure on dialysis, and of 41 patients with nephrolithiasis was determined. The family of a patient with low APRT-activity was further investigated. The enzyme of patients with diminished activity was investigated by measuring heat and cold stability at 55°C and 4°C respectively.

In addition we determined the Michaelis constant for adenine and phosphoribosylpyrophosphate.

RESULTS AND DISCUSSION

APRT-activity of erythrocytes is shown in fig. 1.

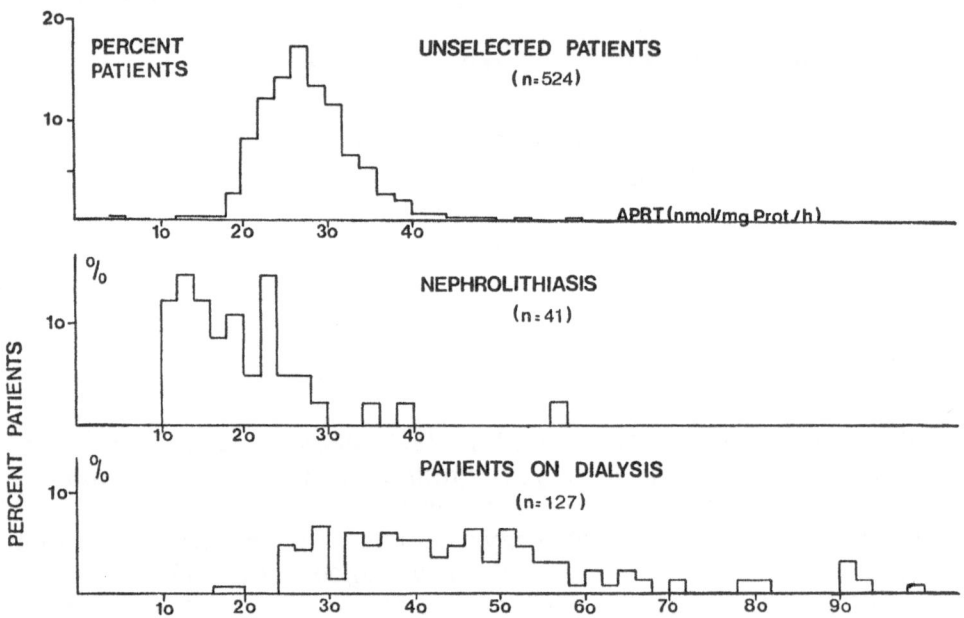

Fig. 1: APRT-activity (nmol/mg/h) in erythrocytes lysates of 524 unselected patients of the Medical Policlinic Munich, 127 uremic patients mostly on dialysis, and 41 patients with nephrolithiasis. Patients with an enzyme activity between two even values (for instance 10 to 11, 9 nmol/mg/h) were combined in one column. The number of patients is shown in percentage figures relating to each group.

The average activity in 524 unselected patients was 28, 9+5, 7 (SD) nmol/mg/h. Four patients (0, 8 %) produced an activity below the 2s-limit. Further analysis is necessary to determine whether there is a subpopulation with low APRT values hidden in the distribution curve.

In the group with nephrolithiasis the average value was lower

(21, 4+12, 6 nmol/mg/h) than that of the control group. 11 patients
(27 %) produced an activity below the 2s-limit. Unfortunately the
nature of the stones is not known in all cases.
In the group with renal failure APRT-activity was higher than in
the control group (46, 4+17, 2 nmol/mg). This is presumably due
to a younger erythrocytes population in patients with uremia
(Becher et al. 1979). A partial APRT-deficiency might therefore
be masked in uremia.

It was possible to investigate the family of one patient (E. F.)
with diminished APRT-activity further (fig. 2).

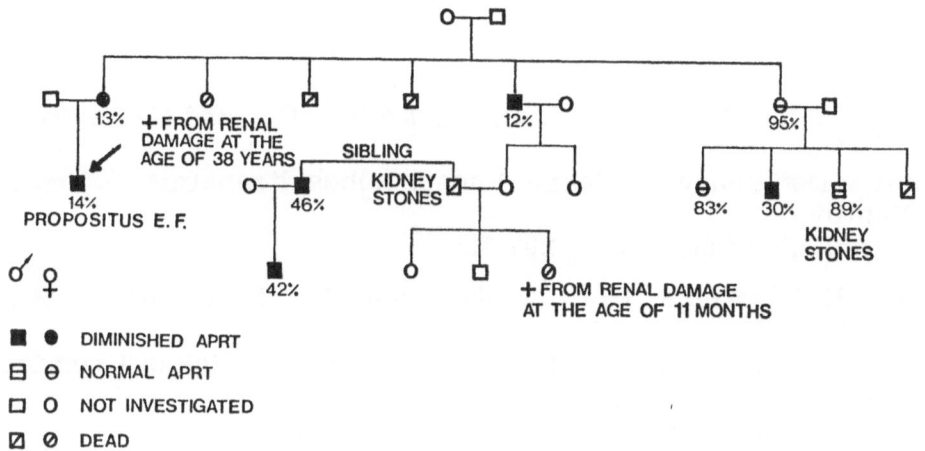

Fig. 2: Pedigrée of a patient (E. F.) with low APRT-activity.
APRT-activity from erythrocytes lysates is shown
in percent of the mean value of the control group
(28, 9 nmol/mg/h).

Two out of 25 investigated persons have died from renal damage
at the age of 38 years and 11 months respectively. Two suffered
from kidney stones (not analysed). In 9 persons APRT-activity
was measured. 6 persons appeared to have a low activity (12 to
46 % of the mean value of the control group). They had no symptoms
except one who suffered from recurrent abacterial dysuria.

Heat and cold stability from patients with diminished enzyme activity was undistinguishable from normals (n=12).
The apparent Michaelis constant for adenine was 25-38 uM (n=9) compared to 11-26 uM in control persons (n=12). The Michaelis constant for phosphoribosylpyrophosphate showed no difference.

In no group did we find a complete APRT-deficiency. Our data suggest that the homozygous form of APRT-deficiency is not a frequent cause for nephrolithiasis and/or renal failure.

ACKNOWLEDGEMENT

We want to thank Mrs. Gamulin-Vucov for excellent technical assistance.

REFERENCES

Van Acker, K.J., Simmonds, H.A., Potter, C., and Cameron, J.S., 1977,
Complete Deficiency of Adenine Phosphoribosyltransferase. Report of a Family,
New. Engl. J. of Medicine, 297:127

Becher, H.J., Weise, H.J., Volkermann, U., Schollmeyer, P., 1980,
Enhanced Purine Nucleotide Synthesis in Erythrocytes of Uremic Patients,
Klin. Wochenschr. 58:1243

Cartier,P., and Hamet, M., 1974,
Une nouvelle maladie métabolique: le déficit complet en adénine phosphoribosyltransférase avec lithiase de 2,8 DHA, C.R.
C.R. Acad. Sci. Sér. D. 279:883

Fox, I.H., Lacroix, S., Planet, G., and Moore, M., 1977,
Partial Deficiency of Adenine Phosphoribosyltransferase in Man,
Medicine, 56:515

Gault, M.H., Simmonds, H.A., Snedden, W., Dow, D., Churchill, D.N., and Penney, H., 1981,
Urolithiasis due to 2,8-Dihydroxyadenine in an Adult,
New Engl. J. Med. 305:1570

Kelley, W.N., Rosenbloom, F.M., Henderson, J.F., and Seegmiller, J.E., 1967,

A specific enzyme defekt in gout associated with overproduction of uric acid,
Proc, Natl. Acad. Sci. U.S. 57:1735

Kuroda, M., Miki, T., Kiyahara, H., Usami, M., Nakamura, T., Kontake, T., Takemoto, M., and Sonoda, T., 1981,
Urologia internat. 36:274

Simmonds, H.A., 1979,
2,8-dihydroxyadeninuria, or when is a uric acid stone not a uric acid stone?
Clin. Nephrol. 12:195

Simmonds, H.A., Cameron, J.S., Dillon, M.J., Barrat, T.M., and Van Acker, K.J., 1981,
"Uric Acid" Stones in Children: Problems of Diagnosis and Treatment in a New Defekt - Adenine Phosphoribosyltransferase Deficiency,
Fortschritte der Urol. und Nephrol. 16:52

NEW PERSPECTIVES IN THE DIAGNOSIS AND TREATMENT OF ADENOSINE DEAMINASE (ADA) DEFICIENCY

R.J. Levinsky, E.G. Davies, H.A. Simmonds,
D.R. Webster and M. Adinolfi

Institute of Child Health and Guy's Hospital
Medical School, London

INTRODUCTION

Severe combined immunodeficiency (SCID) due to adenosine deaminase deficiency (ADA) exhibits autosomal recessive inheritance, and may account for up to 20% of cases (1). Accumulation of intracellular toxic deoxynucleotides and/or S-adenosyl homocysteine particularly in T cells, is considered responsible for the severe lymphoid depletion and dysfunction observed in affected children (2). Unlike other forms of SCID, this variety is amenable to enzyme replacement therapy using regular fresh irradiated red blood cell transfusions as the source of enzyme. Another form of therapy suggested from in vitro studies, is the use of deoxycytidine which theoretically would act as a competitive substrate for deoxycytidine kinase, the enzyme considered responsible for the intracellular accumulation of deoxy-ATP (dATP). This study reports our experience of various treatments in three ADA deficient children.

METHODS

Routine immunological function tests were performed using standard techniques. T cell subsets were estimated by indirect immunofluorescence using monoclonal antibodies. The antibodies were OKT3 which recognises all mature T cells, OKT4 and OKT8 which recognise the helper and suppressor/cytotoxic T cell populations respectively and OKT6 which recog-

nises cortical thymocytes.   OKT10 recognises immature
cells of the lymphoid and granulocyte series (3).   WT1
recognises mature T cells, cortical thymocytes and
precursors of T cells.   Enzyme activity in erythrocyte lysates and intact cells were determined using
radiochemical methods previously described (4).   Nucleotide and nucleoside levels in erythrocytes and lymphocytes were determined by high performance chromatography.

RESULTS

    All three children showed the immunological
parameters typical of SCID due to ADA deficiency with
lymphopaenia, virtually no circulating T cells, a few
B cells but no immunoglobulin production, a poor or
absent PHA response and no response to an intradermal
delayed hypersensitivity test to candida antigen.
The first child S.Y. was initially treated with a calf
thymic hormone extract (TP1 Serono Laboratories) but
this produced no response.   Red cell transfusion
therapy initially given weekly and then every fortnight
was accompanied by improvement in lymphocyte numbers,
T cell numbers and PHA response.

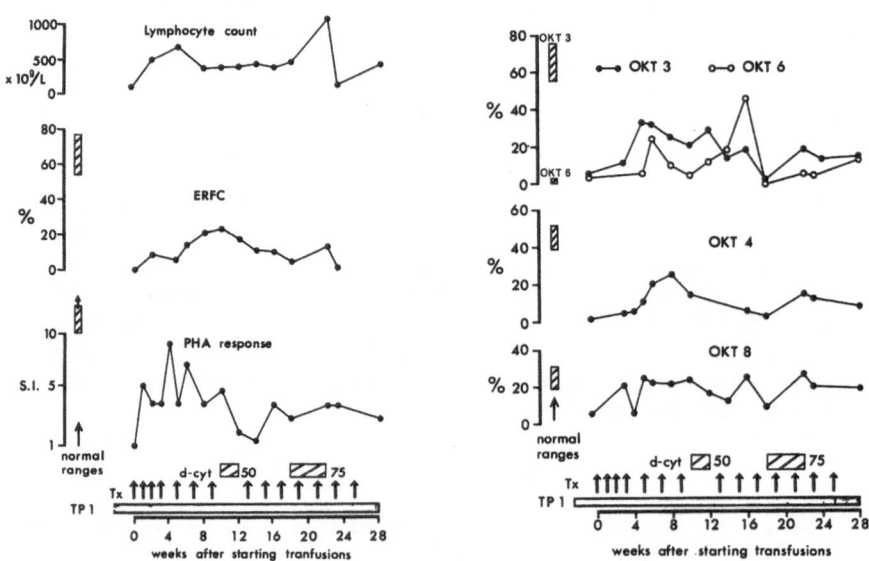

Figures 1 and 2

ABBREVIATIONS   Tx - red cell transfusions
ERFC -   E rosette forming cells
d-cyt   deoxycytidine (mg/Kg)   TP1 - thymic hormone

Monoclonal markers for T cell subsets showed a rise in
cells staining with OKT3, 4, 8 and 6.   Numbers of
cells positive for OKT4 and 8 together often exceeded
numbers of OKT3 positive cells, and this together with
the finding of OKT6 +ve cells in the circulation
suggest that many of these cells were immature.   The
two courses of parenteral deoxycytidine, the first at
50mg/Kg/day for two weeks and the second at 75mg/Kg/
day for four weeks, produced no further benefit.
Table I lists the main biochemical features noted in
the three SCID children, S.Y., K.A. and A.W.   A.W. had
received a blood transfusion prior to referral.   K.A.
who had exceptionally high levels of dATP within red
cells showed no reconstitution of immune function with
repeated red cell transfusions plus thymic hormone
therapy.   The high urinary deoxyadenosine levels and
grossly raised deoxyadenosine nucleotide levels (dATP:
dADP:dAMP in the approximate ratio of 10:1:.1) are
characteristic of ADA deficiency.   S.Y. and K.A. also
showed severe erythrocyte ATP depletion, as noted by us
in a previous case.   Another remarkable feature was the
rapid reduction in dATP levels and reciprocal relationship with ATP post transfusion in K.A. and S.Y.;
particularly in K.A. who showed the highest dATP
levels yet recorded.   The low levels of deoxyadenosine
compounds in urine and red cells of A.W. as well as
the measurable erythrocyte ADA levels are all attribut-
able to the earlier transfusion.   Deoxyadenosine was
found in the red cells and urine of S.Y. and K.A. prior
to transfusion but not after.   It was never detectable
in plasma.   dATP and dADP were detected in pretreatment
blood mononuclear cells of K.A. but not S.Y.   They
were not detected subsequently (Table I).   Erythrocyte
and blood mononuclear cell nucleotide levels in S.Y.
during both courses of deoxycytidine were essentially
unchanged from the post transfusion values listed in
Table I.

     Human amniotic cells were considered as an alter-
native source of enzyme replacement.   They do not
express HLA antigens or β_2 microglobulin and consequently are not rejected when transplanted (5).   They
secrete many lysosomal storage enzymes and in addition
appear to express significant adenosine deaminase
activity.   However, very little immune function was
restored in S.Y. and none at all in K.A. after enzyme
replacement therapy by red cell transfusions, and it
was thought unlikely that they would have provided
any further benefit.   All three children have

subsequently received haplotype mismatched bone marrow transplants from one of their parents. The technique used to prevent fatal graft versus host disease was to remove all T cells by differential agglutination with soy-bean lectin and E rosetting (6). The cells transplanted were greatly enriched for both erythroid and granulocyte colony forming activity. However in S.Y. there was 0.5% T cell contamination and he developed very severe graft versus host disease and died six months later of this. Cyclosporin A was given to K.A. and A.W. in addition to the bone marrow graft. K.A. eight weeks after transplant is showing signs of engraftment but the T cells remain very primitive and are not maturing to fully functional T cells. They are E rosette positive, express early T cell antigens OKT10 and WT1, but not OKT3, 4 or 8. No OKT6 positive cells have been seen. It was thought that this child does not have any functional thymic tissue and therefore thymic hormones have been given subsequently. It is still too early to know if they will effectively mature the T cells. A.W. four weeks post transplant is not yet showing any signs of engraftment.

Table I

Biochemical Parameters in 3 ADA deficient neonates at diagnosis and after therapy

Patient	Treatment	ADA activity (nmol/mg Hb/h)	Adenine based nucleotides						Urine deoxyadenosine (mmol/mmol creatinine)
			ATP	ADP	AMP	dATP	dADP	dAMP	
			Red cells (nmol/ml packed cells)						
SY	Nil	<1.0	323	135	7	946	114	4	0.15
	2 weeks post RBC transfusion	not done	1110	98	10	110	12	2	Not detected
			PBMs (nmol/ 10^6 cells)						
	Nil		3.34	0.98	0.16	-	-	-	
	Post Therapy		3.22	1.84	0.27	-	-	-	
			Red Cells						
KA	Nil	<0.1	666	80	3	1478	84	10	0.14
	2 weeks RBC	not done	1818	170	10	98	20	-	Not detected
			PBMs						
	Nil		6.15	0.93	.09	1.12	.17	-	
	Post RBC		2.76	.76	.29	-	-	-	
			Red Cells						
AW	*	8.6*	1676	214	12	147	12	-	0.02
Control Values		40-100	Red Cells					GTP	GDP
			1278 ±127	114 ±24±	10 ±3	-	-	-	60 ±10	20 ±5
			PBMs						
			3.07 ±0.70	1.35 ±0.37	0.18 ±0.06	-	-	-	0.40 ±0.10	0.23 ±0.07

- indicates below the limits of detection by HPLC method used

* had received blood transfusion

DISCUSSION

There appears to be a spectrum of severity in the immunological defect in ADA deficiency which correlates closely with amounts of residual ADA activity in the lymphocytes. We suggest that erythrocyte dATP levels also provide an accurate guide to the immunodeficiency. It seems that those patients with the most severe forms (such as ours) with a complete deficiency of the enzyme, show the poorest response to therapy, and indeed may not respond at all. Another discrepancy exists between the degree to which we were able to correct the bio-chemical defects (deoxynucleotide levels decreased to less than 10% of pretreatment values) and the immuno-logical improvement obtained. The precise mechanism by which the immunodeficiency is related to the lack of the purine enzyme remains unexplained. The observation in S.Y. that T cells remained immature despite full biochemical restoration emphasises the complexity of the immunodeficiency. It has been suggested that the deoxynucleotides destroy thymic epithelium. Since the differentiation of the T cell series is dependent both on contact with thymic epithelial cells and on hormones secreted by these cells, their absence may account for the lack of maturation. The similar findings of very immature T cells engrafting in K.A. but not maturing further suggests that in very severe depletion of ADA (as in our cases) there is little residual thymic tissue and restoration by either enzyme replacement or bone marrow transplantation must be difficult. It is possible that in these children the additional implantation of foetal thymic cells may be necessary, since neither synthetic nor crude thymic hormones have succeeded in restoring immunity.

In theory deoxycytidine should act as a com-petitive substrate for deoxycytidine kinase, but parenteral deoxycytidine in S.Y. produced no additional benefit. The presence of high dATP and dADP levels in the peripheral blood mononuclear cells in one child but not the other is difficult to explain. As discussed elsewhere (Goday et al, this Symposium) the preparations contained both platelets and nucleated red cells and this may account for the discrepancy.

The optimal form of therapy for these children is undoubtedly bone marrow transplantation. We hope that in time where no donor is available haplotype

mismatched transplants from a parent will be possible
without the development of fatal graft versus host
disease.   However, in these patients with complete
ADA deficiency and with very high levels of dATP a
question mark over functional residual thymic tissue
must remain.

REFERENCES

1.   R. Hirschhorn.   Clinical delineation of adenosine
      deaminase deficiency.   In: Enzyme Defects and
      Immune Dysfunction.   Ciba Foundation Symposium
      68 (New Series)   Excerpta Medica Amsterdam: 35
      (1979)
2.   D.A. Carson, J. Kaye and D.B. Wasson.   The
      potential importance of soluble deoxynucleotidase
      activity in mediating deoxyadenosine toxicity in
      human lymphoblasts.   J. Immunol. 126   (1981)
3.   P.C. Kung, G. Goldstein, E.L. Reinherz and S.F.
      Schlossman.   Monoclonal antibodies defining
      distinctive human T cell surface antigens.
      Science: 206: 347 (1979)
4.   H.A. Simmonds, R.J. Levinsky, D. Perrett and
      D.R. Webster.   Reciprocal relationship between
      erythrocyte ATP and deoxy ATP levels in inherited
      ADA deficiency.   Biochem. Pharmacol. 31: 947
      (1982)
5.   C.A. Akle, M. Adinolfi, K.I. Welsh, S. Leibowitz,
      and I. McColl.   Immunogenicity of human amniotic
      epithelial cells after transplantation into
      volunteers.   Lancet ii: 1003 (1981)
6.   Y. Reisner, N. Kapoor, D. Kirkpatrick, M. Pollack,
      B. Dupont, R.A. Good, R.J. O'Reilly.   Transplant-
      ation for acute leukaemia with HLA-A and B non-
      identical parental marrow cells fractionated with
      soy-bean agglutinin and sheep red blood cells.
      Lancet ii: 327 (1981)

INTRAVENOUS DEOXYCYTIDINE THERAPY IN A PATIENT WITH ADENOSINE

DEAMINASE DEFICIENCY

Morton J. Cowan, David W. Martin, Jr., Diane W. Wara,
and Arthur J. Ammann

Departments of Pediatrics and Medicine, University
of California, San Francisco, California 94143

INTRODUCTION

Adenosine deaminase (ADA) deficiency usually results in severe combined immunodeficiency disease. Without therapy, these children usually die from overwhelming infections. The most successful therapy remains bone marrow transplantation from a histocompatible sibling donor. Efforts at a biochemical approach have focused on enzyme replacement using repeated transfusions from ADA-positive donors (Polmar et al., 1976). Unfortunately, only a relatively few patients with ADA deficiency have histocompatible siblings to provide a bone marrow transplant and less than 50% of patients with ADA deficiency show a significant response to red cell transfusions. In addition, there are significant risks of repeated red cell transfusions, including iron overload, trans-fusion reactions, and viral infections.

Studies of the pathogenesis of ADA deficiency indicate that at least one mechanism for immunodeficiency in this disorder in-volves the inhibition of ribonucleotide reductase by toxic levels of deoxyATP (Cohen et al., 1978). Further, it has been shown in cell culture models of ADA deficiency that a limiting metabolite of ribonucleotide reductase inhibition is deoxyCTP and that deoxy-cytidine in certain cells can bypass the inhibition of ribonucleo-tide reductase by being directly metabolized to deoxyCTP (Ullman et al., 1978).

In this report we describe an 8 year old girl who was diag-nosed as having ADA-negative combined immunodeficiency disease and who was treated with intravenous deoxycytidine in order to deter-mine the potential of this biochemical approach to restore her immunologic function.

CASE REPORT

A.A. is an 8 year old girl who was well until 3 years of age when she had an episode of pneumonia requiring hospitalization. Until that time she had developed normally and was living an uneventful life. Subsequently she developed frequent mild upper respiratory illnesses associated with bronchitis and a clinical picture of asthma. She did not require hospitalization and was referred to us after an initial evaluation by her allergist because of lung disease which seemed more profound than her clinical history warranted as well as a severe lymphopenia. She had one older brother with no history of frequent infections and an otherwise unremarkable family history. Immunologic evaluation revealed severe lymphopenia as well as significantly reduced lymphocyte responses to phytohemagglutinin (PHA) and alloantigen (mixed lymphocyte reaction, MLR). Quantitative immunoglobulins were normal and while there was some specific antibody response to immunization with pneumococcal polysaccharide and keyhole limpet hemocyanin, both were significantly decreased compared to normal. No histocompatible bone marrow donor was available.

METHODS

T Cell Studies

Peripheral blood mononuclear cells (PBMC) were obtained from heparinized peripheral blood samples using a standard Hypaque-Ficoll procedure. Studies of cellular immunity including T cell numbers and lymphocyte responses to PHA and alloantigen were performed using previously described methods (Cowan et al., 1980).

Enzyme Analysis

Adenosine deaminase activity was determined using a modification of the spectrophotometric method of uric acid production as previously described (Cowan et al., 1982).

Measurement of Purine and Pyrimidine Metabolites

Acid soluble extracts (trioctyl-amine freon method) were frozen at -70°C until measurements were made. In those samples in which deoxycytidine was measured tetrahydrouridine (THU) was added to the tube in which the blood was collected to diminish the catabolism of deoxycytidine by deoxycytidine deaminase which is competitively inhibited by THU. Purine and pyrimidine metabolites were measured using high-pressure liquid chromatography as previously reported (Cohen et al., 1978).

RESULTS

ADA Activity and Erythrocyte DeoxyATP Levels

Adenosine deaminase levels in erythrocyte and PBMC extracts
from the patient, both parents, and the male sibling are shown in
Table 1. The patient's erythrocyte ADA activity was undetectable
while she had approximately 1.9% of control levels of ADA activity
in her PBMC. Levels in both parents are compatible with the heter-
zygote carrier state and the brother is normal. Mixing studies
with extracts from the patient and normal control demonstrated no
evidence for an inhibitor to the enzyme present in the patient's
cells (data not shown).

The patient's erythrocytes were processed for acid-soluble
deoxynucleotides and her deoxyATP level was 200 μM (normal <5 μM).
Deoxyadenosine in her urine was found to be significantly elevated.
Deoxycytidine Infusion

Following informed consent from both the patient and her parents deoxycytidine was administered intravenously at a dose of
50 mg/kg/day. During this continuous infusion, a trial of the
deoxycytidine deaminase inhibitor, THU, was done at an intravenous
dose of 150 mg/kg/day. The patient tolerated the infusion without
any difficulty and there was no evidence of toxicity based on
clinical and laboratory evaluation of major organ systems. Plasma
deoxycytidine levels with and without the presence of THU are
shown in Figure 1. Following the initial rapid rise in deoxycytidine levels there was a more gradual increase approaching
steady-state levels after the first 24-36 hours without THU and
sometime after 72 hours in the presence of THU. The levels
achieved with THU were significantly greater over the same time
period (72 μM versus 30 μM, respectively). At the discontinuation

Table 1. Adenosine Deaminase Levels

	A.A.[a]	M	F	B	Control
RBC[b]	0	12.9	23.4	-	39.0 ± 9.9[c]
PBMC[d]	17.2	257	371	1127	897 ± 295

[a] A.A. = patient, M = mother, F = father, B = brother
[b] Red blood cells, nm/hr/ms hemaglobin
[c] Mean ± standard deviation
[d] Peripheral blood mononuclear cells, nm/hr/mg protein

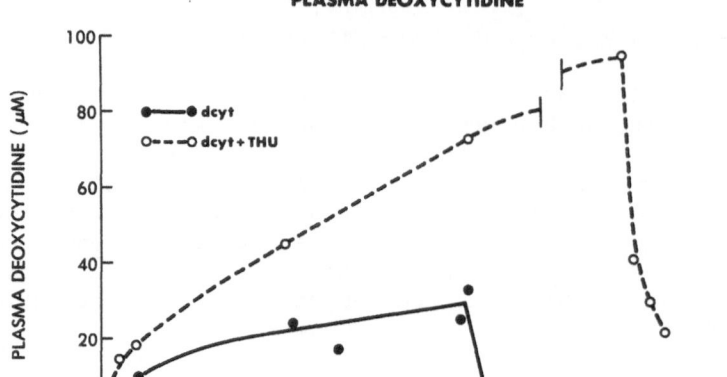

Figure 1.   Plasma deoxycytidine levels.   Arrows indicate cessation
            of infusion.

of the infusion the plasma levels fell rapidly with a half-life of
less than 1 hour.   Erythrocyte deoxyCTP levels were expressed as
a percent of deoxyATP and the pattern with and without THU was
similar to what was found in the plasma.   However, when the infusion was stopped there was no significant drop in deoxyCTP
levels, indicating minimal intraerythrocyte catabolism.

Immunologic and Clinical Response

        Cellular immune studies before and after infusion are shown
in Table 2.   Prior to therapy the absolute number of T cells (ATC)
averaged 232 (normal >720), while the lymphocyte response to PHA
was 4% of normal and to alloantigen (MLR) was 25% of normal.
These studies remained abnormal until 2 weeks following the
infusion when her ATC increased to 530.   At 3 months post infusion, the lymphocyte response to PHA increased to 16% of control
and the MLR became normal.

        Clinically, the patient subjectively felt better within 2 weeks
of the infusion and by several criterion including appetite, weight
gain, endurance and pulmonary symptoms, she remained improved until
4 months post infusion when she developed pneumonia.   Immune studies
showed a return to pre-infusion levels coincident with clinical
deterioration, both of which have persisted for 6 additional months.

Table 2.   Immune Response to Deoxycytidine

	Pre	Post	2 Wks.	3 Mo.	5 Mo.
ATC[a]	232±27[d]	437	530	–	235
PHA[b]	423±199	521	–	1881	634
MLR[c]	1602±456	2124	–	7261	1403

[a] Absolute T cell count, cells/mm^3, normal >720
[b] Phytohemagglutinin, cpm, normal >11750
[c] Mixed lymphocyte reaction, cpm, normal >5800
[d] Mean ± standard deviation, 3 separate studies

DISCUSSION

For the majority of children with T cell immunodeficiency disease there is no effective therapy. In those who have a known biochemical defect, it may be possible to develop effective phar-macologic approaches to bypassing or correcting their deficiency. It is likely that at least one of the mechanisms responsible for the immunodeficiency seen in ADA deficiency is ribonucleotide reductase inhibition by deoxyATP and in vitro cell studies suggest that deoxycytidine can bypass this inhibition (Ullman et al., 1978).

Our patient is similar to 10-15% of patients with ADA defic-iency, in that she presented later in life and has a combined (but not severe combined) immunodeficiency. Also, she has very low ADA activity associated with increased levels of deoxyadenosine in the urine and deoxyATP in erythrocytes. She is different in two respects. One is that she is more mildly affected than pre-viously reported cases with no life-threatening infections during the first 8 years of her life. The other difference is that while her erythrocyte ADA activity is unmeasurable, ADA activity in her PBMC is 1.9% of control. Other reported affected cases have had ADA activities <1%. It is interesting that in 2 reported cases of absent erythrocyte ADA activity but normal immunity, the lympho-cyte ADA was 5% of normal (Jenkins et al., 1976; Reem, 1979). Our patient has a lymphocyte ADA of about 2% and is significantly affected suggesting that the threshhold of ADA activity necessary for normal immunity is between 2 and 5%.

Other attempts at therapy with deoxycytidine have been made without success and one explanation has been that the cells are too severely affected by the time of diagnosis to be salvaged. Also, it has been reasoned that deoxycytidine is too rapidly metabolized in the plasma prior to reaching the target cells. We felt that our patient was a good candidate for a trial of deoxycytidine because of her relatively less severe deficiency. We postulated that she was in an early stage of her immunologic attrition and had evidence for the presence of salvagable stem cells. We have shown that adequate deoxycytidine levels can be achieved safely in plasma and erythrocytes. THU significantly raises the levels and prolongs the plasma half-life. Also, there is no significant intracellular catabolism of deoxyCTP, at least during the time period of our study.

From this single trial of deoxycytidine it is impossible to come to a definitive conclusion as to its effect on immunity in ADA deficiency. Certainly, our clinical and laboratory data suggest an in vivo confirmation of in vitro studies using cell models. However, a repeat trial will be necessary for a more definitive conclusion. If confirmed, our results suggest that at least some patients with ADA deficiency may respond to this biochemical approach.

REFERENCES

Cohen, A., Hirschhorn, R., Horowitz, S., Rubinstein, A., Polmar, S., Hong, R., and Martin, Jr., D., 1978, Deoxyadenosine triphosphate as a potentially toxic metabolite in adenosine deaminase deficiency, Proc. Natl. Acad. Sci. USA, 75:472.

Cowan, M., Fujiwara, P. and Ammann, A., 1980, Cellular immune defect in selective IgA deficiency using a microculture method for PHA stimulation and limiting dilution, Clin. Immunol. & Immunopath., 17:595.

Cowan, M., Fraga, M., Andrew, J., Lameris-Martin, N., and Ammann, A., 1982, Purine salvage pathway enzyme activities in human T, B and null lymphocyte populations, Cell. Immunol., 67:121.

Jenkins, T., Rabson, A.R., Nurse, G.T., Lane, A.B., and Hopkinson, D.A., 1976, Deficiency of adenosine deaminase not associated with severe combined immunodeficiency, J. Pediatr., 89:732.

Polmar, S.H., Stern, R.C., Schwartz, A.L., Wetzler, E.M., Chase, P.A. and Hirschhorn, R., 1976, Enzyme resplacement therapy for adenosine deaminase deficiency and severe combined immunodeficiency, N. Engl. J. Med., 295:1337.

Reem, G.H., Borkowsky, W., and Hirschhorn, R., 1979, Purine
    and phosphoribosylpyrophosphate metabolism of lympho-
    cytes and erythrocytes of an adenosine deaminse defi-
    cient immunocompetent child, Pediatr. Res., 13:649.
Ullman B., Gudas, L.J., Cohen A. and Martin, Jr., D.W., 1978,
    Deoxyadenosine metabolism and cytotoxicity in cultured
    mouse T lymphoma cells: a model for immunodeficiency
    disease, Cell, 14:365.

PROPERTIES OF A NOVEL PEG DERIVATIVE OF CALF ADENOSINE DEAMINASE

Charles Beauchamp, Peter E. Daddona
and David P. Menapace

Department of Internal Medicine, University of
Michigan Medical School and Ann Arbor   Veterans
Administration Medical Center, Ann Arbor, Michigan USA

INTRODUCTION

Human adenosine deaminase (ADA) deficiency has been causally
associated with severe combined immunodeficiency disease[1].
Some success has been achieved in restoring immune function in
these patients by administering the enzyme via red blood cell
transfusions[2]. A derivative of ADA which retains enzymatic
activity and which has a long circulation time and minimal
immunogenicity when injected intravenously could be useful as an
alternate means of enzyme replacement therapy.

Abuchowski et al[3] have described a procedure for covalently
attaching polyethylene glycol (PEG) to protein. Two interesting
characteristics of an intravenously injected PEG-protein adduct
are an increase in circulation time and a decrease in
antigenicity, and immunogenicity as compared to the unmodified
protein[3]. These changes in protein properties upon
modification with PEG can be prerequisites for the use of the
protein as a means of studying disease mechanisms in vivo, or for
the parenteral administration of a non-autologous protein as an
enzyme based therapy[4]. For example, PEG modified Candida
utilis uricase has been shown to experimentally lower uric acid
levels in man[4]. However, Abuchowski et al's method[3] causes a
dramatic decrease in catalytic activity upon the derivatization
of certain enzymes.

In this study a simple and gentle procedure is described for
PEG derivatization of calf ADA which preserves enzymatic
activity. The blood clearance time in mice and the

47

antigenic/immunogenic properties of the derivative are
described. The utility of a PEG derivative of ADA which could be
used for enzyme replacement therapy in man is evaluated.

METHODS

The development of a gentle and efficient method of coupling
PEG to calf ADA is based on a simple method of using
carbonyldiimidazole (CDI)[5,6] as an activator for PEG.
Monomethoxypolyethylene glycol (mPEG) is first activated with CDI
and then the activated mPEG is incubated with protein to form
mPEG-protein derivatives (see Fig. 1) in which a carbonyl group
bonds the hydroxyl group of the mPEG to amino groups of the
protein. Activated mPEG2 and activated mPEG5, average MW of 2000
and 5000 respectively, were formed by the incubation of the PEG
polymers with CDI[6]. A long incubation of activated PEG with
calf ADA was used to prepare the mPEG-calf ADA derivative.
Activated mPEG5 (125 mM) and 2×10^{-5} M calf ADA (from Sigma)
were incubated in 10 mM borate (pH 8.5) at 4° C for 48 h. At
48 h the volume of the incubation was doubled and activated mPEG5
was added to maintain the [mPEG5] = 70mM. At 96 h the [mPEG5]
was increased to 85mM and at 144 h to 100mM. At 192 h the volume
of the solution was again doubled and activated mPEG2 was added
so that the [mPEG2] = 50mM. After an additional 48 h the
solution was dialyzed and concentrated using an XM-50 membrane in
an Amicon ultrafiltration system. The percentage of calf ADA
amino groups which reacted with the activated PEG was determined
using a TNBS titration procedure[7]. The blood clearance half
life of calf ADA and the mPEG derivative was determined in
mice[6]. ADA activity and antigenic properties were determined
as previously described[8,9]. The change in the half-life of the
mPEG derivative of ADA upon repetitive injection into a mouse was
used as a measure[10] of the immunogenicity of the derivative.

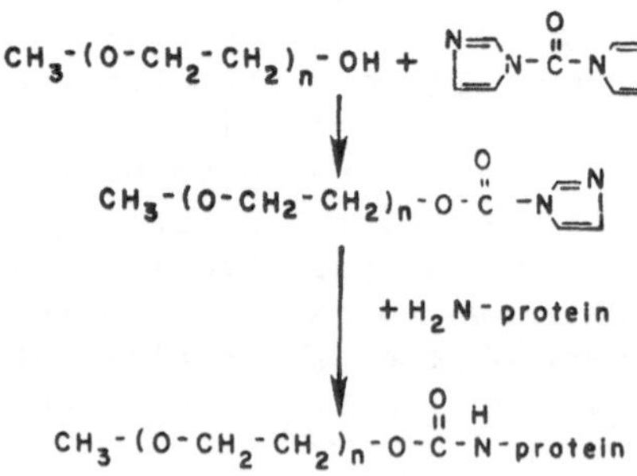

$$CH_3-(O-CH_2-CH_2)_n-OH \quad + \quad \overset{O}{\underset{}{N=\underset{}{\square}N-\overset{\|}{C}-N\underset{}{\square}=N}}$$

$$\downarrow$$

$$CH_3-(O-CH_2-CH_2)_n-O-\overset{O}{\overset{\|}{C}}-N\underset{}{\square}=N$$

$$\downarrow \quad +H_2N-protein$$

$$CH_3-(O-CH_2-CH_2)_n-O-\overset{O}{\overset{\|}{C}}-\overset{H}{N}-protein$$

Fig. 1. Proposed mechanism[6] for the formation of CDI activated
        mPEG and for the reaction of the activated mPEG with
        lysine groups of proteins.

RESULTS

Enzyme Activity

The mPEG—protein adduct used in this study is formed on prolonged incubation of activated mPEG5 and activated mPEG2 with calf ADA. The mPEG—calf ADA derivative retains 76% of the enzymatic activity of the unmodified protein.

TNBS Titration

Only 15% of the available nitrogen groups of calf ADA remain in the mPEG derivative by TNBS titration. These results are consistent with a covalent bond between the mPEG and protein amino groups as diagrammed in Fig. 1.

Blood Clearance

The blood circulation times in mice of intravenously injected calf ADA and its mPEG derivative are shown in Fig. 2.

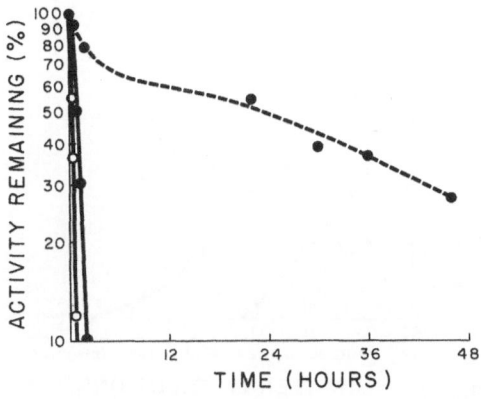

Fig. 2. Blood clearance of calf ADA and its mPEG derivative in mice. Graph shows the clearance of unmodified ADA (—○——○—), PEG modified ADA, 1st injection (- ● - ●), mPEG modified ADA, 14th injection (—●——●). The % ADA activity remaining in the plasma is graphed in a semi-log plot against time.

Less than 15% of enzymatic activity remains in the circulation 60 min after injection of unmodified calf ADA whereas greater than 25% of the activity of the mPEG derivative remains 44 h after the initial injection into a mouse. The unmodified enzyme has a half-life of 13 min and the mPEG derivative has a half-life of 24 h (a ~110 fold increase).

The blood clearance of the mPEG derivative after 14 injections is also given in Fig. 2. The half-life of the mPEG derivative decreases after the 6th injection and by the 14th injection the half-life is 30 min, only a 2.3 fold increase over the half-life of the unmodified protein.

Antigenicity

The relative antigenicity of calf ADA and its PEG derivative is determined by a comparison of the immunoprecipitation curves of the two proteins as shown in Fig. 3. The results illustrated in Fig. 3 indicate that there is a 75% reduction in the antigenicity of the mPEG derivative of calf ADA as compared to the unmodified protein.

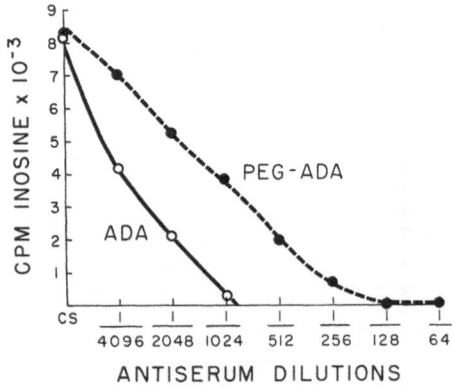

Fig. 3.   Comparison of the antigenicity of calf ADA and its mPEG derivative. Graph shows the activity of ADA remaining in the supernatant after double antibody precipitation using calf ADA antiserum. Unmodified calf ADA (–o——o–), mPEG modified calf ADA (– ● – ●). CS is control serum.

DISCUSSION

Attaining the goal of using non-autologous calf ADA protein to treat patients with the enzyme deficiency state and severe combined immunodeficiency disease requires that the protein have at least the following properties: 1) a long circulating half-life, 2) significant retention of enzymatic activity, and 3) no clinically significant antigenicity or immunogenicity. The mPEG derivative of calf ADA described in this study has 2 of these 3 properties. This derivative retains a significant fraction of the enzymatic activity of the native protein and it has a very prolonged circulation half-life. While the mPEG-calf ADA derivative has a significant decrease in its in vitro antigenicity, the results of the repetitive injection experiment imply that it retains considerable immunogenicity when injected into the mouse. Failure of the mPEG fully to mask antigenic sites on calf ADA is one possible explanation for this finding. Even though 85% of the available amino groups on the enzyme have been blocked by mPEG according to the TNBS titration results, the derivative is still apparently recognized as being immunogenic by the mouse. Thus, the mPEG derivative of calf ADA described herein is not likely to be useful as an enzyme based therapy in man.

The failure to abolish the immunogenicity of calf ADA by our method could be analogous to a similar report of phenylalanine ammonia-lyase[9] derivatized by the procedure of Abuchowski et al[3]. It is possible that a more completely PEG derivatized enzyme would be non-immunogenic and would thus not have a reduced circulation half-life on repetitive injection. The authors are exploring this possibility. This study should not hinder future investigations to determine the applicability of the CDI based procedure for making PEG-protein derivatives of other enzymes which have potential therapeutic benefit in man.

The utility of the CDI based procedure for making PEG-protein derivatives has been demonstrated by the use of PEG-enzyme adducts to determine if the substrates of these enzymes contribute to the mechanism of lung damage in a thermally injured rat[11]. The marked prolongation of the circulation half-life of PEG-enzyme adducts allows them to be used in vivo for delineating disease mechanisms. Other animal models of human disease can undoubtedly be probed using PEG-enzyme adducts produced by this relatively simple technique.

REFERENCES

1. B.S. Mitchell and W.N. Kelley, Purinogenic immunodeficiency
       diseases: Clinical features and molecular mechanisms,
       Ann. Int. Med., 92: 826 (1980).

2. S.H. Polmar, Enzyme replacement and other biochemical
    approaches to the therapy of adenosine deaminase deficiency,
    Ciba Foundation Symposium, 68: 213 (1978).
3. A. Abuchowski, J.R. McCoy, N.C. Palczuk, T. van Es and F.F.
    Davis, Effect of coualent attachement of polyethylene glycol
    on immunogenicity and circulating life of bovine liver
    catalase, J. Biol. Chem., 252: 3582 (1977).
4. S. Davis, Y.K. Park, A. Abuchowski, and F.F. Davis,
    Hypouricemic effect of polyethylene glycol modified urate
    oxidase, Lancet, 2: 281 (1981).
5. G.S. Bethell, J.S. Ayers, W.S. Hancock, and M.T.W. Hearn, A
    novel method of activation of cross-linked agaroses with
    1,1'-carbonyldiimidazole which gives a matrix for affinity
    chromatography devoid of additional charged groups, J. Biol.
    Chem. 254: 2572 (1979).
6. C. Beauchamp, D. Menapace, S. Gonias and S. Pizzo, Manuscript
    in preparation (1982).
7. A.N. Glazer, R.J. Delange and D.S. Sigman, Chemical
    Modification of Proteins, American Elsevier Publishing Co.,
    New York (1975).
8. M.B. van der Weyden, R.H. Buckley and W.N. Kelley, Molecular
    forms of adenosine deaminase in severe combined
    immunodeficiency, Biochem. Biophys. Res. Comm., 57: 590
    (1974).
9. B.E. Chechik, W.P. Schrader and P.E. Daddona, Identification
    of human thymus-leukemia-associated antigen as a
    low-molecular-weight form of adenosine deaminase, J. Natl.
    Cancer Inst., 64: 1077 (1980).
10. K.J. Wider, N.C. Palczuk, T. van Es and F.F. Davis, Some
    properties of polyethylene glycol: phenylalanine
    ammonia-lyase adducts, J. Biol. Chem., 254: 12579 (1979).
11. G.O. Till, C. Beauchamp, W. Tourtellotte Jr., D. Menapace, R.
    Kunkel, K.J. Johnson and P.A. Ward, Oxygen radical dependent
    lung damage following thermal injury of rat skin, J. Clin.
    Invest., manuscript submitted (1982).

PURINE NUCLEOSIDE PHOSPHORYLASE (PNP) DEFICIENCY: A THERAPEUTIC

CHALLENGE

A.R.Watson,H.A.Simmonds,D.R.Webster,L.Layward, and
D.I.K.Evans

Departments of Child Health and Immunology and Purine
Laboratory
Royal Manchester Children's Hospital and Guy's Hospital
Medical School, London

Purine nucleoside phosphorylase (PNP:EC 2.4.2.1) deficiency
appears to result in a predominantly T cell immune defect[1]. Lack
of the PNP enzyme leads to the inability to degrade deoxynucleosides, particularly deoxyguanosine (dGR) resulting in the
intracellular accumulation of deoxyguanosine triphosphate (dGTP)
which is known to inhibit DNA synthesis in vitro[2]. Deoxycytidine
kinase is considered responsible for the initial conversion of dGR
to dGMP[1]. Hence deoxycytidine (dCR), the preferred substrate, should
competitively inhibit dGTP accumulation as has been demonstrated in
vitro[1,2]. However, oral dCR therapy for six months in a PNP
deficient child produced no clinical or immunological improvement,
probably due to degradation in the gut or rapid deamination[3].

We recently demonstrated total PNP deficiency shortly after
birth in the second child of a family in which the first U.K.
homozygote to be identified had died of a lymphoproliferative
disorder[4].

This paper presents clinical, immunological and biochemical
studies over twenty months in which specific therapeutic approaches
were designed to try to prevent the progressive attrition of T cell
function.

CASE HISTORY

A male infant (S.B.) was born at term in June 1980 to healthy
parents who are fourth cousins[4]. The child was physically and
neurologically normal at birth with total PNP deficiency being
confirmed in red and white cell lysates at four days of age. He
has thrived with full recovery from urinary tract, respiratory

syncytial virus and Bordetella Pertussis infections. Oral thrush
infection occurred for the first time at 16 months of age.

Head lag and excessive irritability were noted at three months
of age. Since then the infant has remained profoundly hypotonic and
developmentally delayed with recent hypertonia in the lower limbs.

METHODS

Immunological investigations were carried out using standard
techniques on lymphocytes separated by Ficoll-Triosil gradient.
T-cell subpopulations were detected by OKT4,6 and 8 sera (ortho).
PHA responses, immunoglobulin levels and other standard tests of
lymphocyte and also leucocyte function were measured by established
methods.

Biochemical methods used for purine levels and enzymes are
referenced in detail in a previous publication[5]. Due to the lymphopenia 'lymphocyte' (peripheral blood mononuclear cell: PBM) nucleotide levels were not obtained until 9 months. High pressure liquid
chromatography (HPLC) was used for the estimation of intracellular
nucleotide and nucleoside levels[5].

Therapeutic regimes were divided into four separate periods. The
first comprised deoxycytidine initially intravenously and then subcutaneously in a dose of 15mg/kg on three days a week for three
months (I). During the second period the dCR dosage was increased
to 50mg/kg on five days a week for three months (II). Period III
consisted of dCR 25mg/kg and Tetrahydrouridine (THU) 10mg/kg five
days a week. The fourth period involved oral guanine hydrochloride
in a dose of 10mg - 20 mg/kg/day for three months (IV).

RESULTS

A lymphopenia (1.3×10^9/1) was already evident at four days of
age (Fig.1a). The absolute lymphocyte count rose initially but by
three months there was a sustained lymphopenia (0.5-0.8×10^9/1). This
was not altered significantly by any therapeutic regime except during
the period of proven Bordetella pertussis infection. Despite
initial improvement with dCR alone subsequent values for E rosetting
cells were all below the normal range of 50-75% with specific T-cell
sub-populations OKT4 (helper) between 21-35% (n=30-50%), OKT8
(suppressor) 4-8%(N=15-30%). OKT6 (thymocyte) sera did not identify
any of the peripheral lymphocyte population (Fig.1a). PHA responses,
normal at birth, diminished by six months with transient improvement
during the pertussis infection and at the commencement of oral
guanine (Fig.1b)

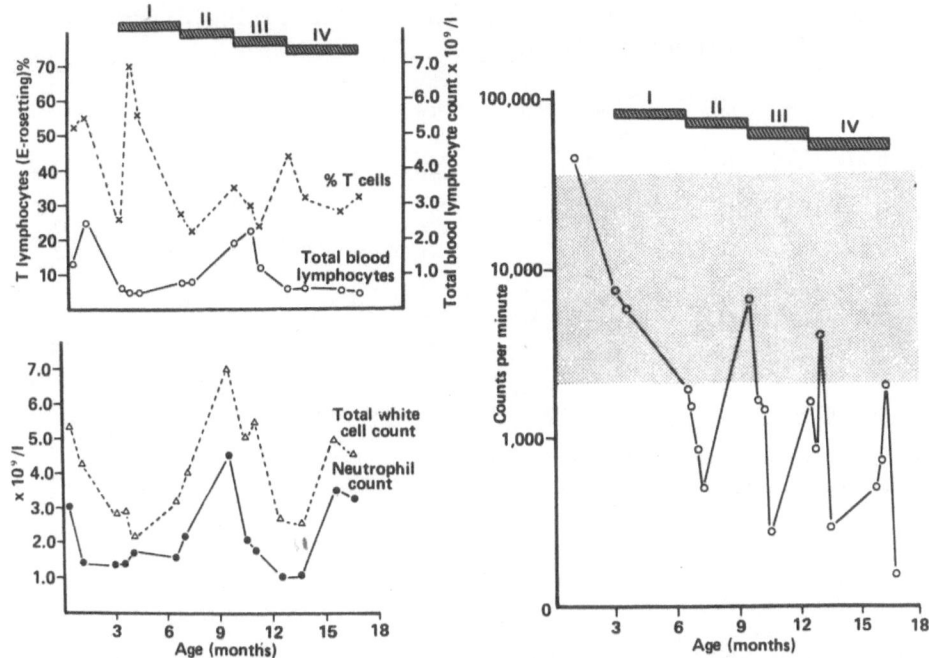

Fig.Ia.Cell counts and percentage of T-lymphocytes prior to and during periods I-IV (see text)

Fig.Ib. Whole blood responses to PHA(2μg/ml) prior to and during periods I-IV (see text). Normal range ± 2SD of controls indicated (hatched area) at same dose level

Although the total B-cell population has been low due to the lymphopenia the %PBM's recognised as B cells remains normal, as have serum and secretory immunoglobulins. Immunoelectrophoresis has shown no monoclonal band. Sero-conversion to respiratory syncytial virus has been noted (Titre 1/10,1/40,1/10 at 6,12, and 16 months respectively. The patient was immunised with diptheria and tetanus toxoid plus inactivated polio vaccine at 4,6 and 12 months. No polio antibodies were detected but following tetanus immunisation 1-2 units/ml tetanus antitoxin were produced. There has been no response to intradermal candida antigen at 6,12, and 15 months of age.

A mild neutropenia has also been evident apart from the whooping cough period. The polymorphonuclear leucocytes appear normal and 3-monthly tests of bacteriocidal killing, mobility, opsonisation and chemiluminescence have been normal.

Table 1.   Purine metabolites in patient SB during different treatment periods.

PERIOD		DATE	PLASMA (μmol/l)					URINE (mmol/24h)					TOTAL *
			UA	HR	GR	dHR	dGR	UA	HR	GR	dHR	dGR	mmol/mmol Cr.
	Nil	22.7.80	3	28	10	4	4	.003	0.51	0.16	0.05	0.07	2.4
		16.9.80	7	40	8	4	3	<.001	0.96	0.29	0.21	0.15	2.9
TREATMENT	I	23.9.80	3	41	11	3	6	<.001	0.76	0.34	0.15	0.22	2.9
		13.1.81	14	63	8	2	1	.022	1.67	0.51	0.18	0.22	4.7
	II	31.3.81	19	69	12	5	1	.007	1.53	0.55	0.30	0.47	3.2
	III	21.4.81	8	17	4	<1	<1	.012	1.29	0.65	0.46	0.41	3.8
		28.4.81	7	41	10	<1	<1	.040	1.08	0.46	0.14	0.18	2.3
		7.7.81	5	14	4	<1	<1	.008	0.97	0.42	0.17	0.28	2.3
	IV	29.7.81	78	31	12	<1	<1	.104	1.18	0.64	0.20	0.18	3.3
		27.10.81	108	31	11	2	2	.256	1.56	0.86	0.31	0.34	2.9

Values given are representative data taken generally after one week, compared with the values after 2-3 months of each treatment.

* =   Total oxypurines - sum of the nucleosides (HR, GR, dHR, dGR) - factored by Creatinine (Cr)

UA = Uric Acid; HR = Inosine; GR = Guanosine; dHR = deoxyinosine

Metabolic Studies

Table I shows that uric acid was almost completely absent from plasma or urine, confirming the severity of the enzyme defect in either direction[5]. No significant increment was noted until guanine therapy commenced. Table I also shows the gross purine overproduction characteristic of this defect with a total purine end product of 2.4-3.8 mmol/mmol creatinine compared with <1.0 for controls.

Inosine (HR) was the principal end product, with lesser amounts of guanosine (GR), and even smaller but comparable amounts of deoxy inosine (dHR) and deoxyguanosine (dGR) in the ratio 60:20:10:10. The ratio changed slightly to approximately 50:20:15:15 but no significant alteration in the total (factored by creatinine on a millimolar basis) was produced by any treatment. All four nucleo-sides were found in the plasma in similar proportions with the deoxy-compounds near and subsequently often below the limits of detection by the method (<1μmol/l) indicating effective renal

Table II. Nucleotide levels in patient SB

| PERIOD | DATE | ATP | ERYTHROCYTES (nmol/ml packed cells) (a) | | | | | NAD+ |
			ADP	AMP	GTP	GDP	dGTP	
Nil	22.7.80	750	92	7.5	1.5	–	2.8	255
	16.9.80	745	111	8.5	4.0	–	6.0	288
I	23.9.80	1140	170	8.0	6.0	–	6.8	257
	13.1.81	1120	160	6.0	4.2	–	2.5	302
II	31.3.81	1234	176	–	24.0	10.6	11.0	227
	21.4.81	1327	103	7.7	14.0	–	8.0	287
III	28.4.81	1505	211	26.0	9.0	*	8.0	378
	7.7.81	1050	299	16.0	9.0	11.0	4.0	277
IV	29.7.81	1230	250	25.0	8.0	10.0	6.0	255
	27.10.81	1240	150	11.0	11.0	–	8.0	381
	Controls	1278	114	10.0	60.0	20	–	75
		±127	±24	±3	±10	±5	–	±14

LYMPHOCYTES (nmol/10^6 cells) (b)

DATE	ATP	ADP	AMP	GTP	GDP	dGTP
31.3.81	3.68	0.91	0.31	0.35	*	0.2
Controls	3.07	1.35	0.18	0.40	0.23	–
	±.70	±.37	±.06	±.10	±.07	

Results are corrected (a) for the haematocrit (RBC) or (b) dilution by cell water (PBMs) using isotope dilution. The term lymphocyte refers to peripheral blood mononuclear cells (PBMs), separated as described.
– below the limits of detection for the method.   * impure peak

clearance. Less than 10% of dCR was found in the urine with the drug alone; but 20-40% was recovered when dCR was combined with THU. The plasma level remained <1µmol/1/ indicating effective inhibition of dCR deaminase but rapid renal excretion. No nucleosides, deoxynucleosides (or bases) were detected in extracts of PNP deficient PBMs or erythrocytes. Both contained low but detectable amounts of dGTP. In addition, severe erythrocyte GTP depletion associated with grossly raised NAD+ levels was noted throughout (Table II). Despite evidence of effective absorption of guanine (≃ 30% as urinary metabolites) no significant increment in red cell GTP levels was noted indicating rapid deamination. GTP, ATP,and NAD+ levels in PBMs appeared similar to controls. However, it was found subsequently that all 'lymphocyte' preparations separated by Ficoll-Triosil are heavily and varyingly contaminated with platelets (see Goday et al this symposium) so that no significance can be placed on the PBM results at present.

DISCUSSION

The clinical, immunological and biochemical data in our patient
are in accord with previous findings by others[1,3,6]. In addition,
we have made several novel observations. The patient's erythrocytes showed severe GTP depletion, with grossly raised NAD^+ levels,
and dGTP was detected in the patients PBMs as well as erythrocytes.
No therapy has had any effect on the erythrocyte or lymphocyte dGTP
levels, or the immunological status of the patient. Unfortunately,
since no pre-treatment PBM levels were available and because of the
heavy platelet contamination no comment can be made regarding nucleotide levels in the latter. The immunological progress of our patient
over 20 months and our inability to reverse this with any regime, is
further confirmation of the selective attrition of T-cell function
that occurs in PNP deficiency.

The enzyme defect in this (the eleventh patient reported) is
clearly severe and similar to that of Stoop et al[3]. In three other
surviving patients the defect appears not so complete since significant amounts of uric acid are reported in plasma and urine[1]. No
satisfactory therapy had been found for these patients but there had
already been a long period of lymphopenia prior to investigation.
Two patients are currently supported by enzyme replacement with
irradiated red cell transfusions[3,6], with only partial restoration
of in vitro T-cell function.

Although therapy with parenteral dCR commenced at three months
there was already evidence of profound lymphopenia with diminishing
PHA responses. The lack of effect of DCR in vivo, even with THU,
on either the immunology or the dGTP levels, contrasts with its
effectiveness in vitro[2]. This suggests that either current
hypotheses are incorrect or we failed to sustain circulating dCR
levels for long enough due to rapid renal excretion, despite THU
inhibiting dCR deamination.

The only metabolic improvement was the increment in erythrocyte
ATP and GTP levels. However, the latter substance never reached
normal values despite evidence of effective guanine absorption
during oral therapy. The NAD^+ levels remained consistently high
and accumulation of this substance may well be central to the
erythrocyte GTP depletion and purine over-production in PNP
deficiency as discussed elsewhere. The gross purine over-production
and low GTP levels may also be related to the neurological problems
of severely PNP deficient children. The profound hypotonia and
developmental delay has assumed greater importance than the
immunodeficiency in our patient at present. A spastic tetraparesis
was noted in our patient's sibling and also in the case of Stoop
et al where there has been no neurological improvement after 4 years

of enzyme replacement therapy with red cell transfusions[3,4]. This
fact combined with the lack of convincing evidence of improvement
in T cell function in vivo following transfusion therapy attracted
us to the recent report of the absence of immunogenicity of human
amniotic epithelial cells[7].   Having demonstrated the ability of
such cells to synthesise PNP in vitro we proceeded to an implant
of human amnion at 19 months of age. No detectable clinical,
immunological or biochemical improvement has yet accrued.

Although successful strategies to abolish the purine
overproduction in our patient have not been defined, our studies
have provided insight into possible control mechanisms for this
phenomenon which could provide a basis for further study.

REFERENCES

1. Enzyme defects and Immunodysfunction. Ciba Symposium 1968
    (new series) Amsterdam. Excerpta Medica (1979).

2. D.A.Carson, J.Kaye,J.E.Seegmiller, Pathogenetic mechanisms
    in deficiencies of adenoside deaminase and purine nucleoside
    phosphorylase. In: Inborn errors of immunity and
    phagocytosis. Gutthler F.,Seakins J.W.T.,Harkness R.A.,eds.
    Lancaster: MTP Press, 129 (1979).

3. J.W.Stoop,B.J.M.Zegers,W.Kuis,C.J.Heijen,J.J.Roord,M.Duran,
    S.K.Wadman,G.E.J.Staal, Purine Nucleoside Phosphorylase
    Deficiency: Long Term Clinical, Immunological and Metabolic
    Follow-up. In: Primary Immunodeficiencies. Elsevier.
    Amsterdam 301, (1980).

4. A.R.Watson,D.I.K.Evans,H.B.Marsden, V.Miller,P.A.Rogers.
    Purine nucleoside phosphorylase deficiency associated with a
    fatal lymphoproliferative disorder. Arch.Dis.Child.
    56 : 7;563 (1981).

5. H.A.Simmonds,A.R.Watson,D.R.Webster,A.Sahota,D.Perrett,
    GTP depletion and other erythrocyte abnormalities in
    inherited PNP deficiency. Biochem. Pharmac. 31, 941 (1982).

6. K.C.Rich,E.Mejians,I.H.Fox. Purine nucleoside phosphorylase
    deficiency: improved metabolic and immunological function
    with erythrocyte transfusion. N.Eng.J.Med.303 : 973 (1980).

7. C.A.Akle,K.I.Welsh,M.Adinolfi,S.Leibowitz,I.McColl.
    Immunogenicity of Human Amniotic Epithelial Cells after
    Transplantation into Volunteers. Lancet ii : 1003 (1981).

THE EFFECT OF DEOXYCYTIDINE AND TETRAHYDROURIDINE IN PURINE NUCLEOSIDE PHOSPHORYLASE DEFICIENCY

J.W.Stoop[1], B.J.M.Zegers[1], L.J.M.Spaapen[1], W.Kuis[1],

J.J.Roord[1], G.T. Rijkers[1], G.E.J.Staal[2], G.Rijksen[2],

M.Duran[1] and S.K.Wadman[1]

1=University Children's Hospital "Het Wilhelmina Kinderziekenhuis", Nieuwe Gracht 137, 3512 LK Utrecht
2=Department of Medical Enzymology, State University
Hospital, Catharijnesingel 101, 3500 CG Utrecht, The
Netherlands

INTRODUCTION

Adenosine deaminase (ADA) and purine nucleoside phosphorylase (PNP) deficiency have been recognized as the primary cause of an associated immune deficiency syndrome. A number of mechanisms have been proposed to explain the predominant effect of these enzyme deficiencies on the development and function of the lymphoid system. One of the mechanisms concerns the phosphorylation of accumulated metabolic compounds i.e. deoxyadenosine (dAdo) in case of ADA-deficiency and deoxyguanosine (dGuo) in case of PNP deficiency in the lymphoid cells and particularly in thymocytes (1). Indeed increased deoxyATP and deoxyGTP levels have been found in the lymphocytes of ADA- and PNP-deficient patients respectively (2,3). These triphosphates may inhibit the enzyme ribonucleotide reductase which leads to a depletion of deoxyCTP and interference with lymphocytic DNA-synthesis (1).

In ADA-deficiency transplantation of bone marrow derived lymphoid stem cells is the method of choice. However, in most instances no suitable bone marrow donor is available. Enzyme replacement therapy with irradiated normal erythrocytes to lower accumulated dAdo and dGuo has resulted in some ADA- or PNP-deficient patients in improvement of immune function (4,5,6,7). Iron overload and

viral infection (e.g. cytomegalovirus, hepatitis B virus) are some
of the complications of this treatment. Administration of deoxycytidine (dCyd), to bypass the ribonucleotide reductase has been suggested as a possibility to overcome the impairment of DNA-synthesis
(1). However, a high activity of (deoxy)cytidine-deaminase is found
in the mucosa of the gut as well as in the liver and as a consequence
orally and intravenously administered dCyd, is rapidly converted to
uridine and uracil before reaching the cells where it is needed.

We treated a PNP-deficient patient, born in January 1975, (8),
with erythrocyte transfusions for more than $5\frac{1}{2}$ years. This regimen
has resulted in a partial restoration of in vitro T cell function.
Orally, and lateron intravenously administered dCyd had no additional benificial effect at all(6). The still existing T cell deficiency is reflected in the clinical condition of the patient: she still
shows a vulnerability to infection(9). We therefore have treated the
patient again with dCyd, this time in combination with tetrahydrouridine (THU), a potent non-toxic inhibitor of (deoxy)cytidine deaminase. We present here the metabolic and immunological results
during this treatment.

CLINICAL PROTOCOL AND METHODS

Clinical protocol:THU, kindly provided by Dr.Ch.W.Young (Me-
morial Sloan-Kettering Cancer Center, New York) was administered
subcutaneously 50 mg/kg bodyweight/day in 3 doses. dCyd was given
from the 3rd day on, 50 mg/kg bodyweight/day in a continuous intravenous infusion, for 3 weeks; during the 4th week the same amount
of dCyd was given in a continuous subcutaneous infusion during 10
hours overnight. This enzyme replacement therapy was continued
during the administration of THU and dCyd.
At regular intervals the immune status of the patient was
assessed as well as the serum content and the urinary excretion of
purine- and pyrimidine metabolites according to methods described
in detail previously (6,8,10). The (deoxy)ribonucleotide content
of the erythrocytes was analyzed by HPLC (11). Perchloric acid
extracts of freshly withdrawn blood were made according to Cohen
et al. (2) with minor modifications. In order to analyze deoxyribonucleotides the neutralized perchloric acid extracts were treated
with sodium periodate according to Garret and Santi (12). 2,3 Di-
phosphoglycerate (2,3-DPG) in the erythrocytes was determined as
previously described (13). Ecto-5'-nucleotidase on intact lymphocytes was determined as described(14). Adenosine deaminase activity
of the lymphocytes was determined essentially according to van Laarhoven et al. (15).

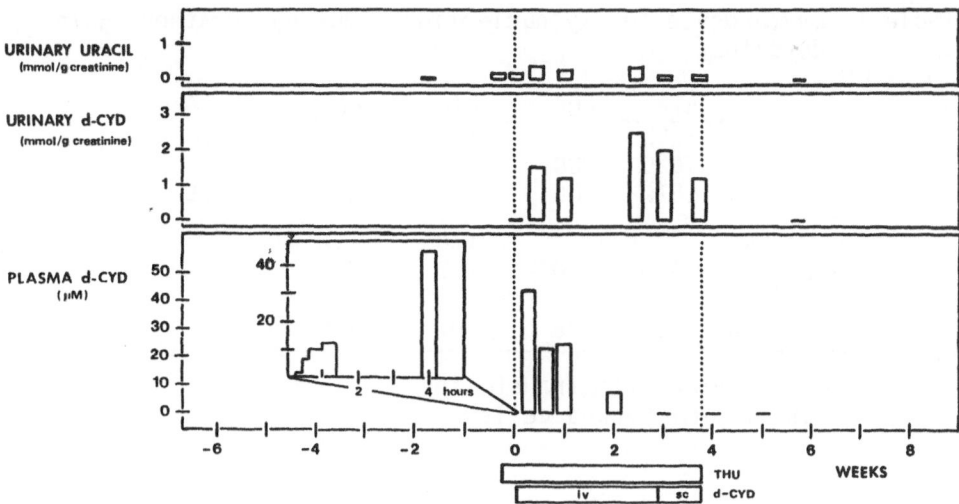

Fig.1. Deoxycytidine (dCyd) and uracil levels in plasma and urine
       during treatment with dCyd/THU.

RESULTS AND DISCUSSION

    An earlier trial of administration of dCyd orally and intrave-
nously did not result in a detectable serum level of dCyd, whereas
the urinary excretion of uracil increased markedly. This time, during
combined administration of dCyd and THU a considerable amount of
dCyd was found in the plasma and in the urine and uracil excretion
did not change (Fig.1). This reflects an effective inhibition of deoxycytidine deaminase by THU and strongly suggests that the compound
is administered under conditions which make adequate uptake by cells
possible. The low dCyd plasma level during subcutaneous administration (see Fig.1) is due to the time interval between the end of the
overnight infusion and the blood sampling.

    Table 1 shows (deoxy)nucleotide levels of the erythrocytes
during and after treatment. It is apparent that during treatment
dCTP levels are about 5 nmol/ml packed cells whereas afterwards the
level fell to less than 1 nmol/ml packed cells. Most interestingly
erythrocyte dGTP levels decreased during treatment; this may suggest
effective competition between dGuo and dCyd for the enzyme deoxycytidine kinase. The observation supports earlier studies that dCyd
does not only bypass the ribonucleotide reductase but also competes
with dGuo for phosphorylation by deoxycytidine kinase (2). The
dCTP levels of the mononuclear cells of the patient could not be
determined appropriately since the level is below the detection
limit of the HPLC system.

Table 1. Erythrocyte (deoxy)nucleotides[*] during treatment with
         dCyd/THU.

	ATP	ADP	GTP	dGTP	dCTP
0.5 week	1750	210	190	6.0	4.6
1 week	1620	200			5.1
2 weeks	1570	200	160	4.6	4.1
→					
3 weeks			170	1.8	
4 weeks	1470	200	150	0.4	3.3
after					
treatment	1200	106	107	3.5	<1

→ erythrocyte transfusion
* in nmol/ml packed cells

        Table 2 shows the ADA and ecto-5'-NT activity of the lymphocy-
tes during dCyd treatment.No significant changes of enzyme activity
were observed as compared to pretreatment values.The high ADA activities and the low ecto-5'-NT activities found are compatible with
the presence of a relatively immature cell population. Low ecto-5'-
NT activities have also been found in lymphocytes of an ADA-deficient patient (16). The 2,3 DPG content in the erythrocytes remained
high as previously shown (13), reflecting the active breakdown of
purine nucleosides by PNP present in the transfused erythrocytes.

        The aim of dCyd treatment to obtain improvement of T cell
function and clinical condition was not achieved during the first
four weeks of treatment. However, one observation presented in
detail in this volume (17) warrants to be mentioned. Already one
year before treatment we observed that freshly isolated lymphocytes of the patient contained about 10% cells which are in S-phase.

Table 2. Lymphocyte ADA and ecto-5'-NT activity[*] during treatment
         with dCyd/THU

	PNP[-] patient	Controls
ecto 5'-NT	0.78 $\pm$ 0.33 (n=3)	13.6 $\pm$ 6.1 (n=23)
ADA	103 $\pm$ 142 (n=6)	83 $\pm$ 18 (n=19)
	range 167-560	

* nmol/hr/10^6 cells

This findings suggested that the lymphocytes of the patients are hampered in vivo in the completion of the cellcycle. During dCyd/THU treatment the S-phase cells disappeared and the absolute number of peripheral blood lymphocytes intermittently increased. This suggests that the cells were able now to complete the cellcycle and that most probably in vivo lymphocyte proliferation has taken place. The observation supports in our opinion the use of dCyd in PNP- and ADA-deficient patients.

Postscriptum

A second course of dCyd/THU was started for a period of two months, five weeks after cessation of the first. Both dCyd and THU were given subcutaneously and in the same dosage as in the former treatment course. Again a slight rise in peripheral blood lymphocyte numbers was observed. This time an important change of in vitro T cell function was observed: the antigen-specific proliferative response of the lymphocytes on the anamnestic antigen Candida Albicans became repeatedly positive and for the first time in the life of the patient. The delayed type skin reaction to Candida Albicans remained negative. We should realize that in vivo reconstitution of T cell function may take a rather long time in this condition.

REFERENCES

1. Martin, D.W. and Gelfand E.W. (1981). Ann.Rev.Biochem.50,845-877.
2. Cohen, A., Gudas, L.J., Ulman.B. and Martin D.W. (1979). In: Enzyme defects and immune dysfunction. Ciba Foundation Series, 68, Excerpta Medica, pp. 101-109.
3. A.Simmonds (1982). Personal communication.
4. Polmar, S.H. (1979). In: Inborn errors of specific immunity. (B.Pollara, K.J. Pickering, H.J. Meuwissen, J.H.Porter, eds.) Academic Press, New York, pp. 343-351.
5. Zegers, B.J.M., Stoop, J.W. (1982). Adv.Clin.Enzymology (in press).
6. Zegers, B.J.M., Stoop, J.W., Staal, G.E.J. , Wadman, S.K., (1979). In: Enzyme defects and immune dysfunction. Ciba Foundation Series 68, Excerpta Medica pp.231-241.
7. Rich, K.C., Mejias, E., Fox, H.J. (1980). N.Engl.J.Med. 303, 973-977.
8. Stoop, J.W., Zegers, B.J.M., Hendricks, G.F.M., Siegenbeek van Heukelom, L.H.Staal, G.E.J., de Bree, P.K., Wadman, S.K., Ballieux, R.E. (1977). New Engl.J.Med. 296, 651-655.
9. Stoop, J.W., Zegers, B.J.M., Kuis, W., Heijnen, C.J., Roord, J.J., Duran,M., Wadman, S.K., and Staal, G.E.J. (1980). Primary Immunodeficiencies INSERM symposium no.16 (Seligmann M. and Hitzig, W.H., eds). Elsevier/North Holland Biomedical Press, Amsterdam, pp. 301 -311.

10. Wadman, S.K., de Bree, P.K., van Gennip, A.H., Stoop, J.W.,
    Zegers, B.J.M., Staal, G.E.J., Siegenbeek van Heukelom, L.H.
    (1977). In: Purine Metabolism in Man II. Regulation of pathways
    and enzyme defects (Müller, M.M., Kaiser, E. and Seegmiller,
    J.E., eds.) Plenum Press, New York, pp.471-476.
11. Van Gennip, A.H., Grift, J., Wadman S.K., de Bree, P.K.(1978).
    In: Biological Biomedical Applications of liquid chromatography
    II. (Hawk, G.L. ed.) Marcel Dekker Inc. New York, Basel. pp.
    337-348.
12. Garret, C. and Santi, D.V. (1979). Anal.Biochem. 99, 268-273.
13. Staal, G.E.J., Stoop, J.W., Zegers, B.J.M., Siegenbeek van Heu-
    kelom, L.H. van der Vlist M.J.M., Wadman, S.K., (1980). J.Clin.
    Inv. 65, 103-108.
14. Bouman, H., Rijksen, G., Hofstede, J., Staal, G.E.J., Zegers,
    B.J.M. and Spaapen, L.J.M., this volume.
15. Van Laarhoven, J.P.R.M., Spierenburg G.Th. and de Bruyn C.H.M.M.
    (1980). J.Immunol.Methods 39, 47-58.
16. Boss, G.R., Thompson, L.F., O'Connor R.D., Ziering R.W., Seeg-
    miller J.E. (1981). Clin.Immunol.Immunopath. 19, 1-7.
17. Rijkers, G.T., Zegers, B.J.M., Spaapen, L.J.M., Rutgers, D.H.,
    Kuis, W., Roord, J.J., and Stoop, J.W., this volume.

LYMPHOCYTE ECTO 5'-NUCLEOTIDASE IN IMMUNODEFICIENCY AND LEUKAEMIA

A.D.B. Webster, T. Shah and T.J. Peters

Divisions of Immunological Medicine and
Clinical Cell Biology, Clinical Research Centre
Harrow, Middx. England

Circulating blood lymphocytes have relatively high activity
of ecto 5'-nucleotidase (EC 3.1.3.5 - 5'N) as compared to
monocytes of neutrophils (Webster et al 1978). This activity
varies between different subpopulations; B lymphocytes having
about four times the activity of 'E' rosetting T lymphocytes (Rowe
et al 1980). 5'N activity also varies between T cell
subpopulations, OKT8 (suppressor) T cells having about half the
activity of OKT4 (helper) cells (Boss et al 1980, Massaia et al
1982). Pro-thymocytes and cortical thymocytes have very low 5'N
activity, medullary thymocytes and cord blood T cells having about
half the activity of circulating T cells (Fig.1). The 5'-N
activity in cortical thymocytes can be increased by incubation
with either thymic epithelial cells or thymosin (Cohen et al (1981).

The active site of the enzyme seems to be superficially
placed on the plasma membrane since its activity can be blocked by
lectins such as Concanavolin A and diazosulphanilic acid
(Williamson et al 1976). The physiological substrate for ecto
5'-N has still not formally been identified. The enzyme has
activity against a variety of 5'-mononucleotides; but has
relatively weak activity against deoxynucleotides as compared to
the cytosol nucleotidase (see below). In practice, adenosine
monophosphate (AMP) is used as substrate in most assays. However,
it is very unlikely that this is the physiological substrate
because AMP is present in less than pmol quantities in plasma
while the Km for AMP is between 10-20 µM (Webster et al 1979;
Edwards and Fox 1979). However, ecto 5'-N may be part of a system
for degrading adenosine diphosphate (ADP) released from platelets
during aggregation. Lymphocytes do have a surface ecto ADPase
which will convert ADP to AMP; the latter then being converted by

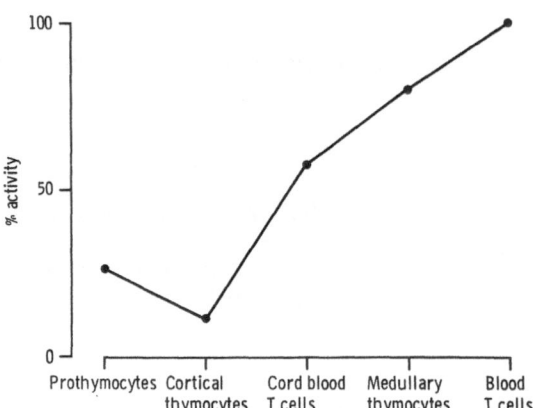

Fig.1.   Taking the 5'-N activity of circulating T cells as 100%,
         this figure shows the relative activities at various stages
         of T cell maturation. The percentages were derived from
         Rowe et al (1980) and Ma et al (1982) and are only approximate
         since the same assays were not used for all the cell types.

ecto 5'N to adenosine (Smith et al 1981). Adenosine deaggregates
platelets and is a potent vasodilator, and this might help the
lymphocyte gain access to inflammatory sites through partially
thrombosed vessels. However the main function of ecto 5'-N may not
involve purine metabolism; and it's function may be to remove the
phosphate from some other ribosylated compound which can then enter
the cell.

Ecto 5'N in immunodeficiency syndromes

     Many groups have now confirmed that the circulating lymphocytes
of patients with both X-linked and late onset hypogammaglobulinaemia

have low 5'-N activity (Johns on et al 1979; Edwards et al 1978).
This is partly due to the very low numbers of B lymphocytes in
some of these patients, and is also due to low activity in their T
cells (Rowe et al 1979; Thompson et al 1979). Early work by
ourselves (Rowe 1980) and others (Edwards et al 1978) suggested
that the enzyme in hypogammaglobulinaemic patients was
structurally normal, with the same Km for AMP as in normal
lymphocytes. However, using lymphocytes from defibrinated blood
and a different assay system at pH 9, Shah et al (1982) have
recently found that the lymphocytes from some patients with late
onset hypogammaglobulinaemia have a Km for AMP which is 10-fold
higher than that in normal lymphocytes. This implies that the
enzyme itself is abnormal and that the defect may be closely
related to the primary cause of the disease. However, the
prevailing view is that the low level of 5'-N in the T and B
lymphocytes of patients with hypogammaglobulinaemia reflects
immaturity of these cell types, and is not directly implicated in
the underlying cause of the immunodeficiency. This is supported
by the finding that αβ-methylene adenosine diphosphate, an
inhibitor of 5'-nucleotidase, has no effect on the in vitro
proliferation and immunoglobulin production of normal B cells
stimulated with pokeweed mitogen (Webster et al 1979). However,
this may be irrelevant since the defect in most patients with
hypogammaglobulinaemia probably lies in a failure of the pre-B
cell to switch isotypes and differentiate into mature B cells.

From the diagnostic point of view, the measurement of
lymphocyte ecto 5'-N has little place in the diagnosis of immunodeficiency syndromes, as these can be diagnosed by other tests.
The level of lymphocyte 5'-N activity can also not be used to
distinguish between the various immunodeficiency syndromes, since
late onset (varied) hypogammaglobulinaemia, X-linked agammaglobulinaemia and severe combined immunodeficiency are all
associated with low levels (Webster et al 1978; Ross et al 1982).
In fact, any condition where there are large numbers of 'immature'
T cells in the circulation will be associated with low total
lymphocyte ecto 5'-N activity.

Thompson et al (1980) provided indirect evidence that the B
lymphocytes of obligate heterozygotes for X-linked agammaglobulinaemia had low 5'-N activity. In their system, the non-T cells
from heterozygotes were stimulated with EB virus to form lymphoblastoid cell lines. Ecto 5'-N was only low if measured in the
first 10 days of culture, probably because after this individual B
cell clones begin to dominate such cultures. This finding is
surprising because the mothers of patients with X-linked agammaglobulinaemia have normal humoral immunity, although detailed in
vitro functional studies of their circulating B cells have not
been reported. However, Matamoros et al (1982) have found that
the circulating T lymphocytes of obligate heterozygotes are

abnormal, having a similar foetal pattern of lactate dehydrogenase
isoenzymes to that in the T cells of homozygotes. Half the total
lymphocytes in heterozygotes should theoretically be abnormal
(Lyon hypothesis) and it is possible that these accumulate in the
circulation while the normal lymphocytes are involved in immune
reactions elsewhere in the spleen, lymph nodes, bone marrow and
gut.

5'-N activity in the Leukaemias

There are striking variations in the level of ecto 5'-N
activity in different leukaemic cells (Table 1). Both acute and
chronic B lymphatic leukaemia cells (B-ALL, B-CLL) have very low
levels, probably representing the undifferentiated nature of these
cells. The more primitive acute T lymphoblastic leukaemia cells
(THY-ALL) have lower 5'-N activity than the common variety (C-ALL)
(Reaman et al 1981; Sylwestrowicz et al 1982). These two
conditions are sometimes difficult to differentiate, and measuring
ecto 5'-N activity can guide the Physician towards the correct
management. The 'blast' stage of chronic granulocytic leukaemia
is another condition where the measurement of ecto 5'-N can be
helpful (Gutensohn & Thiel, 1981). Myeloblasts, like normal
neutrophils, have low ecto 5'-N activity, in contrast to the
relatively high activity in lymphoblasts; and this is important
because these two types of "blast crises" are treated differently.

Table 1. Ecto 5'-Nucleotidase activity in the Leukaemias

Low	Normal	Raised
Thy-ALL	Sezary cells	C-ALL
B-ALL		
B-CLL		
Hairy cell leukaemia		
Waldenstrom macro-globulinaemia		
Myeloblasts in CGL		Lymphoblasts in CGL

There are ethnic and age-related differences in lymphocyte
ecto 5'-N activity, but these are unlikely to cause problems in
the diagnosis of different types of leukaemia. Nevertheless, it
should be remembered that the ecto 5'-N activity of T cells
declines after the age of 40 years (Boss et al 1980) and East
Africans have lower lymphocyte 5'-N than Europeans (Levin et al
1982).

Cytosol 5'-nucleotidase

Carson et al (1981) have shown that B derived lymphoblastoid cell lines have higher activity of a cytoplasmic 5'-N than T derived lines. This enzyme has a lower pH optimum as compared to ecto 5'-N and prefers deoxyadenosine monophosphate to AMP as a substrate. It is suggested that this difference in activity explains why the B cells in ADA deficient patients with severe combined immunodeficiency are partially spared from the toxic effects of high plasma deoxyadenosine levels. However, it is not yet known whether peripheral blood B and T cells have different activities of cytoplasmic 5'-N.

CONCLUSION

There is good evidence that the level of ecto 5'-N activity reflects the stage of lymphocyte maturation. This is seen in the increasing activity of the enzyme as lymphocytes mature within the thymus; and in the association of low activity with the more 'primitive' types of leukaemia. A parallel phenomenon occurs in chick cartilage cells where ecto 5'-N increases as the cells mature (Rodan et al 1977). It is unclear whether the low lymphocyte ecto 5'-N associated with various immunodeficiency syndromes merely reflects a failure of lymphocyte maturation, or is the primary cause of the immunodeficiency. This question is unlikely to be resolved until the physiological function of ecto 5'-N is discovered.

REFERENCES

Boss, G.B., Thompson, L.F., Spiegelberg, H.L., Pichler, W.J. and Seegmiller, J.E., 1980, Age dependency of lymphocyte ecto 5'-nucleotidase, J.Immunol., 125:679-682.

Carson, D.A., Kaye, J. and Watson, D.B., 1981, The potential importance of soluble deoxynucleotidase activity in mediating deoxyadenosine toxicity in human lymphoblasts, J. Immunol., 126:348-352.

Cohen, A., Dosch, H.M. and Gelfand, E.W., 1981, Induction of ecto-5'-nucleotidase activity in human thymocytes, Clin. Immunol.Immunopathol., 18:287-290.

Edwards, N.L., Magilavy, D.B., Cassidy, J.T. and Fox, I.H., 1978, Lymphocyte Ecto-5'-nucleotidase deficiency in agammaglobulinaemia, Science, 201:628-630.

Gutensohn, W. and Thiel, E., 1981, High levels of 5'-nucleotidase activity in blastic chronic myelogenous leukaemia with common ALL-antigen, Leuk.Res., 5:505.

Johnson, S.M., North, M.E., Asherson, G.L., Allsop, J., Watts, R.W.E. and Webster, A.D.B., 1977, Lymphocyte purine 5'-nucleotidase deficiency in primary hypogammaglobulinaemia, Lancet i:168-170.

Levin, A., Jones, M., Shah, T. and Peters, T.J., 1982, unpublished.

Ma, D.D.F., Sylwestrowicz, T.A., Massaia, M., Price, G., Tidman, N.

and Hoffbrand, A.V.,   1982, unpublished.
Massaia, M., Ma, D.D.F., Sylwestrowicz, T.A., Tidman, N., Price, G.
    and Hoffbrand, A.V.,   1982, Enzymes of purine metabolism in human
    peripheral lymphocyte populations, Clin.Exp.Immunol. (in press).
Matamoros, N., Abad, E. and Webster, A.D.B.,   1982, LDH isoenzymes
    in the T lymphocytes of patients with primary hypogamma-
    globulinaemia, Clin.Exp.Immunol., in press.
Reaman, G.H., Blatt, J. and Poplack, D.G.,   1981, Lymphoblast
    purine pathway enzymes in B-cell acute lymphoblastic leukaemia,
    Blood, 58:330-332.
Rodan, G.A., Bourret, L.A. and Cutler, L.S.,   1977, Membrane
    changes during cartilage maturation, J.Cell.Biol., 72:493-501.
Rowe, M., DeGast, C.G., Platts-Mills, T.A.E., Asherson, G.L.,
    Webster, A.D.B. and Johnson, S.M.,   1979, 5'-nucleotidase of B
    and T lymphocytes isolated from human peripheral blood, Clin.
    Exp. Immunol., 36:97-101.
Rowe, M., DeGast, G.C., Platts-Mills, T.A.E., Asherson, G.L.,
    Webster, A.D.B. and Johnson, S.M., 1980, Lymphocyte 5'-nucleo-
    tidase deficiency in primary hypogammaglobulinaemia and cord
    blood, Clin.Exp.Immunol.,   39:337-343.
Shah, T., Webster, A.D.B., Peters, T.J.,   1982, unpublished.
Smith, G.P., Shah, T., Webster, A.D.B. and Peters, T.J.,   1981,
    Studies on the kinetic properties and subcellular localisation
    of adenosine diphosphatase activity in human peripheral blood
    lymphocytes, Clin. Exp. Immunol., 46:321-326.
Sylwestrowicz, T., Piga, A., Murphy, P., Ganeshaguru, K., Russell,
    N.H., Prentice, H.G. and Hoffbrand, A.V.,   1982, The effects of
    deoxycoformycin and deoxyadenosine on deoxyribonucleotide
    concentrations in leukaemia cells, Brit.J.Haemat., 51.
Thompson, L.F., Boss, G.R., Spiegelberg, H.L., Jansen, I.V.,
    O'Connor, R.D., Waldmann, T.A., Hamberger, R.N. and Seegmiller,
    J.E.,   1979, Ecto-5'-nucleotidase activity in T and B
    lymphocytes from normal subjects and patients with congenital
    X-linked agammaglobulinaemia, J. Immunol., 123:2475-2478.
Thompson, L.F., Boss, G.R., Spiegelberg, H.L., Bianchino, A. and
    Seegmiller, J.E.,   1980, Ecto 5'-nucleotidase activity in lympho-
    blastoid cell lines derived from heterozygotes in congenital
    X-linked hypogammaglobulinaemia, J.Immunol., 125:190-196.
Webster, A.D.B., North, M., Allsop, J., Asherson, G.L. and
    Watts, R.W.E., 1978, Purine metabolism in lymphocytes from
    patients with primary hypogammaglobulinaemia, Clin.Exp.
    Immunol.,   31:456-463.
Webster, A.D.B., Rowe, M., Johnson, S., Asherson, G.L. and
    Harkness, A., 1979, Ecto 5'-nucleotidase deficiency in primary
    hypogammaglobulinaemia, in: Ciba Foundation Symposium Series
    No.68, p.135-151, Elsevier, Amsterdam.
Williamson, F.A., James Morre, D. and Shen-Miller, J.,   1976,
    Inhibition of 5'-nucleotidase by Concanavalin A: Evidence for
    localization on the outer surface of the plasma membrane.
    Cell Tiss.Res., 170:477-484.

CLINICOBIOCHEMICAL ANALYSIS OF FOUR CASES OF XANTHINE

OXIDASE DEFICIENCY

K. Nishioka, H. Yamanaka, T. Nishina,
T. Hosoya and K. Mikanagi

Institute of Rheumatology, Department of
Medicine, Tokyo Women Medical College

INTRODUCTION

Xanthinuria is a rare hereditary disease characterized by a
deficiency of xanthine oxidase (E C 1,2,3,2), which metabolized
hypoxanthine and xanthine to uric acid. In Japan, two cases of
xanthinuria have been reported. However, during the past two years,
we have investigated four patients with hypouricemia, hypouricosuria
and xanthinuria. In almost all of these patients, hypouricemia
(less than 1.0 mg/dl) was the most important index information to
make a definite diagnosis of this disorder. After clinical profiles
of these four patients were described, an easy method of detection
of xanthine oxidase activity from the duodenal mucosa was developed
by the authors. Urinary oxypurine was also analyzed by high-pressure
liquid chromatography, and the results are reported here.

MATERIALS AND METHODS

1. Case reports
Case 1: A 50-year-old female with a history of Basedow's disease
since she was 36 years old. She had no history of hematuria nor
colicky pain. She had no special family history, except her sister,
who had chronic thyroiditis. When she was 49, hypouricemia was discovered by chance. Her serum uric acid level was 0.3 mg/dl.
Case 2: A 67-year-old male with renal stones who died from
renal tumors. He had renal stones, but analysis of them was not
undertaken. His uric acid level was 0.1 mg/dl, and urinary uric
acid excretion was 0.16 mg/day.
Case 3: A 39-year-old male whose hypouricemia was discovered
when studying his chronic renal failure. His serum uric acid level

was 0.3 mg/dl, and urinary uric acid excretion was 0.082 g/day.

Case 4: The younger brother of Case 3, who had neither a medical history nor a present illness.

2. Methods

Xanthine oxidase activity of the duodenal mucosa obtained by gastrofiberscopy was determined by the following procedures. Tissue specimens weighing 5-6 mg were homogenized in pyrophosphate buffer solution (pH 8.0, 0.06M) and centrifuged at 10,000 g for 40 minutes.

The supernatants were concentrated with CF 25 cone, washed twice with 1.0 ml buffer, and reacted with a reaction mixture containing 0.5 n mol/l hypoxanthine at 37°C for 30 minutes. Uric acid product was determined by the rate of increased absorption of 293 nm at 37°C measured in a Gilford 2400 type recording spectrophotometer. Protein concentration of the specimens was determined by the Lowry method. Urinary oxypurines were assayed by HPLC as follows. One to two ml of urine samples from 24 hours total excretion were diluted 5-10 times with 0.9% NaCl and filtrated by a 0.45 µm Millipore filter. Three quarters of ml of this sample with 0.25 ml of 0.1 N NaOH and 1.0 ml of ethyo-acetate; n- Butanol = 2 : 1 solution were mixed for one minute.

Construction of HPLC was as follows LS-410K ODS (ID4 300 mn) and eluent 0.037 mol/l phosphate buffer pH 2.3 with 1% CH_3OH. Temperature was ambient, and flow rate was 1.0 ml/min. All of these oxypurine measurements were done at 260 nm.

RESULTS

1. Hypouricemia and Xanthinuria

Serum uric acid level was 0.3 mg/dl, 0.1 mg/dl, 0.3 mg/dl and 0.2 mg/dl, respectively, in the four cases.

Daily uric acid excretion was extremely low. Urinary hypoxanthine and xanthine were increased. Xanthine was 19.6 mg/dl, 11.3 mg/dl, 12.8 mg/dl and 26.1 mg/dl, respectively.

Xanthinuria was definitely diagnosed by these total oxypurines excreted in the urine over a 24-hour period.

Among these urines, the peak of allopurinol was not detected by HPLC. These data suggested that the xanthinuria of these patients was caused by xanthine oxidase deficiency.

DISCUSSION

Xanthinuria has been thought to be a rare hereditary disease involving xanthine oxidase deficiency. A summary of clinical and chemical features from 42 cases of xanthinuria worldwide was reported by Seegmiller.

Seventeen of these 42 patients had clinical symptoms arising from xanthine calculi of the urinary tract.

In this Japanese study series, two cases had clinical involvement and the other were without any clinical problems. Two cases with clinical problems did not relate directly to xanthine oxidase deficiency.

In one case, the patient had a renal stone and renal tumor, but had been asymptomatic for many years.

Arthritis, a muscle symptom of cramps, was not found in any of the cases. The clinical course of the patients with xanthine oxidase deficiency is benign and symptoms are silent. This is one of the reasons why it is difficult to investigate this disorder.

The fact that four cases were found during these two years suggests that this metabolic disorder is not as rare a disease as had been thought in Japan.

Cases 3 and 4 involved hereditary xanthine oxidase deficiency, but in Case 1, her younger sister's xanthine oxidase was 12% of normal activity and serum uric acid level was 2.4 mg/dl.

Case 2 involved "sporadic" xanthinuria.

There was no specific activity of xanthine oxidase in the duodenal mucosa in Cases 1, 3 and 4.

In a normal control group, this enzyme activity was 17.0 n mol/mg protein/hr in males and 15.4 n mol/mg protein/hr in females.

The assay method of xanthine oxidase from the duodenal mucosa using gastrofiberscopy is available for mass screening of this enzyme deficiency. The tissue distribution of this enzyme does not cause a problem in the liver.

There is no difficulty in measuring oxypurine excretion by the HPLC method.

Finally, our routine method for the definite diagnosis of xanthinuria is summarized:
  (1) Hypouricemia (less than 1.0 mg)
  (2) Hypouricosuria (less than 100 mg/24 hr)
  (3) Urinary uric acid, hypoxanthine, xanthine, allopurinol excretion assay by HPLC
  (4) Spectrophotometric assay of xanthine oxidase of the duodenal mucosa

REFERENCES

1. Seegmiller J.E.: Hereditary Xanthinuria. Metabolic Control and Disease (Boncy P.K. and Rosenberg L.E., ed.) P. 842, Philadelphia E.B. Saunders, 1980.

2. Yokoyama M., Suzuki T., Akaoka I: A Xanthine Stone in a Xanthinuri Boy: A Biochemical Case Study. The Journal of Urology 118; 561-653, 1977.

3. Yamanaka H., Nishioka K.: Allopurinol Metabolism in a Patient with Xanthine Oxidase Deficiency. Annals of Rheumatic Diseases, in press.

4. Greenlee L., Handler P.: Xanthine Oxidase. Influence
    of pH on Substrate Specificity. The Journal of
    Biological Chemistry 239; 1090-1095, 1964.

5. Lowry O.H., Rosebrough N.J., Randall R.J.: Protein Measure-
    ment with the Folin Phenol Reagent. Journal of
    Biochemistry 193; 265-275, 1951.

6. Ramsdell C.M., Kelley W.N.: The Clinical Significance
    of hypouricemia. Annals of Internal Medicine 78;
    239-242, 1973.

7. Kelley W.N.: Hypouricemia. Arthritis and Rheumatism
    18; 731-737, 1975.

8. Watts R.W.E., Watts J.E.M., Seegmiller J.E.: Xanthine
    Oxidase Activity in Human Tissues and Its Inhibition
    by Allopurinol (4-hyftocypytszolo 3,4-d pyrimidine).
    Journal of Laboratory & Medicine 66; 688-697, 1965.

HUMAN MYOADENYLATE DEAMINASE DEFICIENCY

William N. Fishbein

Biochemistry Division
Armed Forces Institute of Pathology
Washington, D.C. 20306

This entity was first described, in 5 patients, in 1978, as the consequence of the introduction of an effective histoenzymatic stain[1] for adenylate deaminase in frozen muscle biopsies. That report[2] established the following: (1) The deficiency state was quite common, occurring in 1-2% of muscle biopsies, and accompanied by relatively mild, though persistent, symptoms of exertional myalgia. (2) Application of a sensitive solution assay[3] verified that all 5 cases had less than 5% of the specific activity of adenylate deaminase in normal muscle biopsies, thus validating the histoenzymatic stain as a diagnostic tool. (3) There was no evidence of an enzyme inhibitor, by crossmixing studies of normal and deficient homogenates. (4) Affected patients had normal levels of adenylate deaminase in their red cells, indicating separate genetic control of the two isozymes. (5) Affected patients had normal skeletal muscle architecture and normal levels of other enzymes, indicating that the enzyme deficiency was specific, and was clearly not the cause of muscular dystrophy, as had often been suggested. (6) A simple provocative procedure, the lactate-ammonia-exercise-ratio (LAER test) demonstrated the physiological deficit in affected patients, and could be used for clinical diagnosis. (7) Skeletal muscle adenylate deaminase could be separated and identified, in its native state, by proper application of PAGE and the catalytic stain.

Subsequent reports have corroborated the utility of the histoenzymatic stain[4-12] and the LAER test[7-10], and the high frequency of the deficiency state in both sexes[6,12,13], but have emphasized its frequent coincidence with other muscle diseases[9,12-14], and questioned its clinical significance[10-12].

We have now identified, and verified by solution assay, 32 cases
of myoadenylate deaminase deficiency. Eighteen had no other neuromuscular pathology, and of these, 16 had persistent symptoms of easy
fatigue or weakness after exercise, and/or muscle soreness,
cramping, and tenderness, usually starting in adult life. About
half of the cases had a mild to moderate elevation in the serum
creatine kinase, and/or non-specific abnormalities in the electro-
myogram, prompting the biopsy. Patients were often misdiagnosed
(usually as polymyositis) and subjected to prolonged courses of
steroid therapy, which offer risk, but no benefit. Two of our
patients were biopsied and diagnosed in childhood, with symptoms of
benign congenital hypotonia, as had been suggested in the original
report[2]; another such case has been reported[11]. In most cases
physical examination has documented muscular weakness, rapid
fatigue, soreness and tenderness, and/or mild to moderate atrophy.
In view of these 18 cases, it seems clear that pristine myoadenylate deaminase deficiency is usually symptomatic, and that the view
that symptoms ought not be attributed to the deficiency[10] is
sorely misguided. The same conclusion was reached by Keleman, et
al[6], who found a 1.5% incidence in the muscle biopsy population,
but a 5-6 fold higher incidence in cases coded as exertional
myalgias.

The remaining 14 cases in our series were associated with a
variety of other muscle diseases, principally neuropathies and
polymyositis. The evidence argues that the deficiency state is
neither primary nor secondary to these other diseases, and we
therefore presume that the association is coincidental, and oc-
casioned by the high frequency of the deficiency state, the
auxiliary indication for muscle biopsy, and the now widespread use
of the histoenzymatic stain.

The best available evidence indicates that myoadenylate deaminase deficiency is genetically determined via autosomal recessive
inheritance: (1) Two affected sibling pairs have now been
reported[6,14]. (2) Patients with the muscle deficit have normal
levels of the enzyme in their red cells[2], lymphocytes[15], and
granulocytes[7,15], indicating a separate genetic origin. (3)
Antisera to the human muscle isozyme do not cross-react with the
isozymes of any of the separated human blood cell components[16],
nor with those of other human tissues[17,18]. (4) Both males and
females are affected, although males predominate (19 of 32 in our
series). (5) As shown in Figure 1, three putative carriers (the
mother and father of an affected male, and the son of an affected
father) had specific activities below the 90% tolerance limits for
the population, when correlated with their fiber-type distribution. The normal values show considerable scatter, but the least
squares regression line indicates a nearly 3-fold greater level of
adenylate deaminase in Type 2 than in Type 1 fibers, as was
suggested by the histoenzymatic stain[1]. The carrier levels are

less than 40% of the corresponding normal mean, but are far in
excess of the values for deficient biopsies. The variability is
sufficient to preclude statistical distinction between normals,
carriers, and deficients where the Type 2 fiber area accounts for
less than 25% of the biopsy. (6) Rabbit antisera to the purified

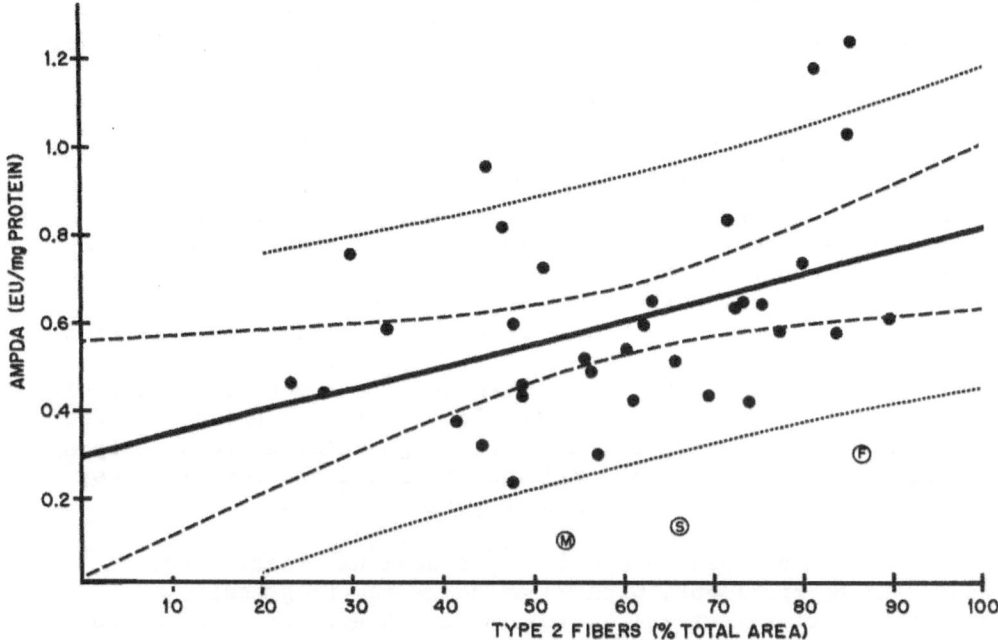

Fig. 1: Distribution of specific activities of adenylate
deaminase in 34 normal human muscle biopsies as a function of the
% area (or volume) occupied by type 2 fibers. Although the data
show considerable scatter, the least-squares-fitted regression
(solid line) indicates a nearly 3-fold higher specific activity in
type 2 than in type 1 fibers. The dashed curves show the 95%
confidence limits for the regression line, while the dotted curves
show the 90% tolerance limits (with $\alpha = 0.25$) for the population,
for use in evaluating individual cases. Shown in circles are the
values for three putative carriers: the mother (M) and father (F)
of a deficient male, and the son (S) of a deficient father. All
three are below the lower tolerance limit for the population at
the appropriate fiber-type distribution, although they are much
higher than the levels encountered in deficient patients ($<.05$
EU/mg protein). Note that normals, carriers, and deficients could
not be discriminated statistically where type 2 fibers account for
less than 25% of the biopsy area. Note also that one carrier (F)
would have been interpreted as normal on the basis of simple mean
specific activity, or if he were assumed to have the usual fiber
type distribution (45-60% type 2).

human skeletal muscle isozyme do not react with the residual
adenylate deaminase activity of deficient biopsies[19], although
they completely remove the enzyme from normal and carrier
biopsies, and cross-react extensively with the skeletal muscle
isozyme from lower animals[16]. This indicates that the residual
activity in such patients is not due to the normal muscle isozyme,
but represents either contaminant blood isozymes, or a fetal
muscle isozyme. Muscle tissue culture from deficient and normal
patients showed equivalent, low activities of adenylate deaminase,
suggesting that a fetal isozyme does exist, as for many other
muscle enzymes, and is the only form expressed in tissue
culture[7]. (7) No evidence could be found by competitive antigen
binding, of catalytically inactive enzyme protein in deficient
muscle biopsies, either in salt or Triton extracts[17]. This
argues that most cases of the deficiency state are due to a
complete gene block. (8) The deficit is physiologic, and not an
artifactual result of freezing damage, since fresh and frozen
biopsies on the same patient showed equivalent deficits[19].
This is also supported by the negative LAER tests on all deficient
cases studied thus far (Figure 2). (9) Kinetic ratios using the
artificial substrate, AMPS, also discriminate the residual activity
in deficient biopsies. Of all the blood cells and tissues tested,
only the skeletal muscle isozyme has a high activity towards this
substrate, and this is preserved in normal and carrier biopsies,
but is absent in those of deficient patients[17].

        The prognosis in myoadenylate deaminase deficiency
appears to be excellent, with no evidence for progressive
debilitation or structural damage in the absence of other disease,
and therapy should be aimed at reassurance, and the avoidance of
hazardous and ineffective long-term drug treatment.

        The plausible physiologic roles of muscle adenylate deami-
nase may be summarily listed in terms of the reactants and products
involved: (1) Removal of AMP may (a) deactivate phosphorylase and
shunt the glycolytic carbon source from glycogen to glucose[20]
(b) enforce myokinase dismutation of all ADP to ATP (and AMP)[21].
This effect may be reinforced at the site of contraction by isozymes bound to actomyosin[4,22,23]. (2) Production of ammonia may
(a) neutralize the acidosis consequent to lactate production (b)
maintain phosphofructokinase, and hence glycolysis, in an activated
state. (3) Accretion of IMP may (a) directly stimulate phosphorylase[24] (b) increase cAMP levels, by inhibiting its phosphodi-
esterase[25], to stimulate phosphofructokinase and phosphorylase b
to a conversion, thus reinforcing the hormonal affects of
epinephrine (c) lead to replenishment of guanine nucleotides[26]
(d) recycle to AMP during[27], or after[28] contractions,
accompanied by conversion of aspartate to fumarate, which fuels
the Krebs cycle for recovery from oxygen debt, and provides a
pathway for muscle amino acid and protein utilization.

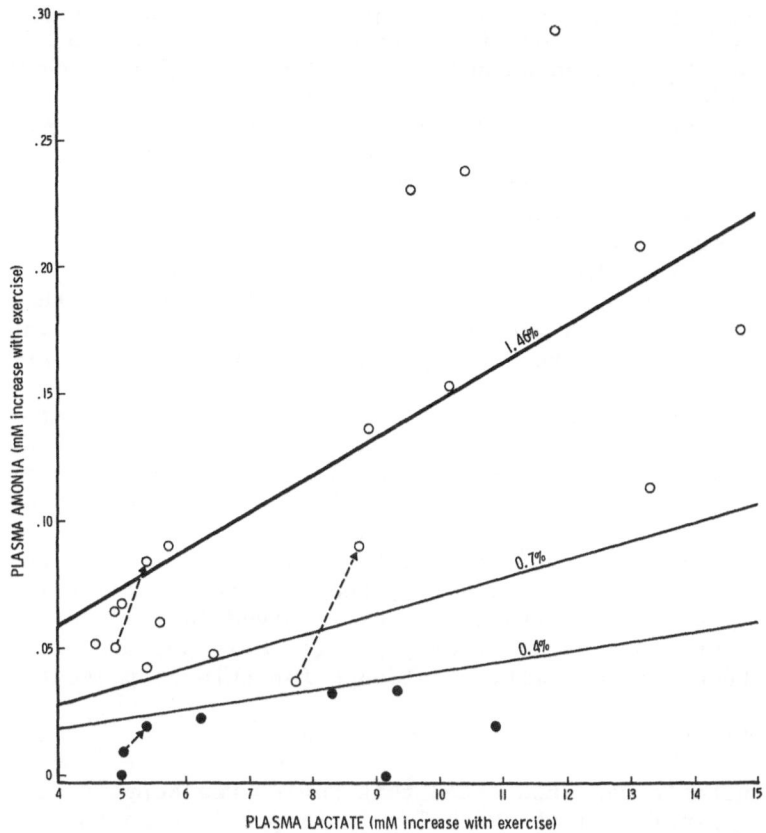

Fig. 2. The lactate-ammonia-exercise-ratio (LAER test) in normal
(o) and deficient (•) patients. After withdrawal of an anticoagu-
lated blood sample from the anticubital vein, a sphygmomanometer
is placed over the other arm, inflated to midway between systolic
and diastolic pressures, and locked. The patient then squeezes a
sponge or rubber ball as rapidly as possible, until extreme
fatigue appears, whereupon a post-exercise sample is withdrawn
before the cuff is released. The plasma samples separated by
centrifugation are used for determination of mM lactate and
ammonia. The exercise-induced increase is obtained by subtraction,
and the ammonia increase as a percent of the lactate increase is
the criterion for diagnosis. The test was repeated on three cases
(see arrows). Normal data show considerable scatter, but the
least-squares-fitted regression line (forced through the origin)
shows an average expected ammonia increase about 1.5% the lactate
increase. Only one normal case fell below the 0.7% line, and was
above it on repeat. All deficient cases lay below the 0.4% line.
A lactate increase > 4.5 mM is essential to indicate sufficient
hypoxic glycolysis[2].

The few deficient patients examined thus far have shown, not the expected accumulation of AMP in exercised muscles, but rather a depletion of all adenine nucleotides and their loss from muscle as nucleosides[8,29]. This suggests a compensatory dephosphorylation by nucleotidase when AMP deaminase is absent, in order to lower AMP levels. Since it is now known that muscle is a major site of de novo purine synthesis[30], exercise in affected patients might accelerate both purine synthesis and breakdown, and predispose them toward hyperuricemia and gout. It may be more than coincidental, therefore, that 1 of our patients has gout, 2 others have had elevated uric acid levels, and another case of the deficiency coexisting with gout has been reported[7]. The evidence at present is too meager to justify other than a cautionary attitude.

The same holds for possible association with malignant hyperthermia, since no untoward anesthetic incidents have been reported, and it is clear that myoadenylate deaminase deficiency is not a prerequisite[31]. The possibility that heterozygotes for myoadenylate deaminase deficiency may also be symptomatic has not been carefully evaluated, and awaits a more accurate delineation of normal enzyme levels. Finally, the ubiquitous plasma inhibitor of the muscle enzyme[32] might also be involved in the symptomatology of certain disease states, particularly polymyositis, although there is no positive evidence for this at present.

ACKNOWLEDGEMENT

The author's research cited this review was supported, in part, by a grant from the Muscular Dystrophy Association, under the auspices of the American Registry of Pathology, Inc. The opinions or assertions contained herein are the private views of the author and are not to be construed as official or as reflecting the views of the Departments of the Army or of Defense.

REFERENCES

1. Fishbein, W.N., J.L. Griffin, and V.W. Armbrustmacher, 1980. Stain for skeletal muscle adenylate deaminase: an effective tetrazolium stain for frozen biopsy specimens. Arch. Path. Lab. Med. 104: 462-466.
2. Fishbein, W.N., V.W. Armbrustmacher, and J.L. Griffin. 1978. Myoadenylate deaminase deficiency: A new disease of muscle. Science. 299: 545-548.
3. Fishbein, W.N. 1979. Indicator enzyme assays. I. Adenylate deaminase: principles and application to human muscle biopsies and blood cells. Biochem. Med. 22: 307-322.
4. Ashby, B., C. Frieden, and R. Bischoff. 1979. Immunofluorescent and histochemical localization of AMP deaminase in skeletal muscle. J. Cell Biol. 81: 361-373.
5. Meyer, R.A., J. Gilloteaux, and R.L. Terjung. 1980. Histo-

chemical demonstration of differences in AMP deaminase activity in rat skeletal muscle fibers. Experientia. 36: 676-677.

6. Keleman, J., D.R. Rice, W.G. Bradley, T.L. Munsat, S. DiMauro, and E.L. Hogan, 1982. Familial myoadenylate deaminase deficiency and exertional myalgias. Neurology. 32: In press.

7. DiMauro, S., A.F. Miranda, A.P. Hays, W.A. Franck, G.S. Hoffman, R.S. Schoenfeldt, and N. Singh. 1980. Myoadenylate deaminase deficiency. Muscle biopsy and muscle culture in a patient with gout. J. Neurol. Sci. 47: 191-202.

8. Sabina, R.L., J. L. Swain, B. M. Patten, T. Ashizawa, W.E. O'Brien, and E.W. Holmes. 1980. Disruption of the purine nucleotide cycle. A potential explanation for muscle dysfunction in myoadenylate deaminase deficiency. J. Clin. Invest. 66: 1419-1423.

9. Mercelis, R., J.J. Martin, I. Dehaene, Th. deBarsy, and G. Van den Berghe. 1981. Myoadenylate deaminase deficiency in a patient with facial and limb girdle myopathy. J. Neurol. 225: 157-166.

10. Hayes, D.J., B.A. Summers, and J.A. Morgan-Hughes. 1982. Myoadenylate deaminase deficiency or not? Observations on two brothers with exercise-induced muscle pain. J. Neurol. Sci. 53: 125-136.

11. Shumate, J. B., K.K. Kaiser, J.E. Carroll, and M. H. Brooke. 1980. Adenylate deaminase deficiency in a hypotonic infant. J. Pediatr. 96: 885-887.

12. Shumate, J.B., R. Katnik, M. Ruiz, K. Kaiser, C. Frieden, M. H. Brooke, and J.E. Carroll. 1979. Myoadenylate deaminase deficiency. Muscle and Nerve. 2: 213-216.

13. Kar, N.C., and C.M. Pearson. 1981. Muscle adenylate deaminase deficiency. Report of six new cases. Arch. Neurol. 38: 279-281.

14. Scholte, H.R., H.F.M. Busch, and I.E.M. Luyt-Houwen. 1981. Familial AMP deaminase deficiency with skeletal muscle type I atrophy and fatal cardiomyopathy. J. Inher. Metab. Dis. 4: 169-170.

15. Fishbein, W.N., J.L. Griffin, K. Nagarajan, J.W. Winkert, and V.W. Armbrustmacher. 1979. Immunologic uniqueness of muscle adenylate deaminase and genetic transmission of the deficiency state. Clin. Res. 27: 274A.

16. Fishbein, W.N., J. I. Davis, K. Nagarajan, J.W. Winkert, and J.W. Foellmer. 1980. Immunologic distinction of human muscle adenylate deaminase from the isozyme(s) in human peripheral blood cells: implications for myoadenylate deaminase deficiency. Arch. Biochem. Biophys. 205: 360-364.

17. Fishbein, W.N., and J.I. Davis. 1982. Immunologic and kinetic evaluation of myoadenylate deaminase deficiency. Fed. Proc. (USA). 41: 902.

18. Ogasawara, N., H. Goto, Y. Yamada, T. Watanabe, and T. Asano. 1982. AMP deaminase isozymes in human tissues. Biochim. Biophys. Acta. 714: 298-306.

19. Fishbein, W.N., V. W. Armbrustmacher, and J.L. Griffin.
    1981. Myoadenylate deaminase deficiency; verification on
    repeat biopsy, fresh or frozen, and origin of the residual
    enzyme. IRCS Med. Sci. 9: 103-104.
20. Davuluri, S.P., F.J.R. Hird, and I.J. Stanley. 1981. On the
    significance of adenylic acid aminohydrolase in skeletal
    muscle of vertebrates. Comp. Biochem. Physiol. 68B: 369-375.
21. Chapman, A.G., A.L. Miller, and D.E. Atkinson. 1976. Role of
    the adenylate deminase reaction in regulation of adenine
    nucleotide metabolism in Ehrlich ascites tumor cells. Cancer
    Res. 36: 1144-1150.
22. Koretz, J.F., and C. Frieden. 1980. Adenylate deaminase
    binding to synthetic thick filaments of myosin. Proc. Natl.
    Acad. Sci. (USA). 77: 7186-7188.
23. Shiraki, H., S. Miyamoto, Y. Matsuda, E. Momose, and H.
    Nakagawa. 1981. Possible correlation between binding of
    muscle type AMP deaminase to myofibrils and ammoniagenesis in
    rat skeletal muscle on electrical stimulation. Biochem.
    Biophys. Res. Comm. 100: 1099-1103.
24. Aragon, J.J., K. Tornheim, and J.M. Lowenstein. 1980. On a
    possible role of IMP in the regulation of phosphorylase
    activity in skeletal muscle. FEBS Lett. 117 Suppl! K56-K64.
25. Liang, C.-M., Y.P. Liu, and B.A. Chabner. 1980. Modes of
    action of hypoxanthine, inosine and inosine 5'-monophosphate
    on cyclic nucleotide phosphodiesterase from bovine brain.
    Biochem. Pharmacol. 29: 277-282.
26. Buchwald, M., B. Ullman, and D.W. Martin, Jr. 1981.
    Biochemical and genetic analysis of AMP deaminase deficiency
    in cultured mammalian cells. J. Biol. Chem. 256: 10346-10353.
27. Aragon, J.J., and J.M. Lowenstein. 1980. The purine
    nucleotide cycle. Comparison of the levels of citric acid
    intermediates with the operation of the purine nucleotide
    cycle in rat skeletal muscle during exercise and recovery from
    exercise. Eur. J. Biochem. 110: 371-377.
28. Meyer, R.A., and R.L. Terjung. 1980. AMP deamination and IMP
    reamination in working skeletal muscle. Am. J. Physiol. 239:
    C32-C38.
29. Holmes, E.W., et al. These Proceedings.
30. Brosh, S., P. Boer, E. Zoref-Shani, and O. Sperling. 1982.
    De novo purine synthesis in skeletal muscle. Biochim.
    Biophys. Acta. 714: 181-183.
31. Fishbein, W.N., V.W. Armbrustmacher, and J.L. Griffin. 1980.
    Skeletal muscle adenylate deaminase, adenylate kinase, and
    creatine kinase in myoadenylate deaminase deficiency and
    malignant hyperthermia. Clin. Res. 28: 288A.
32. Fishbein, W.N., J.I. Davis, K. Nagarajan, and M.J. Smith.
    1981. Specific serum/plasma inhibitor of muscle adenylate
    deaminase. IRCS Med. Sci. 9: 178-179.

MYOADENYLATE DEAMINASE DEFICIENCY:

AN ENZYME DEFECT IN SEARCH OF A DISEASE

E. Joosten[x], C. van Bennekom[xx], F. Oerlemans[xx],
C. De Bruyn[xx], T. Oei[xx], and J. Trijbels[xxx]

Dept. Neurology [(x)], Human Genetics [(xx)],
Pediatrics [(xxx)], Radboud Hospital, Catholic
University, Nijmegen, The Netherlands

INTRODUCTION

In 1978 Fischbein et al,[1] described 5 patients with a myoadenylate deaminase (MAD) deficiency, which were detected during a
histochemical screening of 250 consecutive muscle biopsies. The
deficiency was confirmed by a biochemical assay of the enzyme.
Clinically three of the patients complained of muscle cramping, in
one case associated with postexercise fatigue. CK was only moderately elevated, the myogram showed minor abnormalities, in muscle
there were histologically no or only slight alterations. These
three patients didn't show any muscle weakness or atrophy.
In 1979 they,[2] described 7 other patients, three of them, however,
showing a quite different clinical picture with signs of collagen
vascular disease. In their series of consecutive muscle biopsies
they found a frequency of about 1,5% of MAD deficient biopsies.
Other authors,[3,4,5] described a deficiency in about 1,5-2,0% of the
biopsies. Half of these biopsies came from patients with a clinical
symptomatology of exercise intolerance with muscular pain and/or
fatigue. The other half came from patients displaying a wide diversity of partly well defined clinical entities, among others collagen
vascular disease, neuropathy, amyotrophic lateral sclerosis, dystrophy, cardioskeletal myopathy, spinal muscular atrophy, paroxysmal
myoglobinuria, facio-scapulo-humeral syndrome, infantile hypotonia
and so on. Therefore some authors,[3,5,6] consider the deficiency
state as a normal variant, not associated with any particular clinical picture, others, however, as a cause of exercise intolerance
and exercise induced muscle pain [1,2,4,7,8].
This paper reports a number of MAD deficient patients occurring in
a series of patients with exercise intolerance, muscle pain and/or

fatigue, and in a consecutive series of muscle biopsies.
In a control series and in some deficiency cases together with a
number of their relatives plasma hypoxanthine concentration was
measured in addition to the NH_3, and lactate levels before and
after 2 minutes of ischaemic exercise of the forearm.

METHODS

MAD activity was assayed radiochemically accordingly to Leech
et al,[9]. NH_3 was measured enzymatically by the assay provided by
Boehringer.[3]
Hypoxanthine was estimated using high performance liquid chromatography on prepacked Bondapack-C18 columns (Waters Ass) eluted with
an isocratic system consisting of 0.1 M KH_2PO_4 and 9% (v/v)
methanol pH 5.3.

RESULTS

The results of enzyme assays in four deficiency cases are
summarized in table 1. Information concerning the clinical picture
of these patients is given in table 2.
The case described as case one was discovered during a biochemical
study of 36 consecutive muscle biopsies of patients with various
complaints. Case 2 and 4 were detected by assaying myoadenylate
deaminase in muscle biopsies (partly needle biopsies) of 16 patients
with exercise intolerance with muscle pain and/or fatigue. Case 3
was found by screening 12 patients with these complaints with the
test on NH_3 formation during ischaemic exercise[1].

Table 1.   MAD activity in four deficiency cases and 50
           reference cases (μ mole IMP formed/mg protein·hour)

50 reference biopsies [x] mean 113.0 range 16.4 - 311.3	
Patient 1	6.3
" 2	3.6
" 3	1.5
" 4	0.8

[x] The distribution of the values is skewed. The mode is 70,
the median 95.

Table 2.   Case reports

Patient	Sex	Age	Complaints	CK	EMG
1	M	59	sensory neuropathy since 5 years	normal	denervation
2	M	29	muscular pain and cramping after exercise since 1½ years. In left calf fasciculation.	slightly elevated	minimal signs of de-and reinnervation
3	M	21	longstanding complaints of exercise induced pains and postexercise weakness influenced by cold.	moderately elevated	some small "myopathic" units. Electrically silent cramp.
4	M	28	since 3 years muscular aching after moderate exercise, post-exercise weakness, lifelong muscular stiffness.	normal	normal

Increase in plasma NH_3 versus increase in plasma lactate after ischaemic exercise is plotted in figure 1; increase in plasma NH_3 versus increase in plasma hypoxanthine is plotted in figure 2. As shown, some of the relatives of patient 2 show decreased NH_3 formation, comparable to that of MAD deficient patients.

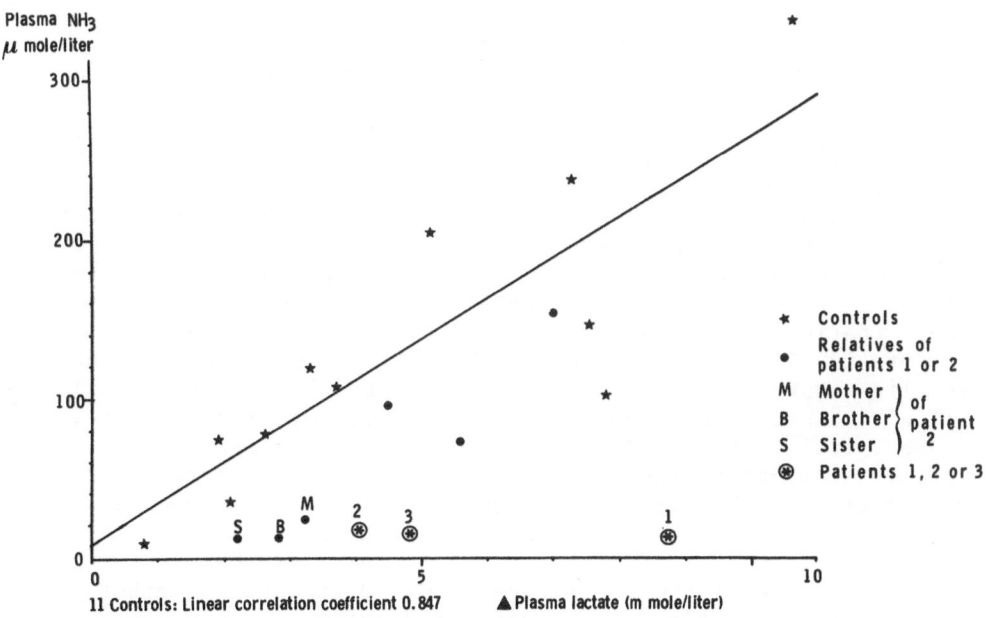

Figure 1.   Increase in plasma NH_3 versus lactate after ischaemic exercise

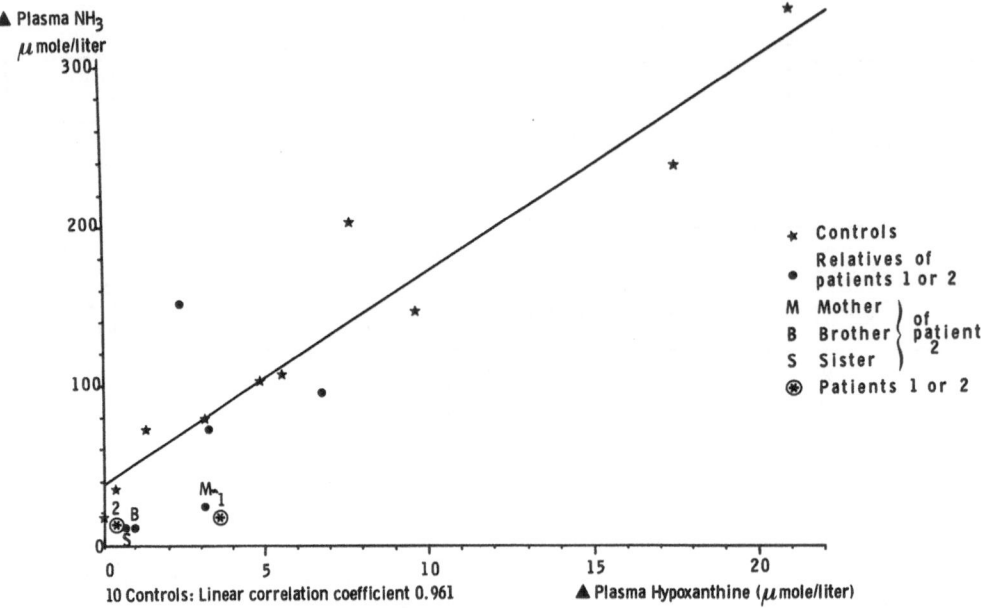

Figure 2.   Increase in plasma NH$_3$ versus hypoxanthine after
            ischaemic exercise

DISCUSSION

In biopsy material from 36 individuals without exercise intolerance 1 was found to be MAD deficient. His residual activity was
the highest among the deficient patients (table 1, patient 1), but
the ischaemic exercise test showed equally decreased NH$_3$ formation
(figure 1,2). In the group of 16 individuals with exercise intolerance, which were not prescreened with the ischaemic exercise test,
2 MAD deficiencies were detected (patient 2,4). Among 12 patients
with these complaints, one showed decreased NH$_3$ formation during
ischaemic exercise. MAD subsequently was found decreased in muscle
biopsy (patient 3). MAD deficiency so could be found in these
patients in a frequency of about 10%. In consecutive biopsy series
from literature,[1,2,3,4,5] a frequency of about 1,8% can be estimated.
Half of the cases in these series display exercise intolerance.
However, it is very probable that much less than half of the
biopsies in these series come from patients with exercise intolerance. Therefore the available evidence from our study and from
literature suggests that myoadenylate deaminase is possibly a
rather frequent cause of exercise intolerance. Further research is
needed to decide if the occurrence of MAD deficiency together with
other often well defined neuromuscular diseases is determined by
chance, or if there exists a rather varying clinical expression of
the deficiency.

Figure 1 confirms that the ischaemic exercise test discriminates between controls and MAD deficient patients, when increase in NH_3 is plotted versus increase in lactate. Figure 2 indicates that discrimination is also possible when NH_3 is plotted versus increase in hypoxanthine. In the control group a better correlation was found between NH_3 and hypoxanthine (linear correlation coefficient 0,961, than between NH_3 and lactate (linear correlation coefficient 0,847). The results of the test show that the defect is familial, but to elucidate the mode of inheritance it will probably be necessary to assay enzyme activity in muscle.

SUMMARY

The frequency of MAD deficiency in cases with exercise intolerance compared with the frequency in series of consecutive muscle biopsies suggests a relation between the deficiency and exercise intolerance. Deficiency cases can be presumed by an impaired NH_3 production during ischaemic exercise. The ischaemic exercise test also gives information concerning the familial character of the deficiency.

REFERENCES

1. W.N. Fischbein, V.W. Armbrustmacher, J.L. Griffin, Science, 200:545-548 (1978).
2. W.N. Fischbein, J.L. Griffin, K. Nagajaran, J.W. Winkert, V.W. Armbrustmacher, Clin. Res., 27:274A (1979).
3. J.B. Shumate, R. Katnik, M. Ruiz, K. Kaiser, C. Frieden, M.H. Brooke, J.E. Carroll, Muscle Nerve, 2:213-216 (1979).
4. N.C. Kar, C.M. Pearson, Arch.Neurol., 38:279-281 (1981).
5. R.R. Heffner, J. Neuropath. exp. Neurol., 39:360 (1980).
6. D.J. Hayes, B.A. Summers, J.A. Morgan-Hughes, J. Neurol. Sci., 53:125-136 (1982).
7. R.L. Sabina, J.L. Swain, B.M. Patten, T. Ashizawa, W.E. O'Brien, E.W. Holmes, J. Clin. Invest., 66:1419-1423 (1980).
8. S. DiMauro, A. Miranda, A.P. Hays, W.A. Franck, G.S. Hoffman, R.S. Schoenfeldt, N. Singh, J. Neurol. Sc. 47:191-20? (1980).
9. A.R. Leech, E.A. Newsholme, Analyt. Biochem., 90:576-589 (1978).
10. W.N. Fischbein, J.L. Griffin, W. Vernon, W. Armbrustmacher, Arch. Pathol. Lab. Med., 104:462-466 (1980).

PHOSPHORIBOSYLPYROPHOSPHATE SYNTHETASE SUPERACTIVITY: DETECTION,

CHARACTERIZATION OF UNDERLYING DEFECTS, AND TREATMENT

Michael A. Becker

Department of Medicine
The University of Chicago
Chicago, Illinois 60637, U.S.A.

Phosphoribosylpyrophosphate (PRPP) synthetase catalyzes the synthesis of PRPP, a critical substrate common to the de novo and salvage pathways of purine nucleotide synthesis.   The PRPP synthetase reaction involves the transfer of the terminal pyrophosphate group of ATP (as a MgATP complex) to the C-1 carbon of ribose-5-P and is dependent on inorganic phosphate (Pi) and Mg^{2+} which serve as enzyme activators as well as cofactors.   Activity of PRPP synthetase is inhibited by a number of phosphorylated compounds including the reaction products (PRPP and AMP), purine, pyrimidine, and pyridine nucleotides and 2,3-DPG.[1] Human PRPP synthetase is composed of a single polypeptide subunit[2] the structural gene for which maps to the long arm of the X-chromasome.[3]   Under appropriate conditions of enzyme and effector concentration in vitro, PRPP synthetase subunits are capable of reversible self-association to aggregates containing 2,4,8,16 and 32 subunits, with only the largest two of these containing significant enzyme activity.[4]

Since the initial report by Drs. Sperling and de Vries and their colleagues,[5] inherited superactivity of PRPP synthetase has become established as an unusual cause of purine overproduction, hyperuricemia, and gout in man.   To date, detailed investigations of 7 families with superactive PRPP synthetase have been published,[5-11] and I am aware of 4 additional families currently under study.   In each family, the index cases have been males, and where studied, the patterns of inheritance of the enzyme aberrations have been consistent with X-linked transmission,[5,6,9-13] reflecting the apparent structural basis of superactivity in each defective enzyme.

Most affected male patients have shown the onset of acute gouty arthritis and/or uric acid urolithiasis in early adulthood.   In these

individuals, the clinical course has been dominated by the usual consequences of excessive uric acid production and excretion, and no
readily apparent associated disease has been identified in these adult patients, the oldest of whom was 63 years of age at the time of
detection of enzyme superactivity. Although excessive daily urinary
uric acid excretion has been found in several heterozygous female carrier relatives of the adult patients, these women have not shown clinical abnormalities.[5,6,11]

Evidence of purine overproduction in childhood has led to detection of superactive PRPP synthetases in 2 families which are of special interest for several reasons.[9,14] First, the hemizygous affected
males in these families show severe sensorineural deafness in addition
to uric acid overproduction. Second, the mothers of these boys share
both the metabolic and hearing abnormalities with their sons, and one
of these women[9] has had both acute gouty arthritis and uric acid urolithiasis. Finally, as discussed below, the functional derangement in
the enzyme of one of the families[9] is unusually marked with more severe metabolic consequences of PRPP synthetase superactivity which
might explain the childhood clinical onset, the development of gout in
the mother, and even the associated deafness.

Structural alteration in PRPP synthetase appears to underlie enzyme superactivity in each family studied in detail,[6-11] Evidence to
support this contention is indirect, since no precise alteration in
enzyme primary structure has been demonstrated. Nevertheless, each
superactive enzyme studied in partially purified or homogenous preparation has shown at least one variant property in either electrophoretic mobility,[8,11,15] thermal stability,[9,11,15] substrate affinity,[8]
inhibitor responsiveness,[7,9] or immunochemical inactivation[9,11,15]
when compared to normal enzyme of comparable purity. Moreover, the
pattern of variant properties of each aberrant enzyme has been distinct indicating that enzyme superactivity can result from a diverse
array of inherited structural changes in PRPP synthetase.

Despite structural diversity in the superactive enzymes of individual families, studies of PRPP and purine metabolism carried out
both in vivo and in cells cultured from affected hemizygous males support the idea that a common mechanism accounts for the association of
PRPP synthetase superactivity with uric acid overproduction.[6-11] Increased intracellular PRPP concentrations and rates of PRPP generation
as well as increased rates of all PRPP-dependent purine nucleotide
synthetic processes are constant accompaniments of enzyme superactivity. These findings suggest a scheme to explain the association of
the enzyme defect with uric acid overproduction: PRPP synthetase superactivity → increased intracellular PRPP generation and concentration → increased rate of purine nucleotide synthesis → excessive uric
acid synthesis.

In addition to heterogeneity in the structural defects leading to

PRPP synthetase superactivity, diversity in the kinetic mechanisms underlying increased PRPP synthesis has been identified. This diversity has important implications for the design of methods for detection of abnormalities of the enzyme. The four categories of kinetic alteration thus far associated with PRPP synthetase superactivity in man are: abnormal catalytic properties (increased maximal reaction veloccity); 2) defective regulatory properties (purine nucleotide feedback resistance); 3) increased affinity for the substrate ribose-5-P; and 4) combined alterations of catalytic and regulatory properties.

In four of the families investigated,[10,11,15] separable structural defects in PRPP synthetase result in increased maximal reaction velocity with no apparent change in substrate affinity or inhibitor responsiveness. No increase in the amount of enzyme in extracts of cells from affected individuals has been found, and excessive PRPP synthetase specific activities are proportionally increased at all Pi concentrations examined (Figure 1A). With crude or dialyzed extracts, Pi activation curves are sigmoidal, paralleling those found with normal extracts and reflecting the normally greater purine nucleotide feedback inhibition of enzyme activity at low than high Pi concentrations.[16] Superactive PRPP synthetases of this type are readily detectable by testing enzyme activities in either dialyzed or chromatographed cell extracts over a broad range of Pi concentrations at saturating concentrations of MgATP, ribose-5-P, and Mg^{2+}.

In contrast to catalytically overactive forms of the enzyme, nucleotide feedback-resistant PRPP synthetases[5,17] show hyperbolic Pi activation with enzyme superactivity clearly demonstrable only at Pi concentrations of less than about 2 mM (figure 1B). Separation of inhibitors from normal PRPP synthetase results in a similar hyperbolic pattern of Pi activation.[7] Forms of PRPP synthetase solely altered in regulatory properties are thus indistinguishable from normal enzyme when Pi activation is examined in chromatographed or purified preparations (Figure 1C). Detection of this of enzyme defect requires either the use of crude or dialyzed extracts retaining endogenous nucleotides or the addition of nucleotide inhibitors (such as ADP or GDP) to the assay mixture when chromatographed or partially purified enzyme preparations are studied. The mutant PRPP synthetase studied by Sperling and de Vries[5] exemplifies this class of alteration. Although no additional example of purely feedback-resistant PRPP synthetase has been encountered to date in man, Green and Martin[17] have demonstrated a TDP feedback-resistant PRPP synthetase in a rat hepatoma cell (HTC) line.

A third class of superactive PRPP synthetase is represented by an enzyme with increased affinity for the substrate ribose-5-P.[8] When assayed at saturating substrate and Mg^{2+} concentrations, activities of this enzyme are entirely normal in all preparations studied. Kinetic analysis at subsaturating ribose-5-P concentrations is required to demonstrate the kinetic derangement. This abnormality suggests that

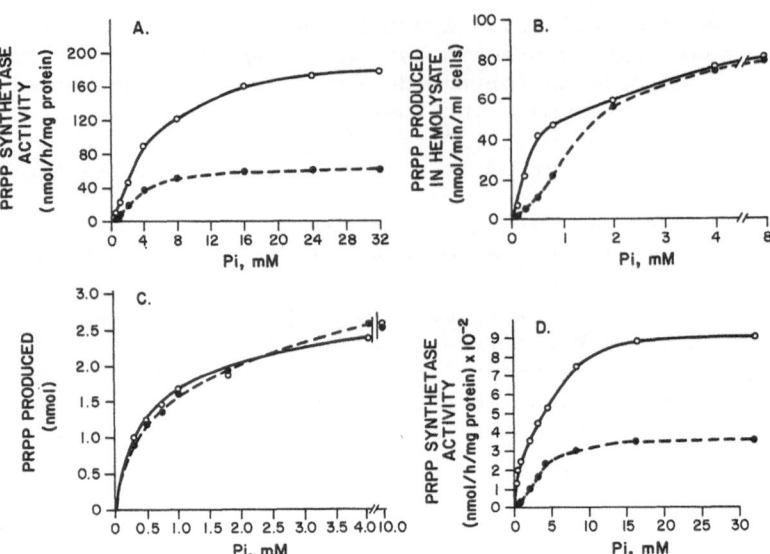

Figure 1. Pi activation of normal (●---●) and mutant (o—o) human
PRPP synthetases. A. Enzyme activities in dialyzed hemolysates.
Sigmoidal activation is seen for both enzymes. Activity of the mutant
enzyme with an increased maximal reaction velocity is, however, increased by a constant proportion at all Pi concentrations. B. Enzyme activities in undialyzed (crude) hemolysates. The mutant enzyme,
deficient in inhibitor responsiveness, shows hyperbolic activation
with increased activity only at Pi concentrations below 2 mM. C. Enzyme activities in partially purified erythrocyte preparations from a
normal individual and the patient with the feedback-resistant PRPP
synthetase studied in B. Note hyperbolic activation of the purified
normal enzyme and the resulting similarity of the Pi activation
curves. D. Enzyme activities in dialyzed fibroblast extracts. The
mutant enzyme, with combined increased maximal reaction velocity and
diminished nucleotide responsiveness, shows both hyperbolic activation
and increased enzyme activity at all Pi concentrations.

screening at subsaturating as well as saturating substrate concentrations must be carried out in order to achieve maximal sensitivity in
the detection of enzyme variants.

    A final category of functional abnormality underlying PRPP syn-
thetase superactivity is represented by the enzyme characterized from
the fibroblasts of the child with purine overproduction and deafness
referred to earlier. Both increased maximal reaction velocity and

decreased nucleotide inhibitor-responsiveness are demonstrable.[9] In
this case, Pi activation curves are hyperbolic in both dialyzed and
chromatographed cell extracts, and enzyme activities are at least
twice normal at all Pi concentrations (Figure 1D). Paradoxically,
PRPP synthetase activities in erythrocyte lysates from this child are
markedly diminished, a reflection of increased lability of the enzyme
in his erythrocytes. The severity of the metabolic consequences of
this abnormality is indicated by the fact that despite substantial diminution of the amount of enzyme in erythrocytes, these cells show excessive rates of incorporation of purine bases into nucleotides at
physiological Pi concentrations.

   The enzyme deficiency in red cell lysates from the above patient
points out one limitation in identifying superactivity of PRPP synthetase in hemolysates. Another limitation in the use of hemolysates is
that identification of heterozygous female carriers is often not possible using erythrocyte lysates.[5,9,11] Although extracts of fibroblasts cultured from obligate carriers have reliably shown enzyme activities and metabolic derangements intermediate between those shown by
extracts from affected male relatives and normal individuals[7,9,11,12]
(as predicted for an X-linked trait by the Lyon hypothesis[18]), enzyme
activities in hemolysates from these women have been quite variable.
Carriers with hemolysate PRPP synthetase activities which are normal,[5,9,11] intermediate,[11] or fully superactive[6] have all been encountered in different families. No ready explanation for this variation
is available.

   For practical reasons, screening for abnormalities of PRPP synthetase will amost certainly continue to utilize hemolysates, with the
use of fibroblast or lymphoblast samples remaining the province of a
relatively few laboratories. From the above discussion, it is clear
that enzyme activities should be measured under a variety of conditions. In our laboratory, screening of PRPP synthetase activity[6,9]
is carried out in dialyzed hemolysates from patients with uric acid
overproduction as follows: 1) as a function of Pi concentration
(range, 0.2-32 mM) at saturating substrate and Mg^{2+} concentrations;
2) at both 1.0 and 32 mM·Pi, as functions of each of three inhibitors
(ADP, GDP, and 2,3-DPG) added to reaction mixtures containing saturating concentrations of substrates and Mg^{2+}; and 3) at both 1.0 and 32
mM Pi, as functions of substrate concentrations varied individually or
in combination from saturating to subsaturating levels. Where differences in activity from control hemolysates are found, more precise
evaluations appropriate to the specific finding are carried out. For
example, a blunted response to inhibitors or hyperbolic Pi activation
is further investigated in chromatographed extracts or partially purified enzyme preparations. When available, fibroblast extracts are
used to corroborate hemolysate findings, to extend the sensitivity of
detection efforts, and to identify female carriers. In addition, PRPP
concentrations and rates of generation and ribose-5-P concentrations
are also measured in fibroblasts, since these determinations are

useful in directing attention to cryptic forms of superactive PRPP synthetase[8] not readily identifiable by the usual screening methods.

Treatment of patients with PRPP synthetase superactivity is currently that appropriate to the treatment of primary uric acid overproduction of any cause. Total purine production and hyperuricemia are diminished with allopurinol, and this agent is beneficial in reducing the number of attacks of gouty arthritis and renal colic and in resolving tophi. Long term administration of agents which specifically diminish intracellular PRPP availability has not been attempted although several chemical agents with this property have been shown to reduce in vivo rates of purine synthesis de novo in brief trials.

ACKNOWLEDGEMENT: The author's studies reported here were supported in part by Grant AM-28554 from the National Institutes of Health and a grant from the Illinois Chapter of the Arthritis Foundation.

REFERENCES

1. Fox, I.H., and Kelley, W.N., J. Biol. Chem. 247:2166 (1972).
2. Becker, M.A., Meyer, L.J., Huisman, W.H., Lazar, C., and Adams, W.B., J. Biol. Chem. 252:3911 (1977).
3. Becker, M.A., Yen, R.C.K., Itkin, P., Goss, S.J., Seegmiller, J.E., and Bakay, B., Science 203:1016 (1979).
4. Meyer, L.J. and Becker, M.A., J. Biol. Chem. 252:3919 (1977).
5. Sperling, O., Boer, P., Persky-Brosh, S., Kanarek, E., and de Vries, A., Rev. Europ. Etud. Clin. Biol. 17:703 (1972).
6. Becker, M.A., Meyer, L.J., and Seegmiller, J.E., Am. J. Med. 55: 232 (1975).
7. Zoref, E., De Vries, A., and Sperling, O., J. Clin. Invest. 56: 1093 (1975).
8. Becker, M.A., J. Clin. Invest. 57:308 (1976).
9. Becker, M.A., Raivio, K.O., Bakay, B., Adams, W.B., and Nyhan, W.L., J. Clin. Invest. 65:109 (1980).
10. Akaoka, I., Fujimori, S., Kamatani, N., Takeuchi, F., Yano, E., Nishida, Y., Hashimoto, A., and Horiuchi, Y., J. Rheumatol. 8: 563 (1981).
11. Becker, M.A., Losman, M.J., Itkin, P., and Simkin, P.A., J. Lab. Clin. Med. 99:945 (1982).
12. Yen, R.C.K., Adams, W.B., Lazar, C., and Becker, M.A., Proc. Natl. Acad. Sci. USA 75:482 (1978).
13. Zoref, E., de Vries, A., and Sperling, O., Adv. Exp. Med. Biol. 76A;287 (1977).
14. Simmonds, H.A., Webster, D.R., Wilson, J., Fairbanks, L.D. and Potter, C.F., Adv. Exp. Med. Biol. (this volume).
15. Becker, M.A., Kostel, P.J., and Meyer, L.J., J. Biol. Chem. 250: 6822 (1975).
16. Hershko, A., Razin, A., and Mager, J., Biochim. Biophys. Acta 184:64 (1969).
17. Green, C.D., and Martin, D.W., Jr., Proc. Natl. Acad. Sci. USA 70:3698 (1973).
18. Lyon, M.F., Nature 190:372 (1961).

EVIDENCE OF A NEW SYNDROME INVOLVING HEREDITARY URIC ACID OVER-

PRODUCTION, NEUROLOGICAL COMPLICATIONS AND DEAFNESS

H. A. Simmonds, D. R. Webster, J. Wilson, C. F. Potter, and L. D. Fairbanks

Purine Laboratory and Department of Neurology, Guy's Hospital and Hospital for Sick Children, Great Ormond Street, London

Hypoxanthine-guanine phosphoribosyltransferase (HGPRT: EC 2.4.2.8) deficiency[1] is an X-linked recessive disorder which may present in early infancy as the Lesch-Nyhan syndrome associated with gross purine overproduction, neurological abnormalities and bizarre self-mutilating behaviour. Less severe forms have been reported in adults with gouty arthritis and/or renal complications but a complete absence of neurological abnormalities[1].

Becker et al[2] recently reported a variant of the purine synthetic enzyme phosphoribosylpyrophosphate (PP-ribose-P) synthetase (EC 2.7.6.1) in a patient with symptoms characteristic of the Lesch-Nyhan syndrome at three years of age, but with normal HGPRT enzyme levels. The clinical manifestations of this X-linked disorder in three previous families in which it had been described had been restricted to gout, uric acid lithiasis and/or renal insufficiency and had not developed until early adulthood.[3]

We report a fifth family which appears to have an even more severe form of the defect than the case of Becker et al[2] in that the symptoms developed neonatally in three siblings.

METHODS

Clinical data: NB was the third child of apparently healthy non-related parents. At four days he was admitted with congestive cardiac failure for repair of coarctation of the aorta. Two previous male siblings had died within the first two years of life. Both the child and his mother have a high tone hearing deficit. One other sibling was also reported to have had the same hearing defect. All three siblings had presented with head drops at 5-7

97

months and showed severe neurodevelopmental retardation.

Biochemical investigations: Studies in NB and both parents were
carried out on a low purine diet. Nucleotide, nucleoside and base
levels were investigated in erythrocytes using high pressure
liquid chromatography (HPLC) and enzyme activities were measured[4,5]
in intact as well as lysed cells using radiolabelled substrates.
PP-ribose-P levels were estimated by the method of Becker et al[2].
The lymphoblast lines from NB and CB were kindly established by
Dr. S. Pereira of the Clinical Research Centre, Northwick Park.

RESULTS

Enzyme activity and PP-ribose-P levels: The erythrocyte lysate
enzyme levels, together with PP-ribose-P levels, are compared
in Table 1.   HGPRT and APRT were both within the normal range.
PP-ribose-P levels were not raised in the erythrocytes of either
NB or his mother (Table 1).   However, lymphoblast PP-ribose-P
levels in NB were eight-fold increased above the control range
(Table 1).

Table 1   Purine metabolites and enzyme levels in family B.

	ERYTHROCYTE ENZYMES (nmol/mgHb/h)		PP-RIBOSE-P LEVELS			AGE	PLASMA Uric Acid mmol/l	URINE (mmol/24h) Uric Acid	Hypo-xanthine	Xanthine	TOTAL mmol/mmol creatinine
SUBJECT	HGPRT	APRT	RED CELLS μmol/l	LYMPHO BLASTS pmol/10⁶	Controls						
					Child	4	0.18	0.87	0.02	0.02	0.58
NB	80	23	4.1	864	Adult ♀	23	0.18	2.58	0.05	0.06	0.28
CB♀	84	26	2.4	-	Adult ♂	25	0.28	3.34	0.06	0.03	0.15
JB♂	81	18	-	-							
Control range	80-130	16-32	0.5-4	51-81	HGPRT						
					* (LN) child	11	0.32	0.51	1.18	0.83	2.7
HGPRT⁻	0.1	40	-	-	(Partial) adult	18	0.57	7.92	0.63	0.14	1.1
HGPRT⁻ (LN)	<0.01	56	20.5	1095							
					NB	1½	0.45	0.98	0.10	0.03	2.2
PNP⁻	111	45	16.2	861	CB ♀	32	0.45	5.84	0.14	0.07	0.37
					JB ♂	33	0.27	3.10	0.03	0.01	0.21

* on allopurinol

Values are representative of repetitive estimations in Family B
compared with a number of controls.
† Uric acid (UA) + hypoxanthine (H) + xanthine (X) factored by
creatinine (Cr).

<u>Plasma and urinary purines</u>: Uric acid, xanthine and hypoxanthine in plasma and urine in Family B are also given in Table 1.   The plasma and urine uric acid levels were greatly increased in NB and his mother (CB) when compared with appropriate controls.   Urinary hypoxanthine levels were excessive in both (Table 1).   The urine purine creatinine ratio confirmed gross purine overproduction in both, that in the mother (CB) being approximately twice normal for a healthy adult female.   Values in the father (JB) were normal for age and sex.

<u>Erythrocyte nucleotide levels</u>: Nucleotide levels are given in Table 2.   ATP:ADP:AMP ratios were essentially in the normal range (10:1:0.1) but ATP was low in CB.   Both NB and CB showed severe GTP depletion.   NAD$^+$ levels were also extremely low in NB and below the normal range in CB.   The results are compared with nucleotide levels in two other defects also resulting in purine overproduction and show that erythrocyte GTP depletion is common to all three (Table 2).

<u>Intact erythrocyte studies</u>: Figure 1 shows that during incubations with {8-^{14}C} adenine or {8-^{14}C} hypoxanthine (not shown) in buffer at either physiological (1mM P_i) or high phosphate (18mM P_i : PP-ribose-P generating conditions) NB cells showed no significant increase in incorporation of label, in contrast to control cells which showed a marked stimulation of rate of incorporation into the nucleotide IMP at 18mM P_i.   Values for the mother were half-way between these two.

Table 2   Erythrocyte nucleotide levels (μmol/l packed cells)

SUBJECT	ATP	ADP	AMP	GTP	GDP	NAD$^+$
Propositus NB	1005	144	5	<1	9	14
♀ CB	620	121	6	8	6	37
♂ JB	1095	150	15	30	8	74
Control range	1278	114	10	60	20	75
	±127	±24	±3	±10	±5	±14
(HGPRT$^-$) LW	1550	219	4	9	14	176
DT	1260	107	10	8	8	360
SB (PNP$^-$)	1319	276	12	5	5	276

HGPRT$^-$: hypoxanthine guanine phosphoribosyltransferase deficiency
PNP$^-$: purine nucleoside phosphorylase deficiency

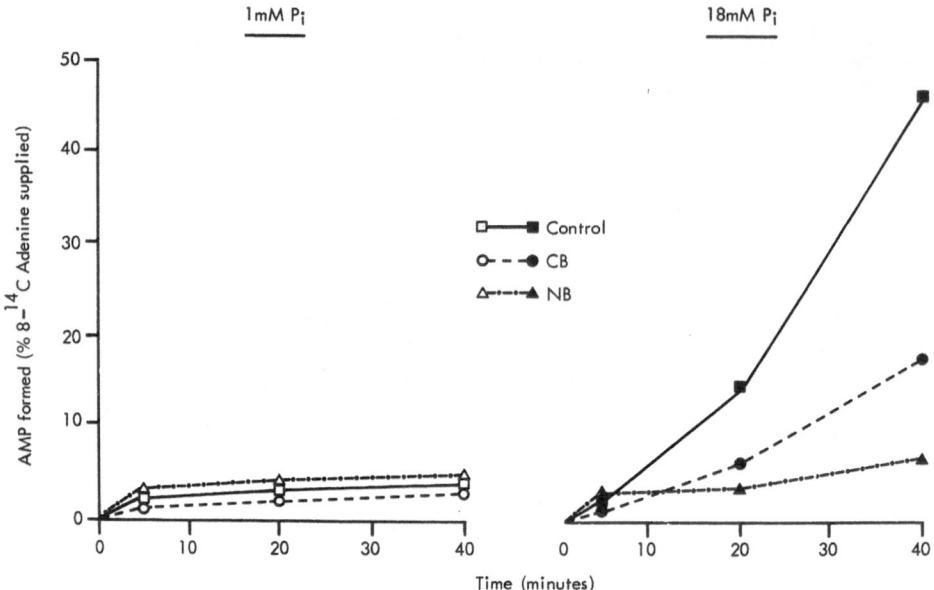

Figure 1   Metabolism of {8-^{14}C}adenine (25μM);intact erythrocytes
from NB △–·–·–▲, the mother CB ○–––● compared with a control □——■
over 40 minutes at 1 and 18mM P$_i$.

DISCUSSION

The family in this report clearly suffers from an X-linked disorder
associated with gross purine overproduction, hereditary deafness
and neurodevelopmental abnormalities.   Two X-linked disorders are
known to be associated with gross purine overproduction in children[1,3].   The finding of normal APRT and HGPRT levels in
erythrocytes lysates would appear to exclude the Lesch-Nyhan
syndrome where low to absent HGPRT activity is associated with
raised APRT levels[1].   Becker et al[2] also found normal APRT and
HGPRT levels in erythrocyte lysates from their patient, and a
similar lack of response to activation by high phosphate, using
intact erythrocytes.   They found that while PP-ribose-P synthetase
activity was increased in fibroblasts it was virtually absent from
all but the youngest red cells and attributed this apparent
paradox to a diminished stability of the PP-ribose-P synthetase
from their patient's erythrocytes.   Although we were unable to
detect raised PP-ribose-P levels in the erythrocytes, the raised

levels in the lymphoblasts and the increased synthetase activity
suggest the propositus is a hemizygote for a variant form of the
purine synthetic enzyme PP-ribose-P synthetase.   Although the
mutant enzyme has not yet been characterised, we have likewise
ascribed the normal red cell PP-ribose-P levels to diminished
stability of the synthetase in our patient's erythrocytes.

The history of two previous male siblings dying in infancy and the
presence of identical but milder symptoms in the mother suggest
that as with the X-linked urea cycle defect, ornithine carbamoyltransferase deficiency (OCT: EC 2.1.3.3), the defect in its
severest form, may be lethal to the hemizygous male and strongly
expressed in the heterozygous female.[5]   The association of the
aberrant enzyme with deafness in a second family also indicates
more than a chance association.   We suggest that as with HGPRT
deficiency the spectrum of manifestations may be broad - with gout
at one end of the scale, neurodevelopmental abnormalities and
deafness at the other.   It is of interest that a large kindred
was reported in 1970 with X-linked hereditary deafness and
hyperuricaemia but no enzyme abnormality could be identified at
that time.[6]

The novel findings in the red cell nucleotides in this child -
extremely low GTP and NAD^+ levels - may provide further insight
into the recognition and treatment of this defect.   We have also
found extremely low erythrocyte GTP levels in a retarded immuno-
deficient child with PNP deficiency[5], and also in children with
the full Lesch-Nyhan syndrome.   Severe hypotonia is characteristic
of all these defects.   Adults with a partial defect of HGPRT
deficiency, as with the mother in this case, do not have such severe
erythrocyte GTP depletion.   They also do not have the associated
neurological problems (Cameron et al - this symposium).   Such
similar findings in three disparate enzyme defects, all associated
with gross purine overproduction and neurodevelopmental abnormalities,
suggest a central role for GTP in both.

The extremely low red cell NAD^+ levels also indicate involvement of
pyridine nucleotide metabolism in the clinical expression of this
disorder.   PP-ribose-P is required for the formation of the
pyridine nucleotides NAD^+ and $NADP^+$; both are vital to the
erythrocyte.   However, high, rather than low NAD^+ levels would
be anticipated and have been found by us in both HGPRT and PNP
deficient red cells; it is possible that the raised NAD^+ in the
latter reflect the raised PP-ribose-P levels found in PNP and HGPRT
deficiency.   PP-ribose-P levels were not raised in NB's red cells
and the NAD^+ values were extremely low.   The latter finding may
also be associated with the observed inability of the intact red
cells to stimulate the re-cycling of hypoxanthine or adenine at
high phosphate and suggest that NAD^+ may be necessary for the
stabilisation of erythrocyte PP-ribose-P.   This would in turn

explain the raised APRT levels (APRT is stabilised by PRPP) in PNP
and HGPRT deficiency but not in this child.   Further work is in
progress in an attempt to resolve these points.

These studies show that diagnosis of the aberrant enzyme may be
impossible from red cells alone but that altered nucleotide levels
together with the raised uric acid values in plasma and urine in
both mother and child, provide useful diagnostic tools.   This
again distinguishes the defect from HGPRT deficiency where mothers
are rarely affected.[1]

The results presented also question whether an aberrant PP-ribose-P
synthetase is the primary disorder or whether it is secondary to an
underlying enzyme defect resulting in reduced guanine nucleotide
and/or NAD^+ levels and inefficient feedback control of synthetase
activity.

REFERENCES

1. W. N. Kelley and J. B. Wyngaarden.   The Lesch-Nyhan syndrome,
in: "The Metabolic Basis of Inherited Disease (4th edition), J. B.
Stanbury, J. B. Wyngaarden, and D. S. Fredrickson, eds., McGraw-
Hill, New York (1978).

2. M. A. Becker, K. O. Raivio, B. Bakay, W. B. Adams and W. L. Nyhan.
Variant human phosphoribosylpyrophosphate synthetase altered in
regulatory and catalytic functions.   J.Clin.Invest. 65: 109 (1980).

3. M. A. Becker.   Abnormalities of PRPP metabolism leading to an
overproduction of uric acid.   in: "Uric Acid".   W. N. Kelley and
I. M. Weiner eds., Springer-Verlag, Berlin (1978).

4. H. A. Simmonds, A. R. Watson, D. R. Webster, A. Sahota and D.
Perrett.   GTP depletion and other erythrocyte abnormalities in
inherited PNP deficiency.   Biochem.Pharmac. 31: 941 (1982).

5. D. R. Webster, H. A. Simmonds, D. M. J. Barry and D. M. O.
Becroft.   Pyrimidine and purine metabolites in ornithine
carbamoyltransferase deficiency.   J.Inher.Metab.Dis. 4: 27 (1981).

6. A. L. Rosenberg, L. Bergstrom, B. T. Troost, B. A. Bartholomew.
Hyperuricaemia and neurological deficits: A family study.   N.Eng.
J.Med. 282: 992 (1970).

DIAGNOSTIC AND THERAPEUTIC APPROACHES IN PYRIMIDINE

5'-NUCLEOTIDASE DEFICIENCY

E.H. Harley and P. Berman

Department of Chemical Pathology
University of Cape Town
Cape Town
South Africa

Hereditary deficiency of erythrocyte pyrimidine 5'nucleotidase results in a chronic haemolytic anaemia. The red cells show
basophilic stippling and contain a markedly increased content of
nucleotides, 3-6 times greater than normal, and 65 to 80% of
nucleotides are pyrimidine in type (Valentine, et al., 1974;
Torrance and Whittaker, 1979). Pyrimidine 5'nucleotidase is
unique amongst the 5'nucleotidases in its strict substrate specificity for pyrimidine nucleoside 5'monophosphates, its pH profile,
and for its cytosolic localisation (Paglia and Valentine, 1975).

A simple screening test is available for providing provi-
sional evidence for pyrimidine 5'nucleotidase deficiency. It is
based on the characteristic difference in the absorption profile
of pyrimidine as opposed to purine nucleotides in perchloric acid
(PCA) extracts of erythrocytes, and provides both quantitative
and qualitative information on the erythrocyte nucleotides
(Valentine et al., 1974). It also provides a means of monitoring
the progress of the disease, since there is evidence that erythrocyte nucleotide levels can fluctuate significantly during attempts
at therapy (Harley, Heaton and Wicomb, 1978). The disease also
shows episodic features which may be linked with fluctuations in
erythrocyte nucleotide levels, but despite the availability of
this simple quantitative procedure no studies addressing this question have been reported. Such studies could be particularly useful in providing information as to the source of the accumulated
nucleotides.

Confirmation of the diagnosis is by assay of pyrimidine
5'nucleotidase in haemolysates (Valentine et al., 1974). The sen-

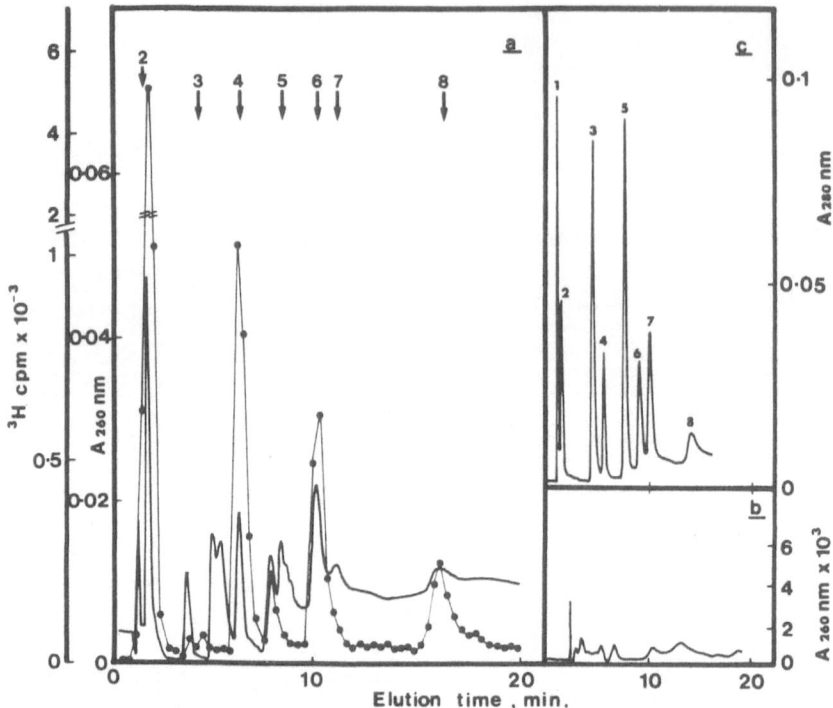

Fig. 1. HPLC of PCA extracts equivalent to approximately 2 x 10⁷ erythrocytes from a) a patient with pyrimidine 5' nucleotidase deficiency and b) a control. Marker nucleotides and nucleosides are shown in c), and are 1, cytidine; 2, uridine; 3, CMP; 4, UMP; 5, CDP; 6, UDP; 7, CTP; 8, UTP. In a) the fresh washed red cells were labelled for 2.5 hours with 10 µCi/ml 5-³H uridine in saline containing phosphate and glucose both at 6 mmol/l, pH 7.4. After washing once, incubation was continued for a further 12 hours. Nucleotides were separated on a Spectra-Physics Lichrosorb AN10 anion exchange HPLC column using a 5 to 100% linear gradient of 0.5 M NH₄H₂PO₄, pH 4.0, and a sweep time of seven minutes after a 1 minute delay. The flow rate was maintained at 3 ml/min and fractions taken for counting every 20 secs.

sitivity of the assay can be improved by measuring the phosphate released by the method of Itaya and Ui (1966). Alternatively, the other product of the reaction, uridine (or cytidine, if CMP is the substrate) can be accurately and rapidly quantitated by high pressure liquid chromatography (HPLC). UMP and uridine are easily

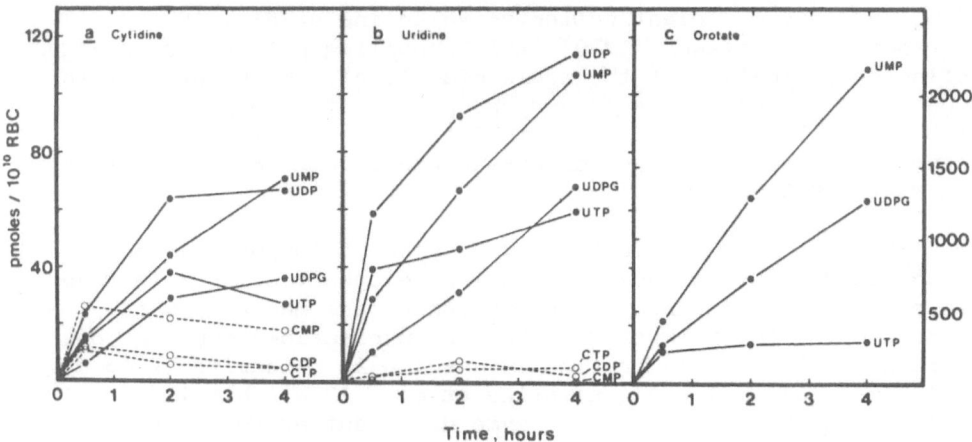

Fig. 2.   Erythrocyte nucleotide formation from exogenous pre-
          cursors. 0.5 ml aliquots of normal washed erythrocytes
          from fresh heparinised blood was labelled for the times
          indicated with a) 20 μCi 5-^{3}H uridine (1.6 μmoles/l),
          b) 20 μCi 5-^{3}H cytidine (1.4 μmoles/l) or c) 0.5 μCi
          2-^{14}C orotate (16 μmoles/l), and labelled nucleotides
          in the PCA extracts quantitated after separation on HPLC.

separated by isocratic elution with 1 mM KH_2PO_4, pH 6.0, on a
reverse phase column (Spectra-Physics Spherisorb ODS).

     HPLC also provides a rapid and effective means of characteris-
ing the intracellular nucleotides in erythrocyte PCA extracts, and
this information can be augmented by incubating the cells with a
labelled pyrimidine nucleotide precursor before PCA extraction.
Fig. 1 shows such a separation in a PCA extract of ^{3}H-Uridine
labelled erythrocytes from a patient with pyrimidine 5'nucleotidase deficiency. Peaks corresponding in elution time to uridine
and cytidine mono-, di-, and tri-phosphates can be seen. Label is
present in all three uridine nucleotide peaks as well as in a
fourth peak which corresponds in elution time in these and under
other separation conditions with UDP-glucose (UDPG).

     The accumulation with time of pyrimidine nucleotides from
labelled precursors in normal erythrocytes is illustrated in Fig
2. With either cytidine, uridine or orotate as the labelled precursor, UDPG shows the greatest relative accumulation with time,
and with still longer periods of label (not shown) becomes the
predominately labelled species. With cytidine as the labelled
precursor, only after short periods of labelling are cytidine
nucleotides the predominant labelled species. Thereafter counts
accumulate in uridine nucleotides. Since cytidine nucleotides are
the predominant pyrimidine nucleotides accumulating in pyrimidine

5'nucleotidase deficient patients (Valentine et al., 1974;
Torrance and Whittaker, 1979), this labelling pattern presumably
reflects deamination at the nucleoside level. With the orotate
label, Fig. 2 (c), there is a disproportionately high ratio of UMP
to UTP labelling when compared to Fig 2 (a), and the possible
reasons for this have been discussed previously (Harley, Zetler
and Neal, 1980).

There is no specific therapy at present for pyrimidine
5'nucleotidase deficiency. Logical development of an effective
therapy depends primarily on establishing the mechanisms responsible for the bulk of the abnormal nucleotide accumulation.
These nucleotides were originally assumed to be derived from degradation of ribosomal RNA (Valentine et al., 1974), but the results
of labelling both normal and enzyme deficient erythrocytes with
labelled orotate or uridine showed that much if not most of the nucleotide accumulation was from circulating pyrimidine nucleotide
precursors (Harley, Heaton and Wicomb, 1978). Assuming that biosynthetic pathways do contribute significantly to the nucleotide
accumulation, therapy might be feasible by searching for drugs
which would block uptake or metabolism of either orotate or
uridine (plus cytidine) in erythrocytes. There is no a priori
contraindication to such therapeutic approaches, since a) deficiency of OPRT in hereditary orotic aciduria is compatible with normal
health if the patient's diet is supplemented with uridine and
b) uridine salvage by uridine kinase may be merely a supplemen-
tary rather than an essential pathway for maintaining pyrimidine
nucleotide levels in cells other than the erythrocyte. The question then is which of the two possible biosynthetic pathways contribute most to erythrocyte nucleotide accumulation, that from
orotate via OPRT, or that from uridine via uridine kinase. Data
presented elsewhere in this volume (Berman and Harley) estimates
the Km (apparent) for the conversion of orotate to uridine nucleotides to be about 30 μmole/litre in the erythrocyte, a value over
an order of magnitude lower than that for the conversion of
uridine to uridine nucleotides (700 μmole/l) (Harley, Heaton and
Wicomb, 1978), and the Vmax for orotate uptake and conversion to
uridine is at least 25 pmoles/min/10^{10} RBC.
These factors would necessarily keep circulating orotate concentrations very low. Measurement of actual fluxes in vivo are not
possible without a measurement of the rate of entry of orotate
into the circulation. Orotate could enter the circulation from
two sources: 1) the diet; milk and milk products containing more
than 0.1% orotate, and 2) from de novo synthesis in certain
tissues, especially the liver, which is the only known tissue,
other than kidney and erythrocytes, to have an effective transport
system for orotate (Ord and Stoken 1973). Evidence is presented
elsewhere in this volume (Berman and Harley) demonstrating the
feasibility of a pyrimidine flux from liver to erythrocyte (as
orotate) and from erythrocyte to peripheral tissue (as uridine).

An attempt at therapy of a patient with pyrimidine 5'nucleotidase deficiency by administration of allopurinol, reported previously (Harley Heaton and Wicomb, 1978), gave results compatible with this concept. The rationale for this approach was the hope that the allopurinol induced inhibition of erythrocyte OPRT might decrease any flux of pyrimidines through this pathway into erythrocytes and lower the pyrimidine nucleotide levels. In practice the opposite occurred, and erythrocyte nucleotide levels increased by over 60%, and were maintained at the elevated level until the cessation of allopurinol therapy. An interpretation for these unexpected results which is more likely than that originally proposed (Harley, Heaton and Wicomb, 1978) i.e. that allopurinol stabilises red cell OPRT (Beardmore, Cashman and Kelley, 1972) is that allopurinol inhibition of OPRT could give rise to a greatly increased efflux of orotate from the liver into the circulation. This would present the erythrocyte with increased orotate availability with concomitant increase in uptake by the erythrocytes, and trapping of the uridine nucleotides formed on account of the deficiency of the 5'nucleotidase.

In summary, it is proposed that the most hopeful prospect for therapy of this condition would be to search for a specific inhibitor of orotate transport across cell membranes. This is likely to affect the metabolism of pyrimidines in only two tissues, liver and erythrocytes, and, unless it gives rise to toxic accumulations of orotate in the liver, could help to reduce pyrimidine nucleotide levels in pyrimidine 5'nucleotidase deficient erythrocytes. A trial of a specific inhibitor of uridine kinase might also be appropriate, and the final choice of a therapeutic approach will depend on the demonstration of which of these two alternative pathways amenable to therapeutic intervention contributes most to erythrocyte pyrimidine nucleotide accumulation.

REFERENCES

Beardmore, T. D., Cashman, J.S., and Kelley, W. N., 1972, Mechanism of allopurinol-mediated increase in enzyme activity in man, J. Clin. Invest., 51:1823.
Berman, P., and Harley, E. H., 1982, Orotic acid uptake and metabolism by human erythrocytes, Adv. Exp. Med. Biol. this volume.
Harley, E. H., Heaton, A., and Wicomb, W., 1978, Pyrimidine metabolism in hereditary erythrocyte pyrimidine 5'-nucleotidase deficiency, Metabolism, 27:1743.
Harley, E. H., Zetler, P., and Neal, S., 1980, Kinetics and compartmentation of erythrocyte pyrimidine metabolism, Adv. Exp. Med. Biol., 122B:217.
Itaya, K., and Ui, M., 1966, A new micromethod for the colorimetric determination of inorganic phosphate, Clin. Chem. Acta 74:361.

Ord, M. G., and Stocken, L. A., 1973, Uptake of orotate and thymi-
    dine by normal and regenerating rat livers, Biochem. J.,
    132:47.
Paglia, D. E., and Valentine, W. N., 1975, Characteristics of a
    pyrimidine-specific 5'-nucleotidase in human erythrocytes, J.
    Biol. Chem., 250:7973.
Torrance, J. D., and Whittaker, D., 1979, Distribution of eryth-
    rocyte nucleotides in pyrimidine 5'-nucleotidase deficiency,
    Brit. J. Haematol., 43:423.
Valentine, W. N., Fink, K., Paglia, D. E., Harris, S. R., and
    Adams, W. S., 1974, Hereditary hemolytic anaemia with human
    erythrocyte pyrimidine 5'-nucleotidase deficiency, J. Clin.
    Invest., 54:866.

NEW DEFECTS OF PYRIMIDINE METABOLISM

S.K. Wadman, F.A. Beemer, P.K. de Bree, M. Duran,
A.H. van Gennip*, D. Ketting, and F.J. van Sprang

University Children's Hospital "Het Wilhelmina Kinderziekenhuis", Nieuwe Gracht 137, 3512 LK Utrecht, The
Netherlands
* Children's Hospital "Het Emma Kinderziekenhuis", Spinoza-
straat 51, 1081 HJ Amsterdam, The Netherlands

INTRODUCTION

The number of inherited defects of the pyrimidine metabolism
described so far is small, compared to that of the purine metabolism. Combined deficiency of orotate phosphoribosyltransferase
(OPRT) (EC 2.4.2.10) and orotidine 5'-monophosphate decarboxylase
(ODC) (EC 4.1.1.23), designated as type I hereditary orotic aciduria, presents with characteristic clinical features such as hypo-
chromic anemia with a megaloblastic bone marrow and crystalluria.
Only six patients have been described and, as far as we know, new
cases have not been discovered recently. ODC deficiency with similar clinical phenomena and leading to increased urinary excretion
of orotate and orotidine has been detected in only one patient (1).
A third defect, a deficiency of pyrimidine 5'-nucleotidase (Py-5N)
(EC 3.1.3.5.) in erythrocytes, is associated with chronic hemolytic anemia and prominent basophylic stippling of the erythrocytes
due to accumulated pyrimidine nucleotides. An increasing number
of patients have been reported, their detection being facilitated
by the typical phenomena. We do not know whether the urinary pyrimidine profile in this condition is abnormal.
A difficulty in finding inborn errors of pyrimidine metabolism by
metabolite screening procedures is the absence of a typical end-
product, like uric acid is in the purine metabolism. Moreover,
pyrimidine metabolism is not easily accessible for simple chromatographic screening techniques. Nevertheless, with more complicated methods we are able to evaluate patterns of urinary pyrimidine
bases and nucleosides. With routine gas-liquid chromatography
(GLC) as is used for urinary organic acid analysis strongly increased uracil and thymine concentrations can be discovered.

Other possibilities are automated column chromatography on ion-exchange resins, high performance liquid chromatography (HPLC), or high resolution two-dimensional thin-layer chromatography (TLC). Purine bases and nucleosides are simultaneously detected.
With our routine GLC-system for urinary organic acid analysis in use for the screening for inborn errors of metabolism we detected persistent uraciluria in a few children. One of these patients will be described briefly here. Another patient showed a persistent excretion of thymine and uracil both. This child has been studied more in detail; the results will be given in the present paper.

CASE REPORTS

E.v.d.B (m), with uraciluria, is the first child of healthy, unrelated parents. He was dysmature at birth and he had a prolonged jaundice. Breast feeding was not possible due to poor sucking. He was admitted for clinical investigation and urinary metabolite screening when he was 4 y. Main abnormalities were: psycho-motor retardation, severe speech retardation, sparse fair hair, hyper-laxation of ligaments and hypotonia. He had a macrocephaly.

R.E. (m), with thymine-uraciluria, is the third child of healthy parents. The first child is healthy but the second one died of perinatal asphyxia. A few hours after birth the patient developed cyanosis with a mild respiratory distress. On the third day he developed also a pneumothorax. Furthermore there was a hyperbilirubinemia. Treatment was successful and at $3\frac{1}{2}$ weeks he was discharged. Psycho-motor development was normal until the age of $1\frac{1}{2}$ years. Then he developed petit mal seizures which were treated. After one month treatment could be discontinued. The parents noticed that behavioural changes occurred after the onset of the seizures. Speech did not develop and his behaviour became solitary. At admission for evaluation of his developmental problems no physical abnormalities were seen. Psychological investigations revealed a normal intelligence with autistic features and the absence of auditory defects.

METHODS

Gas-liquid chromatography and GLC combined with mass spectrometry (GLC-MS) were performed as described previously (2). Quantitative analysis of urinary pyrimidines and purines was done with automated cation-exchange chromatography according to (3) and with HPLC (4), (5), (6). For high resolution two-dimensional TLC the method described previously (7) with minor modifications (8) was used. Synthetic dihydrouracil and dihydrothymine (Sigma) were hydrolyzed in 6M HCl for 48h at 150° and determined as β-alanine (β-Ala) and β-aminoisobutyric acid (β-AIB) resp. by automated amino acid chromatography on a cation-exchange resin. Urinary dihydrouracil and dihydrothymine were comprised in fractions 3 and 4 obtained with the prefractionation procedure (7). These fractions were hydrolyzed and β-Ala and β-AIB were determined in the hydrolysate. R- and S-enantiomers of β-AIB were determined as described in (9).

RESULTS AND DISCUSSION

    Uraciluria in E.v.d.B. The first urine sample was presented for metabolite screening when the child was 3 10/12 y. The gas chromatogram showed a moderate amount of uracil. Cation-exchange column chromatography revealed 0.93 mmol/l - 1.29 mmol/g creatinine, which is significantly elevated (mean value in 6 controls 0.13, range 0.07 - 0.30). Thymine was not detected. Orotic acid and orotidine were not clearly elevated. One month later urinary uracil was 0.61 mmol/l - 1.03 mmol/g creatinine. Orotic acid: 12 μmol/l - 20 μmol/g creatinine. Mean value in 6 controls 20, range 8.5 - 42. At 4 2/12 y. uracil was 0.89 mmol/l - 1.29 mmol/g creatinine. There was no hyperammonemia, fasting blood ammonia being 16 and 20 μmol/l. On loading with protein, 4 g/kg b.w., blood ammonia remained normal, 25 and 30 μmol/l and there was no increase in urinary uracil: 0.48 mmol/l - 1.30 mmol/g creatinine. Herewith we excluded hyperammonemia as the underlying cause of the uraciluria. The absence of overflow of uridine and orotic acid is in agreement with this conclusion Oral loading of the patient with uracil (intake 15 mmol - 1 mmol/kg b.w.) resulted in a urinary excretion of 2.73 mmol uracil during the first 24 hours. This makes a block at the level of dihydropyrimidine dehydrogenase (diHPyDH) (EC 1.3.1.2.) improbable. No thymine was excreted, which we should expect in case of a partial deficiency of this enzyme. To date we have no explanation for the persistent uraciluria in this patient and we do not know whether this metabolic abnormality is associated with the developmental retardation of the patient. In this connection we should mention another patient (A.G.) with extensive neurological damage, who also had an excessive excretion of uracil. But this patient showed periodic spells of hyperammonemia of unknown etiology and during an attack she excreted also an excessive amount of orotic acid and some uridine.
Possible hypotheses for the defect in E.v.d.B. are: 1. Decreased reutilisation of uracil at the level of uridine phosphorylase or uridine kinase. At low uracil levels reutilisation will predominate over catabolism via the diHPyDH reaction and at high uracil concentrations the reverse will be the case presumably. 2. Increased synthesis of pyrimidines due to a fluctuating regulation at the level of aspartate transcarbamylase and/or ornithine transcarbamylase. It has to be presumed that attacks of periodic hyperammonemia have not occurred during the limited time of investigations, with the phenomena of patient A.G. in mind.

Thymine-uraciluria in R.E. The gas chromatogram of the tri-
methylsilylated organic acids displayed two medium size abnormal
peaks, which turned out to be due to uracil and thymine according
to GLC-MS analysis. The two-dimensional thin-layer chromatogram
showed except uracil and thymine a third abnormal spot. Isolation
of this compound from a two-dimensional chromatogram, followed by
trimethylsilylation and GLC-MS showed the compound to be 5-hydroxy-
methyluracil. This compound is not extracted with ethylacetate and
is therefore absent in the gas chromatogram.
The elevated excretion of uracil,thymine and 5-hydroxymethyluracil
appeared to be persistent during the four years of follow-up.
Initially the excretions expressed as mmol/g creatinine in the 24h
urine on a free diet were higher than later on.

	16-8-78	7-4-82		16-8-78	7-4-82
uracil	4.6	2.4	pseudouridine	0.6	0.6
thymine	3.0	2.3	uric acid	3.6	6.3
5-OH Me uracil	0.2	0.4			

Normal pseudouridine and uric acid values were found, indicating
that overall tissue breakdown is not enhanced.
Oral loading of the patient with uracil, 1 mmol/kg b.w., resulted
in a 24h excretion of 74% of the intake. On oral loading with thymine 73% was excreted in 24 hours. These results suggest an enzyme
deficiency at the level of diHPyDH. Subsequently we loaded the
patient with dihydrouracil. Only 6.5% of the intake was excreted;
no free β-Ala was detected in the urine. Normal metabolism of dihydrouracil is to be expected because this compound is the first
metabolite beyond the block. We also loaded the patient with synthetic dihydrothymine, 1 mmol/kg b.w. On analysis by the procedure
described in the section methods the synthetic dihydrothymine appeared to be racemic. Only 9% of the load was excreted, indicating
again an intact dihydropyrimidinase (EC 3.5.2.2.). Loading with
dihydrothymine resulted in an excessive excretion of β-AIB: 0.9
mmol/24h, being 6.4% of the intake on a molar basis. Apparently
the R-β-aminoisobutyrate amino transferase (EC 2.6.1.22) was saturated. It could be shown that 95% of the β-AIB had the R-configura-
tion.
The results described prove that we deal with an in vivo diHPyDH-
deficiency, presumably of an inherited character.
As far as we know persistent thymine-uraciluria has been described
once before, in a child with a fatal medulloblastoma (10). It
seemed likely that the brain tumor caused the high excretion of
uracil and thymine, because there was a correlation between their
excretion rates and the clinical progress of symptoms, remission
after therapy and relapse. In our patient, however, no signs of
malignancy have developed during the four years of observation.
Therefore an inherited enzyme defect is held as the most probable
cause.

REFERENCES

1. W.N. Kelley and L.H. Smith Jr., Hereditary orotic aciduria, in 'The Metabolic Basis of Inherited Disease', eds. J.B. Stanbury, J.B. Wijngaarden, D.S. Frederickson, McGraw-Hill Book Co., New York (1978), pp 1045-1071.

2. J.P. Kamerling, M. Brouwer, D. Ketting and S.K. Wadman, Gas chromatography of urinary N-phenylacetylglutamine, J. Chromatogr. 64: 217-221 (1979).

3. S.K. Wadman, P.K. de Bree, A.H. van Gennip, J.W. Stoop, B.J.M. Zegers and G.E.J. Staal, Urinary purines in a patient with a severely defective T cell immunity and a purine nucleoside phosphorylase deficiency, in "Purine Metabolism in Man-II: regulation of pathways and enzyme defects", M.M. Müller, E. Kaiser and J.E. Seegmiller, Plenum Publishing Corporation, New York (1977) pp 471-477.

4. A.H. van Gennip, E.J. van Bree-Blom, J. Grift, P.K. de Bree and S.K. Wadman, Urinary purines and pyrimidines in patients with hyperammonenia of various origins, Clin. Chim. Acta, 104: 227-239, (1980).

5. A.H. van Gennip, E.J. van Bree-Blom, S.K. Wadman, M. Duran and F.A. Beemer, HPLC of urinary pyrimidines for the evaluation of primary and secondary abnormalities of pyrimidine metabolism, in "Biological/biomedical Applications of Liquid Chromatography III, ed. G.L. Hawk, Marcel Dekker, Inc. New York and Basel (1982), pp 285-296.

6. A.H. van Gennip, J. Grift, P.K. de Bree, B.J.M. Zegers, J.W. Stoop and S.K. Wadman, Urinary excretion of orotic acid, orotidine and other pyrimidines in a patient with purine nucleoside phosphorylase deficiency, Clin. Chim. Acta, 93: 419-423 (1979).

7. A.H. van Gennip, D.Y. van Noordenburg-Huistra, P.K. de Bree and S.K. Wadman, Two-dimensional thin-layer chromatography for the screening of disorders of purine and pyrimidine metabolism, Clin. Chim. Acta, 86: 7-20 (1978).

8. A.H. van Gennip, Screening for disorders of purine and pyrimidine metabolism. A chromatographic approach. Thesis, Utrecht (1982).

9. A.H. van Gennip, J.P. Kamerling, P.K. de Bree and S.K. Wadman, Linear relationship between R- and S-enantiomers of β-aminoisobutyric acid in human urine, Clin. Chim. Acta 116: 261-267 (1981)

10. G. Berglund, J. Greter, S. Lindstedt, G. Steen, J. Walden-
    ström and U. Wass, Urinary excretion of thymine and uracil
    in a two-year-old child with a malignant tumor of the brain,
    Clin. Chem. 25: 1325-1328 (1979).

CLINICAL SURVEY OF 200 PATIENTS WITH GOUT

A. Spaccarelli, A. Giacomello, M. L. Sorgi,
and A. Zoppini

Institutes of Rheumatology and of Probability Calculus
University of Rome, Rome, Italy

A retrospective computer assisted analysis of 200 men with primary gout has been carried out. Diagnosis was based on clinical observations, X ray findings and laboratory data[1].

The results obtained are summarized in the tables.

Age at onset, frequency of family history of gout, hypertension urate crystals in joint fluid during the acute attack, are in good agreement or within the range of previous surveys[1,2]. Many patients

Table 1. Mean, Standard Deviation (S.D.) and Range of the Variables
Listed on the Left

Variable	Mean	S.D.	Range
Age of the patients (years)	53.47	12.36	24 – 82
Age at onset of gout (years)	45.34	13.04	16 – 79
Duration of gout (years)	8.13	8.07	0 – 52
Serum urate concentration (mg/100 ml)	9.15	2.56	3.4 – 16
Joints involved for each patient (No)	6.84	3.17	1 – 14
Age at onset of nephrolithiasis (years)	43.76	11.72	7 – 71

Table 2. Percentage Distribution of Positive Responses Among Men
         with Primary Gout

Family history of gout	15.62 % of 160 patients
Minimum average daily intake of 1 l of wine	50.05 % of 200 patients
Nephrolithiasis	32.00 % of 200 patients
Hypertension (diastolic blood pressure of 90 mm Hg or more)	33.50 % of 200 patients
Vascular diseases (coronary artery disease as evidenced by a history of myocardial infarction or angina, or by suggestive electrocardiogram changes; cerebrovascular disease; other evidence of severe atheroma)	30.50 % of 200 patients
Urate crystals in joint fluid during the acute attack	87.50 % of 64 patients
Urate crystals in synovial punch biopsy	71.43 % of 21 patients
Subcutaneous tophi	31.50 % of 200 patients
First metatarsophalangeal involvement	63.58 % of 162 patients
Periarticular osteopenia	78.86 % of 194 patients
Bone erosions	73.71 % of 194 patients
Heberden's nodes	19.50 % of 200 patients
Tests for rheumatoid factor	8.54 % of 199 patients
Renal failure (a blood urea of more than 50 mg/100ml) and/or proteinuria	13.50 % of 200 patients

were regular drinkers and an alcoholic polyneuropathy was present in
two subjects. The mean duration of disease was higher in patients
with, than in those without hypertension (9.66 ± 7.86 and 7.36 ± 8.12
years respectively).

The patients had a much greater frequence of nephrolithiasis,
crystalline urate deposits in synovial punch biopsies and were somewhat more tophaceous than expected.[1,2] The severity of their gout
was confirmed by the high incidence of radiographic finding of bone
erosions and by the high number of joints involved for each patient.

In good agreement with the literature data,[1] the prevalence of
nephrolithiasis increased with the height of the serum urate concentration reaching 43.75 % at serum urate values above 11 mg/100 ml.

Table 3. Relative Frequency of Joint Involvement. First Attack and
Any Attack

	First Attack		Any Attack	
Foot	80/175	45.71%	181/200	90.50%
Ankle	43/175	24.57%	143/200	71.50%
Knee	51/175	29.14%	181/200	90.50%
Hip	3/175	1.71%	18/200	9.00%
Hand	15/175	8.57%	96/200	48.00%
Wrist	10/175	5.71%	97/200	48.50%
Elbow	6/175	3.43%	82/200	41.00%
Shoulder	3/175	1.71%	44/200	22.00%

Table 4. Summary of "Chi Square" Test of Indipendence between Variables Listed on the Left and on the Head of the Table[a]

	Serum Urate concentration	Duration of gout	Age of the patients	Age at onset of gout
Bone erosions	−	++	−	−
Periarticular osteopenia	−	−	−	−
Vascular disease	−	−	++	++
Hypertension	−	++	−	−
Heberden's nodes	−	−	−	−
Renal failure and or proteinuria	−	−	−	−
Rheumatoid factor	++	++	−	−
Tophi	−	++	++	−
Nephrolithiasis	+	−	−	−

[a] Significance level: (−) Significance level > 5%; (++) Significance
level < 1%; (+) 5% > Significance level > 1%

In 47.62 % of the patients stone preceded the development of gouty
arthritis by more than a decade in 14.30 percent. The mean duration
of gout was higher in patients with subcutaneous tophi (10.71 ± 9.17
years) or with bone erosions (9.19 ± 8.49 years) than in those without tophi (6.94 ± 7.27 years) or without erosions (5.43 ± 6.41 years).

Although the preponderance of lower extremity involvement during
acute episodes of gout was confirmed, the frequency of acute attacks
at the first metatarsophalangeal was less than expected and knee and
ankle involvement was more common than anticipated[1,2].   Only 64.20 %
of patients had monoarticular gout in the first episode. The pattern
of joint involvement was independent of duration of disease.

Rheumatoid factor (RF) appeared in a small but definite group
of patients with gout confirming previous observations[2,3]. The mean
duration of disease was lower in the RF positive patients than in the
RF negative patients (5.06 ± 3.19 and 8.35 ± 8.33 years respectively).
The mean serum urate concentration was also lower in the RF positive
patients than in the RF negative patients (6.80 ± 1.60 and 9.37 ± 2.54
mg/100 ml respectively). Ten of 62 patients (16.13 %) with chronic
tophaceous gout were RF positive, compared with 5.11 % (7 of 137) of
those with acute arthritis only.

The frequency of periarticular osteopenia was striking and might
be related to disuse and/or to an active synovitis. The frequency of
Heberden's nodes was more common than anticipated and was independent
of age of the patient and duration of gout.

Aseptic necrosis of bone has been observed in two patients[4].

REFERENCES

1. J. B. Wyngaarden and W. N. Kelley, Gout and Hyperuricemia, Grune
   & Stratton, New York (1976)
2. S. L. Wallace, H. Robinson, A. T. Masi, J. L. Decker, D.J. McCarty,
   T. F. Yu, Selected Data on Primary Gout, Bull. Rheum. Dis.,29:
   992 (1979)
3. D. Gigante, A. Giacomello, Rheumatoid Factors in the Serum of
   Gouty Patients, Arthritis Rheum. 23: 379 (1980)
4. M. Castagnoli, A. Giacomello, R. Scaffidi Argentina, A. Zoppini,
   Kienbock's Disease in Gout, Arthritis Rheum., 24: 974 (1981).

INCORPORATION OF ^{15}N FROM GLYCINE INTO

URIC ACID IN GOUT: A FOLLOW-UP STUDY

Ts'ai-fan Yü and John Roboz

Mount Sinai School of Medicine
The City University of New York
New York, N.Y. 10029 U.S.A.

INTRODUCTION

The incorporation of ^{15}N-glycine into uric acid was studied three times in one gouty patient in an interim of 19 years. The objective of repeated studies was to see the effects of long-term therapy, aging, and changing life style on the nature of the metabolic aberration of uric acid metabolism in gout using ^{15}N-glycine.

MATERIAL AND METHODS

Patient SK had recurrent gouty attacks for 5 years when first seen at 32 years of age in 1963. He was nontophaceous with no history of renal calculi. His plasma urate was > 10 mg/dl and urinary uric acid was > 1000 mg/day. His renal function was normal. Shortly after the first study using ^{15}N-glycine, he received daily colchicine and allopurinol from 1964-75. A second isotope study was made in 1976 after discontinuing allopurinol for 10 months. Allopurinol was resumed thereafter until 1981. A third isotope study was conducted in 1982 after allopurinol had been discontinued for 8 months. Throughout the period of 19 years, his renal status remained normal, with no tophi or renal calculi.

His diet was essentially the same during the three studies. ^{15}N-glycine (60 atom % excess), 100 mg/kg body weight was given with 250 ml milk. Thereafter, urine was collected daily for one week, using toluene as preservative. Uric acid was determined using the uricase method. Uric acid was isolated and crystallized. Isotope enrichment was analyzed by standard mass spectrometer techniques as previously described (1).

RESULTS

In 1963, there was overwhelming incorporation of ^{15}N-glycine into uric acid. Maximal abundance reached 0.351 and 0.357 atom % excess in Days 1 and 2, with cumulative incorporation of 0.527 % in a week (2). In 1976, the incorporation was much less with max-ima at 0.163 and 0.158 atom % excess in Days 2 and 3; and cumula-tive incorporation of 0.216 % of the total dose administered in one week. In the present study, the atom % excess was not much different from that in 1976. The isotope enrichment was 0.164 and 0.145 atom % excess in Days 2 and 3, and the cumulative incorpora-tion was 0.205 % in one week. Although the rate of ^{15}N incorpora-tion in Studies 2 and 3 was much less than that of Study 1, he still incorporated more than the nongouty subjects (Tables 1 and 2). Also, his daily excretion of uric acid, and the ratios of urinary uric acid-N to total nitrogen remained higher than those of the nongouty subjects. His serum urate became lower in Study 2, and normal in Study 3 (Table 3).

Table 1   Daily ^{15}N Atom % Excess in Urinary Uric Acid

Day	1	2	3	4	5	6	7
Study 1	0.351	0.357	0.303	0.255	0.255	0.193	0.164
Study 2	0.139	0.163	0.158	0.148	0.131	0.123	0.116
Study 3	0.138	0.164	0.145	0.138	0.125	0.133	0.113

Table 2 Cumulative % of ^{15}N Incorporation as Urinary Uric Acid

Day	1	2	3	4	5	6	7
Study 1	0.107	0.212	0.298	0.359	0.430	0.385	0.527
Study 2	0.030	0.064	0.099	0.133	0.164	0.190	0.216
Study 3	0.033	0.071	0.099	0.129	0.159	0.181	0.205

Table 3 Serum and Urinary Uric Acid At Different Studies

Year Studied	Age (Yr)	Weight (kg)	Serum Urate (mg/dl)	Urine Uric Acid (mg/day)	T.N.* (gm/day)	UA-N/T.N.** (%)
1963	32	109	11.3	1047±99	14.9±1.1	2.55±0.20
1976	45	107	9.5	790±32	12.9±1.1	2.60±0.20
1982	51	116	6.2	853±73	13.0±1.1	2.20±0.13

* T.N. = Total Nitrogen
** UA-N/T.N. = Uric Acid Nitrogen/Total Nitrogen

DISCUSSION

There is distinct alteration in the metabolic aberration of uric acid metabolism in this patient. His life style has not changed, although he has aged by 19 years, the interval between Studies 1 and 3. The long-term therapy of allopurinol is most likely the cause of the observed changes. Allopurinol effectively blocks the enzyme xanthine oxidase which converts hypoxanthine and xanthine to uric acid. This effect can apparently be quite long-lasting after the use of allopurinol for years. Even after its discontinuance for 8-10 months, the rate of uric acid production remained suppressed.

SUMMARY

Over-incorporation of ^{15}N-labeled glycine into uric acid indicates over-production of uric acid by de novo purine biosynthesis. This metabolic aberration, though considered to be inborn (3), may be modified by changing of life style, aging and long-term therapy. In the patient under study, the protracted use of allopurinol seems to have played the most important role. Aging contributed to a certain extent, and changing life style was the least significant factor.

REFERENCES

1. Yü TF, Roboz, J   (1981)   Am J Med 70:797-802
2. Gutman AB, Yü TF   (1963)   Am J Med 35:820-831
3. Gutman AB, Yü TF   (1963)   Trans Assoc Am Physicians 76:141-151

HYPERURICAEMIA IN YOUNG NEW ZEALAND MAORI MEN

T. Gibson, R. Waterworth, P. Hatfield, G. Robinson,
and K. Bremner

Guy's Hospital, London, U.K. and Wellington Public
Hospital, New Zealand

INTRODUCTION

Many Polynesian peoples exhibit a susceptibility to hyperuricaemia and gout.[1,2] This predisposition is a feature of widely
distributed populations and suggests a common genetic role.[3] The
mechanism has not been clearly established. Maori men are reported
to have a prevalence of gout in excess of 10%. They are also prone
to obesity, diabetes, and hypertension.[3] It is not known whether
renal dysfunction is an additional feature of this disease spectrum.
The present study examined the current prevalence of hyperuricaemia
and gout in a male Maori population of working age. An attempt was
made to reassess the relationship of obesity and hypertension with
hyperuricaemia; and to determine the prevalence of renal impairment. The renal excretion of uric acid was also estimated.

METHODS

All Maori males at a car assembly plant were asked to participate. Those unwilling to volunteer completed a questionnaire to
determine whether they or family members had gout or renal disease.
Subjects were interviewed and supine and standing blood pressures
measured with a large cuff. Upper arm circumference, height and
weight were also recorded. Two consecutive 24h. urine collections
and simultaneous venous blood samples were obtained. No dietary
restrictions were imposed. Alcohol was discouraged during the
period of urine collection. Serum and urine uric acid were
estimated by a uricase method; urea and creatinine by standard
autoanalyser techniques. Statistical comparisons were made by
Student's t test and the chi square test.

RESULTS

There were 115 who agreed to take part and 16 who declined. Of the latter, one had gout and six (37%) had a family history of gout. Ten of the participating subjects had a history of gout, three of whom were receiving hypouricaemic treatment. Hyperuricaemia was defined as a serum uric acid in excess of 0.44 mmol/l. A normal range for males was derived from 200 healthy blood donors, 8% of whom were of Polynesian extraction. On this basis, 26 (23%) had asymptomatic hyperuricaemia. Regular alcohol consumption was defined as a daily intake of one litre or more of beer and hypertension as a supine or standing diastolic blood pressure in excess of 100mm Hg (Korotkov IV). Clinical details are shown in Table 1.

TABLE 1. Details of Maoris in survey

	History of gout	Asymptomatic Hyperuricaemia	Normo- uricaemic	Total
No.	10 (8%)	26 (23%)	79 (69%)	115
Mean age (± S.D.)	45 ± 8	28 ± 10	27 ± 10	-
Family History gout	5 (50%)	10 (38%)	18 (23%)	33 (28%)
Regular alcohol	5 (50%)	10 (38%)	28 (35%)	43 (37%)
Hypertension	4 (40%)	3 (11%)	12 (15%)	19 (16%)
Mean wt (± S.D.) Kg.	93 ± 20	85 ± 10	79 ± 15	-
Mean Ponderal Index – ht/3√wt	11.4 ± 0.9	11.9 ± 0.7	12.2 ± 0.7	-
Mean arm circumference (± S.D.) cm	32 ± 3.9	31 ± 2.0	29.5 ± 2.7	-

Those with a history of gout were significantly older than both the asymptomatic hyperuricaemic subjects (t = 4.5; p < 0.001) and the normouricaemic. A family history of gout, regular alcohol intake and hypertension were more frequent in the gouty group but not significantly. The gouty were heavier than the asymptomatic hyperuricaemic and the latter heavier than the normouricaemic but the differences were not significant. Ponderal index (height in ins. divided by cubed root of weight in pounds) was significantly lower (t = 2.14; p < 0.05) and arm circumference significantly greater (t = 2.16; p < 0.05) in the asymptomatic hyperuricaemic compared with the normouricaemic. These measurements were not significantly different between the gouty and asymptomatic hyperuricaemic. In the survey as a whole, serum uric acid correlated with weight (r = 0.27; p < 0.01), arm circumference (r = 0.29; p < 0.01) and most strikingly with ponderal index (r = - 0.31; p < 0.001).

Serum and urine uric acid and indices of renal function are outlined in Table 2. Mean serum creatinine and urea were slightly higher in the gouty group and creatinine clearance was lower but the differences were not significant.

TABLE 2. Mean ($\pm$ S.D.) serum uric acid, urate excretion and renal function

	History of gout	Asymptomatic hyperuricaemic	Normouricaemic
Serum uric acid (mmol/l)	0.52 $\pm$ 0.1	0.48 $\pm$ 0.03	0.38 $\pm$ 0.04
Urate excretion (mmol/24h)	2.3 $\pm$ 1.0	3.3 $\pm$ 1.3	3.2 $\pm$ 1.2
Serum creatinine (mmol/l)	93 $\pm$ 10	89 $\pm$ 9.0	88 $\pm$ 10
Blood urea (mmol/l)	6.1 $\pm$ 1.8	5.6 $\pm$ 1.0	5.6 $\pm$ 1.5
Creatinine clearance (ml/min/1.73m^2)	95 $\pm$ 27	106 $\pm$ 28	113 $\pm$ 22

Urate clearance and fractional urate clearance of the asymptomatic hyperuricaemic were significantly lower than those of the normouricaemic (t = 3.0; p < 0.05). Values of the gouty and asymptomatic hyperuricaemic were not significantly different. A comparison with results of a recent survey of British gouty and normouricaemic subjects[4] seemed to indicate that Maoris are in general more obese (Table 3). Normouricaemic Maoris were heavier and their ponderal index was significantly lower than the British normouricaemic subjects, despite the greater age of the British (t = 2.01; p < 0.05). The ponderal index of the gouty British, who have previously been considered obese[4], was identical to that of

TABLE 3. Comparisons of mean ($\pm$ S.D.) and fractional urate
clearance within the Maori population and a comparison
with British gouty and normouricaemic subjects

	No.	Age	Body wt (kg)	Ponderal index	Urate clearance (ml/min/ 1.73m^2)	Urate clearance / GFR x100
Maori normo- uricaemic	79	27 ± 10	79 ± 15	12.2 ± 0.7	5.5 ± 2.1	4.9 ± 1.5
British normo- uricaemic	51	47 ± 12	75 ± 11	12.46 ± 0.6	8.3 ± 2.3	8.1 ± 3.2
Maori gout	10	45 ± 8	93 ± 20	11.4 ± 0.9	3.2 ± 1.0	3.6 ± 1.3
Maori asympto- matic hyper- uricaemic	26	28 ± 10	85 ± 10	11.9 ± 0.7	4.1 ± 1.7	3.9 ± 1.4
British gout	51	48 ± 12	81 ± 12	12.2 ± 0.6	5.0 ± 1.5	5.4 ± 1.7

the normouricaemic Maori population. Urate and fractional urate
clearance were lower in both the Maori gouty and asymptomatic
hyperuricaemic subjects compared with British gout patients. Urate
and fractional urate clearance were significantly lower in the Maori
normouricaemic population compared with their British counterparts
(t = 6.23 and 6.5; p < 0.001).

DISCUSSION

The study confirmed the very high prevalence of hyperuricaemia and
gout amongst Maoris and it was clear that those who participated
were representative of the survey population as a whole. It is
acknowledged that even amongst teenagers, Maoris tend to have a
higher blood uric acid level than New Zealand Europeans.[5] The
prevalence of gout in this study was approximately ten times that
amongst males of similar age in American and European populations.[6,7]

The correlation of serum uric acid with indices of obesity in
this and earlier studies[8] suggest that this may be a contributory
factor, even amongst Maori teenagers.[5] The comparisons with British
normouricaemic and gouty subjects emphasises the tendency toward

obesity amongst the Maori people. The frequent family history of
gout, especially amongst the gouty and asymptomatic hyperuricaemic
Maoris, was not surprising and may imply an inherited factor.
Alcohol consumption was no higher amongst the asymptomatic hyperuricaemic compared with the normouricaemic population and by itself
could not therefore be of paramount aetiological significance.
Undoubtedly, alcohol may have accentuated hyperuricaemia in some
cases.

The prevalence of hypertension was high amongst those with
gout. Two subjects were receiving hypotensive treatment. Amongst
the remaining hyperuricaemic and normouricaemic subjects, the
prevalence of mild hypertension appeared high for the age range
under investigation but comparisons with similar age groups in
other populations were not possible. In a survey of New Zealand
adults, presumably mainly Eurpoeans, the prevalence of hypertension
using similar criteria was 20%, a value which is in accord with
several Western surveys.[9]

Renal function was not obviously impaired in those with
asymptomatic hyperuricaemia. Mean values of creatinine clearance
of both groups were similar to GFR results of other normouricaemic
subjects of similar age.[10,11] The relatively low creatinine clearance of the small gouty population may have been partly due to their
age, hypertension and the slowly progressive kidney dysfunction
associated with gout.[4] In the age group examined, asymptomatic
hyperuricaemia appeared to have no adverse effect on the kidneys.

Urate clearance, even when expressed as a percentage of GFR
(fractional urate clearance) was significantly lower in the gouty
and asymptomatic hyperuricaemic subjects. This is a common
characteristic of many gouty patients and may be one mechanism of
hyperuricaemia.[4] Nevertheless, it was evidence that mean urate
clearance was lower than that of British subjects with gout. Urate
clearance of normouricaemic Maoris was also considerably less than
that of normouricaemic British males, despite the greater age of
the latter. Differences of methodology in the separate studies
could have accounted for this wide disparity. It is nevertheless
notable that the fractional urate clearance values of normouricaemic
men obtained in this study were also lower than those previously
published for other healthy populations.[10,11,12,13] This would
suggest that Maori males are relatively less efficient at excreting
uric acid. A similar comparative deficiency has been described
amongst subjects from the Philippines[14] and gives credence to the
notion that populations from the Pacific area share a similar,
inherited defect which increases their susceptibility to hyperuricaemia. Amongst the Maoris, this trait is probably accentuated
by a common tendency to obesity.

ACKNOWLEDGEMENT

The authors thank the Arthritis and Rheumatism Council of Great
Britain for financial support and the workforce at Todd Motors,
Porirua, for their magnificent cooperation.

REFERENCES

1. I.A. Prior. Metabolic maladies in New Zealand Maoris. Brit.
        Med. J. 1: 1065 (1964)
2. P.Z. Zimmet S. Whitehouse, L. Jackson, K. Thoma. High
        prevalence of hyperuricaemia and gout in an urbanised
        Micronesian population. Brit. Med. J. 1: 1237 (1978)
3. I.A. Prior, B.S. Rose, H.P. Harvey, F. Davidson. Hyperuricaemia,
        gout and diabetic abnormality in Polynesian people.
        Lancet 1: 333 (1966)
4. T. Gibson, J. Highton, C. Potter, H.A. Simmonds. Renal impair-
        ment and gout. Ann. Rheum. Dis. 39: 417 (1980)
5. J.M. Stanhope and I.A. Prior. Uric acid, joint morbidity and
        streptococcal antibodies in Maori and European teenagers.
        Ann. Rheum. Dis. 34: 359 (1975)
6. W.M. Mikkelsen, H.J. Dodge, I.F. Duff, I.H. Kato. Estimates of
        the prevalence of rheumatic disease in the population of
        Tecumseh, Michigan. J. Chron. Dis. 20: 351 (1967)
7. J. Zalokar, J. Lellouch, J.R. Claude. Goutte et uricémie dans
        une population de 4663 hommes jeunes actifs. Sem. Hop.
        Paris 57: 664 (1981)
8. G.W. Brauer and I.A. Prior. A prospective study of gout in
        New Zealand Maoris. Ann. Rheum. Dis. 37: 466 (1978)
9. B.W. Christmas and A.S. Turner. Prevalence of high blood
        pressure treated and untreated in an urban adult New
        Zealand population: Napier 1973. New Zealand Med. J. 86:
        419 (1977)
10. T.H. Steele and R.E. Rieselbach. The renal mechanism for urate
        homeostasis in man. Am. J. Med. 43: 868 (1967)
11. M.L. Snaith and J.T. Scott. Uric acid clearance in patients
        with gout and normal subjects. Ann. Rheum. Dis. 30: 285
        (1971)
12. M.A. Ogryzlo, M.B. Urowitz, H.M. Weber, J.B. Houpt. Effects of
        allopurinol on gouty and non-gouty uric acid nephropathy.
        Ann. Rheum. Dis. 25: 673 (1966)
13. C.A. Nugent and F.H. Tyler. The renal excretion of uric acid
        in patients with gout and in non-gouty subjects. J. Clin.
        Invest.   38: 1890 (1959)
14. L.A. Healey and P.S. Bayani-Sioson. A defect in the renal
        excretion of uric acid in Filipinos. Arthritis Rheum.14:
        721 (1971)

LEAN DRY GOUT PATIENTS

L. G. Darlington

Epsom District Hospital
Dorking Road, Epsom, Surrey, U.K.

Hypertriglyceridaemia is common in gout (1,2). There is good evidence that the hyperprebetalipoproteinaemia of gout patients may be associated with obesity and excessive alcohol consumption (3). It remains uncertain, however, whether obesity and alcohol, either alone or in combination, are sufficient to explain the hyperlipidaemia in all cases.

The aim of the present study was to measure lipid levels in non-obese gouty patients with little or no alcohol intake to determine whether the hyperprebetalipoproteinaemia frequently found in gouty patients occurred in such a lean and abstemious group.

All patients were of desirable weight or less for their age and frame by Metropolitan Life Insurance Company tables (4) and the mean weight of the gouty group was 2.7% below the mean of their desirable weight range. No patient drank more than one pint of beer per day (or its equivalent) and their average alcohol intake was less than half a pint of beer or its equivalent daily. 3 of the patients were receiving allopurinol therapy and 4 were untreated.

"Lean, dry" gout patients are rare and only 7 were found in 4 years from a busy gout clinic. Fasting sera were analysed for lipid and lipoprotein concentrations at a laboratory with its own control population (5). Serum uric acid levels were determined for the gout patients but uric acid data for the control population, unfortunately, were not available.

RESULTS

Results are given for gout patients in Table 1 with control data in Tables 2 (a) and (b).

129

The readings for serum cholesterol, triglyceride, beta and pre-beta lipoprotein levels in the gout patients lay within $\pm$ 2 standard deviations of the corresponding control mean. This means that they were within the 95% confidence limits for the control population and it is, therefore, unlikely that there is any real difference between gouty and control patients.

It is not possible entirely to exclude an effect of allopurinol therapy on the lipid levels of the three patients receiving treatment but levels were certainly not lower in treated patients.

DISCUSSION

The question arises whether the hypertriglyceridaemia associated with gout is due to a metabolic link between uric acid and triglyceride or, alternatively, due to the effects of obesity and/or of excessive alcohol intake. Indeed a combination of both mechanisms may occur or either mechanism may apply in various cases.

The theory of a metabolic link was favoured by Mielants et al. in 1973 and perhaps supported by the work of Trevaks and Lovell (6) and of Aronow et al. (7) who described the shared hypolipidaemic and uricosuric properties of clofibrate and halofenate respectively. Support for a metabolic link is also given by the occasional enhancement of triglyceride removal during allopurinol therapy (8). Furthermore, in 1981, MacFarlane et al. (9) reported reduced apoprotein C11 in gouty subjects which might account for their hypertriglyceridaemia.

The aetiological effects of obesity and of alcohol intake have long been considered to be important but, unfortunately, have usually been considered together with resulting confusion.

Wiedemann et al. (10) and Darlington and Scott (1) could not implicate obesity in the genesis of gouty hyperlipidaemia.

Furthermore, we have considered the effects of alcohol on serum lipids (Darlington and Scott, unpublished work) and noted a non-significant reduction in lipid and uric acid levels in gout patients after four weeks of confirmed alcohol abstention. This suggested that alcohol was only one aetiological factor in the pathogenesis of hyperlipidaemia.

In 1979, however, Gibson et al. (11) felt that hyperuricaemia and hypertriglyceridaemia were simply linked by means of alcohol excess and obesity and not by a direct mechanism and, indeed, our family study of gout patients (12) did not reveal hypertriglyceridaemia among first-degree relatives of these patients - a result which

Table 1.   Lipid, Lipoprotein and Uric Acid Data From Gout Patients

Name	Serum cholesterol (mg/dl)	Serum triglyceride (mg/dl)	Agarose electrophoresis (%)			Serum uric acid (μmol/l)	High Density lipoprotein cholesterol by precipitation (mg/dl)
			Beta	pre-β	Alpha		
F.T.	264	90	48.2	17.9	33.9	400*	56
N.P.	265	169	50.6	33.5	14.1	333*	49
J.C.	221	74	42.9	35.7	21.4	252	52
S.T.	231	86	37.4	16.0	46.6	468	-
P.H.	297	161	40.3	38.9	20.8	326*	-
M.N.	220	99	56.0	23.1	20.9	-	-
R.D.	188	63	58.3	10.0	31.7	490	-

* on hypouricaemic therapy

Table 2.   Fasting Control Data for Lipids and Lipoproteins in White Men and White Women Not on Oral Contraceptives (5)

A

Age (years)	Cholesterol (mg/dl)				Triglyceride (mg/dl)			
	Men		Women		Men		Women	
	No.	Mean $\pm$ 1 SD	No.	Mean $\pm$ 1 SD	No.	Mean $\pm$ 1 SD	No.	Mean $\pm$ 1 SD
18-29	213	191.7 $\pm$ 39.2	50	184.8 $\pm$ 28.7	211	96.2 $\pm$ 47.8	50	66.6 $\pm$ 29.3
30-39	182	213.6 $\pm$ 42.9	54	199.7 $\pm$ 35.9	182	110.5 $\pm$ 56.9	54	78.3 $\pm$ 39.4
40-49	267	232.5 $\pm$ 43.4	142	227.0 $\pm$ 38.5	264	133.7 $\pm$ 71.2	142	95.0 $\pm$ 42.5
50-59	237	230.7 $\pm$ 42.7	203	247.2 $\pm$ 43.8	235	115.9 $\pm$ 53.0	202	109.4 $\pm$ 49.8
60-69	97	236.0 $\pm$ 46.5			95	124.4 $\pm$ 67.5		

B

Age (years)	β- Lipoprotein (%)				Pre- β- lipoprotein (%)				α- Lipoprotein (%)			
	Men		Women		Men		Women		Men		Women	
	No.	Mean $\pm$ 1 SD	No.	Mean $\pm$ 1 SD	No.	Mean $\pm$ 1 SD	No.	Mean $\pm$ 1 SD	No.	Mean $\pm$ 1 SD	No.	Mean $\pm$ 1 SD
18-29	195	49.0 $\pm$ 6.8	48	49.0 $\pm$ 5.1	195	20.3 $\pm$ 7.3	48	13.1 $\pm$ 5.6	195	30.5 $\pm$ 7.6	48	37.8 $\pm$ 6.2
30-39	173	50.8 $\pm$ 7.2	50	49.7 $\pm$ 6.5	173	20.6 $\pm$ 7.8	50	15.2 $\pm$ 5.5	173	28.4 $\pm$ 7.2	50	35.0 $\pm$ 6.6
40-49	249	50.0 $\pm$ 7.4	135	50.2 $\pm$ 6.1	249	23.0 $\pm$ 9.2	135	16.4 $\pm$ 6.6	249	26.8 $\pm$ 6.6	135	33.3 $\pm$ 7.6
50-59	227	51.5 $\pm$ 6.6	191	50.7 $\pm$ 7.5	227	19.9 $\pm$ 7.8	191	17.7 $\pm$ 7.6	227	28.5 $\pm$ 7.1	191	31.4 $\pm$ 7.2
60-64	93	52.6 $\pm$ 6.8			92	21.3 $\pm$ 9.0			93	26.4 $\pm$ 6.4		

makes less probable a familial or genetic cause for the hyperlipidaemia.

CONCLUSION

On balance, therefore, it would appear that there is no strong
evidence for a familial cause for the hyperlipidaemia associated with
gout although a metabolic abnormality, possibly involving apoprotein
C11, remains a possibility. Furthermore, alcohol and obesity, either
singly or together, are important in the aetiology of the hypertriglyceridaemia at least in some patients.

In the present study, in spite of the small numbers of these
rare, "lean, dry" patients, the results revealed no intrinsic
hyperlipidaemia in gouty subjects when obesity and an excess of
alcohol were removed as causes of hypertriglyceridaemia.

REFERENCES

1. L.G. Darlington, and J.T. Scott, Plasma lipid levels in gout.
Ann. Rheum Dis. 31: 487-489 (1972).
2. H. Mielants, E.M. Veys and A.de Weerdt, Gout and its relation
to lipid metabolism. I. Serum uric acid, lipid and lipoprotein levels
in gout. Ann Rheum Dis. 32: 501-505 (1973).
3. T. Gibson and R. Grahame, Gout and hyperlipidaemia, Ann Rheum
Dis. 33: 298-303 (1974).
4. Metropolitan Life Insurance Company Tables. Build and Blood
Pressure Study, Society of Actuaries, Chicago, Illinois,U.S.A.(1959)
5. J. Slack, N. Noble, T.W. Meade, and W.R.S. North. Lipid and
lipoprotein concentrations in 1,604 men and women in working
populations in N.W.London. Br. Med. J. 2: 353-356 (1977).
6. G. Trevaks and R.R.H. Lovell. Effect of Atromid and its components on uric acid excretion and on gout. Ann Rheum Dis. 24:
572-575 (1965).
7. W.S. Aronow, P.R. Harding, M. Khursheed, J.S. Vangrow and
N.P. Papageorges. Effect of halofenate on serum uric acid. Clin.
Pharmacol. Therap. 14: 371-373 (1973).
8. R. Bluestone, B.Lewis and L. Mervart. Hyperlipoproteinaemia in
gout. Ann. Rheum Dis. 30: 134-137 (1971)
9. D.G. MacFarlane, C.A. Midwinter, P.A. Hopes, B. Slade,P.A.Dieppe
Lipoproteins, gout, and vascular disease. Ann Rheum Dis 41: 200 (1982)
10. E. Wiedemann, H.G. Rose and E. Schwartz,. Plasma lipoproteins,
Glucose Tolerance and Insulin response in Primary Gout. Amer. J.
Med. 53: 299-307 (1972).
11. T. Gibson, K. Kilbourn, I. Horner and H.A. Simmonds. Mechanism
and treatment of hypertriglyceridaemia in gout. Ann Rheum Dis. 38:
31-35 (1979).
12. L.G. Darlington, J. Slack and J.T. Scott. Family study of lipid
and purine levels in gout patients. Ann. Rheum. Dis. In press.

URATE DEPOSITS WITHOUT GOUTY ARTHRITIS

J. T. Scott, P. Hollingworth, and H. C. Burry

Kennedy Institute of Rheumatology

Bute Gardens
London W6 7DW, England

Wellington Clinical School of Medicine, Wellington, N.Z.

The sequence of events from asymptomatic hyperuricaemia to acute gout, intercritical gout and finally chronic tophaceous gout is well established, one fifth of patients who suffer acute gout eventually developing tophaceous deposits if untreated with hypouricaemic drugs. Deposition of tophi in the absence of acute gouty arthritis is extremely rare: the cases described here invite speculation on the mechanism of inhibitory factors involved in acute gout.

Patients 1-5 (Table) developed tophi without any history of acute gouty arthritis. All were taking drugs, for some other condition, which may have contributed to hyperuricaemia. Cases 1-4 showed varying degrees of renal impairment and Case 5 had co-existent rheumatoid arthritis, a tophaceous deposit being discovered in a renal biopsy carried out because of penicillamine nephropathy with a tophus on the ear being observed at the same time.

Case 6 is of related interest. His attacks of podagra ceased abruptly with the onset of classical polyarticular seropositive rheumatoid arthritis, although hyperuricaemia and a crystal-proven tophus on the ear were demonstrated several years later.

Case 7 is also of particular interest. Arthroscopy and synovial biopsy in a patient with S.L.E., who had a mild chronic synovitis but no history of acute arthritis, showed, unexpectedly, urate crystal deposition.

133

Table

Case	Sex	Age	Urate deposit	Other diseases	Drugs	Plasma urate µmols/l	Creatinine clearance mls/min
1	Male	69	Finger	Generalized osteoarthrosis	Aspirin	590	40
2	Male	70	Finger, toes & olecranon bursa	Recurrent pulmonary emboli	Frusemide	550	60
3	Male	75	Finger pulp	Chronic pyelonephritis	Bendrofluazide	510	24
4	Female	72	Fingers	Generalized osteoarthrosis	Aspirin	580	62
5	Male	46	Ear & kidney	Rheumatoid arthritis	Aspirin Prednisolone	600	126
6	Male	68	Ear	Rheumatoid arthritis	Aspirin	710	110
7	Male	64	Synovial membrane	S.L.E.	Prednisolone	610	70

Previous reports have drawn attention to the very infrequent occurrence of acute gouty arthritis in association with chronic renal failure[1] or inflammatory disorders of connective tissue such as rheumatoid arthritis and S.L.E.[2] It has been suggested that duration of untreated renal impairment may be insufficient for the miscible pool of urate to reach a critical level for crystal deposition to occur[3], while the age and sex incidences of rheumatoid arthritis and S.L.E. are very different from those of gout.

Our admittedly anecdotal experience, however, suggests that hyperuricaemia and crystal deposition may indeed occur in such situations, but that an acute inflammatory response may not be elicited.

Factors possibly involved in such inhibition of inflammation are:-

1.   Suppression of phagocytic activity, as has been reported in both uraemia[4] and rheumatoid arthritis[5].

2.   An alteration of the protein coating of crystals[6]: the protein coat acquired by urate crystals in rheumatoid synovial fluid may be less inducive to crystal phagocytosis than that present on the crystal surface in non-rheumatoid individuals.

3.   Synovial hypocomplementaemia, as has been shown in rheumatoid arthritis and S.L.E.[7]

REFERENCES

1.   S C Wallace, O Bernstein. The relationship between gout and the kidney. Metabolism 12: 440 (1963)

2.   B A Wall, C A Agudelo, M E Weinblatt, R A Turner. Acute gout and systemic lupus erythematosus: report of 2 cases and literature review. J. Rheumatol.9(2): 305 (1982)

3.   C B Sorenson. The pathogenesis of gout. Arch.Int.Med.109:55 (1962)

4.   W W Buchanan, J R Klinenberg, J E Seegmiller. The inflammatory response to injected microcrystalline monosodium urate in normal, hyperuricaemic, gouty and uraemic subjects. Arthr. & Rheum. 8: 361 (1965)

5.   R A Turner, H R Schumacher, A R Myers. Phagocytic function of polymorphonuclear leukocytes in rheumatic diseases. J.Clin.Invest.52:1632 (1973)

6.   F Kozin, D J McCarty. Protein adsorptions of monosodium urate,
      calcium pyrophosphate dihydrate and silica crystals:
      relationship to the pathogenesis of crystal induced
      inflammation. <u>Arth. & Rheum.</u> 19: 433 (1976)

7.   P Hasselbacher. Immunoelectrophoretic assay for synovial
      fluid C_3 with correction for synovial fluid globulin.
      <u>Arthr. & Rheum.</u> 22: 243 (1979)

POSTHEPARIN PLASMA LIPOPROTEIN LIPASE AND HEPATIC TRIGLYCERIDE LIPASE ACTIVITIES IN GOUT

Y. Nishida, T. Miyamoto, T. Kodama*, M. Okazaki** and
I. Hara**
Department of Medicine and Physical Therapy, *Third
Department of Internal Medicine, University of Tokyo,
7-3-1, Hongo, Bunkyo-ku, Tokyo, Japan 113
**Tokyo Medical and Dental University

Many studies have reported that gouty patients were associated frequently with hypertriglyceridemia(1). In the present study, postheparin plasma lipoprotein lipase (LPL) and hepatic triglyceride lipase (H-TGL) activities in gouty subjects were measured.

MATERIALS AND METHODS

10 male patients with primary gout were studied after an overnight fast. Before injection of heparin, a 3 ml blood sample was collected for the determination of triglyceride, total cholesterol and the quantitation of lipoproteins. Lipoprotein fractions were separated by high performance liquid chromatography(2). Sodium heparin at 10 IU per Kg of body weight was then injected. 3 ml of blood collected 10 min later was immediately cooled on ice and centrifuged at 2000 g for 10 min at 4°C. Postheparin plasma lipolytic activities were assayed using the radioisotope method. Two lipases, i.e. LPL and H-TGL, in this postheparin plasma were separately measured using the antiserum prepared against H-TGL(3). Plasma samples from 12 age-matched control subjects were used for the lipoprotein analysis. Normal values of postheparin plasma lipase activities were those for 13 normo-lipidemic male subjects reported elesewhere(3). The serum concentration of apolipoprotein A-I, A-Ⅱ and apolipoprotein B were measured by a immunodiffusion technique using specific anti-serum.

RESULTS AND DISCUSSION

Plasma triglyceride, cholesterol in the lipoprotein fractions of patients with gout and control subjects are shown in Table I.

As expected, the triglyceride in the VLDL fraction in gouty
subjects (82.5±48.2 mg/100ml) was significantly elevated compared
with controls (35.1±23.2 mg/100ml). No significant difference was
observed in the concentration of total cholesterol between gouty
patients and controls. However, gouty patients had significantly
lower, concentrations of HDL_2 cholesterol than controls (12.7±4.5
vs 19.4±9.1 mg/100ml) respectively.

Table I   Lipid contents in serum lipoproteins determined by HPLC
method

		VLDL	LDL	HDL_2	HDL_3
	TC	28.8±12.2	123.5±27.1	12.7±4.5*	25.1±7.8
gout	PL	34.9±19.9	74.0±13.9	29.6±10.3	54.4±11.3
	TG	82.0±48.2**	54.0±12.1	11.4±5.1	
	TC	13.1±5.1	123.9±15.6	19.4±9.1	28.2±7.1
control	PL	14.1±7.9	83.8±10.3	30.4±5.3	38.9±8.0
	TG	35.1±23.2	36.6±12.0	11.4±3.2	

(mg/100ml)   **$p<0.01$ *$p<0.05$

Apoprotein A-I, A-II and B concentration in gouty subjects
had mean values of 138.6±20.3, 34.3±7.0 and 88.2±15.3 mg/100ml
respectively, which were not significantly different from those of
controls.

The LPL and H-TGL activities of gouty patients are shown in
Table 2. Postheparin plasma LPL activities in these patients were
normal, whereas H-TGL activities were significantly low.

Table II   Postheparin plasma LPL and H-TGL activities in gouty
subjects

	LPL	H-TGL
gout	6.5±2.2	6.4±1.9*
normal values	6.1±2.1	9.9±1.8

(μ mol FFA/ml/hr)   *$p<0.01$

Patients with primary gout showed hypertriglyceridemia characterized by an increase of VLDL-TG and a decrease of HDL_2-cholesterol. In these patients, LPL, a key enzyme for the removal of VLDL-TG, in postheparin plasma was normal. We are interested in the finding that postheparin plasma H-TGL activities were low in gouty patients. It is not known whether this is a primary defect in gout or whether it is a defect secondary to the metabolic abnormalities due to gout. Physiological function of H-TGL are not yet clear. Kuusi etal(4) reported a negative correlation between postheparin plasma H-TGL activity and HDL_2-cholesterol, suggesting a role for this enzyme in the metabolism of HDL_2. Contrary to this observation, our patients with low H-TGL activity showed low HDL_2-cholesterol levels. Recently, Nicoll and Lewis(5) have suggested from the data on patients with functional LPL deficiency that H-TGL plays a role in intermediate density lipoprotein (IDL) metabolism. In our present study, IDL was included in LDL fraction and it was impossible to detect the changes of IDL. More detailed studies are needed to demonstrate a possible defect in remnant lipoprotein catabolism in gouty patients.
(Postheparin plasma lipases were kindly measured by Dr. Murase.)

REFERENCES

1. Y.Nishida, I.Akaoka, T.Nishizawa, and T.Yoshimura,
   Hyperlipidaemia in gout.
   Clin. Chim. Acta. 62: 103 (1975).
2. M.Okazaki, N.Hagiwara, and I.Hara,
   Quantitation method for choline-containing phospholipids in human serum lipoproteins by high performance liquid choromatography.
   J. Biochem 91: 1381 (1982)
3. T.Mrase, N.Yamada, N.Ohsawa, K.Kosaka, S.Morita, and S.Yoshida,
   Decline of postheparin plasma lipoprotein lipase in acromegalic patients.
   Metabolism. 29: 666 (1980).
4. T.Kuusi, P.Saatinen, and E.A.Nikkilä,
   Evidence for the role of hepatic endothelial lipase in the metabolism of plasma high density lipoprotein$_2$ in man.
   Atherosclerosis 36: 589 (1980).
5. A.Nicoll and B.Lewis,
   Evaluation of the roles of lipoprotein lipase and hepatic lipase in lipoprotein metabolism: in vivo and in vitro studies in man.
   Europ. J. Clin. Invest. 10: 487 (1980).

HDL-CHOLESTEROL LEVELS IN GOUTY PATIENTS

A. Carcassi, S. Boschi, P. Macri, and S. Mondillo

Institute of Clinical Medicine, University of Siena
(Head: Prof. A. Caniggia)
Piazza Selva 7, 53100 Siena, Italy

INTRODUCTION

Lipoprotein abnormalities are common in gout. The most frequent is hypertriglyceridemia, which has been reported in three-quarters of patients with primary gout (1, 2). This alteration may or may not be related to obesity, alcohol consumption, and glucose intolerance (3). Patients with gout secondary to lead nephropathy do not have hypertriglyceridemia (4).

Total cholesterol levels are elevated in some gouty patients, but mean values do not differ significantly from controls (5) and no correlation has been shown between serum urate and cholesterol values (2).

The incidence of coronary heart disease is twice as high in gouty subjects as in a male population with normal serum urate or in subjects with hyperuricemia but without clinically overt gout (6).

Recently an inverse relationship has been observed between HDL-cholesterol levels and coronary heart disease and mortality (7,8).

To investigate these new aspects of lipoprotein metabolism in primary gout, we studied a group of 32 patients and a group of 32 controls comparable in age and body weight, but with normal serum uric acid.

141

MATERIALS AND METHODS

Studies were carried out on:

- 32 patients with primary gout (all males);
- 32 controls comparable in age and body weight (all males).

Almost all were out-patients on a free diet, and nobody received any drug affecting uric acid or lipoprotein metabolism for at least 15 days.

Blood samples were withdrawn in patients fasting for 12 hours. Standard methods were used for measurement or uric acid (9); total cholesterol was determined using an enzymatic method (Coles-Cinet Scalvo, Siena, Italy). Triglyceride levels were determined using an enzymatic method (Boehringer). HDL-cholesterol was determined using a Dextran Mg sulfate method (Chol-HDL Sclavo, Siena, Italy).

RESULTS

The results obtained are reported in Table I and Figures 1 and 2.

Table I. Behavior of Various Parameters in 32 Controls (A) and in 32 Patients with Primary Gout (B)

		A		B		t
		mean ± SE		mean ± SE		
Age	years	57.4		57.7		
Body weight	Kg	73.1		80.1		
Height	cm	161.7		168.9		
Uric acid	mg%	5.3	0.1	7.5	0.2	p < 0.01
Cholesterol	mg%	205	7.3	214	7.7	n.s.
Triglycerides	mg%	145	10.5	200	16.2	p < 0.05
HDL-cholesterol	mg%	47.7	2.3	46.7	2.0	n.s.
Cholesterol/HDL	—	4.5	0.2	4.7	0.2	n.s.
LDL	mg%	130	6.2	125	8.1	n.s.

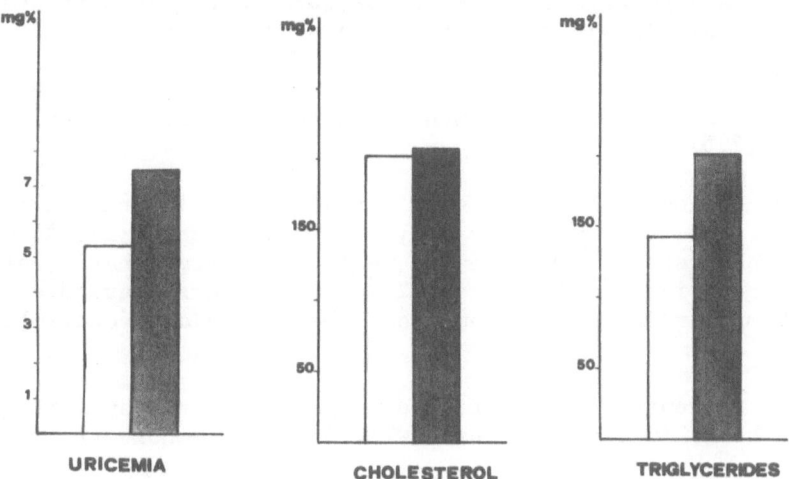

Fig. 1.   Mean values of uric acid, cholesterol, and triglycerides
          in controls (○) and in gouty patients (●).

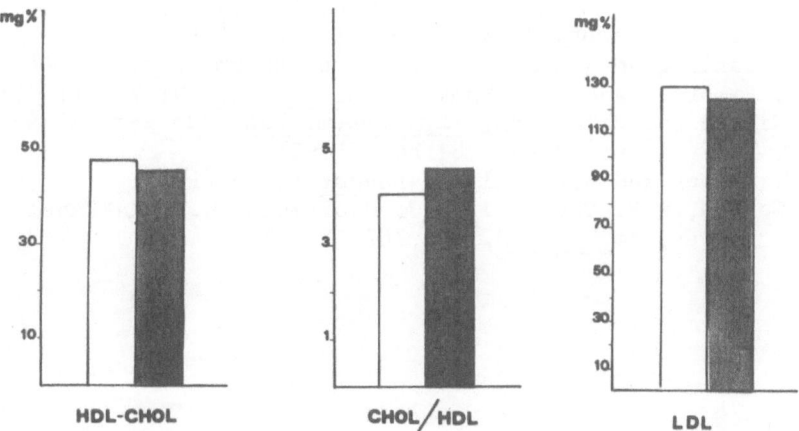

Fig. 2.   Mean values of HDL-cholesterol, cholesterol/HDL, and LDL
          in controls (○) and in gouty patients (●).

As shown in Table I, a significant difference was observed in uric acid and triglyceride levels between gouty patients and controls. No difference was observed in mean values of total cholesterol.

These data are in agreement with previous results in the literature (1,2,5).

HDL-cholesterol levels were not different in the two groups of subjects, but if we consider only the 6 patients (= 17.6%) with low HDL-cholesterol levels (< 35 mg%), we observe that 3 patients had hypertensive heart disease, 2 had electrocardiographic evidence of past myocardial infarction, and 1 had chronic coronary heart disease.

The patients with moderate arterial hypertension without hypertensive heart disease had normal levels of HDL-cholesterol.

These data show that also in gouty patients low levels of HDL-cholesterol correlate with presence of coronary artery disease.

ACKNOWLEDGMENTS

The authors are grateful to M.S. Campagna B.D. and to A.L. Avanzati and A. Pallassini for their excellent technical assistance.

REFERENCES

1. E.B. Feldman and S.L. Wallace, Circulation 29: 508 (1964)
2. T.G. Benedeck, Ann. Intern. Med. 66: 851 (1964)
3. L.G. Darlington and J.T. Scott, Ann. Rheum. Dis. 31: 487 (1972)
4. B.T. Emmerson and B.R. Knowles, Metabolism 20: 721 (1971)
5. T. Gibson and R. Graham, Ann. Rheum. Dis. 33: 298 (1974)
6. A.P. Hall, Arthr. Rheum. 8: 846 (1965)
7. J.G. Miller and N.E. Miller, Lancet 1: 16 (1975)
8. T. Gordon, W.P. Castelli, M.J. Hjortland, W.B. Kakkanel, and T.R. Dowler, Am. J. Med. 62: 707 (1977)

URINARY cAMP EXCRETION IN GOUTY PATIENTS WITH AND WITHOUT NEPHROLITHIASIS

A. Carcassi, M. Galli, S. Boschi and G. Di Cairano

Institute of Clinical Medicine, University of Siena
(Head: Prof. A. Caniggia)
Piazza Selva 7, 53100 Siena, Italy

INTRODUCTION

The prevalence of nephrolithiasis in patients with primary gout is very high, ranging from 10 to 40%, except in Israel where 75% of gouty subjects are afflicted with renal calculi (1).

Hyperuricosuria with increased urinary concentration of uric acid and undue low pH of urine are probably the main but not the only factors in the pathogenesis of nephrolithiasis in gouty patients (1).

Calculi are frequently of pure uric acid but in some cases also of calcium oxalate or mixed.

To investigate other possible metabolic factors we studied in blood and/or in 24/hours urine:
- calcium, phosphate,oxalate, hydroxyproline (HOP), creatinine, immunoreactive parathyroid hormone (iPTH) and cyclic adenosine monophosphate (cAMP), in two groups of patients affected with primary gout without and with renal stones.

MATERIALS AND METHODS

studies were carried out on:
- 16 gouty patients without nephrolithiasis (all but one male)(A);
- 19 gouty patients with nephrolithiasis (all males)(B).

Almost all were out-patients and nobody received any drug affecting uric acid metabolism for at least 15 days. Blood samples were withdrawn in patients fasting for 12 hours.

Uric acid was determined by the method of Archibald (2). Calcium was determined using atomic absorption spectrophotometry (atomic absorption spectrophotometer Perkin Elmer 107).

Standard methods were used for measurements of creatinine (Creatinine Test Sclavo Siena Italy), phosphate (3), serum alkaline phosphatase (4), urinary hydroxyproline (HOP) (5) and oxalic acid (6) (normal range in adult males in our laboratory is 16.4 - 42.1 mg/24 hours) (7).

Urinary cAMP was analyzed by mean of cAMP assay kit (Radio-chemical Centre Amersham England). Samples were added to a liquid scintillant mixture (Instagel, Packard) and counted in a liquid scintillation system Unilux I° Nuclear Chicago C 6850. (Normal range in our laboratory is 1440 - 4500 nmoles/24 hours) (8).

Immunoreactive parathyroid hormone (iPTH) was analyzed by mean of radioimmunoassay method, utilizing specific antibody for C-terminal fragment (ISO - TEX Diagnostics, Friedswood Texas, U.S.A.). (Normal range in our laboratory is 0.4 - 1.4 ng/ml).

RESULTS

The results obtained are reported in Table I and in Figures 1 and 2.

No differences were observed in levels of the biochemical para-meters studied in the blood of the two groups of gouty patients with and without nephrolithiasis.

Levels of alkaline phosphatase and also of iPTH were normal in both the groups of patients.

Urinary excretion of uric acid, oxalic acid, calcium, phosphate, hydroxyproline and cAMP was higher in patients with renal stones than that observed in patients without nephrolithiasis, but the difference was not statistically significant.

Table I. Biochemical data in gouty patients without (A) and with
         nephrolithiasis (B).

		A		B		t
		mean ± SD		mean ± SD		
Uricemia	mg%	7.5	1.5	7.5	1.6	n.s.
Calcemia	mg%	9.5	0.6	9.6	0.5	n.s.
Phosphoremia	mg%	3.2	0.4	3.6	0.4	n.s.
Alkaline Phosphatase U.		12.4	3.6	11.6	3.9	n.s.
iPTH	ng/ml	1.3	1.0	1.0	0.5	n.s.
Uricosuria	mg/24 h	868	330	1033	311	n.s.
Calciuria	mg/24 h	113	77	154	107	n.s.
Phosphaturia	mg/24 h	666	265	850	312	n.s.
Oxaluria	mg/24 h	23	8.5	28	11.8	n.s.
HOP	mg/24 h	21	11.3	27	13.8	n.s.
cAMP	nmol/24 h	4299	1871	5655	2403	n.s.
cAMP / Creatinine	μmol/g	3.1	1.0	3.5	1.2	n.s.

   This fact does not necessarily mean that the difference has not
clinical importance in the pathogenesis of renal stones, because
elevated concentration of "active" ions in urine contribute to the
"formation product" at which crystals form (9).

   A significant correlation has been observed between uric acid
and cAMP excretion ($p < 0.01$). Further studies are necessary to
better understand this result of the research.

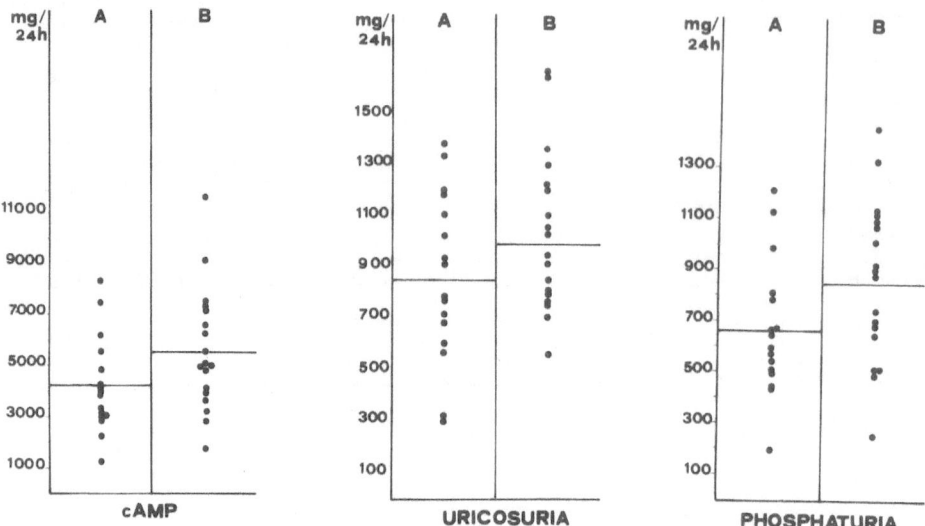

Fig. 1. Urinary excretion of cAMP, uric acid and phosphate in gouty patients without (A) and with nephrolithiasis (B).

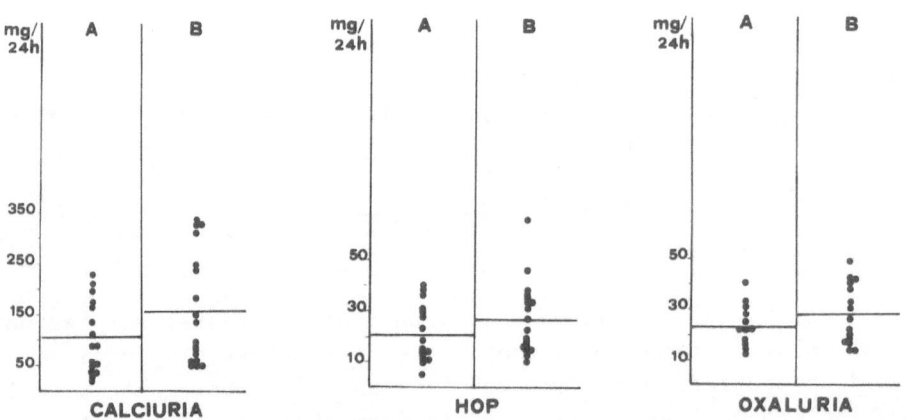

Fig. 2. Urinary excretion of calcium, hydroxyproline and oxalic acid in gouty patients without (A) and with nephrolithiasis (B).

ACKNOWLEDGEMENTS

The authors are grateful to M.S. Campagna B.D., M.B. Franci B.D. M.T. Martorelli B.D. and to A.L. Avanzati and A. Pallassini for their excellent technical assistance.

REFERENCES

1. J.B. Wyngaarden and W.N. Kelley "Gout and Hyperuricemia" Grune § Stratton ed. New York (1976)
2. R.M. Archibald Clin. Chem. 3: 102,(1957)
3. C.H. Fiske and Y. Subbarow J. Biol. Chem. 66: 375,(1925)
4. E.J. King Br. Med. Bull. 9: 160,(1953)
5. R.E. Neuman and M.A. Logan J. Biol. Chem. 184: 299,(1950)
6. A. Hodgkinson and A. Williams Clin. Chim. Acta 36: 127,(1972)
7. F. Loré and G. Manasse Communication IV Symp. CEMO "Diphospho- nates and bone" Nyon (Switzerland) November 1-4,(1981)
8. M. Galli, R. Nuti, M.S. Campagna, B. Franci and A. Caniggia Proc. 2nd Intern. Symp. on Calciotropic Hormones, Taormina 1980 A. Albertini and M. Cecchettin ed. pp. 36-51, Brescia (1981)
9. C.Y.C. Pack Physical Chemistry of Stone formation in "Calcium Urolithiasis, Pathogenesis, Diagnosis and Management" Alvioli L.V. ed., Plenum, New York (1978)

HYPERURICEMIA IN PRIMARY HYPERPARATHYROIDISM:

INCIDENCE AND EVOLUTION AFTER SURGERY

José M. Castrillo, Manuel Díaz-Curiel and
Aurelio Rapado

Unidad Metabólica, Fundación Jiménez Díaz
Universidad Autónoma, Madrid, Spain

INTRODUCTION

Hyperuricemia was first found to be associated with primary hyperparathyroidism (PHP) by Mintz (1) in 1961. He detected it in 50 per cent of his patients. Later on Scott et al. (2) related the hyperuricemia mainly to renal failure, and also to a tubular disorder in nephrocalcinosis, a frequent finding in their PHP cases. Posterior series reported by Mallette (3) and Christensson (4) describe a high incidence of hyperuricemia in patients with PHP, a close relation between the levels of calcium and uric acid being observed. However no changes in the uricemia levels were seen after parathyroidectomy.

Although a clear connection between the excretion of uric acid and the infusion of parathormone (5) or calcium (6) has not been demonstrated, the renal origin of hyperuricemia in certain patients with PHP might be explained by the increased renal synthesis of uric acid through the cyclic nucleotides stimulated by the parathormone (7).

In this paper we study the incidence of hyperuricemia in PHP, its evolution after parathyroid surgery and its possible causes and mechanisms.

MATERIAL AND METHOD

We have studied 73 patients (25 males and 48 females whose ages ranged from 14 to 70 years) with PHP demonstrated by surgery. Serum uric acid and creatinine values were assessed before surgery in every case.

In patients with high serum uric acid levels we tried to rule out other causes of hyperuricemia. We assessed 24-hour uricosuria, uric acid and creatinine clearances and estimated the uric acid clearance/creatinine clearance ratio /Cur/Ccr).

The values of serum uric acid and creatinine after surgery were determined in 72 patients. Both parameters were correlated in the postoperative course, both in absolute values and in the percentage variation regarding the preoperative values in each subject.

Uric acid was assessed by the Caraway method (8) creatinine by the Owen technique (9).

Hyperuricemia was defined by values above 7.2 mg/dl in males and above 6.8 mg/dl in females.

RESULTS

Twenty-three cases of hyperuricemia were found among the 73 patients with PHP (31%). Of them, 10 were males (43 per cent of the total number of males) and 13 females (27 per cent of the total number of females). Fig. 1 shows the histogram of the mean values of uric acid in the PHP population, and fig. 2 the same data after parathyroidectomy.

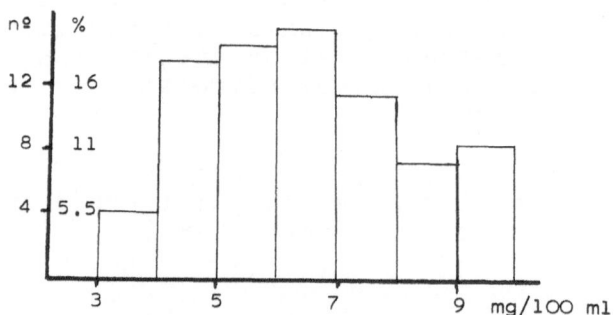

Fig. 1.Basal values of serum uric acid in 73 patients with hyperparathyroidism.

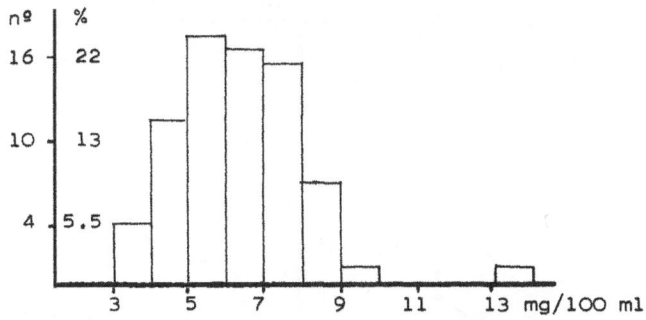

Fig. 2. Postoperative values of serum uric acid in 72
patients with hyperparathyroidism.

        The changes in uricemia and serum creatinine during
the postoperativa follow-up are shown in figures 3 and
4. A close relation between the late increase of serum
uric acid and the values of serum creatinine is observed.

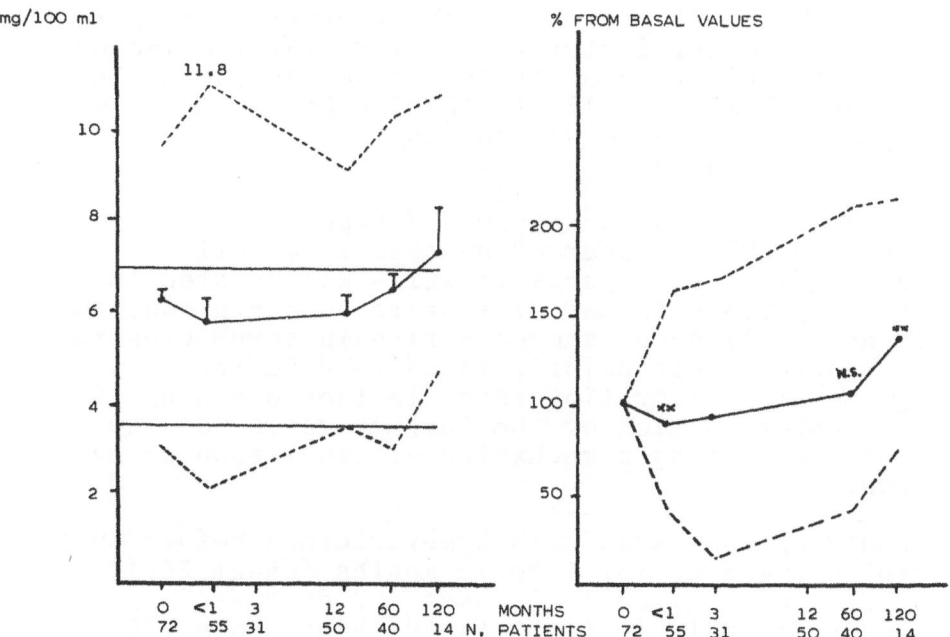

Fig. 3. Serum uric acid changes after parathyroidectomy.
        ** = $p < 0.01$
        n.s. = not significant

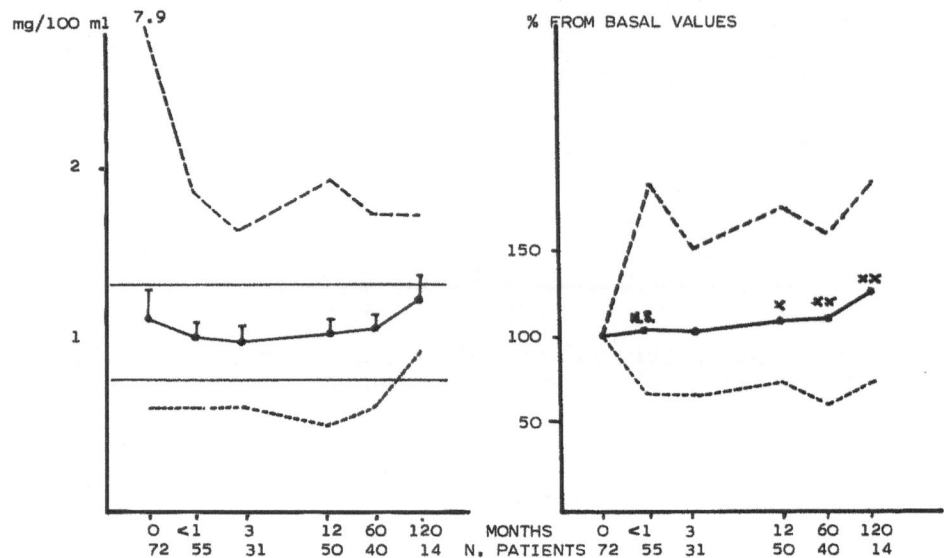

Fig. 4. Serum creatinine changes after parathyroidectomy.
    * = p < 0.05   ** = p < 0.01
    n.s. = not significant

In 20 of the 23 patients with hyperuricemia there was at least another factor responsible for the increase in serum uric acid (Table 1). The cause was unique in 8 cases and multiple in 12. In the 3 other male patients with hyperuricemia there was not any other known cause but their PHP disease.

In analyzing the mechanisms of hyperuricemia in the 23 cases with PHP we observed an associated uricosuria above 800 mg/24 h in 7 patients which was related to an increase of purines in the diet or to an overproduction of uric acid. Six cases showed a rise in serum creatinine, the hyperuricemia being related to a decrease of the renal glomerular filtration rate. In the remaining 10 cases the determination of the Cur/Ccr ratio was lower than 0.05, suggesting a mechanism of renal tubular hypo-excretion.

Twenty of the cases with hyperuricemia before surgery were followed-up for 6 to 72 months (mean: 28.1). The uricemia normalized in 9 cases and decreased, without becoming normal, in 5. Of these 14 patients (70 per cent of the total), 9 showed that the decrease of the uricemia was accompanied by an increase of the Cur/Ccr ratio. The hyperuricemia did not modify or increase in six other cases.

Table 1. Factors contributing to hyperuricemia in 23 cases of primary hyperparathyroidism

Factor	Sex M	Sex F	Total
Arterial high blood pressure	4	8	12
Obesity	3	7	10
Medication	1	5	6
Serum creatinine above 1.2 mg%	2	4	6
Hypercholesterolemia	1	1	2
Volume depletion	-	3	3
Diabetes	-	2	2
Cushing	-	1	1
Myxedema	1	-	1
Cystinuria	-	1	1

Of the three hyperuricemic patients with no other cause that their PHP disease, the uric acid values became normal in 2, showing an increase in Cur/Ccr. The uricemia increded in the other case in relation to a deteriorated renal glomerular filtration rate.

DISCUSSION

Hyperuricemia was found in one third of our patients with PHP. In most cases there existed other factors that could explain the high values of serum uric acid. However 3 cases did not have any other cause save the hyperparathyroidism.

The mechanisms producing the hyperuricemia were found to be either an increase of the uric acid entering the extracellular fluid or a decrease of its renal excretion because of tubular or glomerular changes.

The parathyroidectomy resulted in the early and significant decrease of serum uric acid values in our series (fig. 3). These results are in contradiction with

others reported in the literature where uricemia did not
modify after surgery (1, 2, 3).

In the patients followed-up for more than 5 years
we have observed a progressive increase of serum uric
acid, probably due to the deterioration of the renal
glomerular function (as shown in the rise of serum creatinine), to the advanced age of some of the patients, or
to some other factors (arterial hypertension, medication
with thiazide diuretic drugs, etc.).

The systematic study of our hyperuricemic patients
after the parathyroidectomy showed an overall improvement of the mean values of serum uric acid in 70 per cent
of the cases, the mechanism being explained by an increase of the renal tubular excretion of uric acid.

The early decrease of uricemia following surgery
observed in the whole group, together with the decrease
of the uric acid in the hyperuricemic patients advocates
for the possible effect of the parathormone and/or the
hypercalcemia upon the tubular handling of uric acid in
primary hyperparathyroidism. However this improvement is
blunted by the progressive increase in serum uric acid
resulting from the decreasing renal glomerular function
observed in the long-term follow-up of the patients who
underwent surgery.

REFERENCES

1. D. H. Mintz, J. J. Canary and G. Carreon, Hyperuricemia in hyperparathyroidism. New Engl J Med
   265:112-115 (1961)
2. J. T. Scott, A. St. Dixon and E. G. L. Bywaters,
   Association of hyperuricemia and gout with hyperparathyroidism. Br Med J 5390:1070-1073 (1964)
3. L. E. Mallete, J. P. Bilezikian, D. A. Heath, and
   G. D. Aurbach, Primary hyperparathyroidism:
   clinical and biochemical features. Medicine 53:
   127-146 (1974)
4. T. Christensson, Serum urate in subjects with hyper-
   calcemic hyperparathyroidism. Clin Chem Acta 80:
   529-533 (1977)
5. W. D. Shelp, T. H. Steele and R. E. Rieselbach, Comparison of urinary phosphate, urate and magnesium
   excretion following parathyroid hormone administration to normal man. Metabolism 18:63-70 (1969)
6. C. G. Duarte and J.H. Bland, Uric acid, calcium and
   phosphorus clearances in normal subjects on a low
   calcium, low phosphorus diet. Uric acid, calcium

and phosphorus after calcium infusion in normal and gouty patients. Metabolism 14:203-210 (1965)

7. R. Conlson, Metabolism and excretion of exogenous adenosine 3:5'monophosphate and guanine 3:5' mono- phosphate. Studies in the isolated perfused rat kidney and in the intact rat. J Biol Chem 251: 4958-4967 (1976)

8. M. T. Caraway and H. Morable, Comparison of carbon- ate and uricase carbonate methods for the deter- mination of uric acid in serum. Clin Chem 12: 18-23 (1966)

9. J. A. Owen, B. Iggo, F. J. Scandrett and C. P. Ste- ward, The determination of creatinine in plasma, or serum and in urine: a critical examination. Biochem J 58:426-437 (1954)

PILOT STUDY OF BLOOD COAGULATION IN GOUT PATIENTS

D. A. Lane and L. G. Darlington

Charing Cross Hospital, London and
Epsom District Hospital, Epsom, Surrey
United Kingdom

Gout patients have certain vascular risk factors such as hyperlipidaemia, hypertension, 'type A' personality and, possibly, hyperuricaemia, obesity and glucose intolerance.

Uric acid activates clotting Factor XII and hyperuricaemia in rats is associated with increased platelet aggregation (1) with increased adenosine diphosphate-induced pulmonary platelet thrombosis. Platelet survival is decreased and platelet turnover correspondingly increased in some primary gout patients and this abnormality is reversed during uricosuric therapy with sulfinpyrazone (2,3).

Studies of coagulation and of platelet function in gout have yielded variable results to date (2,4,5) but, in recent years, tests have become increasingly sophisticated and the present study was designed to use modern methods to seek any abnormalities of coagulation in patients with primary gout.

METHOD

Twelve, fasting, male patients with primary gout were investigated. Ten of the twelve patients were untreated, one had ceased allopurinol therapy two weeks before the tests and one was taking 400 mgs cimetidine daily. All patients were requested not to take any medication and, in particular, any aspirin for the two weeks preceeding the tests. Every attempt was made to bleed the patients when they were relaxed, resting, after a fast of at least fourteen hours and at the same time in the morning to minimise effects on coagulation of emotion, exercise, food and diurnal variation. No patient was included within three months of trauma and/or surgery and an atraumatic venepuncture technique was used with a large needle.

159

Assays were performed to fibrinopeptide A (FpA) (6) with modifications (7), to fibrinopeptide B β 1-42 (FpB β 1-42) (8) with modifications (7) and to β-thromboglobulin (βTG) (9) with modifications (7). Serum cholesterol and triglycerides were measured directly (10) while high density lipoprotein cholesterol (H.D.L.C.) was estimated after differential precipitation of β and pre β-lipoproteins (11) followed by measurement of the cholesterol component of the α lipoprotein (10).

RESULTS

Data from gout patients (Table I) were compared with control data from normal subjects in the same age range (Table II).

No significant differences between gouty patients and controls were found for any of the tests performed although the FpA assay revealed an elevation among gouty patients which might have achieved significance with larger numbers of subjects ($0.1 > p > 0.05$).

Since one patient was taking cimetidine and another had only ceased allopurinol therapy two weeks before the tests all data were reanalysed after excluding these patients but results were unchanged.

Spearman Rank correlation coefficients were calculated between FpA, FpB β 1-42 and βTG and lipid levels, alcohol consumption and weight since the latter data tend not to be distributed normally. Results are shown in Table III. Once again, inclusion of the two treated patients had no effect upon the correlation results.

It was not possible to calculate actual r values for the correlation with smoking since most patients were non-smokers, but simple assessment of the data did not suggest any relationship with smoking.

Correlation results must be considered with care since this was a pilot study on small numbers of patients. Furthermore, it must be remembered that abnormalities of coagulation detected in vitro are not necessarily present in vivo nor do they necessarily have a causal relationship with thrombosis.

In the Framingham Study (12) Hall found an increased risk of vascular disease in hyperuricaemic patients although our own study of vascular mortality among gout patients (13) did not confirm the relationship.

Certainly, in the present study, it was concluded that modern methods to assess coagulation and platelet function did not reveal any significant abnormalities in a pilot study among patients with primary gout.

Table I Coagulation Data from Gouty Patients

Name	Age	FpA (pmol/l)	FpBβ1-42 (pmol/l)	βTG (pmol/l)	Platelet Aggregation.
1) B.W.	15.1.06	1.63	4.43	0.64	Secondary aggregation at 0.5 μm ADP (slightly enhanced)
2) A.S.	20.11.39	2.67	2.21	2.14	Normal
3) P.G.	17.8.27	1.77	1.90	0.69	-
4) J.S.	9.2.39	0.39	1.84	0.97	Normal
5) G.E.	18.6.22	-	7.14	7.17	-
6) K.S.	11.6.22	2.35	2.49	1.31	Secondary aggregation at 1.0 μm ADP (slightly enhanced)
7) R.A.	20.3.33	1.31	1.11	0.58	Normal
8) R.H.	11.10.49	0.39	1.81	0.72	-
9) G.S.	7.10.13	1.41	3.43	1.28	-
10) A.B.	27.8.35	2.27	-	1.33	-
11) G.T.	20.4.12	-	-	1.31	-
12) G.G.	25.1.48	1.14	-	0.83	-

Table II Comparison of Mean Data from Gouty Patients and Controls (mmol/l)

		Controls	Gout Patients	Significance
FpA (pmol/l)	n	13	10	N.S.
	mean	0.98	1.53	
	s.d.	0.29	0.77	
FpBβ 1-42 (pmol/l)	n	10	9	N.S.
	mean	1.65	2.92	
	s.d.	0.63	1.85	
βTG (pmol/l)	n	9	12	N.S.
	mean	1.00	1.58	
	s.d.	0.28	1.81	

N.S. = Not Significant

Table III Correlation Table

	FpA (pmol/l)	FpBβ 1-42 (pmol/l)	βTG (pmol/l)
Cholesterol (mmol/l)	n = 9 r = 0.68 p = <0.05*	n = 8 r = 0.62 p = N.S.	n = 11 r = - 0.19 p = N.S.
Triglycerides (mmol/l)	n = 9 r = 0.35 p = N.S.	n = 8 r = - 0.60 p = N.S.	n = 11 r = - 0.24 p = N.S.
H.D.L.C. (mmol/l)	n = 8 r = 0.12 p = N.S.	n = 7 r = 0.86 p = <0.05*	n = 10 r = - 0.02 p = N.S.
Alcohol Consumption (drinks/day)	n = 10 r = 0.23 p = N.S.	n = 9 r = - 0.05 p = N.S.	n = 12 r = 0.18 p = N.S.
Weight (lbs)	n = 10 r = - 0.03 p = N.S.	n = 9 r = - 0.10 p = N.S.	n = 12 r = 0.32 p = N.S.

* p <0.05   N.S. = not significant

REFERENCES

1. Winocour, P.D., Turner, M.R., Taylor, T.G., and Munday, K.A. Gout and Cardiovascular Disease,. Lancet i, 959-960. (1977).
2. Mustard, J.F., Murphy, E.A., Ogryzlo, M.A., and Smythe, H.A. Blood Coagulation and Platelet Economy in Subjects with Primary Gout. Can Med Assoc J. 89, 1207-1211. (1963).
3. Smythe, H.A., Ogryzlo, M.A., Murphy, E.A., and Mustard, J.F., The Effect of Sulfinpyrazone (Anturan) On Platelet Economy and Blood Coagulation in Man. Can Med Assoc J. 92, 818-821. (1965).
4. Darlington, L.G., Scott, J.T. and Shaw, S. Platelet adhesiveness in gout. Postgrad Med J. 49, 24-26. (1973).
5. Bluhm, G.B., and Riddle, J.M. Platelets and Vascular Disease in Gout. Semin Arthritis Rheum. 2, 355-366 (1973).
6. Nossel, H.L., Yudelman, I., Canfield, R.E., Butler, V.P., Spanondis, K., Wilner, G.D., Qureshi, G.D. Measurement of Fibrinopeptide A in Human Blood. J. Clin Invest. 54, 43-53. (1974).
7. Lane, D.A., Ireland, H., Wolff, S., Boots, M., Pegrum, G.D. Simultaneous Measurement of Thrombin and Plasmin Activities and Platelet Releasing stimuli in Plasma. Br J Haematol. 47, 630. (1981).
8. Nossel, H.L., Wasser, J., Kaplan, K.L., La Gamma, K.S., Yudelman, I., Canfield, R.E. Sequence of Fibrinogen Proteolysis and Platelet Release after Intra uterine Infusion of Hypertonic Saline. J Clin Invest. 64, 1371-1378. (1979).
9. Bolton, A.E., Ludlam, C.A., Moore, S., Pepper, D.S., Cash, J.D. Three Approaches to the Radio immuno assay of Human β-Thromboglobulin. Br J Haematol. 33, 233-238. (1976).
10. Smith, L., Lucas, D., and Lehnus, G. Automated Measurement of Total Cholesterol and Triglycerides, in "Tandem ", on the Discrete Sample Analyser Guildford System 3500. Clin Chem. 25, 439-442. (1979).
11. Viikari, J. Precipitation of Plasma Lipoproteins by PEG-6000 and Its Evaluation with Electrophoresis and Ultracentrifugation. Scand J Clin Lab Invest. 36, 265-268. (1976).
12. Hall, A.P. Correlations among hyperuricaemia, hypercholesterolaemia, coronary disease and hypertension. Arthritis Rheum. 8, 846-852, (1965).
13. Darlington, L.G., Slack, J., and Scott, J.T. Vascular Mortality in Gout Patients and their Families. Ann Rheum Dis. In Press. (1982).

CHARACTERIZATION OF AN URATE BINDING α_2-GLOBULIN FROM HUMAN SERUM

M.R. Mazzoni[§], M.L. Ciompi[§], G. Pasero[§],
C. Martini[¶], D. Segnini[¶], and A. Lucacchini[¶]

Servizio di Reumatologia, Istituto di Patologia medica[§]
Istituto Policattedra di Discipline biologiche[¶]

University of Pisa, 56100 Pisa, Italy

INTRODUCTION

Since urate binding to serum protein may influence the deposition in joints and the renal excretion, several authors have studied this interaction with different results[1,2].

Previously we reported the use of affinity chromatography to study urate binding proteins in the human serum[3]. This technique was useful to demonstrate the presence of at least two different serum proteins which recognize urate: albumin and an α-globulin. More recently we described the use of 8-amino-2,6-dihydroxy-purine coupled with Sepharose to purify the urate binding globulin from human serum[4]. This protein which migrates as an α_2-globulin is devoid of enzymatic activity and shows an apparent molecular weight of 70,000±4,000.

In this paper we describe further characterization of this urate binding globulin.

MATERIALS AND METHODS

Lyophilized fraction III, according to Cohn, and γ-globulin were purchased from Farmaceutici Biagini (Italy). Allopurinol was obtained from Schoum S.p.A. (Italy). All other purine derivatives were from Sigma Chemical Co. (USA). ^{14}C-uric acid was purchased from Amersham/Searle Corp. (England). All other chemicals were reagent grade and were used as obtained.

Affinity chromatography procedure was carried out on the specific adsorbent prepared by us, as previously described.

Assay for ^{14}C-uric acid binding: ^{14}C-uric acid (26.5 µC1/mM) was added to the solution of purified protein to obtain a final concentration of 10 µM in a total volume of 200 µl of Tris buffer 20 mM, pH 8.0. After incubation at 22°C for 45 min. 0.5 ml of γ-globulin (1 mg/ml) and 2 ml of ammonium sulphate (80%) were added and then filtered under low vacuum through HAMK filters (0.45 Millipore, USA). After washing with 2 ml of ammonium sulphate (60%) the filters were placed in vials containing 10 ml of scintillation liquid HP (Beckman, USA) and counted. Specific binding is defined as total ^{14}C-uric acid binding minus binding obtained in the presence of 1 mM uric acid.

A saturation analysis was carried out using 1 to 80 µM ^{14}C-uric acid. Protein concentration was determined by BioRad protein assay.

The amino acid analysis was carried out after acid hydrolysis in 6 N HCl at 110°±1°C, using a Beckman Multichrome AA. Analyzer.

RESULTS

The urate binding α_2-globulin purified to homogeneity as previously described was further characterized. The ^{14}C-uric acid binding properties of this protein were investigated. Fig. 1 shows a Scatchard plot obtained with the purified protein. The straight line plot indicates the presence of a single class of saturable binding sites. A Kd = 13 µM was obtained. The binding specificity of the urate binding site was investigated by inhibition studies with some purine derivatives or aspirin. These compounds were tested by incorporation in the binding assay at concentration between 5 and 500 µM. The IC 50 values for these compounds are reported in Table 1. The amino acid composition and the glycidic content are reported in Table 2. It is interesting to note the high content of carbohydrates in the protein, which suggests that α_2-globulin is a glycoprotein.

Table 1 - Inhibition of ^{14}C-uric acid binding

Compound	IC 50 (µM)	Compound	IC 50 (µM)
Allopurinol	8	2-NH$_2$-6,8-OH-purine	50
Hypoxanthine	9	Adenine	>500
Xanthine	16	Caffeine	25
8-NH$_2$-xanthine	15	Theophylline	13
6-NH$_2$-2,8-OH-purine	>500	Aspirin	100

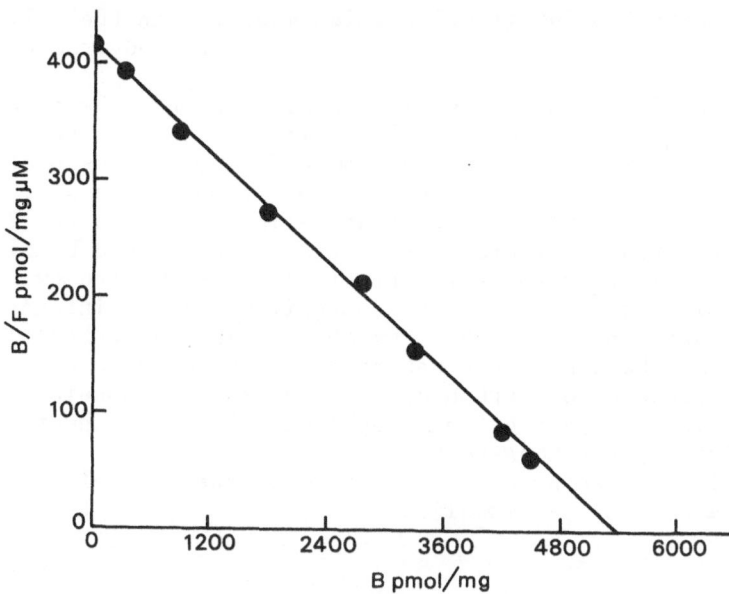

Fig. 1 – Scatchard plot for binding of ^{14}C-uric acid to purified
α₂-globulin.

Table 2 – Amino acid composition of the α₂-globulin
The carbohydrate content, with the anthrone method,
is 20%.

Residue	Residues/100	Residue	Residues/100
Lys	3	Ala	6
His	7	Cys	2
Arg	3	Val	6
Thr	5	Met	2
Ser	12	Ile	4
Glu	14	Leu	6
Pro	4	Tyr	3
Gly	7	Trp	4
Asp	7	Phe	5

CONCLUSION

The urate binding globulin which migrates in the cellulose a-
cetate electrophoresis as an α_2-globulin, purified from human serum fraction III by affinity chromatography, presents a binding
site with a Kd = 13 μM. This binding site seems to be specific for
the 6-position of the purine moiety. In fact adenine and 6-amino-
2,8-dihydroxypurine seem to have low affinity for the urate binding site. On the contrary the imidazole ring of the purine seems
to have less structural requirement: allopurinol and the 8-amino-
derivatives are good inhibitors of the ^{14}C-uric acid binding. The
nature of inhibition for allopurinol was investigated by Linewea-
ver-Burk plot and appears to be competitive. It is interesting to
note that aspirin is an inhibitor of the binding, suggesting that
this fact may be one of the causes of the uricosuric effect of the
drug. The presence of carbohydrates suggests the possibility of
purifying this protein by ConA-Sepharose. However, the protein is
not retained by the ConA-Sepharose.
Studies are in progress to determine the physiologic role of
this urate binding α_2-globulin.

REFERENCES

1. Alvsaker J.O.: Urate-binding protein interaction. Scand.J.Clin.
   Lab.Invest. 30:345 (1972)
2. Kovarsky J., Holmes E.W., Kelley W.N.: Absence of significant
   urate binding to human serum proteins. J.Lab.Clin.Med. 93:85
   (1972)
3. Ciompi M.L., Lucacchini A., Segnini D., Mazzoni M.R.: Urate-
   binding proteins in plasma studied by affinity chromatography.
   Purine Metabolism in Man-III. Part B. Plenum Press, N.Y., 395,
   1980
4. Mazzoni M.R., Martini C., Segnini D., Ciompi M.L., Lucacchini
   A.: Isolation of an urate-binding protein by affinity chromatography. Appl.Biochem.Biotech. (1982), in press.

METABOLIC STUDIES OF HIGH DOSES OF ALLOPURINOL IN HUMANS

Donald J. Nelson and Gertrude B. Elion

Wellcome Research Laboratories
Burroughs Wellcome Co.
Research Triangle Park, NC, USA

INTRODUCTION

Allopurinol has been used for 16 years at doses ranging from 200 to 800 mg/day for the control of primary and secondary hyperuricemia. At the most commonly used doses of allopurinol (300-400 mg/day) about 70% of the allopurinol is oxidized to oxipurinol, which is excreted in the urine. Urinary allopurinol and allopurinol riboside each account for about 10% of the dose. Since the degree of xanthine oxidase inhibition is dose-related, not only the oxidation of hypoxanthine and xanthine to uric acid, but also the oxidation of allopurinol to oxipurinol might be expected to be strongly inhibited at high doses of allopurinol. This would lead to increased levels of allopurinol, as well as allopurinol riboside, in plasma and urine. The extent to which this phenomenon occurs was investigated in several laboratory animal species and in man.

RESULTS

Allopurinol Metabolism in Laboratory Animals - In CD-1 mice given allopurinol at doses of 100 mg/kg, i.p. or i.v., the ratio of urinary allopurinol (HPP) to oxipurinol (DHPP) in a 24 hr period was 0.5, indicating that oxidation of HPP to DHPP is not markedly inhibited even at this high dose. In rats the effect of increasing the dose of allopurinol in the diet upon the metabolism of that compound was evident between 28 and 67 mg/kg/ day (Table 1). At the higher dose, the percentage excreted as DHPP decreased, and that of HPP and allopurinol riboside (HPPR) dramatically increased. Early experiments in dogs showed that the urinary HPP/DHPP ratios were 0.3, 3.6 and 3.8 following intravenous doses

of allopurinol of 5, 25 and 100 mg/kg, respectively. The pronounced
difference between the degree of oxidation of allopurinol at
5 mg/kg and 25 mg/kg in the dog was further reflected in the
plasma half-life of allopurinol: $t_\frac{1}{2}$ = 1.25 hours at 5 mg/kg,
$t_\frac{1}{2}$ = 4.5 hours at 25 mg/kg or 100 mg/kg (allopurinol riboside was
not measured at that time).   In dogs given allopurinol orally at
30, 90 and 270 mg/kg/day on a chronic basis, the HPP/DHPP ratio in
the urine was 0.7, 2.1 and 4.1, respectively.   However, since
absorption appears to have been incomplete at these high doses,
only the general trend rather than the absolute dose at which the
ratio changed is valid.   The species differences in the dose at
which significant inhibition of allopurinol oxidation occurs
probably reflects quantitative differences in the amount of xanthine
oxidase and aldehyde oxidase present in these species.   Rodents
are known to have high levels of these enzymes whereas dog and man
have lower levels (Krenitsky, 1974).   There may also be some
qualitative differences in the xanthine oxidase of these species.

Table 1.   Allopurinol Metabolites in Urine of SD Rats
Given Allopurinol Mixed into the Diet

Allopurinol Dose†	Concentration in Urine; μmoles/mg creatinine			
	HPP	HPPR	DHPP	SUM
0.05% Diet, 13 mg/kg/day	0.24 (6%)	0.19 (5%)	3.43 (89%)	3.86
0.10% Diet, 28 mg/kg/day	0.86 (10%)	0.62 (8%)	6.92 (82%)	8.41
0.25% Diet, 67 mg/kg/day	5.06 (30%)	4.09 (25%)	7.39 (45%)	16.54

†The dose was calculated from the amount of food eaten by each
group averaged over four weeks, with six animals per group. Molar
percent of the recovered drug is shown in parenthesis (%)

Allopurinol Metabolism In Man - There are scattered reports of the
use of allopurinol at doses of 900 mg (~12 mg/kg) per day or
higher (Sweetman, 1968; Rundles, 1966).   The proposed use of high
doses of allopurinol to improve the chemotherapeutic index of
5-fluorouracil (Schwartz, 1980) and also to treat the parasitic
disease, leishmaniasis, gave impetus to the present metabolic
study.

Plasma and urine samples were analyzed by HPLC with an ODS-
reversed phase column.   Three cancer patients given 900 mg of
allopurinol excreted pyrazolopyrimidines in the proportion of 28%
HPP, 17% HPPR and 55% DHPP, on the average, in urine.   The highest

plasma levels detected were: oxipurinol 24 μg/ml, allopurinol
8 μg/ml, and allopurinol riboside 2.4 μg/ml, after three days of
therapy.

Another individual took 1200 mg (17 mg/kg) divided 4 x daily
for 5 days. Serum allopurinol reached 3.6 μg/ml, allopurinol
riboside was as high as 2.2 μg/ml and oxipurinol averaged 31 μg/ml
on days 3 to 5. On day five 45% of the dose was excreted as
oxipurinol, 14% as allopurinol riboside and 22% as allopurinol.

A hyperuricemic individual who had taken 1500 mg of allopurinol (21 mg/kg) as a single daily dose for over one year showed
peak plasma levels of allopurinol of 6.9 μg/ml, and of allopurinol
riboside of 9.1 μg/ml at 3 hrs; oxipurinol varied between 16 and
29 μg/ml over the course of a day. Urine contained 15% of the
dose as allopurinol, 42% as allopurinol riboside, and 37% as
oxipurinol.

Orotate Excretion - Mice and rats given allopurinol excreted
orotate in proportion to the dose (Hitchings, 1975).   The maximum
excretion reached was 16 μmoles orotate/mg creatinine in the mouse
and 1.4 μmoles/mg creatinine in the rat after an oral dose of
70 mg/kg allopurinol. This was 1/3 the level of the maximum orotate

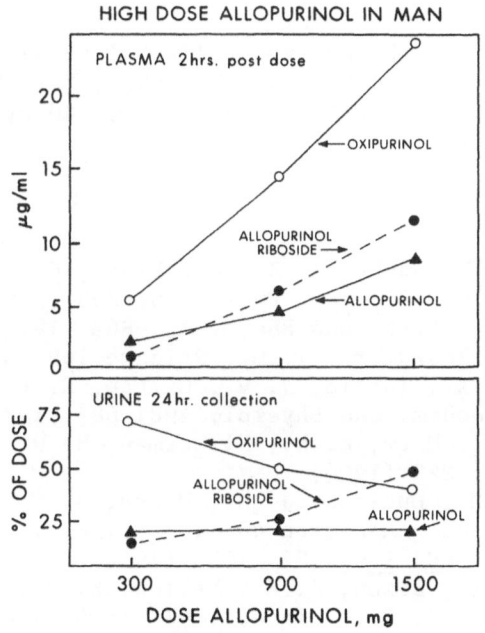

Figure I

excretion produced by pyrazofurin. In man the orotate excretion increased with dose of allopurinol. At 300 mg of allopurinol, orotate excretion was approximately 30 mg/day (Kelly, 1970) whereas at 900-1500 mg doses of allopurinol, orotate excretion was variable between 60 and 370 mg/day. The total daily orotate production by the body is estimated to be about 600 mg/day (Weissman, 1962).

SUMMARY AND CONCLUSION

In animals and in humans given high doses of allopurinol, the oxidation of allopurinol to oxipurinol is inhibited, resulting in a higher proportion of unchanged allopurinol and of allopurinol riboside in plasma and urine than is seen at low doses. The dose which produces this inhibition of allopurinol oxidation is higher in rodents than in man or in the dog. Urinary orotate and orotidine increased in proportion to the dose of allopurinol. These increased levels of orotate would be expected to compete more effectively with 5-fluorouracil for conversion to a nucleotide by orotate phosphoribosyltransferase. Since allopurinol and allopurinol riboside are active against leishmaniae in vitro, it may be possible to attain therapeutic levels of allopurinol and allopurinol riboside in vivo by using high doses of allopurinol.

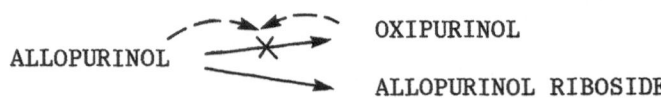

ACKNOWLEDGEMENTS: For the numerous clinical samples thanks go to Drs. P. Schwartz and R. Handschumacher, Yale University, Dr. R. Collont, V.A. Hospital, Durham, N.C., and Dr. J.J. Marr, St. Louis University.

REFERENCES

1.   Elion, G. B., Kovensky, A. K., Hitchings, G. H., Metz, E., Rundles, W., Biochem. Pharm. 15: 863 (1966).
2.   Hitchings, G., Arth. and Rheum. 18: 863 (1975).
3.   Kelly, W. N., Beardmore, T. D., Science 169: 388 (1970).
4.   Krenitsky, T. A., Tuttle, J. V., Cattau, E. L., Wang, P., Comp. Biochem. and Physiol. 49B: 687 (1974).
5.   Rundles, R. W., Metz, E. N., Silberman, H. R., Annals of Int. Med., 64: 229 (1966).
6.   Schwartz, P. M., Dunigan, J. M., Marsh, J. C. Handschumacher, R. E., Cancer Research 40: 1885 (1980).
7.   Sweetman, L., Fed. Proc. 27: 1055 (1968).
8.   Weissman, S. M., Eisen, A. Z., Fallow, H. J., Lewis, M., Karon, M. J. Clin. Invest. 41: 1546 (1962).

GOUT RESISTANT TO ALLOPURINOL: POOR COMPLIANCE OR NON-RESPONSE

H. A. Simmonds, T. Gibson, G. J. Huston, D. R. Webster,
A. V. Rodgers, and J. Munro

Purine Laboratory, Department of Rheumatology and
Metabolic Ward, Department of Medicine
Guy's Hospital, London, U.K.

Patients with chronic gouty arthritis unresponsive to allopurinol
are generally dismissed as poor compliers.   This paper presents
studies in a middle-aged male with a 20-year history of gout in
whom allopurinol in varying doses and combinations had proved
ineffective in controlling either plasma uric acid levels or the
gouty arthritis.   The patient was studied on two separate occasions
under Metabolic Ward conditions because the initial study on low
dose allopurinol had failed to reduce plasma uric acid levels -
they had in fact increased.

Patient: LT was a 51 year old male who presented with a mono-
arthritis in the left ankle in 1961, confirmed by a raised blood
uric acid level.   Over the past 20 years he had suffered intermittent attacks affecting most joints.   Anti-inflammatory drugs,
colchicine, uricosurics and allopurinol - in varying doses and
combinations, at times with steroids - had proved ineffective in
controlling the gouty arthritis or the plasma uric acid.   He had
a long history of pain in the right foot for which talar fusion
had been performed in 1972 since when he had been on crutches and
wore a plaster cast on the leg.   One brother had a history of gout.
Alcohol intake was reported to be maximally 2 pints per week.
He was referred to Guy's for assessment and treatment of his gouty
arthritis in 1980, and was admitted with swelling and tenderness
of the left knee, the right elbow and MTP joint of the left foot.
Urate crystals were identified in synovial fluid.   He had no
evidence of gouty tophi.   Plasma urate was 0.60 and climbed to
0.95mmol/l (Vickers method).   Liver biopsy to investigate
persistently raised alkaline phosphatase, SGOT and 5'nucleotidase
was normal.

171

Metabolic studies: LB was investigated twice in a Metabolic Ward
on a constant low purine (LP) diet [1,2] After four days on the
basal diet the patient was given allopurinol (200 mg/day - Study 1;
300 and then 600mg/day - Study 2) and investigated for a further
four days. Blood was taken at the same time each day and blood[2]
and urine analysed for drug metabolites, uric acid and purines,[1,2]
as well as creatinine, pH and osmols by standard techniques.
Details of sample handling as well as biochemical[2] methods used and
the enzyme assays employed have been published.[1,2]

RESULTS

Plasma uric acid (enzymic method[1]) levels were greatly elevated for
a LP diet (Fig. 1a - all values are the mean for the period indicated), and were increased even further by allopurinol in both
studies (maximally 0.82 mmol/l, Study 1 ; 0.89mmol, Study 2).
Mean fractional uric acid excretion (C_{ur}/C_{cr} x 100) was low and
decreased even further on allopurinol at all doses (Fig. 1b)
indicating that the increase in plasma urate was due to a decrease
in uric acid clearance in both studies. Creatinine clearance
(uncorrected), was not considered abnormal for age (Study 1:
87-97ml/min; Study 2: 72-90ml/min). The patient showed the
characteristic fixed urine pH of the gouty but was unable to
concentrate the urine to the degree noted for controls or primary
gout patients[3] (not shown).

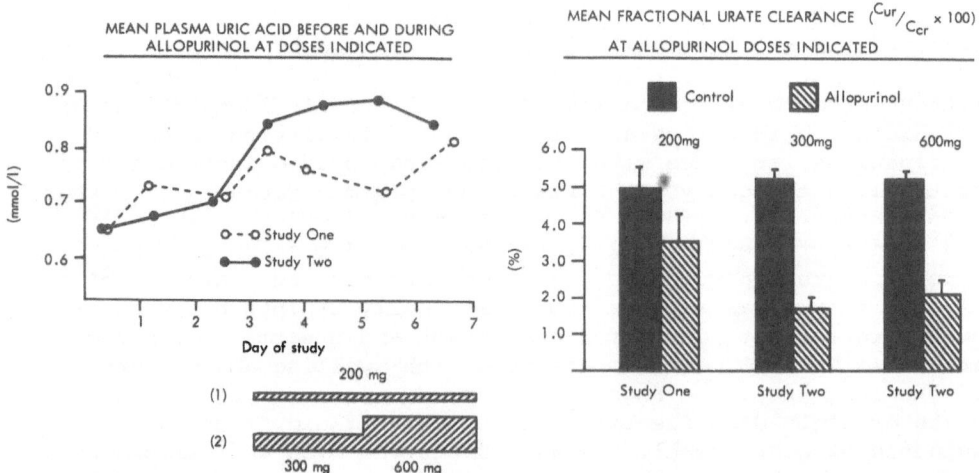

Figure 1. Effect of allopurinol on plasma uric acid (a) and
uric acid clearance relative to creatinine (b) in patient LT.

Table 1.   Purine, allopurinol metabolite and enzyme levels in LT

STUDY	DRUG	URIC ACID			Total Oxypurines mmol/24h	H/X ratio	Drug Metabolites % dose	Plasma Oxipurinol µmol/l
		Plasma mmol/l	Urine mmol/24h	Clearance ml/min				
1	Nil	0.70	4.7	4.7	4.96	3.3	-	-
	Allopurinol 200 mg 1.47mmol	0.76	3.4	3.1	3.92	1.5	25.7	Not done
2	Nil	0.67	4.21	4.4	4.30	3.5	-	-
	Allopurinol 300mg 2.2mmol	0.86	1.50	1.2	1.74	2.4	8.1	17
	Allopurinol 600mg 4.4mmol	0.85	1.92	1.6	2.22	1.6	7.0	50

ERYTHROCYTE STUDIES

	ENZYMES (nmol/mg Hb/h)			Nucleotide Levels (nmol/ml packed cells)						
	APRT	HGPRT	MTAP	ATP	ADP	AMP	GTP	GDP	NAD+	PP-ribose-P
Patient LT	20.1	(H) 81 (G) 86	9.1	1336	210	11	38	14	80	5.6
Controls (n-28)	16 -32	(H) 80- 130	5- 13	1278 ±127	114 ±24	10 ±3	60 ±10	20 ±5	75 ±14	0.5 -4.0

H hypoxanthine) substrate
G guanine   )

APRT adenine   ) phosphoribosyl
HGPRT hypoxanthine guanine) transferase
MTAP methylthioadenosine   phosphorylase

Urinary hypoxanthine/xanthine (H/X) ratios on the basal diet (>3.0) were consistently raised compared with normal (<1.5), as were hypoxanthine levels (Table 1), suggesting increased de novo purine synthesis.   Urinary hypoxanthine and xanthine levels increased slightly, but did not show the dramatic increase in xanthine resulting in reversal of the H/X ratio to values well below unity as usually seen in primary gout or healthy controls on allopurinol. Small but measurable amounts of adenine were found in the urine (<0.02mmol/24h) in some but not all samples in each study.   Mean drug metabolite recovery was low (25%, 8% and 7% on the different doses respectively), but plasma oxipurinol levels were similar to those for subjects on 300mg allopurinol per day[4] and doubled when the dose doubled (Table 1).

APRT, and HGPRT (hypoxanthine or guanine as substrate) were normal[2] as were the levels of PP-ribose-P synthetase and methylthioadenosine phosphorylase (MTAP) in lysed red cells[5].   PP-ribose-P levels were also normal, excluding any of the recognised enzyme defects assoc- iated with purine overproduction in man[6] (Table 1).

DISCUSSION

This patient clearly differs from either the middle-aged male with primary gout or patients homozygous for any of the recognised

inherited defects leading to purine overproduction and gout in man
(partial HGPRT deficiency; PP-ribose-P superactivity)[6]. Levels
of these enzymes were normal. The excretion of significant
amounts of adenine (normally undetectable) suggested possible
overactivity of the polyamine pathway. The latter is now known to
be the source of endogenous adenine in man.[5] However both APRT and
MTAP (the enzymes responsible for the removal and production of
adenine as a by-product of polyamine synthesis) were normal.[5]

Gout patients presenting in early adulthood, as did this man, have
been described with dominant defects associated with an inability
to concentrate the urine but this type of familial gout is associated with rapidly fatal renal disease affecting young men and women
equally.[2] Renal function for age was not significantly impaired,but
the patient did show the reduced fractional urate clearance characteristic of the patient with primary gout as well as the excretion
of a urine of a fixed low pH.[6] However, he differed from the
latter in exhibiting uric acid overproduction on a low purine diet.

The lack of effect of allopurinol in this patient and the poor
recovery suggested the drug had not been taken,despite careful
surveillance. However plasma oxipurinol levels were comparable
to patients on similar doses and doubled when the dosage doubled.[4]
These results argue against poor compliance and suggest plasma
oxipurinol levels should be investigated where this is suspected.
The defect in this patient remains undefined. He obviously belongs
to a group with a defective absorption of allopurinol. Much higher
drug doses may thus be necessary to achieve satisfactory control
of uric acid levels and hence the gouty arthritis.

REFERENCES

1. H. A. Simmonds. Urinary excretion of purines, pyrimidines and
pyrazolopyrimidines in patients treated with allopurinol and oxipurinol. Clin.Chim.Acta. 23: 353 (1969).
2. H. A. Simmonds, D. J. Warren, J. S. Cameron, C. F. Potter and
D. A. Farebrother. Familial gout and renal failure in young
women. Clin.Nephrol. 14: 176 (1980).
3. T. Gibson, J. Highton, H. A. Simmonds and C. F. Potter.
Hypertension, renal function and gout. Postgrad.Med.J. 55:21 (1979).
4. S. Sved and L. Wilson. Analysis of allopurinol and oxipurinol
in plasma and its application to metabolic studies. Biopharma-
ceutics and Drug Disposition, 1: 111 (1980).
5. H. A. Simmonds, J. S. Cameron, M. J. Dillon, T. M. Barratt and
K. J. Van Acker. 'Uric acid' stones in children: problems of
diagnosis and treatment in a new defect - adenine phosphoribosyltransferase deficiency. Fortschritte der Urol und Nephrol. 16: 52
(1981).
6. J. B. Wyngaarden and W. N. Kelley. Gout and Hyperuricaemia.
Grune and Stratton, New York (1976).

URICOSTATIC EFFECT OF ALLOPURINOL IN THE ALLANTOXANAMIDE-TREATED

RAT: A NEW METHOD FOR EVALUATING ANTIURICOPATHIC DRUGS

Max Hropot, Roman Muschaweck and Erik Klaus

Hoechst AG, Postfach 80 03 20
D-6230 Frankfurt (M) 80, FRG

INTRODUCTION

In most mammals, uric acid is converted into allantoin by the enzyme uricase, which does not occur in man and apes. Thus, the urinary end product of purine metabolism in man and apes is uric acid, whereas it is allantoin in most mammals. An attempt to produce an animal model for research into hyperuricemia was made by Johnson[1], who blocked the activity of hepatic uricase in rats by a selective enzyme inhibitor, oxonic acid (oxonate). Particularly the amide of oxonic acid (allantoxanamide) has been described as an effective inhibitor of uricase activity in vitro and in vivo. Hropot[2] demonstrated that the prolonged duration of action of allantoxanamide as compared to oxonic acid is due to its elimination kinetics. Allantoxanamide was eliminated by the kidneys with a half-life of 25.4 min, which was double that of oxonate during the first 30 min after administration. The purpose of this study was to establish an appropriate screening method for evaluating antiuricopathic drugs in the allantoxanamide-treated rat.

METHODS

Male Sprague-Dawley rats with a body weight of 250 $\pm$ 15.5 g were used. Allantoxanamide (2,4-dioxo-1,2,3,4-tetrahydro-1,3,5-triazine-6-carboxylic acid amide, Calbiochem-Behring) was injected intraperitoneally at a dose of 250 mg/kg as a suspension in 0.25 % methylcellulose (control group). In the experimental groups, the following drugs suspended in 2 % starch solution were administered orally at the same time as allantoxanamide: allopurinol at doses of 10, 25, 50 and 100 mg/kg, probenecid and tienilic acid at 100 mg/kg. Blood samples were taken retroorbitally

4, 6 and 24 hours after administration of allantoxanamide and antiuricopathic drugs. The blood samples were assayed for uric acid
concentration according to the urica-quant method.

RESULTS

The serum uric acid concentration of normal rats was about
20 μmol/1. Intraperitoneal injection of 250 mg/kg of allantoxanamide significantly increased the serum uric acid level 4 and 6
hours after the administration as compared to untreated controls
(Fig. 1). The serum uric acid was still elevated 24 hours after
administration, but not significantly.

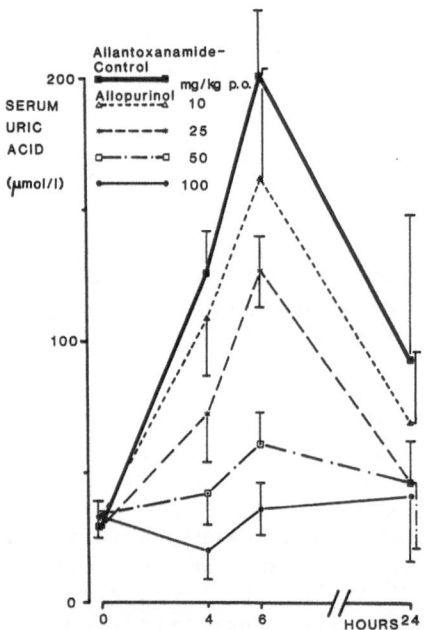

Fig. 1. The time course of dose-dependent inhibition of allan-
       toxanamide-induced hyperuricemia by allopurinol in rats.
       Each point represents the mean ± S.D. of 6 individual
       values.

Simultaneous oral administration of allopurinol in doses of 10,
25, 50 and 100 mg/kg dose-dependently decreased allantoxanamide-
induced elevation of serum uric acid concentration. Allopurinol
inhibited the allantoxanamide-induced elevation of serum uric
acid, the ID_{50} being about 32 mg/kg orally. These observations
are consistent with those reported by Smith[3] in the oxonate-pretreated rat.

The excretion of uric acid in the urine under various experimental conditions is displayed in Fig. 2. The open bars show the excretion of uric acid in the control group receiving allantoxanamide alone. In the collection period of 1 – 6 hours, there was no major difference in uric acid excretion between control group and rats having simultaneously received allantoxanamide and allopurinol in doses of 10, 25, 50 and 100 mg/kg.

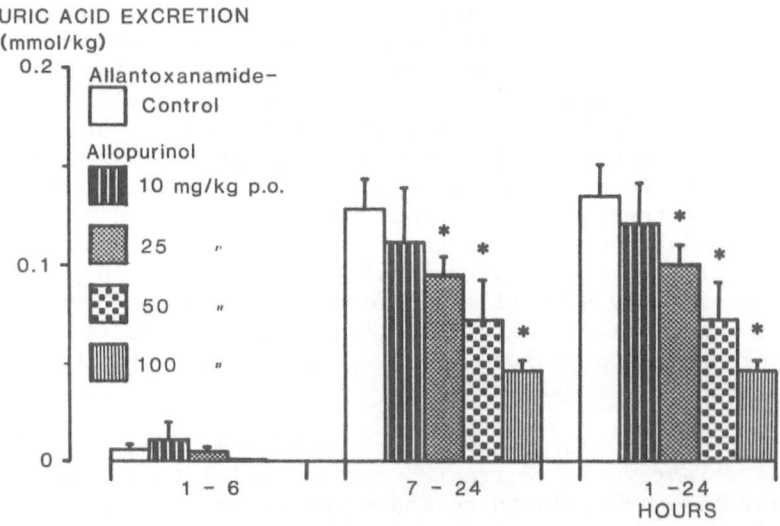

Fig. 2. The excretion of uric acid in the collection periods 1 – 6, 7 – 24 and 1 – 24 hours after allantoxanamide, 250 mg/kg i.p., as compared to rats receiving allantoxanamide and allopurinol simultaneously.

Starting at 6 hours after the administration of allantoxanamide alone, there was an enormous increase of uric acid excretion (140 $\pm$ 15 versus 6.1 $\pm$ 3 μmol/kg). At the same time, i.e. in the collection period of 7 – 24 hours, the excretion of uric acid dose-dependently decreased in the animals receiving simultaneously both drugs. The effect was significant at 25 mg/kg allopurinol and higher doses as compared to the allantoxanamide control group.

The model of the allantoxanamide-treated rat might also be used to test uricosurics. Two known uricosurics, probenecid and tienilic acid, were administered orally, both in a dose of 100 mg/kg given simultaneous with allantoxanamide. However, only tienilic acid increased the uric acid excretion as compared to control.

DISCUSSION

The lack of suitable animal models for testing potential antiuricopathic compounds led to the use of uricase inhibitors. Johnson[1] and Iwata[4] have shown that allantoxanamide and 2,8-diaza-hypoxanthine are the most potent uricase inhibitors, but only the former revealed a prolonged duration of action. The inhibition of uricase activity by allantoxanamide in rats led to significant hyperuricemia and hyperuricosuria. Allopurinol, a potent inhibitor of xanthine oxidase and therefore of uric acid synthesis, dose-dependently decreased allantoxanamide-induced serum uric acid elevation. Smith[3] has shown the same pattern of effects using oxonate; however, oxonate was less potent and had a shorter duration of action than allantoxanamide. As a consequence of allopurinol treatment in the allantoxanamide-treated rat, the excretion of uric acid was dose-dependently decreased. In contrast, tienilic acid, an uricosuric diuretic, increased uric acid excretion in the allantoxanamide-treated rat, whereas probenecid had no effect in this animal model. Smith[3] used the oxonate-pretreated rat as a model for evaluating the hyperuricemic effects of antihypertensive drugs. However, only diazoxide and furosemide produced hyperuric-aemia in these animals. Therefore, it is questionable if this model is appropriate to exclude new drugs possibly producing a clinically relevant hyperuricemia.

On the basis of our results, it can be concluded that the model of the allantoxanamide-treated rat is useful to screen uricostatics but less suitable to screen uricosurics.

ACKNOWLEDGEMENTS

The authors are grateful to Mrs. U. Schwarzer and Mr. P. Hainz for their excellent technical assistance.

REFERENCES

1. W. J. Johnson and André Chartrand, Allantoxanamide: A potent new uricase inhibitor in vivo, Life Sciences, 23: 2239 (1978)
2. M. Hropot, F. Sörgel, B. v. Kerékjártó, H. J. Lang, and R. Muschaweck, Pharmacological effects of 1,3,5-triazines and their excretion characteristics in the rat, Purine Metabolism in Man-III, 122A: 269, Plenum Press, New York (1980)
3. R. D. Smith, A. D. Essenburg, and H. R. Kaplan, The oxonate-pretreated rat as a model for evaluating hyperuricemic effects of antihypertensive drugs, Clin. exp. Hypertens. 1: 487 (1979)
4. H. Iwata, I. Yamamoto, I. Ghoda, E. K. Morita, M. Nakamuro, and K. Sumi, Potent competitive uricase inhibitors - 2,8-diaza-hypoxanthine and related compounds, Biochem. Pharmacol. 22: 2237 (1973)

THE EXCRETION OF ^{14}C-HYPOXANTHINE AND ITS METABOLITES IN RATS FOLLOWING ADMINISTRATION OF URICOSTATIC DRUGS

Béla v. Kerékjártó and Max Hropot

Hoechst AG, Postfach 80 03 20

D-6230 Frankfurt (M) 80, FRG

INTRODUCTION

In vivo xanthine oxidase inhibition results in a reduction of uric acid (and allantoin) in urine and in an increase in the urinary excretion of hypoxanthine and xanthine as described by Elion[1] for allopurinol. The dose-dependence of the excreted amounts of hypoxanthine and xanthine is a relevant consideration when using these parameters to measure the action of uricostatics in the rat. The purpose of our studies was to obtain a simple and rapid method to detect the uricostatic quality of hypouricemic compounds.

METHODS

Male Wistar rats (company's own breeding) with a body weight of 120 - 150 g were used. The last 16 hours preceding the experiment the animals were deprived of food but drinking water remained unlimited. Two rats of the same body weight were placed in one metabolism cage. All animals were taken from the same population (identical age and any other conditions). Each assay was based on the evaluation of the urine samples taken from two or three metabolism cages. The experiment started with the oral administration of the drug in starch gel; the control rats received the vehicle alone. One hour later the animals were given 50 ml tap water/kg b.w. by stomach tube; 40 µCi/kg b.w. ^{14}C-hypoxanthine were simultaneously injected intravenously. Immediately thereafter urine collection was started and samples corresponding to the 1st and 2nd hours were separated. The total radioactivity was measured in 1 - 3 µl of the urine fraction by liquid scintillation. The distribution of the radioactivity was determined

following separation by high pressure liquid chromatography. Using
Lichrosorb RP 18 on a 250 x 4 mm column, 50 μl of urine containing
5 - 20 nCi radioactivity were sampled and eluated at room temperature with either 0.01 M ethanolamine phosphate at pH 3.9, or with
0.01 M potassium dihydrogen phosphate at pH 4.0. The flow was
2.0 ml/min. Radioactivity was detected with a solid scintillator
(Berthold), the obtained peaks were recorded and integrated for
three determinations of each sample. The retention times were.
1.6 min for allantoin, 4.3 min for uric acid, 5.3 min for hypoxanthine and 6.3 min for xanthine. The results presented are the
mean values of a total of two to five experiments.

RESULTS

Allantoin and uric acid plus traces of hypoxanthine and/or
xanthine were excreted in the urine of control groups of rats
which had received radioactivity in the form of hypoxanthine. The
reproducibility of the data and the variation between samples of
corresponding cages are demonstrated in Table 1. The form in which

Table 1. Total radioactivity excreted in form of allantoin, uric
acid and hypoxanthine in urine 1 hour after i.v. injection of ^{14}C-hypoxanthine. The values represent mean $\pm$ S.D.
of three measurements per sample.

| Metabolites | % of excreted radioactivity | | | overall mean |
	Cage 1	Cage 2	Cage 3	Cages 1 - 3
allantoin	87.1+3.3	85.3+0.3	84.4+0.6	85.6+2.1
uric acid	6.5+1.0	10.2+1.2	9.1+0.9	8.6+1.9
hypoxanthine	1.0	0.9	0.6	0.8+0.5
sum	94.6	96.4	94.1	95.0

radioactivity was excreted agrees with earlier findings of Greger[2]
who studied the renal excretion of allantoin and urate in rats.
The biological and analytical deviation of the mean values in
Table 1 also applies for the results in Tables 2 - 4. The sum of
the radioactivity found in the compounds detected in the urine represents 95 % of the total excreted radioactivity.

42 % of the administered ^{14}C-activity was eliminated in the
urine within 2 hours. Table 2 demonstrates the effect of differ-
rent doses of allopurinol on the elimination and distribution of
^{14}C-derived from hypoxanthine.

Table 2. The effect of allopurinol on the urinary recovery of the administered radioactivity and its distribution in form of the metabolites.

Compound	Dose mg/kg	Recovery of administered radioactivity %	A allantoin	U uric acid	H hypox.	X xanthine
			% of excreted metabolites			
control	–	42	85.9	12.0	tr	–
allopurinol	3	25	65.6	–	25.2	9.2
allopurinol	10	15	35.8	–	48.2	16.4
allopurinol	25	14	9.9	–	73.6	13.4

The data confirm the dose-dependent increase in the appearance of hypoxanthine and xanthine[1]. The rate of excreted radioactivity decreases simultaneously while uric acid falls below detectable limits. The differences between the mean values of the controls and those of 3 mg/kg b.w. allopurinol are significant.

In the results shown in Table 2, allopurinol was given one hour prior the administration of radioactivity. This delayed interval can be extended as shown in Table 3. Because of the short plasma half-life of allopurinol, the prolonged delay corresponds

Table 3. The time course of the effect of allopurinol on the urinary recovery of the administered radioactivity and its distribution in form of the metabolites.

Compound	Dose mg/kg	Time h	Recovery of administered radioactivity %	A + U	H + X
				% of excreted metabolites	
control	–		42	97.9	tr
allopurinol	0.3	0.5	46	100	
allopurinol	3.0	0.5	17	70.5	29.5
allopurinol	3.0	1	17	65.6	34.4
allopurinol	10	1	15	35.8	63.6
allopurinol	10	2	28	77.5	22.5
allopurinol	10	4	33	91.0	9.0
allopurinol	10	6	52	97.4	2.6
allopurinol	25	1	14	9.9	87
allopurinol	25	4	30	78	22

to an actual decrease in allopurinol concentration. Thirty to
sixty minutes seem to be adequate to produce optimal effect of
allopurinol. No effect was obtained with the dose of 0.3 mg/kg at
30 min or with 10 mg/kg at 6 hours.

The results represented in Table 2 and 3 are compatible with
the expectation of obtaining similar findings with other xanthine
oxidase inhibitors of comparable potency, while uricosuric drugs
may have no action on the amount of excreted metabolic precursor
of uric acid. As demonstrated in Table 4 thiopurinol, a less active
compound than allopurinol in small animals, results in a significant response with the dose of 25 mg/kg b.w.

Table 4. The effect of thiopurinol and tienilic acid on the urin-
         ary recovery of the administered radioactivity and its
         distribution in form of the metabolites.

Compound	Dose mg/kg	Time h	Recovery of administered radioactivity %	A	U	H + X
				% of excreted metabolites		
control	–	–	56	81.9	18.1	
thiopurinol	25	1	42	75.1	9.5	15.4
control	–	–	39	86.8	13.2	
tienilic acid	100	1	40	87.9	12.1	
	100	2	44	87.4	12.6	
tienilic acid	200	1	46	90.8	9.2	
	200	2	36	86.1	13.9	

Tienilic acid, on the other hand, does not change the distribution of eliminated radioactivity in comparison to the controls
in dosages of 100 or 200 mg/kg b.w.

DISCUSSION

With the use of labeled hypoxanthine for the detection of reduced or intact in vivo xanthine oxidase activity we assume that
nearly constant relationship between the velocity of catabolism
and the velocity of other relevant metabolic events exists. In our

experiments the amount of oxidized metabolites, allantoin and urate, was approximately 50 % of the injected radioactivity. During the inhibition of xanthine oxidase, a greater part of the injected hypoxanthine flows through non-oxidative pathways and the urinary excretion of radioactivity is accordingly decreased. The diminished excretion of radioactivity is less specific than the changes in the distribution of the radioactivity. The most important signs of the activity are the significantly elevated amounts of hypoxanthine plus xanthine versus urate plus allantoin.

The advantages of the method are rapid performance and the lack of interference by foreign compounds other than the test drug. The necessary high potency of drugs to obtain positive findings may be less favorable for screening purposes. Without a reference point for plasma half-life the assay must be performed in consideration of the time dependence of drug action. Essential in any case is the necessity for reproducible diuresis in the test animals.

ACKNOWLEDGEMENTS

The authors wish to thank Mrs. H. Keller and Mr. A. Schilling for their technical assistance.

REFERENCES

1. G. B. Elion, Allopurinol and other inhibitors of urate synthesis, in: Uric acid, W. N. Kelley, I. M. Weiner, ed. Springer-Verlag, Berlin (1978)

2. R. Greger, Purine excretion by the rat kidney, in: Amino acid transport and uric acid transport, S. Silbernagl, F. Lang, R. Greger, ed. Georg Thieme Publishers Stuttgart (1976)

PULSE RADIOLYSIS STUDY OF

HYPOXANTHINE AND XANTHINE OXIDATION

[*]J.Santamaria,[*]C.Pasquier,[*]C.Ferradini and J.Pucheault[†]

[*]Laboratoire de Recherches Organiques de l'ESPCI,Univer-
sité P.& M. Curie,75005,Paris
[*]Hopital Cochin,Institut de Rhumatologie,75014,Paris
[*]Laboratoire de Chimie Physique,Université R. Descartes
75006,Paris

INTRODUCTION

The oxidation of hypoxanthine (Hyx) into xanthine (X) and xanthine into uric acid (U) plays an important role in diseases like
gout and xanthinuria.A good knowledge of this oxidation would allow
to understand the mechanism of action of drugs like pyrazolopyrimidines (allopurinol and thiopurinol) which reduce uric acid synthesis.

It is well known[1,2] that in presence of xanthine oxydase (XO)
the oxidation of oxypurines by oxygen occurs with concomitant formation of superoxide anion (O_2^-), product of the univalent reduction
of molecular oxygen ($^3O_2$).

At this time possible transients have not been observed yet.
Consequently monoelectronic oxidations have been carried out using
pulse radiolysis technique (Febetron 708) and fast kinetics absorption spectrophotometry.

In a first time, the oxidation in aqueous solutions of hypo-

185

xanthine and xanthine by $^{\bullet}$OH radicals has been investigated (it is
known that under nitrous oxide atmosphere, hydrated electrons, produced by water radiolysis, lead to the formation of $^{\bullet}$OH radicals;
thus G($^{\bullet}$OH)=5.4 after less than a microsecond). It produces transient
radicals which have been characterized spectrophotometrically and
kinetically.

   Some experiments have been realized on aerated solutions in
order to display the part of molecular oxygen and superoxide anion.

1. HYPOXANTHINE-XANTHINE SYSTEM

   Irradiation of aqueous solutions of hypoxanthine (under N_2O,
(Hyx)=$2x10^{-3}$M, ($^{\bullet}$OH)=$1.5x10^{-4}$M, buffer phosphate pH=7.8) leads to a
spectrum which reaches a plateau 0.8µs after the pulse(Fig.1,curveI).
This spectrum may be attributed to the radical produced by the oxidation of hypoxanthine by $^{\bullet}$OH radicals. It presents two maxima at
310 and 340 nm and a shoulder towards 475 nm.

   The effect of hypoxanthine concentration shows that at $2x10^{-3}$M,
the uptake of $^{\bullet}$OH radicals is complete and thus, one is able to calculate the differential molar extinction coefficients ($\varepsilon_{R_1^{\bullet}}^{340}$=2200
$M^{-1}cm^{-1}$) (Fig.1, right scale).

   From the results obtained for the lower concentrations of hypoxanthine, the calculation of the two competitive reactions:

$$^{\bullet}OH + ^{\bullet}OH \xrightarrow{2k_1} H_2O_2 \quad \text{and}$$

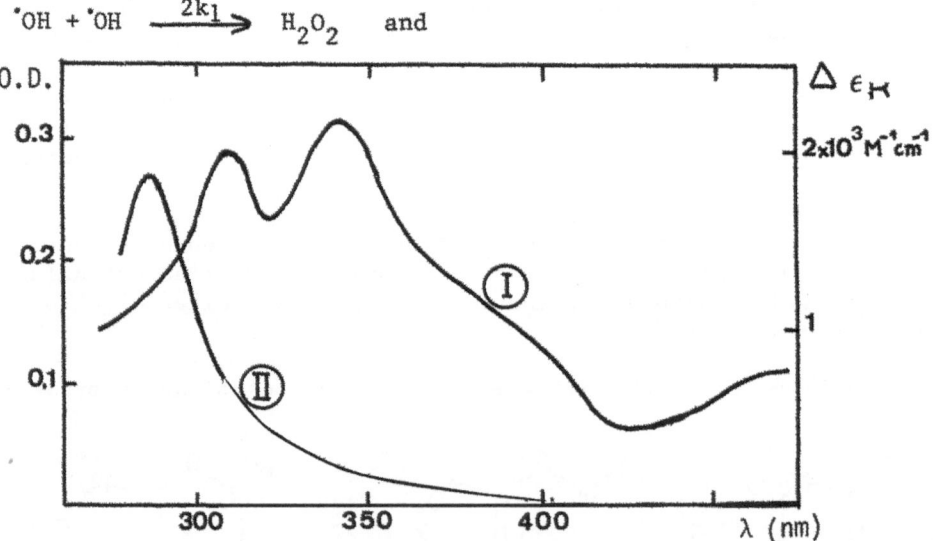

Fig.1. Differential absorption spectra for different delays after
       the pulse.
   I.0.8µs after the pulse(correspondence to $\varepsilon_{R^{\bullet}}$ on the right scale)
II.800µs after the pulse.

$$\text{Hyx} + {}^{\bullet}\text{OH} \xrightarrow{\quad k_2 \quad} R_1^{\bullet}$$

leads to $k_2 = 6.5 \times 10^9 M^{-1} s^{-1}$ with $2k_1 = 10^{10} M^{-1} s^{-1}$.

The $R_1^{\bullet}$ radical decreases by a second order reaction and leads directly to the formation of xanthine (800μs after the pulse) which is characterized by absorption peak at 286 nm (Fig.1, curve II) and by reverse phase HPLC method from the final mixture. Its yield is G(X)=2.26 molec./100 eV. Thus we can assume that $R_1^{\bullet}$ radical disappear by a disproportionation:

$$R_1^{\bullet} + R_1^{\bullet} \xrightarrow{\quad 2k_3 \quad} X + \text{Hyx} \qquad\qquad 2k_3 = 1.3 \times 10^8 M^{-1} s^{-1}$$

2. XANTHINE- URIC ACID SYSTEM

The oxidation of xanthine in aqueous solutions by ${}^{\bullet}$OH radicals (under N_2O, $(X)=10^{-3}M$, $({}^{\bullet}OH)_o=1.4\ 10^{-3}M$, pH=7.8) gives rise to a transient radical $R_2^{\bullet}$ like in the previous system. Its formation is maximum 0.8μs after the pulse. Its differential absorption spectrum versus xanthine (Fig.2, curve I) shows a maximum at 340 nm and as previously, its differential molar extinction coefficients have been calculated ($\varepsilon_{R_2^{\bullet}}^{340} = 4400\ M^{-1} cm^{-1}$) (Fig.2, right scale) and so its formation rate $R_2^{\bullet}$ ($k_2' = 5.2 \times 10^9\ M^{-1} s^{-1}$).

At the end of the reaction (4s after the pulse) the absorption peak at 300 nm (curve IV) may be attributed to uric acid.

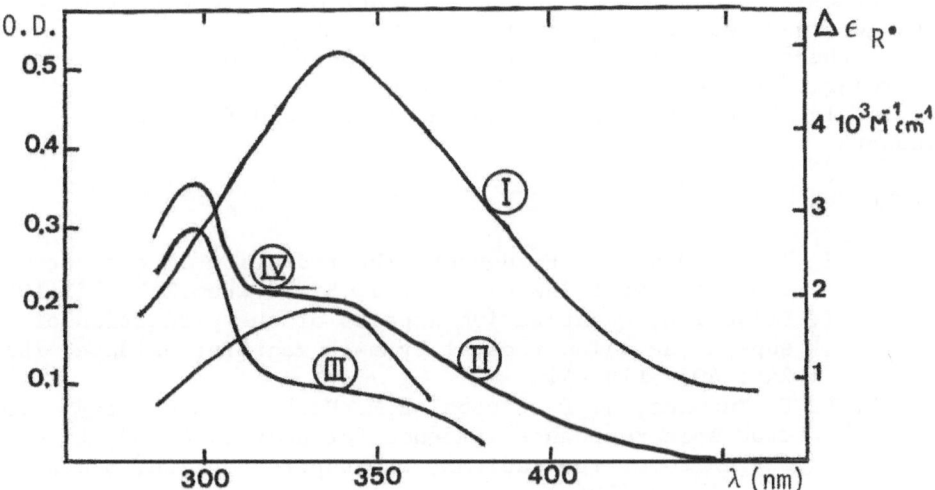

Fig.2 Differential absorption spectra for different delays after the pulse.
I. 0.8μs after the pulse (correspondence to $\hat{\varepsilon}_{R^{\bullet}}$ on the right scale)
II. 400μs ; III. 160 ms ; IV. 4s.

Nevertheless, one can see that the decrease of R_2 radical does not lead directly to uric acid but the formation of another transient species (Fig.2, curve II). This decrease being too, of a second order, we can assume that this species is the dimer D.

$$R_2^{\bullet} + R_2^{\bullet} \xrightarrow{\quad 2k_3' \quad} D \qquad\qquad 2k_3' = 2.0 \times 10^8 \ M^{-1} s^{-1}$$

This dimer may react throught two paths:
- the first path following first order kinetics ($k_4 = 23 \ s^{-1}$) leading to xanthine and uric acid G(U) = 1.4 molec/100 eV (curve III and IV, $\lambda = 300$ nm). Furthermore xanthine and uric acid were identified from the final mixture by reverse phase HPLC method. It has been calculated that this way is responsible of 57 % of the disappearance of dimer D.
- the second (43%) would give an unknown intermediate A 160 ms after the pulse (curve III, absorbance about 330 nm) which could lead, following also first order kinetics ($k_6 = 0.53 \ s^{-1}$) to the product B (curve IV, absorbance about 340 nm) stable under N_2O atmosphere. So, the kinetic scheme is

$$D \quad
\begin{array}{l}
\xrightarrow{\quad k_4 \quad} X + U \qquad 57\% \\
\xrightarrow{\quad k_5 \quad} A \xrightarrow{\quad k_6 \quad} B \qquad 43\%
\end{array}$$

Preliminary experiments carried out with oxygen lead to the following findings:
- superoxide anion $(O_2^{-\bullet})$ does not react on hypoxanthine nor on xanthine.
- radicals formation ($R_1^{\bullet}$ and $R_2^{\bullet}$) is not modified even though their decrease would slightly be.
- in the case of xanthine, the dimer appears too and decrease according to the same distribution between the two ways described previously. Neverless, the second way leads to different final products.

REFERENCES

1. J. Mc. Cord and I. Fridovich, The reduction of cytochrome c by milk xanthine oxidase, J. Biol. Chem.243:5753(1968).
   I. Fridovich, Quantitative aspects of the production of superoxide anion radical by milk xanthine oxidase, ibid. 245: 4053 (1970).
2. P. F. Knowles, J. F. Gibson, F.M. Pick and R.C. Bray, Electron spin resonance evidence for enzymatic reduction of oxygen to a free radical, the superoxide ion, Biochem. J. 111: 53 (1969).
   R. C.Bray, F. M. Pick and D. Samuel, Oxygen- 17 hyperfine splitting in the electron paramagnetic resonance spectrum of enzymically generated superoxide, Eur. J. Biochem. 15: 352 (1970).

URATE CRYSTAL-INDUCED SUPEROXIDE RADICAL PRODUCTION

BY HUMAN NEUTROPHILS

Costantino Salerno, Alessandro Giacomello,
and Egisto Taccari

Institutes of Biological Chemistry and Rheumatology
University of Rome, and Center of Molecular Biology
National Research Council, 00100 Rome, Italy

It has been established that during acute attack of gout crystals of monosodium urate, to which serum proteins (particularly IgG) are avidly adsorbed, are engulfed by polymorphonuclear leukocytes[1]. Within leukocytes, after the fusion of the crystal-containing phagosome with lysosomes, the crystals are stripped enzymatically of protective proteins and become capable of membrane lysis[2]. Recent papers have stressed the potential pathological role of oxygen radicals, produced by exposure of granulocytes to suitable particulate or soluble stimuli, in mediating damage of tissues and of phagocytic cells themselves[3,4]. The oxygen radical-generating system is an integral part of the plasma membrane, activated by membrane perturbance and incorporated in the phagosomal membrane by invagination of the plasmalemma[4]. In the present paper, superoxide production by human neutrophils in the presence of urate crystals of different dimensions, coated and not coated with serum proteins has been studied.

EXPERIMENTAL PROCEDURES

Human polymorphonuclear leukocytes were obtained from heparinized venous blood of healthy donors according to published procedures[5]. Microcrystalline monosodium urate was prepared according to the method reported in reference[6]. Oxygen radical generation was measured[7] in the presence of saturating luminol concentration (0.2 mg/ml). Under these conditions it has been demonstrated that, at all times, the intensity of the luminescence response was proportional to the rate of production of oxygen radicals[7]. Therefore the intensity of the light emission could be considered a measure of the concentration of the activated radical-producing enzyme system located on the plasma membrane. All experiments were carried out in buffered saline[8] at 37°C. The cells were handled in plastic containers.

RESULTS

Fig. 1 shows that the addition of microcrystalline monosodium urate (4 mg/ml) to a suspension of leukocytes in phosphate buffered saline saturated with urate gives rise to a chemiluminescence response after a latency period of about 20 sec. The height of the chemiluminescence peak is at least 18 times lower than that obtained by adding latex particles (1 μm diameter, 5 mg/ml) to a suspension of leukocytes in phosphate buffered saline or in phosphate buffered saline saturated with urate. The chemiluminescence peak obtained with urate crystals is sharper than that obtained using latex particles. After 4 min, the chemiluminescence response of urate activated cells disappears while light emission for more than 10 min is observed using latex particles.

Using latex beads it has been shown that the height of the chemiluminescence peak is a linear function of leukocyte concentration and a hyperbolic function of particle concentration[7] . In increasing urate crystal concentration from 4 mg/ml to 12 mg/ml, no increase in the height of the chemiluminescence peak was observed suggesting that, in the experiment illustrated in fig. 1,

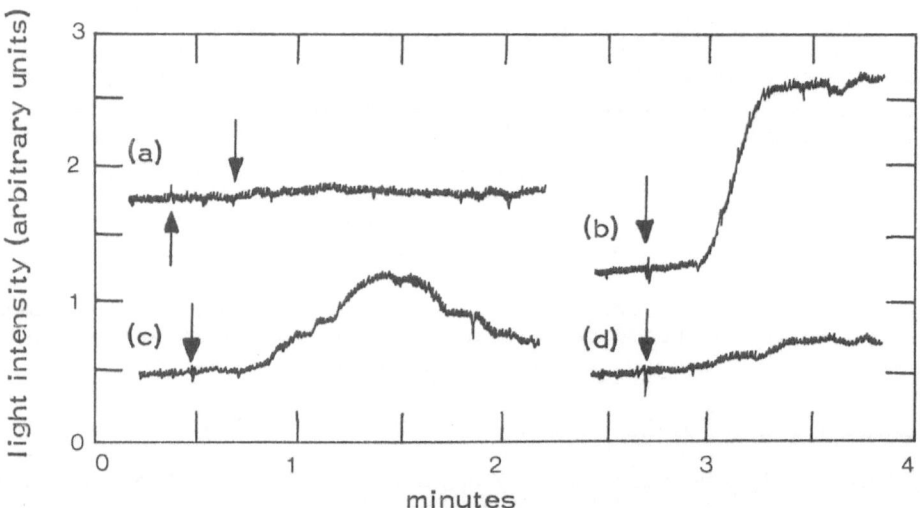

Fig. 1. Chemiluminescence records obtained at saturating luminol concentration by stimulating human polymorphonuclear leukocytes (22.4 x 10^6 cells/ml in a,c,d; 2 x 10^6 cells/ml in b) with urate crystals of different length (2 μm in a,c; 10 μm in d) or latex particles (b) (mean diameter: 1 μm) in the presence (a) or absence (b,c,d) of IgG (1.3 mg/ml). Urate crystal concentration was 4 mg/ml. Latex concentration was 3 mg/ml. Arrows indicate the time of addition of the particulate stimulus (↓) or of IgG (↑). Other experimental conditions are as described in the text.

urate crystal concentration is saturating. As shown in the same
figure, the peak of the luminescence curve is a function of particle
dimension. The height of the chemiluminescence peak is higher using
urate crystals about 2 μm than crystals about 10 μm in length. Latex
particles with a mean diameter of 0.1, 0.4, 0.6, 0.8, 1.0, and 16 μm
have been employed. The highest chemiluminescence peak was obtained
with 1.0 μm particles (data not shown).

When granulocytes are stimulated by latex particles (1 μm diameter, 5 mg/ml) coated with IgG[8] the height of the chemiluminescence
peak is at least 20 times lower than that obtained using uncoated
latex particles[7]. The same results are obtained when latex particles
opsonized with whole serum[9] are used instead of latex particles coated
with IgG. Serum proteins and in particular IgG inhibit the chemiluminescence response of granulocytes by coating latex particles and
not by reacting with granulocytes or with the oxidizing agents produced by activated granulocytes. In fact the addition of 5 mg/ml
IgG at the peak of the chemiluminescence response to 4 mg/ml uncoated latex particles has no appreciable effect on the rate of luminol consumption whereas the same amount of IgG added before uncoated latex particles prevents the chemiluminescence respose[7].
Moreover the chemiluminescence response obtained using uncoated latex
particles is not appreciably affected by incubating granulocytes with
5 mg/ml latex particles coated with IgG or whole serum for up 3 min[7].
Similar results are obtained when urate crystals are used instead of
latex particles: 1.3 mg of IgG or 0.2 ml of serum added to 1 ml of a
suspension of granulocytes before the stimulation of the cells with
4 mg/ml urate crystals lead to a marked decrease in the height of
the chemiluminescence peak.

DISCUSSION

Naked monosodium urate crystals stimulate oxygen radical release
by human polymorphonuclear leukocytes. Very recently, two papers[9,10]
dealing with superoxide anion generation by human neutrophils exposed
to monosodium urate have been published. From the comparison of the
data derived from nitroblue tetrazolium and cytochrome c reduction,
it has been suggested that radical production in response to urate
crystals is compartmentalized and occurs predominantly in the intracellular space[9]. Microcrystalline sodium urate-induced oxygen consumption and nitroblue tetrazolium reduction were present for at
least 15 min after stimulation of the cells. To enhance extracellular radicals neutrophils were converted to secretory non phagocitic
cells by use of cytochalasin B[10].

Taking into account these observations, the rapid disappearance
of the luminol enhanced chemiluminescence response after stimulation
of human leukocytes by urate crystals could be at least in part
attributed to the phagocytosis of the crystals with incorporation of
the radical-producing plasma membrane into phagosome and to the fact

that the luminol-enhanced chemiluminescence method detects only oxygen radicals released extracellularly. The absorption of serum proteins on urate crystals lead to a decrease in the production of luminol-reacting radicals. Using cytochalasin B-treated leukocytes, an increase in O_2^- production was observed[10]. The apparent discrepancy between these results could be resolved by taking into account that cytochalasin B inhibits while IgG absorbed to the crystals accelerates phagocytosis[11]. Grinding monosodium urate crystals to produce smaller crystals lead to an increase in the luminol-reacting radicals. This increase in luminol-enhanced chemiluminescence response cannot be attributed to a decreased phagocytosis since small crystals are phagocytized more avidly than large crystals[11]. Oxygen-derived free radicals must be added to the list of mediators that participate in gouty inflammation. The dimension of the crystals and the absorption of serum proteins can modify the extent of production of these cytotoxic radicals.

REFERENCES

1. F. Kozin, M. H. Ginsberg, and J. L. Skosey, Polymorphonuclear leukocyte responses to monosodium urate crystals: modification by absorbed serum proteins, J. Rheumatol. 6:519 (1979).
2. N. S. Mandel, The structural basis of crystal-induced membranolysis, Arthritis Rheum. 19:439 (1976).
3. L. Simchawikz and I. Spilberg, Evidence for the role of superoxide radicals in neutrophil-mediated cytotoxicity, Immunology 37:301 (1979).
4. D. Roos, Molecular events during phagocyte stimulation, Scan. J. Rheumatol. Suppl. 40:46 (1981).
5. M. A. Trush, M. E. Wilson, and K. VanDyke, The generation of chemiluminescence by phagocytic cells, Methods Enzymol. 57:462 (1978).
6. E. Seegmiller, R. R. Howell, and S. E. Malawista, The inflammatory reaction to sodium urate: its possible relationship to the genesis of acute gouty arthritis, JAMA 180:469 (1962).
7. A. Giacomello, C. Salerno, and E. Taccari, Activation of human granulocytes by latex particles, in press (1982).
8. R. C. Allen, R. L. Stjernholm, and R. H. Steele, Evidence for the generation of an electronic exctitation state(s) in human polymorphonuclear leukocytes and its participation in bactericidal activity, Biochem. Biophys. Res. Comm. 47:679 (1972).
9. S. Abramson, S. T. Hoffstein, and G. Weissmann, Superoxide anion generation by human neutrophils to monosodium urate, Arthritis Rheum. 25:174 (1982).
10. L. Simchowitz, J. P. Atkinson, and I. Spilberg, Stimulation of the respiratory burst in human neutrophils by crystal phagocytosis, Arthritis Rheum. 25:181 (1982).
11. H. R. Schumacher, P. Fishbein, P. Phelps, R. Tse, and R. Krauser, Comparison of sodium urate and Ca pyrophosphate crystal phagocytosis by polymorphonuclear leukocytes, Arthritis Rheum. 18:783 (1975).

RENAL HANDLING OF URIC ACID IN NORMAL SUBJECTS: ITS BEHAVIOUR WITH RESPECT TO DIFFERENT FILTERED LOADS

F. Mateos, J.G. Puig, E. Prieto, E. Herrero, A. Muñoz, and J. Ortiz Vázquez

Departments of Internal Medicine and Clinical Biopatholo-gy, Ciudad Sanitaria La Paz, U. Autónoma, Madrid, Spain

INTRODUCTION

Serum uric acid concentration in man is the result of complex biosynthetic, catabolic, and excretory processes. Recent investigations of the urate transport system in the human kidney have provided evidence for a four-component model: glomerular filtration, presecretory reabsorption, tubular secretion, and postsecretory reabsorption[1].

The different phases that influence the renal excretion of uric acid have been studied in normouricemic controls[2], but we ignore their physiological behaviour over a wide range of filtered urate in order to mantain serum uric acid concentrations under normal limits.

MATERIAL AND METHODS

The investigation was conducted with the voluntary cooperation of 10 normal males (mean age, 33 years) who were placed on an essentially purine-free diet during the study. Renal handling of uric acid was examined by means of the pyrazinamide (PZA) and probenecid (PB) tests[3], performed in three uricemia states: normouricemia 3.6 to 6.4 mg/dl), allopurinol-induced hypouricemia (under 3.5 mg/dl), and hyperuricemia after oral administration of RNA monosodium salt (over 6.5 mg/dl).

RESULTS

The correlation between uricemia and uricosuria, for the range of plasma urate studied, showed an excellent statistical significance (r=0.90; p < 0.01).

Table 1 shows daily uric acid excretion, serum urate levels, and the different tubular phases of the renal handling of uric acid under the three uricemia states. Presecretory reabsorption of filtered uric acid was always above 99%. Tubular secretion of uric acid expressed as the percentage of filtered urate, was significantly higher in hyperuricemia with respect to the hypouricemia state. Tubular reabsorption of secreted uric acid was similar in normouricemia and hypouricemia, but in the sate of hyperuricemia a significant diminution of uric acid postsecretory reabsorption could be evidenced.

Table 1. Uricemia, uricosuria, presecretory reabsorption, tubular secretion, and postsecretory reabsorption of normal males in the states of hyperuricemia, normouricemia, and hypouricemia.

	Pur mg/dl	Uur mg/day	Presecretory[a] reabsorption (% of UA_F)	Tubular secret. (% of UA_F)	Postsecretory reabsorption (% of UA_S)
HYPERURICEMIA					
Mean	10.1[b]	1415[b]	99.7	42.6[c]	68.4[b]
$\pm$ M.S.E.	0.4	109	0.1	1.8	1.8
NORMOURICEMIA					
Mean	5.1	520	99.5	40.4	76.6
$\pm$ M.S.E.	0.1	18	0.2	2.4	1.4
HYPOURICEMIA					
Mean	2.7[b]	176[b]	99.6	37.2	75.7
$\pm$ M.S.E.	0.1	14	0.1	1.8	1.7

[a]UA_F: Uric acid filtered. UA_S: Uric acid secreted.
[b]$p < 0.01$, with respect to the normouricemia state.
[c]$p < 0.05$, with respec to the hypouricemia state.

CONCLUSIONS

1. The correlation between serum and urinary uric acid levels lends
   support to the assessement that when renal function is normal,
   the kidney is the principal organ which mantains uricemia under
   normal limitis.

2. In normal man, filtered uric acid undergoes almost complete
   tubular reabsorption, indicating that presecretory reabsorption
   does not modulate serum uric acid levels, and that urate excretion
   depends on the balance tubular secretion-postsecretory reabsorption.

3. When plasma urate concentrations increase to around 10 mg/dl,
   tubular secretion and postsecretory reabsorption modulate serum
   uric acid levels by a slight increment of the first and a
   marked diminution of the second, thus increasing uricosuria and
   lowering uricemia. The inverse mechanism was not evident in the
   hypouricemia state, possibly because low urate filtrable load is
   undoubtedly less pernicious than hyperuricemia.

REFERENCES

1. H. S. Diamond, and J. S. Paolino, Evidence for a postsecretory
   reabsorptive site for uric acid in man, J. Clin Invest. 52: 1491
   (1973).
2. D. J. Levinson, and L. B. Sorensen, Renal handling of uric acid
   in normal and gouty subjects: evidence for a 4-component system,
   Ann. Rheum. Dis. 39: 173 (1980).
3. A. de Vries, and O. Sperling, Inborn hypouricemia due to isolated
   renal tubular defect, Biomedicine, 30: 75 (1979).

EVIDENCE OF ABNORMAL RENAL HANDLING OF URIC ACID IN PATIENTS WITH

NEPHROLITHIASIS AND HYPERURICOSURIA

F. Mateos, J.G. Puig, E.Mª. Martínez, G. Gaspar,
E. Herrero, and J.A. Martínez Piñeiro

Departments of Internal Medicine and Clinical Biopathology, Ciudad Sanitaria La Paz, U. Autónoma, Madrid, Spain

INTRODUCTION

A group of patients forming calcium oxalate stones are hyperuricosuric and it is thought that their excessive urate excretion contributes to calcium-stones formation[1]. The pathomechanisms invoked
are dietary purine excess and endogenous uric acid overproduction,
being defective tubular reabsorption of urate "unattractive" because uricemia was found to be normal in patients with recurrent calcium nephrolithiasis (RCN) and hyperuricosuria. Current studies
were undertaken to define the incidence, role of diet, abnormalities
of the renal handling of urate, and associated metabolic disturbances in patients with RCN and hyperuricosuria.

MATERIAL AND METHODS

Fifty patients (22 males and 28 females; mean age 46 years)
were consecutively referred to our Metabolic Unit for the evaluation
of recurrent nephrolithiasis. Diagnosis was made on the basis of
spontaneous emission or chirurgical extraction of two or more calculi with an interval superior to one year. In every patient we performed a metabolic study[2] and the results were compared with those
obtained in 20 controls (10 males and 10 females; mean age 33 years)
Hyperuricosuria was defined as daily uric acid excretion above 800
mg for men and 750 mg for woman, while on a purine-free diet. Renal
handling of uric acid was evaluated by means of pyrazinamide (PZA)
and probenecid (PB) tests[3].

RESULTS

Among 50 patients with recurrent nephrolithiasis, 42 showed

calcium in their stone composition. Nine of them were hyperuricosuric
(6 males and 3 females; 21,4%) in the absence of clinical gout.
Table 1 shows the results of the basal metabolic study in 9 patients
with RCN and hyperuricosuria and in the control group. Six patients
had idiopathic hypercalciuria, one patient frank hyperoxaluria and
3 marginal hyperoxaluria. In four patients the results indicate renal
leak of phosphates.

Table 1. Basal metabolic study in patients with calcium nephrolithiasis and hyperuricosuria, and control subjects.

		SERUM			URINE				
		Ca	P	Ur	Ca	P	Ur	Ox	Ccr	Cur/Ccr
		mg/dl			mg/day				ml/min	%
Patient nº	Sex									
1	M	9.1	8.3	6.4	437	919	939	32	104	9.72
2	M	9.5	2.7	5.1	301	676	823	37	122	9.18
3	M	8.7	2.5	4.8	148	793	822	38	108	9.94
4	M	9.7	2.4	6.0	312	748	994	34	130	8.81
5	M	9.1	3.8	5.2	371	997	954	42	174	7.35
6	M	8.9	2.2	6.3	193	892	974	38	108	9.94
7	F	8.6	2.8	2.0	295	604	850	31	126	23.50
8	F	8.7	2.6	1.8	210	560	765	28	106	29.50
9	F	8.9	2.5	3.5	370	850	952	35	124	15.20
CONTROLS										
Males (n=10)										
$\overline{X}$		8.9	3.2	5.2	169	603	518	29	124	6.24
S.D.		0.5	0.4	0.6	51	119	90	5	26	1.17
Females (n=10)										
$\overline{X}$		8.9	3.5	4.3	136	566	522	29	118	9.36
S.D.		0.4	0.5	0.4	49	96	76	5	15	2.88

Data concerning renal handling of uric acid appears in Table
2. The administration of an essentially purine-free diet normalized
uricosuria in 4 patients. PZA and PB tests were within normal
limits in these patients, suggesting that hyperuricosuria could be
related to purine overingestion. Five patients remained hyperuricosuric even under an essentially purine-free diet, and 2 were hypo-

uricemic. A presecretory reabsorption defect of filtered uric acid was excluded by means of the PZA test. The maximum uricosuric response to PB was decreased in 4 patients and augmented in one, allowing us to conclude that hyperuricosuria was due to a postsecretory reabsorption defect and increased tubular secretion of urate, respectively.

Table 2. Uricemia, uricosuria, fractional excretion of uric acid and rate of tubular phases that govern the renal excretion of uric acid in patients with calcium urolithiasis and hyperuricosuria and control subjects, under purine-free diet.

GROUP	Sex	Pur $\frac{mg}{dl}$	Uur $\frac{mg}{day}$	Cur/Ccr %	Presecretory reabsorption (% of UA_F^a)	Tubular secret. (% of UA_F)	Postsecretory reabsorption (% of UA_S)
Patient							
1	M	4.7	823	10.2	99.9	32.1	68
2	M	4.2	508	6.7	99.5	40.2	84
3	M	4.6	655	7.7	99.5	39.7	82
4	M	6.2	912	12.8	99.8	51.7	76
5	M	3.9	666	8.5	99.6	44.2	82
6	M	4.4	664	8.7	99.9	37.2	77
7	F	2.1	818	15.0	99.1	89.3	72
8	F	1.6	743	25.8	99.2	54.6	41
9	F	3.5	996	33.3	99.1	39.2	56
Controls							
Males in normouricemia (n=10)							
$\overline{X}$		4.8	634	7.8	99.5	38.9	82
S.D.		0.6	91	1.4	0.5	7.4	3
Females in normouricemia (n=7)							
$\overline{X}$		3.4	563	10.5	99.5	39.8	75
S.D.		0.4	140	2.2	0.5	4.5	5
Females in hypouricemia (n=5)							
$\overline{X}$		1.6	217	9.7	99.5	38.3	76
S.D.		0.6	87	4.0	0.5	8.8	7

[a]UA_F: Uric acid filtered. UA_S: Uric acid secreted.

CONCLUSIONS

1. Hyperuricosuria is a frequent finding in patients with recurrent
calcium nephrolithiasis (RCN) and is ussually associated with
other metabolic abnormalities. The pathomechanism which may link
hyperuricosuria and RCN could not be definitely established, but
a precise disgnosis concerning the pathogenesis of hyperuricosuria
may contribute to our understanding of renal stone disease.

2. Dietary habits and tubular transport defects of urate explain the
excessive uric acid excretion of patients with RCN. An alteration
of the renal handling of uric acid may be suspected when hyperuricosuria concurs with hypouricemia; normal serum uric acid
does not exclude a tubular transport defect of urate.

REFERENCES

1. F. L. Coe, Nephrolithiasis: pathogenesis and treatment. Year
   book of Medicine. Year book Publishers Inc. Chigago, London.
   (1978).
2. C. Y. C. Pak, F. Briton, R. Peterson, P. P. Ward, C. Northcurtt,
   M. A. Breslau, J. McQuire, K. Sakhace, S. Bush, M. Nicar,
   D. Norman, and P. Peters, Valoración ambulatoria de la nefroli-
   tiasis: Clasificación, formas de presentación clínica y crite-
   rios diagnósticos, Am. J. Med. (spanish edition) 12: 11 (1980).
3. A. de Vries, and O. Sperling, Inborn hypouricemia due to
   isolated renal tubular defect, Biomedicine, 30: 75 (1979).

RENAL HANDLING OF URIC ACID IN GOUT BY MEANS OF THE PYRAZINAMIDE

AND PROBENECID TESTS

J.G. Puig, F. Mateos, A. Muñoz, G. Gaspar,
T. Ramos, and J. Gijón Baños

Departments of Internal Medicine and Clinical Biopatholo-
gy, Ciudad Sanitaria La Paz, U. Autónoma, Madrid, Spain

INTRODUCTION

The protagonism of the kidney in the pathogenesis of primary
gout is essential because at least three-fourths of all patients
have normal urate production and require an elevated plasma urate
concentration to obtain a normal uric acid excretion[1,2]. Recent
investigations of the urate transport system in the human kidney
have provided evidence for a 4-component model: glomerular filtration, proximal or presecretory reabsorption, tubular secretion,
and postsecretory reabsorption[3]. According to this model, renal
hypouricemia might result from diminished glomerular filtration
rate (GFR), augmented either presecretory or postsecretory reabsorption, or diminished tubular secretion of uric acid. In an
attempt to determine the site of the renal abnormality resulting
in decreased clearance of urate, we examined the tubular phases
that govern uric acid excretion in a gouty population in the states
of hyperuricemia and hypouricemia.

MATERIAL AND METHODS

Renal handling of uric acid was examined in 30 patients with
primary gout (mean age, 44 years), and in 10 normal controls (mean
age, 33 years). A normal GFR, as assessed by means of endogenous
creatinine clearance, was a pre-requisite for being included in the
study. Pyrazinamide (PZA) and probenecid (PB) tests were performed
on each subject in the sate of hyperuricemia (controls were fed
with RNA monosodium salt), and in the sate of hypouricemia (allopurinol, 900 mg/day, for five days). Each pharmacological test was
done in the morning following an overnight fast, and after 24 h
urine collection for uric acid and creatinine clearances. Uric acid

Table 1. Uricemia, uricosuria, and tubular handling of uric acid
        in the states of hyperuricemia and allopurinol-induced
        hipouricemia.

	Pur[a] mg/dl	Uur mg/day	Presecretory reabsorption (% of UA_F)	Tubular secretion (% of UA_F)	Postsecretory reabsorption (% of UA_S)
HYPERURICEMIA		Normal subjets (n=10)			
Mean	10.1	1415	99.7	42.6	68.4
± M.S.E.	0.4	109	0.1	1.8	1.8
Gouty patients with abnormal renal handling of urate (n=26)					
Mean	7.9	487	99.6	27.9[b]	78.2
± M.S.E.	0.2	28	0.1	1.0	1.2
Gouty patients with normal renal handling of urate (n=4)					
Patient nº 1	8.4	603	99.1	40.6	70.5
2	5.8	949	99.1	37.9	79.4
3	6.5	934	99.9	38.7	81.0
4	8.1	995	99.3	43.8	83.5
HYPOURICEMIA		Normal subjects (n=10)			
Mean	2.7	176	99.6	37.2	75.7
± M.S.E.	0.1	14	0.1	1.8	1.7
Gouty patients with abnormal renal handling of urate (n=26)					
Mean	3.0	157	99.7	21.2[b]	73.1
± M.S.E.	0.1	18	0.1	1.0	1.0
Gouty patients with normal renal handling of urate (n=4)					
Patient nº 1	2.9	145	99.8	39.5	70.0
2	3.3	231	99.9	33.1	76.8
3	3.1	196	99.9	35.4	81.1
4	3.4	210	99.7	40.1	84.9

[a]Pur: serum uric acid. UA_F: Filtered uric acid. UA_S: secreted urate.
[b]p 0.01 , with respect to controls and gouty patients (n=4).

and creatinine determinations in serum and urine were assayed enzimatically.

RESULTS

Assuming that PZA completely suppresses tubular secretion of urate, the uric acid filtered load that undergoes proximal reabsrption was in every patient over 99% in both uricemia states, thus excluding the possibility that renal hyperuricemia could be due to increased presecretory reabsorption. The maximum uricosuric response to PB, equated with the minimum secretory rate, was diminished in 26 gouty patients, and was similar to that obtained in controls,in 4 patients, in the states of hyperuricemia and hypouericemia. The percentage of secreted uric acid that is reabsorbed distally in the tubule was normal in each gouty patient in both uricemia states.

Three patients with normal PB tests evidenced, at least once, a 24 h urate excretion over 900 mg/day, but one patients with normal tubular secretion showed a urinary uric acid excretion of 603 mg/day similar to that obtained in other patients with reduced uricosuric response to PB.

CONCLUSIONS

1. When GFR is normal, hyperuricemia of most gouty patients is not due to augmented tubular reabsorption of uric acid (either presecretory or postsecretory reabsorption), but to diminished tubular secretion of urate.

2. The maximum uricosuric response to PB allows a physiopathological classification of primary gout in underexcretors (diminished tubular secretion of urate) and normoexcretors (normal tubular secretion of uric acid).

3. Twenty-four hour urinary urate excretion, when expressed in mg/day or mg/Kg, is valuable to ascertain whether a patient underexcretes uric acid or not, but in a particular case may not differentiate patients with normal renal handling of uric acid from those with reduced tubular secretion of urate.

REFERENCES

1. P. A. Simkin, Uric acid excretion in patients with gout, Ann. Intern. Med. 92: 98 (1980).
2. D. J. Levinson, and L. B. Sorensen, Renal handling of uric acid in normal and gouty subjects: evidence for a 4-component system, Ann. Rheum. Dis. 39: 173 (1980).
3. H. S. Diamond, and J. S. Paolino, Evidence for a postsecretory reabsortive site for uric acid in man. J. Clin. Invest. 52: 1491 (1973).

BEHAVIOUR OF SERUM URATE IN RENAL DISEASE OF VARYING ETIOLOGY

Marco Gonella and Giuliano Mariani

Division of Nephrology of the Hospital of Pisa, and 5th Medical Pathology of the University of Pisa, Pisa (Italy)

INTRODUCTION

Renal failure has long been known to lead to hyperuricemia, which, however, has been reported not to be strictly proportional to the reduction of the glomerular filtration rate, due to the adaptative mechanisms of tubular resorption and to increased intestinal urico-lysis (Rieselbach and Steele, 1975).

Among the renal patients hospitalized in our Unit, a wide range of serum urate levels (sUr) was observed to be associated with similar degrees of renal function impairment. Thus, a retrospective study was undertaken in order to elucidate the various types of correlation between sUr and serum creatinine concentrations (sCr) in patients affected by renal diseases of different etiology.

MATERIAL AND METHODS

The clinical charts of a total of 128 patients affected by renal disease were reviewed. Fifty-eight patients had primary chronic glo-merulonephritis (CGN), and the remaining 70 patients were diagnosed to have a chronic interstitial renal disease (ID) (17 with polycystic kidney, 24 with nephroangiosclerosis, and 29 cases of chronic pyelo-nephritis). The patients of both groups were further divided into three subgroups depending of the degree of renal function impairment: 1) with normal sCr (0.8-1.2 mg/dl); 2) with sCr ranging between 1.3 and 3.9 mg/dl; and 3) with sCr higher than 4 mg/dl. Diagnosis of CGN

had been made in all cases by histological examination of renal
biopsies. A total of 33 patients hospitalized in the same period
and resulting not to be affected by any renal disease served as the
control group. None of the patients considered in this study was
receiving any drug known to interfere with the renal urate excretion.
Sex and age distribution of the patients studied is reported in the
Table.

The sUr and sCr levels were measured by routine autoanalyzer
procedures, with normal values ranging respectively between 2.5-5.5
mg/dl and 0.8-1.2 mg/dl.

RESULTS

Sex distribution was similar in the CGN and ID groups considered
as a whole, even though some differences were present among the subgroups considered (see Table 1). The mean age resulted to be significantly higher in the ID patients than in the CGN and the control
patients, while there were no significant differences as concerns
age between CGN patients and controls.

The mean values of sUr, when related to those of sCr, resulted
to always significantly higher in the CGN group than in the ID group
and in the control patients. In particular, the CGN patients with
normal renal function had sUr significantly higher than both the
controls and the ID patients ($P<0.001$ to $P<0.01$). Serum urate increased sharply with increasing sCr in the CGN patients, while the serum
urate increase in the ID patients with respect to the controls was
significant, in the 1.3-3.9 mg/dl sCr range, for women only ($P<0.001$)
or above 4 mg/dl sCr for both sexes. With correspondingly impaired
renal function, serum urate levels were always significantly higher
in the CGN patients than in the ID patients ($P<0.001$).

In both renal disease groups sUr was directly correlated with
sCr. However, while in the ID patients the correlation was linear,
it was of the logarithmic type in the CGN patients, with sUr increasing steeply and early with respect to the sCr increase (see Fig. 1).
The respective equations and the correlation parameters for the two
groups are as follows: $Y = 6.1594 + 2.1344 \ln X$ ($r = 0.70$; $P<0.001$)
for the CGN patients, and $Y = 3.7351 + 0.6777 X$ ($r = 0.78$; $P<0.001$)
for the ID patients.

Table 1. Sex and age distribution, and corresponding sCr and sUr levels in the patients under study (all data are given as mean values ± 1 Standard Deviation).

Patients and diagnosis	Sex Men	Women	Age (years)	sCr (mg/dl)	sUr (mg/dl)
CGN					
0.8 - 1.2 mg/dl sCr	10	10	32.5 ± 13.9	1.01 ± 0.14	6.12 ± 1.72
1.3 - 3.9 mg/dl sCr	19	13	41.6 ± 14.0	2.15 ± 0.71	7.72 ± 1.12
sCr higher than 4 mg/dl	4	2	36.8 ± 14.7	5.07 ± 0.99	9.50 ± 1.57
Overall CGN patients	33	25	37.9 ± 14.5	2.05 ± 1.32	7.35 ± 1.73
ID					
0.8 - 1.2 mg/dl sCr	3	5	35.9 ± 13.4	1.00 ± 0.19	3.71 ± 0.97
1.3 - 3.9 mg/dl sCr	25	14	52.8 ± 13.1	2.66 ± 0.77	5.70 ± 1.34
sCr higher than 4 mg/dl	7	16	52.1 ± 13.4	6.06 ± 1.54	7.84 ± 1.25
Overall ID patients	35	35	50.6 ± 14.1	3.59 ± 2.10	6.18 ± 1.83
Control patients	13	20	41.3 ± 16.0	0.96 ± 0.15	4.37 ± 1.18

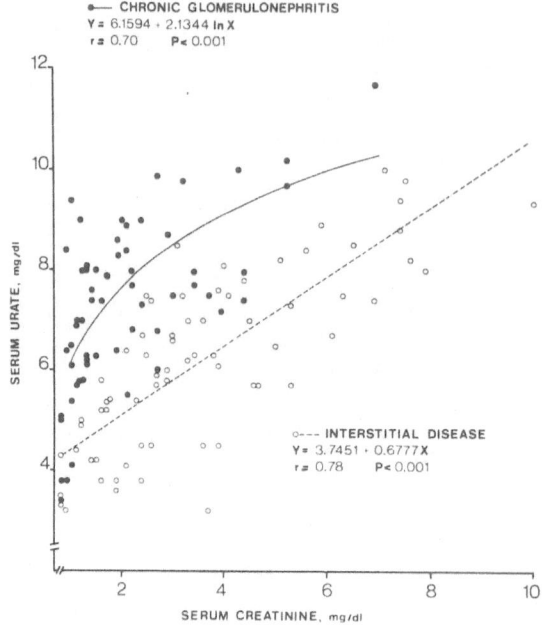

Fig. 1 - Correlations observed between the sCr and the sUr values,
         respectively in the CGN patients and in the ID patients.

DISCUSSION

     Some hypotheses can be put forward to account for the results
obtained in this study. A diffuse tubular damage, as predominantly
occurs in the course of ID, would increase the amount of excreted
urate (by impairing its active resorption at the proximal tubule),
thus preventing in the early stages of kidney involvment the prompt
increase in sUr following the overall reduction of renal function.
The fact that an impaired tubular function occurs also in normal or
slightly impaired renal function may explain the lower sUr values
observed in the ID patients with normal sCr than in the control patients.

     Alternatively (or as an associated mechanism), an urate over-
production at the kidney site in the course of CGN (possibly due to
the glomerular cell hyperplasia) may be postulated. This hypothesis
is consistent with the observation of increased sUr values in CGN
patients with normal renal function, and with the sharp sUr increase
in the early stages of renal function impairment in the patients of
this group.

Some findings previously reported in pre-eclampsia and in eclampsia seem to support this hypothesis. In fact, it is well known that in these conditions sUr levels frankly higher than normal are usually observed; besides a reduced urate tubular excretion due to various factors, a direct correlation has been reported between the degree of the glomerular damage, as evaluated histologically, and the severity of hyperuricemia (Emmerson and Ravenscroft, 1975).

In conclusion, the findings obtained in this study showed that in CGN the values of sUr are higher than those observed in patients with comparable degrees of renal function impairment, but suffering from kidney disorders other than CGN. Further investigations are required to study uric acid metabolism in CGN and in other renal diseases, with the main purpose of clarifying some possible role of hyperuricemia in the sequence of events involved in the pathogenesis of chronic glomerulonephritis.

REFERENCES

Emmerson, B.T., and Ravenscroft, P.J., 1975, Abnormal renal urate homeostasis in systemic disorders, Nephron, 14: 62.
Rieselbach, R.E., and Steele, T.H., 1975, Intrinsic renal disease leading to abnormal urate excretion, Nephron, 14: 81.

HYPOURICEMIA AND MEDULLARY CARCINOMA OF THE THYROID

J.G. Puig, A.F. Mateos, G. Gaspar, E.Mª. Martinez,
T. Ramos, and A. Lesmes

Departments of Internal Medicine and Clinical Biopathology
Ciudad Sanitaria La Paz, Universidad Autónoma, Madrid

INTRODUCTION

Hypouricemia could be the result of diminished uric acid production, excess urate excretion, or a combination of these mechanisms. Recently, hypouricemia has been noted in a few patients with
neoplastic diseases, particularly Hodgkin's disease and pulmonary
tumors[1]. In many cases, renal leak of uric acid was responsible for
the low serum uric acid values due to structural tubulopathies,
inappropriate ADH secretion, or an hypothetical production of a
tumoral uricosuric substance that still remains speculative[2].

We have studied a patient with thyroid carcinoma and renal
hypouricemia that could be attributed to increased calcitonin
secretion.

Case report

A 44-year-old woman had a partial thyroidectomy 14 years-ago
at another Hospital evidencing a thyroid carcinoma of unknown hysto-
llogical type. One year later, left supraclavicular and bilateral
tracheal adenopathies developed and was treated with I^{131}, 5-fluor-
uracil and telecobaltotherapy. Physical examination revealed a left
supraclavicular mass. A biopsy of this mass showed connective tissue
infiltrated by a malignant epitelial tumor. Under normal diet,
serum and 24 h urine sodium, potassium, chloride, calcium, phosphorus
magnesium and osmolarity were normal. Routine urianalysis and ammi-
noaciduria were also normal. Plasma levels of T4, TSH, LH, FSH, and
insulin were normal. Vanilmandelic acid, catecholamines, cyclic AMP,
17-Hydroxysteroids, 17-Ketosteroids, and cortisol in 24 h urine
were normal. Basal plasma ACTH was 1.8 mU/ml (normal values:
2-4 mU/ml).

211

The patient was placed on an essentially purine-free diet and received no medication for 5 days. Uric acid metabolism disclosed: uricemia 2.0 mg/dl; uricosuria 510 mg/day; Cur 15.4 ml/min; Ccr 91 ml/min; fractional excretion of uric acid (Cur/Ccr) 16.9%. Basal plasma calcitonin was 168 pg/ml (normal values: undetectable). A pentagastrin bolus injection[3] of 0.5 g/Kg elevated plasma calcitonin over 1000 pg/ml. Simultaneously, serum uric acid decreased from 2.0 mg/dl to 1.3 mg/dl, and Cur/Ccr increased from 16.9% to 25.7%. A pentagastrin test in two control subjects did not make plasma calcitonin levels detectable, nor did it modify uric acid excretion. Her clinical course was progressively down-hill and she died after several bronchoneumonic episodes and massive tracheal hemorrage. Pyrazinamide and probenecid tests could not be done. Permission for autopsy was denied.

DISCUSSION

Hypercalcitoninaemia is a sine qua non for diagnosing medullary carcinoma of the thyroid (MCT). Pentagastrin injection in patients with MCT usually result in a ten-fold plasma calcitonin increment compared with basal levels. However, elevated calcitonin concentrations have also been observed in neoplasms other than MCT (essentially lung tumors), but the average rise in calcitoninaemia induced by pentagastrin is much lower in these neoplasms (average 20%), allowing a differential diagnosis[4].

The malignant disease of our patient could not be accurately established. Nevertheless, her basal calcitonin level was very high and pentagastrin raised calcitoninaemia, at least, 495%. This fact consistently supports the diagnosis of MCT, instead of ectopic calcitonin secretion.

Several syndromes of renal tubular dysfunction, including hypouricemia, have previously described in patients with malignant diseases. A Fanconi-like syndrome has been observed in several cases of multiple myeloma in association with Bence-Jones proteinuria. It has also been suggested that hypervolemia secondary to inappropriate ADH secretion was responsible for the increased uric acid clearance in two patients with pulmonary tumors. In other cases of paraneoplastic hypouricemia, tumoral secretion of an hypothetic uricosuric substance has been related to disease activity, but the existence of this uricosuric substance remains speculative[2].

Our patient had basal hypouricemia due to renal urate overexcretion. Renal leak of other substances and inappropriate secretion of ADH were ruled out. Pentagastrin injection elevated plasma calcitonin with simultaneous increment of urate fractional excretion and diminution of serum uric acid, lending support to the hypothesis of calcitonin-induced renal hypouricemia. The possible uricosuric action of pentagastrin could not be demostrated in two normal subjects; the fractional excretion of uric acid did not change significantly and calcitoninaemia remained undetectable. These findings can best be interpreted to mean that renal hypouricemia was related to exagerated calcitonin secretion by a MCT, thus causing renal urate overexcretion and low serum uric acid.

According to the four-component model for the renal handling of urate, renal uric acid wasting might result from defects either of of presecretory or postsecretory reabsorption, or enhaced secretion of urate in the tubule[5]. Future studies are required in order to delineate the tubular phases that may account for the uricosuric action of calcitonin.

REFERENCES

1. W. N. Kelley, Hypouricemia, Arthr. Rheum. 18: 731 (1975).
2. A. M. Sanz, F. J. Barbado, J. Mª. Peña, F. Arnalich, J. G. Puig, and J. J. Vázquez, Hipouricemia paraneoplásica en el linfoma de Hodgkin por defecto en la reabsorción postsecretora de ácido úrico, Med. Clin. 76: 307 (1981).
3. R. K. Rude, and F. R. Singer, Comparison of serum calcitonin levels after a 1-minute calcium injection and after pentagastrin injection in the diagnosis of medullary thyroid carcinoma, J. Clin. Endocrinol. Metab. 44: 980 (1977).
4. H Mulder, and W. H. L. Hackeng, Ectopic secretion of calcitonin, Acta Med. Scand. 204: 253 (1978).
5. Y. Tofuku, M. Kuroda, and R. Takeda, Hypouricemia due to renal urate wasting: Two types of tubular transport defects, Nephron. 30: 39 (1982).

PURINE SALVAGE ENZYMES

IN LEISHMANIA DONOVANI AND TRICHOMONAS VAGINALIS

Thomas A. Krenitsky and Richard L. Miller

Wellcome Research Laboratories
Research Triangle Park, NC 27709 USA

INTRODUCTION

The purine salvage pathways of pathogenic protozoa are of special interest because most of these organisms lack the ability to synthesize purines de novo. This absence of an alternative to salvage is the basis for the view that these pathogens might be particularly susceptible to selective chemotherapy with purine analogues and their nucleosides. The preceding paper provides some examples to illustrate the validity of this view. The central point of this paper is the comparison of purine salvage enzymes of two not so distantly related pathogenic protozoa. The basic lesson is that not only do pathogenic protozoa differ from mammals in their purine salvage enzymes, but dramatic diversity among the protozoa themselves can be found.

Protozoa are divided into several major groups. We will be concerned with one group--the flagellates; and more specifically, with two species from this group, Leishmania donovani and Trichomonas vaginalis. The enzyme assays were performed with Sephadex-treated extracts[1] of the free living forms of both parasites. In the case of L. donovani (grown in the laboratory of J. Marr), the free living form (promastigote) is the insect-vector stage in the life cycle. With T. vaginalis (grown in the laboratory of D. Linstead), this form lives in the human host.

RESULTS AND DISCUSSION

Figure 1 compares the levels of phosphoribosyltransfer activities in extracts of Rhesus monkey liver, L. donovani, and T. vaginalis. Magnesium 5-phosphoribosyl-1-pyrophosphate (PRPP) was the phosphoribosyl donor. Leishmania had high levels with adenine, hypoxanthine, guanine, and xanthine relative to those in monkey

215

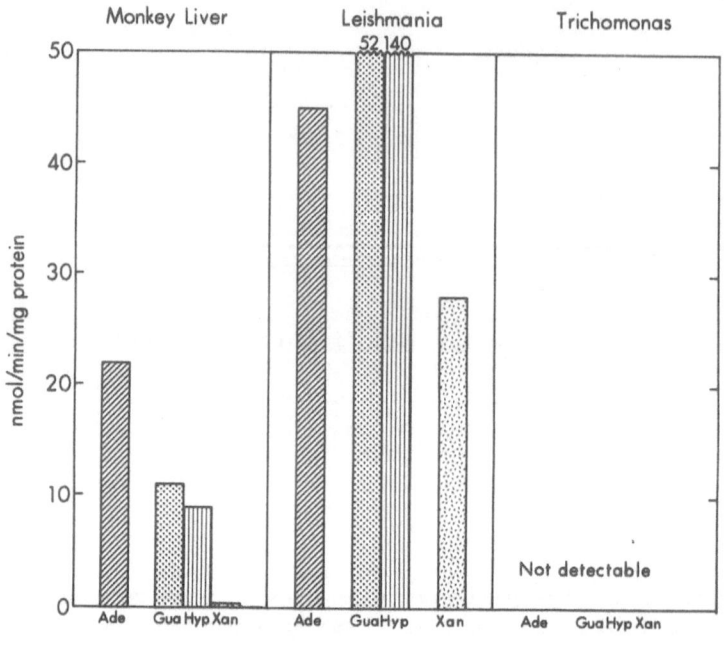

PHOSPHORIBOSYLTRANSFERASES

Figure 1

liver. No activity was detectable with any of these purines in
the trichomonas extracts. However, we were somewhat reluctant to
conclude that trichomonas had no phosphoribosyl transferases. The
negative enzyme data would be more convincing if evidence of an
alternative pathway was obtained. Consequently, the enzymes of
the phosphorylase-kinase pathway were studied. Figure 2 compares
the levels of ribosyltransferase (nucleoside phosphorylase) activities from these sources. Ribose-1-phosphate was used as the
ribosyl donor. With the monkey liver extracts, very little activity
toward adenine but appreciable levels with hypoxanthine, guanine,
and xanthine were detected. With L. donovani the converse was
found, the only appreciable activity was with adenine. With T.
vaginalis, the activities with adenine and guanine were very high.
The activity with hypoxanthine was moderate and none was detectable
with xanthine.

It was previously reported[2] that L. donovani had high levels
of nucleoside hydrolase activities toward all the common purine
nucleosides except adenosine. It was clear that if the 'phosphorylase-kinase' pathway was the only route of purine salvage in
trichomonas the presence of such hydrolases would be difficult to
explain. Figure 3 shows the nucleoside hydrolase levels of leishmania and trichomonas. With leishmania, the levels are high with

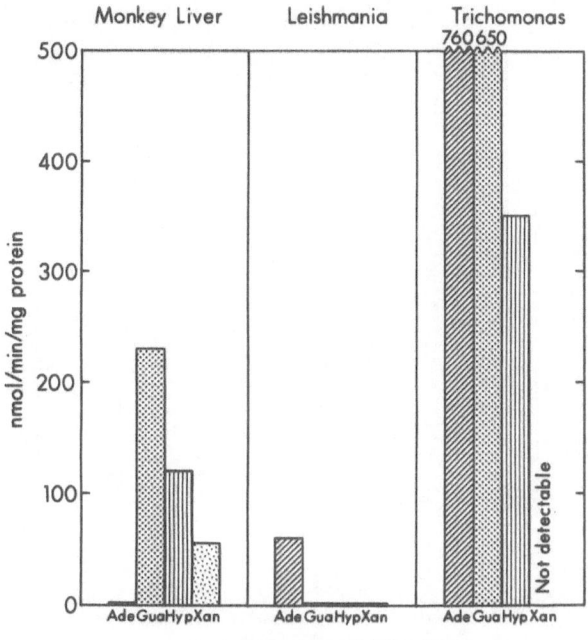

RIBOSYLTRANSFERASES

Figure 2

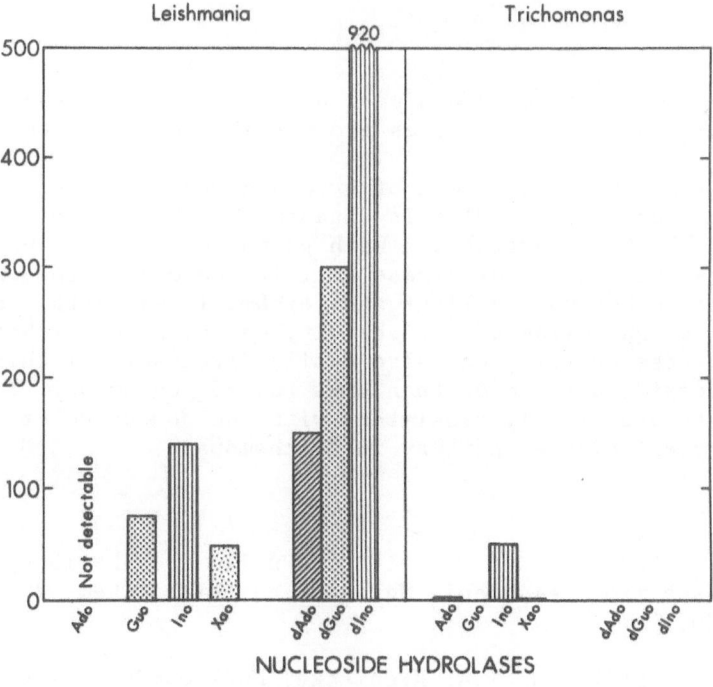

NUCLEOSIDE HYDROLASES

Figure 3

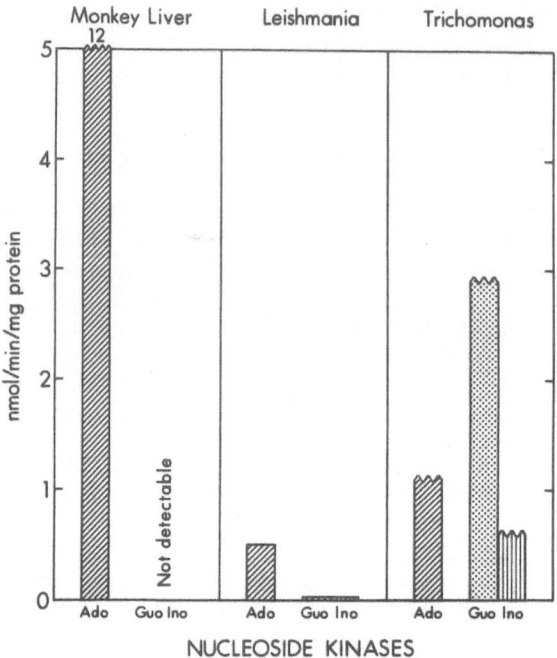

Figure 4

all nucleosides except adenosine. With trichomonas, the levels of
nucleoside hydrolase activities are very low except with inosine.

The last essential piece of evidence was the presence of
kinases (Figure 4). The levels of adenosine kinase were appreciable in all three extracts. With guanosine and inosine as
substrates, the levels of kinase activity were very low or not
detectable in the monkey liver and leishmania extracts. However,
the level of guanosine kinase activity was high in trichomonas.
Inosine kinase activity was also easily detectable in this extract
but was considerably lower than that toward guanosine. These
enzyme data are clearly consistent with the dominance of the
'phosphorylase-kinase' pathway in trichomonas.

REFERENCES

1.   J.V. Tuttle and T.A. Krenitsky, Purine Phosphoribosyltrans-
      ferases from Leishmania donovani, J. Biol. Chem. 255:909
      (1980).

2.   G.W. Koszalka and T.A. Krenitsky, Nucleosidases from
      Leishmania donovani, J. Biol. Chem. 254:8185 (1979).

HYPOXANTHINE METABOLISM BY HUMAN MALARIA INFECTED ERYTHROCYTES:

FOCUS FOR THE DESIGN OF NEW ANTIMALARIAL DRUGS

Horace K. Webster*, William P. Wiesmann,
Marvin D. Walker, Teresa Bean, and June M. Whaun

*Dept. of Immunobiology, Armed Forces Research Institute
of Medical Sciences, Bangkok, Thailand. Dept. of Hematology, Walter Reed Army Institute of Research, Washington, D.C. 20012

INTRODUCTION

It is the blood stage, or intraerythrocytic (IE) form, of the
malaria parasite that produces the major clinical and pathological
features of malaria infection.

Recently, we began the systematic study of purine metabolism
of the IE form of the major human malaria pathogen, Plasmodium
falciparum, using newly developed techniques for continuous
erythrocyte culture (1,2) and novel chromatographic procedures
that permit comprehensive quantitative examination of purine
intermediary pathways (3). Purines are essential to the rapidly
proliferating malaria parasite for nucleic acid synthesis, energy
metabolism and as enzyme cofactors (4).

In this study we report on the metabolism of hypoxanthine by
parasitized erythrocytes in vitro. Emphasis was placed on identified differences in host and parasite purine pathways (5) and on
how through use of specific inhibitors a rational approach to the
selective destruction of the IE malaria parasite can be achieved.

METHODS

The FCR-3 strain of P. falciparum was maintained in continuous erythrocyte culture using described techniques (5,6).

Biochemical studies were done as follows. The contents of

48 hour growth cultures were pooled and aliquots transferred to individual flasks to provide a uniform set of test cultures containing 3 ml of parasitized RBC suspension with an adjusted hematocrit of 12%. Inhibitors (5×10^{-5} M) and (^{3}H) hypoxanthine (specific activity 3 Ci/mmol) were added to the test cultures and incubation continued for 2.5 hours. Perchloric acid (PCA) extracts were prepared on the culture contents to yield acid-soluble (purines) and acid-insoluble (nucleic acid) fractions. Details of these methods have been described elsewhere (5,6).

Purine compounds were quantitated by high-performance liquid chromatography (HPLC). These procedures permited the simultaneous measurement of concentration and radioactivity of a purine compound separated by either reversed-phase (nucleosides and base) or anion-exchange (nucleotides) gradient HPLC. These HPLC procedures have been described in detail (3).

RESULTS

Incorporation of (^{3}H) Hypoxanthine into Purine Nucleotides of Malaria Infected Erythrocytes

Previous studies in our laboratory have demonstrated the pathways for purine salvage and interconversion associated with purine nucleotide synthesis in P. falciparum infected erythrocytes (PRBC) (5). The malaria parasite cannot synthesize purines de novo (4). Hypoxanthine appears to be the preferred substrate for synthesis of both adenosine and guanosine nucleotides, although adenine and guanine can be used (5). Figure 1 shows a representative nucleotide profile for PCA extracted PRBC following incubation with (^{3}H) hypoxanthine. Label from hypoxanthine was incorporated into IMP and then into both adenylates and guanylates.

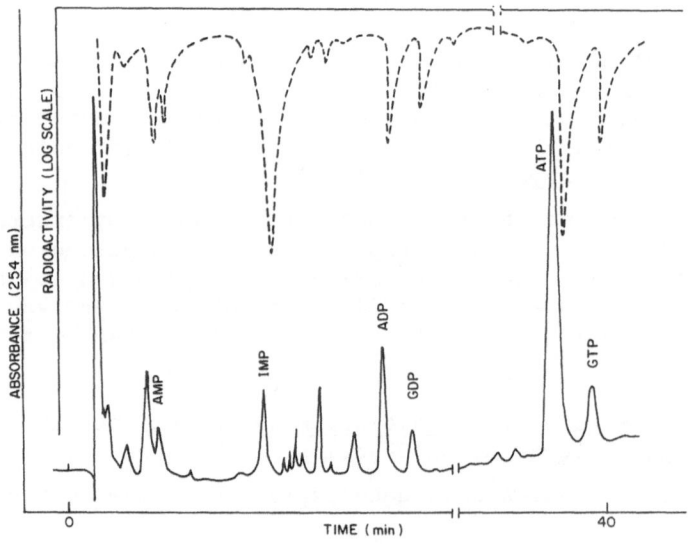

Figure 1

PRBC synthesized pyridine nucleotide (NAD) from the labelled adenylate pool. Unparasitized red cells (RBC) synthesize IMP from hypoxanthine at a slower rate than PRBC (34 pmoles/min/ml RBC compared to 196 pmoles/min/ml RBC for PRBC, HCT = 12%, 4.8% parasitemia). Control RBC cannot synthesize adenylates from hypoxanthine and produce only small amounts of guanylates (5,6).

Effect of Specific Inhibitors on Purine Nucleotide Synthesis

Table 1 shows the effect of various purine inhibitors on incorporation of (^{3}H) hypoxanthine into purine nucleotides of PRBC. Control cultures and inhibitor treated cultures were all prepared from a common pooled source of malaria infected RBC and, therefore, have the same initial parasitemia (4.8% PRBC) and hematocrit (12%).

Hadacidin (N-formyl hydroxy-aminoacetic acid), which is known to inhibit adenylosuccinate synthetase (7), blocked synthesis of adenosine nucleotides from IMP (Table 1). There was a significant decrease (83%, p < .001; Student's T-test for unpaired values) in newly synthesized ATP and in total adenylates (ΣA). The concentration of ATP and the level of the adenylate pool were not decreased. There was a decrease in labelled guanylate in PRBC exposed to hadacidin.

Alanosine (L-2-amino-3-(N-hydroxy, N-nitrosamino) propionic acid), another inhibitor of adenylosuccinate synthetase (7), did not appear to interfere with synthesis of adenosine nucleotides by malaria infected erythrocytes (Table 1). There was, however, an unexpected decrease in labelled GTP and total guanylates (ΣG).

Table 1. Effect of Inhibitors on Incorporation of (^{3}H) Hypoxanthine Into Purine Nucleotides of Human Intraerythrocytic Malaria Parasites

CONDITION		ATP	GTP	IMP	ΣA[f]	ΣG[g]	ΣA/ΣG
CONTROL	nmoles[a]	1562 ± 31[d]	93 ± 2	211 ± 5	1783	109	16
	radioactivity[b]	4840 ± 126	316 ± 12	38170 ± 202	6620	524	13
	specific activity[c]	3	3	181			
HADACIDIN	nmoles	1493[e]	89	342	1701	106	16
	radioactivity	848	108	42514	1448	166	9
	specific activity	0.6	1	124			
ALANOSINE	nmoles	1606	99	256	1827	114	16
	radioactivity	4652	154	31460	6282	312	20
	specific activity	3	2	123			
BREDININ	nmoles	1603	81	301	1846	93	20
	radioactivity	5474	64	39772	7062	70	101
	specific activity	3	0.8	132			
MYCOPHENOLIC ACID	nmoles	1712	82	235	1969	96	21
	radioactivity	3840	48	38990	6162	80	77
	specific activity	2	0.6	166			

[a]Nanamoles per ml RBC.

[b]Integrated radioactive counts (area) per chromatography peak.

[c]Specific activity = radioactivity per nmole.

[d]Values are mean ± SEM, n = 3.

[e]Values are averages of duplicate experiments in which the average deviation of individual measurement from the mean was less than 8 percent.

[f]ΣA (total adenylates) = AMP + ADP + ATP + NAD.

[g]ΣG (total guanylates) = GMP + GDP + GTP.

Bredinin (4-carbamoyl-1-β-D-ribofurnyosyl-imidazolium-5-olate), and inhibitor of IMP dehydrogenase (8), interferred with synthesis of guanosine nucleotides from IMP by PRBC (Table 1). There was a decrease (80%, p < .001) in newly synthesized GTP and in ΣG. The decrease in GTP concentration was also significant (13%, p < .005). There was no apparant effect of bredinin on adenylates of PRBC.

Mycophenolic acid (6-(4-hydroxy-6-methoxy-7-methyl-3-oxo-5-pathalanyl)-4-methyl-4-hexenoic acid) also shown to inhibit IMP dehydrogenase (9) effectively interferred with the synthesis of guanosine nucleotides from IMP by PRBC (Table 1). The decrease (85%) in labelled GTP and total guanylates was significant (p < .001). There was also a moderate (p < .01) decrease in the concentration of GTP and in the level of the guanylate pool. Unlike bredinin mycophenolic acid produced a decrease in newly synthesized ATP (21%, p < .005).

Effect of Inhibitors on Nucleic Acid Synthesis

Table 2 shows the effect of each purine inhibitor on incorporation of (^{3}H) hypoxanthine into nucleic acids (acid-insoluble PCA fraction) of malaria infected erythrocytes. Two clinically proven antimalarial drugs, chloroquine (7-chloro-4-(4'-diethylamino-methyl-butylamino)-quinoline) and mefloquine (α-(2-piperidyl)-2, 8-bis (trifluoro-methyl)-4-quinolinemethenol hydrochloride), were included for comparison. Hadacidin, bredinin and mycophenolic acid all produced significant (p < .005) decreases in nucleic acid synthesis by PRBC as measured by incorporation of (^{3}H) hypoxanthine. Alanosine had no demonstrable effect.

Table 2.   Effect of Various Inhibitors on Incorporation of (^{3}H) Hypoxanthine Into Nucleic Acids of Malaria Infected Erythrocytes

Inhibitor[a]	Inhibition (%)
Control[b]	–
Hadacidin	73
Alanosine	0
Bredinin	43
Mycophenolic Acid	67
Chloroquine	79
Mefloquine	76

[a]All inhibitors (5 x 10^{-5} M)

[b]Control (11496 dpm, n = 3)

DISCUSSION

These studies confirm the importance of hypoxanthine as a precursor for synthesis of both adenosine and guanosine nucleotides by malaria infected erythrocytes.

Hadacidin, bredinin and mycophenolic acid, each shown to effect specific purine enzymes in other types of cells acted predictably to disrupt purine nucleotide synthesis by PRBC. The lack of a detectable effect on adenylate synthesis by alanosine may be due to an inability to form the active metabolite, L-alanosyl-AICOR, which requires an active de novo purine pathway (10).

Each effective purine inhibitor through disruption of nucleotide synthesis interferred with malaria parasite nucleic acid synthesis and thus parasite growth. We have shown in detailed studies with bredinin that the disruption in nucleic acid synthesis due to a specific block in guanosine nucleotide production is directly correlated with parasite killing (6). Other investigators have also established that radiolabelled nucleic acid precursors are reliable indicators of malaria parasite growth (11,12).

These studies identify biochemical targets associated with the malaria parasites' metabolism of hypoxanthine which are essential for synthesis of purine nucleotides and nucleic acids. Specific focus on these unique features of parasite purine metabolism and the classes of inhibitors effective against them could lead to the design of new antimalarial drugs.

REFERENCES

1. Trager W, Jensen J (1976). Science 193:673-675.
2. Haynes J, Diggs C, Hines F, Desjardins R (1976). Nature 263:767-769.
3. Webster H, Whaun J (1981). Journal of Chromatography 209: 283-292.
4. Sherman I (1979). Microbiological Review 43:453-495.
5. Webster H, Whaun J (1981). Progress of Clinical and Biological Research 55:557-573.
6. Webster H, Whaun J (1982). Journal of Clinical Investigation (In press)
7. Gale R, Smith A (1968). Biochemical Pharmacology 17: 2495-2498.
8. Sakaguchi K, Tsuijino M, Yoshigaiva M, Mizuno K, Haynao K (1975). Cancer Research 35:1643-1648.
9. Franklin T, Cook J (1969). Biochemical Journal 113:515-524.
10. Tyagi A, Cooney D (1980). Cancer Research 40:4390-4397.
11. McCormick G, Canfield C (1972) Proceeding Helminthological Society (Washington) 39:292-297.
12. Desjardins R, Canfield C, Haynes J, Chulay J (1979). Antimicrobial Agents and Chemotherapy 16:710-718.

ADENOSINE DEAMINASE IN MALARIA INFECTION: EFFECT OF

2'-DEOXYCOFORMYCIN IN VIVO

Horace K. Webster*, William P. Wiesmann, and
Charles S. Pavia

*
*Department of Immunobiology, Armed Forces Research
Institute of Medical Sciences, Bangkok, Thailand
Department of Hematology, Walter Reed Army Institute of
Research, Washington, D.C.

INTRODUCTION

We have observed a dramatic increase in erythrocyte (RBC)
adenosine deaminase (ADA: E.C. 3.5.4.4) activity during malaria
infection of the rhesus monkey. Previous studies in our laboratory
using P. knowlesi infected rhesus monkeys showed cyclic changes in
RBC purine metabolites—particularly ATP and GTP—that were
associated with the stage of parasite schizogony (1).

Purine nucleotides are required by the rapidly proliferating
malaria parasite primarily for nucleic acid synthesis and energy
metabolism. The malaria parasite cannot synthesize purines de novo
and depends for its intraerythrocytic (IE) growth and development
on salvage of purine bases from the host RBC and extracellular
environment (2). We have shown with P. falciparum, in vitro, that
hypoxanthine is an essential purine base precursor for parasite
synthesis of adenosine and guanosine nucleotides (3). Whether
hypoxanthine is the malaria parasites preferred substrate in vivo
is not known.

An increase in ADA activity, however, is an obvious means for
production of IE hypoxanthine. Increased availability of hypoxanthine would be a natural consequence of adenosine catabolism in the
mature erythrocyte (viz:adenosine $\xrightarrow{ADA}$ inosine $\xrightarrow{PNP}$ hypoxanthine)
since this cell lacks the enzyme xanthine oxidase (4). Conversely,
inhibition of ADA activity could act to deprive the rapidly
growing IE malaria parasite of a readily accessible hypoxanthine
pool for purine nucleotide synthesis.

It was, therefore, of interest to determine whether specific inhibition of adenosine deaminase activity in vivo using the tight binding inhibitor, 2'-deoxycoformycin, interferred with the malaria parasites' IE growth and development.

METHODS

Adult male rhesus monkeys (Macaca mulatta) were experimentally infected with the simian malaria parasite, Plasmodium knowlesi. Parasitemias were expressed as percent of parasitized erythrocytes (% PRBC). Samples of heparinized whole blood were collected at selected times during the IE infection cycle and at various levels of parasitemia. Each animal served as its own uninfected control. Lysates were prepared from washed RBC and perchloric acid extracts from whole blood by described methods (1).

The ADA assay was done by a radiochemical method using an automated HPLC system (5). ADA activity was measured by the conversion of (^{14}C) adenosine to (^{14}C) inosine. The reaction was linear with time and protein concentration. Specific activity was expressed as nanomoles/min/mg of protein.

Purine nucleotides were determined by an anion-exchange gradient HPLC method (6). This method separates all major purine nucleotides as well as the ribonucleotide from the deoxyribo-nucleotide form.

Administration of 2'-deoxycoformycin (Warner-Lambert Pharmaceuticals) to malaria infected rhesus monkeys was by a single 250 ug/kg (i.v.) dose. Evaluation of the drugs effect on parasitemia and parasite morphology was by microscopic examination of Giemsa stained thin blood smears.

RESULTS AND DISCUSSION

Table 1 shows the change in erythrocyte ADA level of P. knowlesi infected rhesus with moderate parasitemias. There was a 3.6 fold increase in RBC ADA activity of infected rhesus with a mean parasitemia of 6.2% PRBC. Several animals were serially sampled at increasing parasitemias and ADA levels were observed to increase in direct proportion to the number of RBC parasitized. A similar observation has been made for P. falciparum in culture (7). Malaria infected RBC have been shown by starch gel electrophoresis to contain a distinct parasite ADA (7,8). At high parasitemia (> 15% PRBC) considerable amounts of ADA activity were observed in the infected animals circulation (Monkey L901 showed a 17 fold increase in erythrocyte ADA activity with a 23% parasitemia).

It is apparant, therefore, that IE malaria parasite growth and proliferation is associated with an increase in PRBC ADA activity.

Table 1. Adenosine Deaminase Activity in Malaria (P. Knowlesi) Infected Erythrocytes of Rhesus Monkeys

Condition	Adenosine deaminase activity (nanomoles/min/mg. protein)
Uninfected (control) Erythrocytes	1.91 ± 0.18[1]
Infected[2] Erythrocytes	6.95 ± 0.70 $(p < .005)$[3]

[1] Mean $\pm$ SEM (n = 12)

[2] Parasitemia (6.2% $\pm$ 0.9; n = 12)

[3] Student's T-test for paired values

Table 2 shows that a single i.v. dose of 2'-deoxycoformycin effectively inhibited erythrocyte ADA activity in malaria infected monkeys at 6 and 24 hours. The parasitemia was decreased at 6 hours and continued to fall over the ensuing 24 hours. Microscopic examination of 6 hour 2'-deoxycoformycin treated PRBC revealed parasite nuclear and cytoplasmic deterioration. By 24 hours the majority of PRBC contained degenerate parasite forms.

2'-Deoxycoformycin thus appears to produce a potent antimalarial effect in vivo.

Table 2. Adenosine Deaminase Activity in Malaria (P. Knowlesi) Infected Rhesus Monkeys Treated with Deoxycoformycin

Monkey	Adenosine deaminase (nanomoles/min/mg. protein) Uninfected	Infected	Deoxycoformycin treated[1] (6 h)	(24 h)
M 032	1.47	6.26 (8.2)[2]	0.12 (4.3)	0.69 (1.8)[3]
M 025	1.07	6.36 (6.8)	0.65 (3.0)	0.62 (2.0)[3]
M 189	2.13	8.19 (6.1)	0.37 (1.6)	0.23 (0.1)[3]
Mean $\pm$ SEM (n = 3)	1.56 ± 0.31	6.94 ± 0.63 (6.7 ± 0.89)	0.38 ± 0.15[4] (3.0 ± 0.78)	0.51 ± 0.14[4] (1.3 ± 0.60)

[1] Single i.v. dose of deoxycoformycin (250 ug/kg)

[2] Value in parenthesis is parasitemia (% PRBC)

[3] Remaining malaria infected erythrocytes showed parasite nuclear and cytoplasmic deterioration

[4] $(p < .01)$

There are three plausible mechanisms to explain malaria parasite killing following in vivo deoxycoformycin inhibition of ADA. First, there could be an IE deficiency of the catabolite hypoxanthine which is needed by the parasite for nucleotide synthesis. Second, there may be direct toxicity to critical parasite enzymes from accumulated adenosine or deoxyadenosine. Increased levels of 2'-deoxyadenosine, for example, have been shown to irreversibly inactivate S-adenosylhomocysteine hydrolase due to accumulation of S-adenosylhomocysteine which inhibits methyltransferase reactions (9). Third, there could be an accumulation of deoxyadenosine triphosphate, dATP, such as has been observed in ADA associated severe combined immunodeficiency disease (SCID) (10). DeoxyATP could act to inhibit ribonucleotide reductase and thereby interfere with DNA synthesis (11). In line with this third mechanism we observed an accumulation of both dADP and dATP in nucleotide profiles (24 hr) of PRBC following in vivo 2'-deoxycoformycin treatment of the P. knowlesi infected rhesus monkeys.

Finally, the mechanism for deoxycoformycin's antimalarial activity in vivo could be a combination of all three stated mechanisms.

It is apparant that parasite adenosine deaminase is a potential metabolic target for the design of new antimalarial chemotherapy. Work is currently under way in our laboratory to more fully understand the action of 2'-deoxycoformycin on the malaria parasite and the possible role of ADA in cellular development and proliferation.

REFERENCES

1. Webster HK, Haut MJ, Martin LK, Hildebrandt PK (1982). International Journal for Parasitology 12:75-79.
2. Sherman IW (1979). Microbiological Reviews 43:453-495.
3. Webster HK, Wiesmann WP, Walker MD, Bean T, Whaun JM (1982). Published in this symposium.
4. Bishop C (1960). Journal of Biological Chemistry 235:3228-3232.
5. Webster HK, Wiesmann WP (1982). Manuscript in preparation.
6. Webster HK, Whaun JM (1981). Journal of Chromatography 209: 283-292.
7. Prichard JS, Haynes JD, Haut MJ (1979). Clinical Research 27:480A.
8. Thaithong S, Sueblinwong T, Beale GH (1981). Transaction of the Royal Society of Tropical Medicine and Hygiene 75:268-273.
9. Hershfield MS (1979). Journal of Biological Chemistry 254:22-25.
10. Donofrio J, Coleman MS, Hutton JJ, Daoud A, Lampkin B, Dyminski J (1978). Journal of Clinical Investigation 62:884-886.
11. Cohen A, Hirschhorn R, Horowitz SD, Rubinstein A, Polmer SH, Hong R, Martin DW (1978). Proceedings National Academy of Sciences USA 75:472-476.

In conducting the research described in this report, the investigators adhered to the "Guide for Laboratory Animal Facilities and Care", as promulgated by the Committee on the Guide for Laboratory Animal Facilities and Care of the Institute of Laboratory Animal Resources, National Academy of Science-National Research Council.

PYRAZOLOPYRIMIDINE METABOLISM IN LEISHMANIA: AN OVERVIEW

J. Joseph Marr

Department of Medicine
University of Colorado
Health Sciences Center
4200 East 9th Avenue
Denver, Colorado 80262

The hemoflagellates that infect man have proven recalcitrant
to effective chemotherapy. The specificity of antimicrobial
therapy, so useful in the treatment of diseases caused by bacteria,
is lacking in diseases caused by these protozoans which have adapted
themselves to survival within mammalian cells. This is evinced by
the significant toxicity of the agents used in their treatment. A
high therapeutic index for a given antimicrobial usually requires a
specific difference in metabolism between the host and the parasite.
Such a difference appears to exist in the ability of organisms of
the genera Leishmania and Trypanosoma to metabolize the pyrazolopyrimidine nucleus as though it were a purine base.

PARASITOLOGY

There are three disease syndromes included under the general
title of leishmaniasis. These are visceral, cutaneous, and muco-
cutaneous leishmaniasis. They are clinically distinct and each has
a definite geographical distribution, but all are associated with
the parasite of the genus Leishmania.

The life cycle of these organisms involves an alternate
existence in a vertebrate host and an insect vector. The natural
hosts, other than man, include domestic and wild mammals. The in-
vertebrate vectors are sand flies, of the genus Phlebotomus, which
take up the parasites during a blood meal on the mammalian host.
The organisms taken up by the insect develop in the midgut and in
the infective forms migrate forward to the pharynx and buccal cavity.

231

During subsequent attempts of the sand fly to ingest blood they are
regurgitated into the bite wound. Having regained access to a mamm-
malian host, the organisms are taken up by the macrophages. These
amastigotes reproduce within the macrophages, cause the cells to
rupture, and the free parasites invade other macrophages or are taken
up by a sand fly to perpetuate the cycle.

   Much of our investigation has been done using the promastigote
form of three species of leishmania: <u>L. donovani</u>, <u>L. braziliensis</u>,
and <u>L. mexicana</u>. These can be grown <u>in vitro</u> in liquid culture and
are analogous to the form found in the insect vector. Most metabolic
data apply to the intracellular form (amastigote) as well.

PYRAZOLOPYRIMIDINE COMPOUNDS

   Pyrazolopyrimidines are purine analogues in which the nitrogen
and carbon of the imidazole ring are inverted. The best known member
of this group, allopurinol (4-hydroxypyrazolo(3,4-d)pyrimidine; HPP),
is a structural analogue of hypoxanthine. Its major metabolic conversions in mammalian cells are to oxipurinol, <u>via</u> xanthine oxidase,
and to allopurinol ribonucleoside (HPPR)[1].

<u>Biological and Biochemical Effects of Allopurinol in Leishmania</u>
   The principal feature which distinguishes the metabolism of
allopurinol in leishmania is the rapid formation of allopurinol ribonucleoside monophosphate (HPPR-MP). This can be demonstrated by grow
ing <u>L. donovani</u> in the presence of $(6-^{14}C)$-allopurinol and analyzing
the perchloric acid-soluble extracts by high performance liquid
chromatography (HPLC). It is the predominant radioactive material in
the elution profile. The chromatogram reveals three other radiolabeled compounds. The first radioactive metabolite, which appears
ahead of HPPR-MP, is 4-aminopyrazolopyrimidine ribonucleoside monophosphate (APPR-MP). The second is between ADP and GDP and the third
between ATP and GTP. These are 4-aminopyrazolopyrimidine ribonucleoside diphosphate (APPR-DP) and triphosphate (APPR-TP), respectively.
When the RNA from these cells is analyzed, 4-aminopyrazolopyrimidine
(APP) can be found in the RNA[2].

   The enzymology of these interconversions was suggested by the
metabolic studies; that is, activation of the base to the ribonucleotide of allopurinol and subsequent amination of that ribonucleotide
to 4-aminopyrazolopyrimidine ribonucleotides. The activation of
allopurinol to allopurinol ribonucleotide is mediated by the hypoxanthine-guanine phosphoribosyltransferase (HGPRTase). Although the
K'_m of the HGPRTase for allopurinol is much higher than for its
normal substrate (K'_m Hyp $= 8$ uM; K'_m HPP $= 230$ uM), allopurinol is
acted upon by this enzyme since it is concentrated almost 10-100-
fold into the parasite; it has an intracellular concentration of
about 0.5-1mM[2].

Amination of HPPR-MP

Subsequent investigations of the adenylosuccinate synthase of
L. donovani, have demonstrated that HPPR-MP, in the concentrations
in which it is found within the parasite, can serve as a substrate
for this enzyme[3]. Although the kinetic constant for HPPR-MP is very
high (K'_m = 340 uM; V'_{max} = 1%), it is well within the concentration
achieved in the cell and is about 100-fold higher than the concentration of inosinic acid (IMP), the normal substrate for this enzyme
(K'_m = 12 uM; V'_{max} = 100%). The intracellular S/K'_m ratio of 3-6
for HPPR-MP is similar to that for IMP (S/K'_m = 3) and suggests that
HPPR-MP can compete favorably with the normal substrate. The enzymological data are consistent with the metabolic studies which show that
a large concentration of HPPR-MP (1-2 mM) is found within the cell;
APPR-MP is formed very slowly and in concentrations about one-tenth
that of its precursor. The rate-limiting step is the formation of
succino-APPR-MP by the adenylosuccinate synthetase. This conversion
does not occur in mammalian tissues[4]. The same metabolic sequences
have been shown to exist in the intracellular parasitic form of L.
donovani[5], in the in vitro culture form, the blood stream and intracellular forms of Trypanosoma cruzi[6,7], the causative agent of American trypanosomiasis, and in the African trypanosomes T. brucei
and T. rhodesiense[8].

Metabolism of Allopurinol Ribonucleoside

Since allopurinol is converted rapidly in man to oxipurinol
(60 per cent of the total pyrazolopyrimidines) and allopurinol-1-
ribonucleoside (10 per cent), we tested these compounds for their
antileishmanial effects. Oxipurinol was relatively ineffective but
HPPR was very active against all three Leishmania sp [9]. The promastigotes of L. braziliensis and L. donovani are about 100-fold more
sensitive to HPPR than to HPP. For L. mexicana, the toxicity of
HPPR and HPP are approximately the same.

Leishmania promastigotes convert HPPR to allopurinol ribonucleotide. This compound is converted further to the APP ribonucleotides
as described previously (Fig. I). The enzyme involved is a nucleoside
phosphotransferase[9]. The stability of HPPR is an important factor
in its direct phosphorylation to the nucleotide. Natural nucleosides are rapidly degraded to their respective bases [10].

OTHER PYRAZOLOPYRIMIDINES

A review of the information obtained to this point in the investigation indicated there were several promising avenues for
exploration: the pyrazolopyrimidine structure was accepted by
enzymes of purine metabolism in these organisms; the pyrazolopyrimidine bases could be activated by the appropriate PRTases; the
ribonucleoside could be activated by another route; the ribonucleoside was stable to nucleoside cleaving activities of these cells.
Thus, both pyrazolopyrimidine bases and ribonucleosides could be
tested. Many analogues were investigated for their activities

against leishmania and three emerged as promising compounds. They
are the thio- analogues of HPP and HPPR, 4-thiopyrazolo(3,4-d)-
pyrimidine (TPP) and its ribonucleoside (TPPR), and 4-hydroxypyrazolo-
(4,3-d)pyrimidine ribonucleoside (formycin B;FORB).

TABLE I

Activities of Pyrazolopyrimidines against Leishmania in vitro

Organism	HPP	TPP	Compound (uM) HPPR	TPPR	FORB
L. donovani	300	100	0.3	2	2
L. braziliensis	50	20	3	4	NT
L. mexicana	40	110	40	30	NT

The concentrations listed are those which inhibit growth by 90%.

Metabolism of TPP and TPPR
        TPP and its ribonucleoside are effective in vitro against the
intracellular and extracellular forms of L. donovani and the extracellular forms of L. braziliensis and L. mexicana. These thio-
analogues have about the same activities as HPP and its ribonucleoside, HPPR, except in the case of L. mexicana where HPP and HPPR
are of equal potency (Table I).

        The promastigotes of L. donovani were incubated for 24 hours
at a concentration of about 10^8 cells/ml in the presence of
$(4-^{35}S)$TPP, $(4-^{35}S)$TPPR, and $(U-^{14}C-ribose)$TPPR. A radioactive
metabolic product was identified which had a retention time and
ultraviolet absorption spectrum identical to authentic 4-thio-
pyrazolo(3,4-d)pyrimidine ribonucleoside-5'-phosphate (TPPR-MP).
There was no radioactivity in the di- or triphosphate region of the
chromatogram, indicating that the organisms did not phosphorylate
TPPR-MP to aminopyrazolopyrimidine ribonucleotides[11].

        Enzymatic investigations were undertaken to clarify the meta-
bolism of these compounds. An earlier study showed that TPP was a
substrate for the HGPRTase[12]. Kinetic studies performed with adenylosuccinate synthetase revealed that the possible product, succino-
APPR-MP, was not formed. The inability of TPPR-MP to serve as a substrate of this enzyme corroborated the in vivo studies which did not
detect the formation of APPR-MP. The ribonucleoside of thiopurinol
was phosphorylated by a nucleoside phosphotransferase partially
purified from promastigotes of L. donovani[11]. The kinetic data were
similar to those found previously for the ribonucleoside of allopurinc

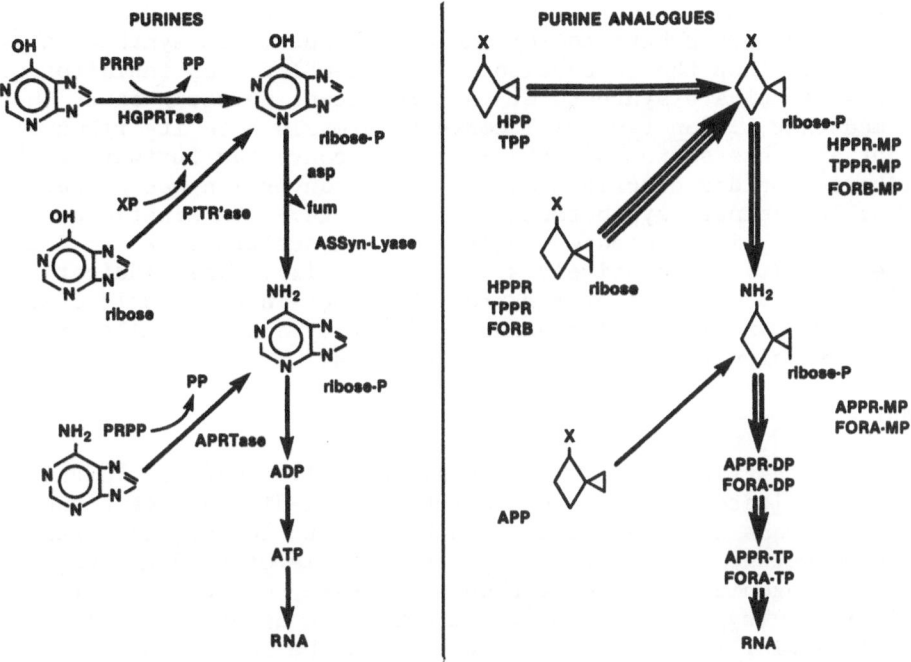

Figure I. Comparison of purine and pyrazolopyrimidine metabolism in
          Leishmania

The left panel indicates normal purine salvage pathways. The panel
on the right summarizes the metabolism of pyrazolopyrimidine analogues.
The X in the 4 position indicates the possibility of substituting
the following groups: -OH, -SH, -NH^2. The pyrazolopyrimidine
abbreviations are given in the text. The enzymes which mediate the
pyrazolopyrimidine conversions are those listed in the left panel.
The abbreviations are as follows: HGPRTase = hypoxanthine-guanine
phosphoribosyltransferase; P'TR'ase = nucleoside phosphotransferase;
APRTase = adenine phosphoribosyltransferase; ASSyn-Lyase = adenylosuccinate synthetase-lyase.

Metabolism of Formycin B

Another pyrazolopyrimidine, formycin B, has been shown to be active against L. donovani (Table I) and L. mexicana[13]. Carson and Chang demonstrated that this compound was phosphorylated to the ribonucleotide but was not converted to any APP ribonucleotides since it did not serve as a substrate for the adenylosuccinate synthetase[13]. They did show that the ribonucleotide (FORB-MP) was an inhibitor of the adenylosuccinate synthetase. A more recent investigation[14] has confirmed that L. donovani will convert formycin B to its ribonucleotide but also has shown that the latter is converted further to the APP ribonucleotides described above. This conversion was mediated by the adenylosuccinate synthetase at a slow rate. The kinetic constant for FORB-MP is high, compared to its natural substrate (K'_m = 26uM), and the velocity is relatively slow (V'_{max} = 1%). This indicates that formycin B does not differ from the other inosine analogues, HPPR and TPPR, in its antileishmanial activity and is similar to HPPR in its conversion to the APP-ribonucleotides.

SUMMARY

The demonstration of this unusual metabolic sequence in these organisms indicates that there are substantial differences between these protozoans and their mammalian hosts with respect to pyrazolopyrimidine metabolism. The intracellular forms of L. donovani, which are the pathogenic agents of the human disease, metabolize allopurino identically. Trypanosoma cruzi, the causative agent of American trypanosomiasis, metabolizes allopurinol in the same manner as leishmania; in addition, the intracellular and bloodstream forms of this organism, the agents of the disease in man, carry out the same reaction sequence. The African trypanosomes, T. rhodesiense and T. brucei also convert allopurinol to the same metabolic products. Thes metabolic sequences appear to be common to most of the pathogenic hem flagellate parasites that infect man and this opens a new avenue of research in purine metabolism.

ACKNOWLEDGEMENT

Although this overview is reported by the single individual, the work described is the result of many who contributed their time, physical and intellectual efforts, and friendship. The results belon to all of them. They are: R. L. Berens, G.B. Elion, W.R. Fish, T. A.Krenitsky, S. W. LaFon, D. L. Looker, R. L. Miller, D. J. Nelson T. Spector.

REFERENCES

1.   Hitchings, G.H., Arthr. and Rheum., 18, 863 (1975).
2.   Nelson, G.J., Bugge, C.J.L., Elion, G.B., Berens,R.L., and
          Marr, J.J., J. Biol. Chem. 254, 395 (1979).

3.   Spector, T., Jones, T.E., and Elion, G.B., J. Biol. Chem.
         254,, 8422 (1979).
4.   Spector, T. and Miller, R.L., Biochim. Biophys. Acta. 445,
         509 (1976).
5.   Berens, R.L., Marr, J.J., Nelson, D.J., and LaFon, S.W.,
         Pharmacol. 29, 2397 (1980).
6.   Marr, J.J., Berens, R.L., and Nelson, D.J., Science 201,
         1018 (1978)
7.   Berens, R. L., Marr, J.J., Cruz, F.S., and Nelson, D.J.,
         Submitted for publication.
8.   Berens, R.L., Marr, J. J., and Brun, R., Mol. Biochem.Parasitol.
         1, 69 (1980).
9.   Nelson, D. J., LaFon, S.W., Tuttle, J.V., Miller, W.H.,
         Miller, R.L., Krenitsky, T.A., Elion, G.B., Berens, R.L.
         and Marr, J. J., J. Biol. Chem., 254, 11544 (1979).
10. Koszalka, G. W., and Krenitsky, T.A., J. Biol. Chem., 254,
         8185 (1979).
11. Marr, J.J., Berens, R.L., Nelson, D.J., Krenitsky,T.A.,
         Spector, T., LaFon, S.W. and Elion, G.B., Biochem.
         Pharmacol., 31, 143 (1982).
12. Tuttle, J.V. and Krenitsky, T.A., J. Biol. Chem., 255, 909 (1980).
13. Carson, D. A. and Chang, K-P., B.B.R.C., 100, 1377 (1981)
14. Nelson, D.J., LaFon, S.W., Spector, T., Berens, R.L. and
         Marr, J.J. Submitted for publication.

SINGLE CELL CLONING OF LEISHMANIA PARASITES IN PURINE-DEFINED MEDIUM: ISOLATION OF DRUG-RESISTANT VARIANTS

David M. Iovannisci and Buddy Ullman

University of Kentucky

Lexington, Kentucky

Abstract: A simple technique for the isolation of Leishmania donovani and Leishmania tropica promastigote clones derived from a single cell involves the use of semi-solid agar. Both species of Leishmania form discrete colonies at high efficiency in completely defined medium lacking serum. Visible colonies appear between eight and fourteen days. Viability of colonies transferred from semi-solid agar to liquid suspension culture is 100%. Using these techniques, we have isolated clonal populations of cells resistant to tubercidin and formycin b.

INTRODUCTION

The ability to isolate clonal populations of parasites is a prerequisite to thorough genetic analysis. Present culture methods for the propagation of parasitic protozoa involve the continuous growth of extracellular parasite forms in liquid medium or the cultivation of intracellular forms inside mammalian cells. Most parasitic protozoa are cultivated in undefined growth media (Castellani et al., 1967; Mattei et al., 1977). However the genus Leishmania is one of the few protozoan parasites which can be cultivated in semi-defined or completely defined tissue culture media (Steiger and Steiger, 1976; Berens and Marr, 1977).
Despite the fact that Leishmania and other trypanosomatids have been single cell cloned (Keppel and Janovy, 1980), no method exists for isolation of colonies derived from single cells in completely defined conditions. We have modified previous cloning techniques by adapting Leishmania donovani and Leishmania tropica to completely defined growth conditions and report the high efficiency cloning of both Leishmania species in semi-solid agar using bacteriological techniques.

MATERIALS AND METHODS

Leishmania donovani and Leishmania tropica were routinely
propagated in commercially available Dulbecco's Modified Eagle
(DME) tissue culture medium (Gibco, Grand Island, NY) supplemented
with $NaHCO_3$ (3.7 g/l), hemin (5 mg/l), dextrose (1.5%), Hepes,
pH 7.3 (40 mM), adenosine (0.1 mM), Tween-80 (40 mg/l) and 0.3%
bovine serum albumin, Fraction V (Sigma). The liquid component
was prepared by filter sterilizing DME medium as a five-fold
concentrate, adding the remaining filter sterilized components and
diluting to a two-fold concentrate. Agar at 2% in H_2O was
autoclaved separately. The two-fold growth medium concentrate and
the liquid agar were combined in equal proportions before pouring
into 100 X 15 mm Petri dishes. Approximately 20-25 ml of complete
medium in 1%, Difco-Bacto agar were placed in each Petri dish and
allowed to cool overnight at room temperature. Medium in
semi-solid agar could be stored at 4° for at least two weeks prior
to plating of the parasites.

Before inoculating the Petri dishes with parasites, the
plates were allowed to pre-equilibrate for one hour in a
humidified 5% CO_2 atmosphere. Exponentially growing
promastigotes were diluted to a concentration of 2000 cells/ml
into freshly prepared liquid medium and 0.1 ml spread onto plates
with a flame sterilized bacteriological spreader to a final
expected density of 200 colonies/plate. Cells were counted either
in a model ZB1 Coulter Counter (settings were 1/amp = 1/4;
1/aperture current = 1) or manually using a hemacytometer. The
plates were incubated in an uninverted position in a water
jacketed incubator with 5% CO_2 atmosphere at 26°C. With final
agar concentrations of 1%, Petri dishes could also be maintained
in an inverted position with no loss in cloning efficiency. L.
donovani colonies were visible after 10 days, while L. tropica
could be visualized after 14 days.

Selective clonings in the presence of either tubercidin or
formycin b were performed similarly except that the cytotoxic drug
was added to the semi-solid medium, 10^8 promastigotes in a
volume of 0.1 ml were plated, and Petri dishes were maintained in
an inverted position.

RESULTS

Cells of Leishmania donovani and Leishmania tropica when
plated on semi-solid complete growth medium containing Difco-Bacto
agar yielded translucent, slightly mucoid colonies resembling
Gram-negative enteric bacteria (Fig. 1). Although plates
containing 0.5% agar yielded higher plating efficiencies as well
as larger colony sizes for L. donovani the agar was found to be
too soft for routine use. The medium was easily dislodged from
the plates and colonies often grew into each other after longer

incubation periods. In addition, because the cells are motile
(promastigote stage) on the semi-solid medium, it was ambiguous
whether low agar concentrations provided a sufficient barrier
against cell swarming. By contrast, media containing 1.0% agar
provided a sufficiently solid surface to promote the growth of
discrete colonies while still allowing workable plating
efficiencies (Fig 1.). The cloning efficiency of L. tropica
promastigotes also decreased as a function of increasing agar
concentration. Nevertheless, media containing 1% agar again proved
to be the most efficaceous for the production of large discrete
colonies.
     That the cells of both Leishmania species remain motile when
grown on even the highest agar concentrations could be
demonstrated by removing cells from a colony with a sterile pipet
or metal loop and examining them under the light microscope.
Cells isolated from mature colonies grown on 1% agar were
flagellated and motile immediately after suspension in phosphate
buffered saline, indicating they had retained their promastigote
phenotype. Non-motile, non-flaggelated, spherical cells could be
observed in visible colonies sampled from 1.5% or 2% agar shortly
after plating, but they appeared to regain their promastigote
phenotype.

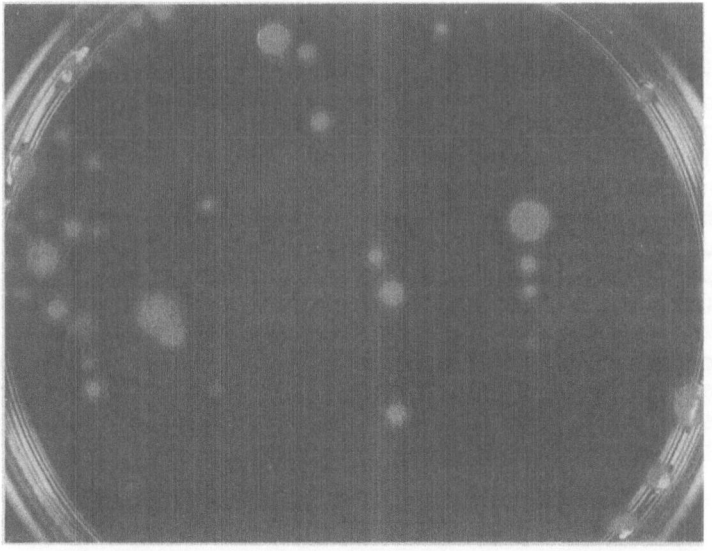

Fig. 1. L. donovani colonies 15 days after plating on
purine-defined semi-solid medium containing 1% agar.

DISCUSSION

Early cloning experiments with leishmanial promastigotes were attempted in this laboratory using agarose as a solidifying agent. Because of the motility of these parasites, the agarose (Sea-Kem ME) concentrations (0.5%) necessary to prevent colony swarming proved to be cytotoxic. This problem was reduced by substituting agar as the solidifying agent. Although L. donovani and L. tropica responded differently to varying agar concentrations, both species formed discrete macroscopic colonies at high efficiency in DME based semi-solid media. Media containing 1% agar maximized the parameters required for high efficiency cloning of both species of Leishmania. Lower agar concentrations (0.5%) increased the cloning efficiency slightly for L. donovani and shortened the interval between plating and appearance of visible colonies of both species. However, the viscosity of 0.5% agar was too low for discrete colony formation and did not impede the migration of parasites from one colony to another.

Allowing freshly poured plates to stand overnight at room temperature promoted the evaporation of excess surface moisture prior to inoculation of the parasites. This further reduced the risk of cell migration (swarming) from one discrete colony to another but necessitated pre-equilibration of the plates in 5% CO_2 to attain the proper pH conditions. Maintenance of plates in an inverted position prevents the possibility that accumulated moisture droplets might fall onto the growing colonies. Furthermore, Petri dishes inoculated with parasites can be maintained in an inverted position with no loss of cloning efficiency.

Colonies of either L. donovani or L. tropica could be transferred by either micropipet or by a flame sterilized bacteriological loop from the semi-solid to liquid medium with 100% viability. The organisms were all motile and retained their flagella and promastigote phenotype.

Growth of L. donovani in completely defined medium has facilitated the isolation of clonal variants resistant to cytotoxic purine analogues In these experiments 10^8 promastigotes were plated under selective conditions, either 5μM formycin b or 1μM tubercidin, and from one to about fifty resistant colonies appeared per plate. The isolation of such clonal drug-resistant variants from large populations of parasites is a necessary prelude to the genetic dissection of the purine salvage pathways unique to these parasites.

REFERENCES

Berens, R. L. and Marr, J.J. 1978. An easily prepared defined medium for cultivation of Leishmania donovani promastigotes. J. Parasitol. 64:160.

Castellani, O., Ribeiro, L. V. and Fernandes, J. F. 1967. Differentiation of Typanosoma cruzi in culture. J. Parasitol. 14:447-451.

Iovannisci, D. M., and Ullman, B. 1982. A simple defined medium for the continuous cultivation of Leishmania promastigotes. Manuscript in preparation.

Keppel, A. D. and Janovy, J., JR. 1980. Morphology of Leishmania donovani colonies grown on blood agar plates. J. Parasitol. 66(5): 849-851.

Mattei, D. M., Goldberg, S., Morel, C., Azevedo, H.P., and Roitman, I. 1977. Biochemical strain characterization of Trypanosoma cruzi by restriction endonuclease cleavage of kinetoplast DNA. Federation of European Biochemical Societies Letters. 74:264-268.

Steiger, R. E. and Steiger, E. . 1976. A defined medium for cultivating Leishmania donovani and L. braziliensis. J. Parasitol. 62:1010-1011.

SELECTIVITY OF ANTIVIRAL EFFECTIVENESS DERIVED FROM DIFFERENCES OF

HERPES SIMPLEX VIRUS-CODED THYMIDINE KINASES

James A. Fyfe

Wellcome Research Laboratories
Research Triangle Park, NC   27709
USA

INTRODUCTION

   Herpes simplex viruses are human pathogens that cause oral
and ocular lesions (HSV-1) or genital lesions (HSV-2). These
viruses code for enzymes with substrate specificities different
from those of the host cell. The pyrimidine metabolism of the
cell is augmented by a virus-coded deoxythymidine (dThd) kinase
(TK). This enzyme has a relatively broad phosphate acceptor
specificity which permits the phosphorylation of several antiviral
nucleoside analogs, the first step in their activation. These
analogs are selective inhibitors of DNA synthesis in virus-
infected cells, in part, because they are phosphorylated to their
triphosphate derivatives only in infected cells.

   While it is known that the TK induced by HSV-1 has the unusual property of bifunctionality, little has been reported about
the specificity or importance of the second enzyme activity, the
deoxythymidylate (dTMP) kinase. The following report compares its
role in the activation of the nucleoside analogs, acyclovir (9-(2-
hydroxyethoxymethyl)guanine and BrVdUrd ((E)-5-(2-bromovinyl)-2'-
deoxyuridine) in both HSV-1 and HSV-2 infected cells.

RESULTS AND DISCUSSION

   The virus-coded dThd kinases from HSV-1 and HSV-2 infected
cells were purified by previously described methods[1,2]. The
enzyme from HSV-1 was previously shown to be bifunctional[3], but
this had not been reported with the enyzme from HSV-2. Figure 1
shows the co-elution of dThd and dTMP phosphorylating activities
(peak 4) from a cytosolic extract of Vero cells infected with

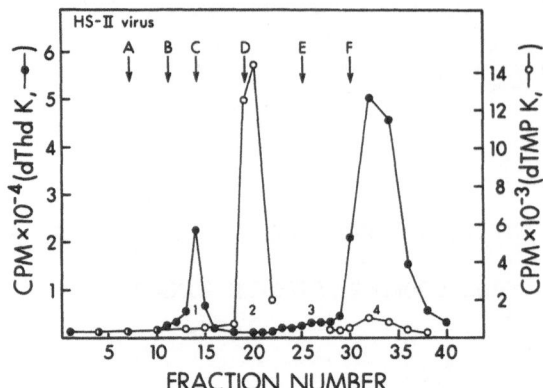

Fig. 1.   Kinase activity profiles from dThd-agarose affinity
          column. The sample added (arrow A) was from a cytosol
          fraction of cells infected with HSV-2. The elution
          solutions were: B, 20 mM Tris (pH 7.5), 2 mM dithio-
          threitol, and 10% glycerol; C, 800 mM Tris (pH 6.8),
          2 mM dithiothreitol, and 10% glycerol; D, solution "C"
          with 1 mM dTMP; E, solution "C" with 1 mM ATP, 0.2 mM
          dThd and albumin (0.5 mg/ml); F, solution "E" except
          with 0.5 mM dThd. Fractions were assayed for dThd
          kinase (— 0 —) or dTMP kinase (— 0 —) activity with
          100 μM [14C]dThd or dTMP. Other details are described
          elsewhere[5].

HSV-2. Both activities were biospecifically eluted from a dThd-
agarose affinity column by 0.5 mM dThd. The ratios of activities
with dThd versus dTMP across the peak were constant. Host cell
dTMP kinase (fractions 19 and 20) was also biospecifically eluted
from the column with dTMP. The two dTMP phosphorylating activities could be distinguished by their susceptibilities to dThd
inhibition. Only the virus-coded enzyme was inhibited.

     Although dTMP was phosphorylated by the purified HSV-2 TK,
the ratio of activities (rate with dTMP/rate with dThd) from HSV-1
was substantially greater than that from HSV-2. These were 1.3
and 0.016, respectively. Relative amounts (per cell volume) of
extractable virus-coded dTMP phosphorylating activities were
calculated to be about 50:1.

     Both purified dThd-dTMP kinases catalyzed the phosphorylation
of BrVdUrd and acyclovir to their respective monophosphate deriva-

tives.   The phosphorylation of acyclovir monophosphate to its
diphosphate is catalyzed efficiently only by host cell guanylate
kinase[4].   Since the catalyst for BrVdUrd monophosphate (BrVdUMP)
phosphorylation was unknown, the viral dThd-dTMP kinase was tested
for this ability.

BrVdUMP was synthesized enzymatically and characterized[5].
This nucleotide was first incubated with ATP·Mg and cytosolic
fractions from cells infected with HSV-1 and HSV-2.   Samples after
incubation were analyzed by high pressure liquid chromatography[5].
A product with the appropriate spectral, chromatographic and
enzymatic properties for the diphosphate (BrVdUDP) was observed
with the incubates from HSV-1 but not from HSV-2.   Next, the
formation of BrVdUDP was quantitated with the purified enzymes
(Fig. 2).   The rate with the HSV-1 dThd-dTMP kinase was about
0.09 pmol/min/dTMP kinase unit; with the HSV-2 enzyme or with the
host cell dTMP kinase, no product was detected (<0.002 and
<0.0002 pmol/min/dTMP kinase unit, respectively).

It appears, then, that only with HSV-1 infection does the
amount of viral dThd-dTMP kinase and/or the specificity of the
enzyme (BrVdUMP compared with dTMP) permit formation of BrVdUDP.

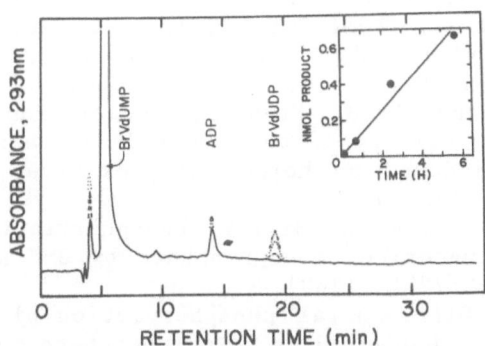

Fig. 2.   Quantitation of product formation with purified HSV-1
           dThd-dTMP kinase.   The reaction mixture (30°) contained
           0.5 mM BrVdUMP,   10 mM ATP·Mg,   75 mM Tris (pH 7.5),
           albumin (0.5 mg/ml), and 280 pmolar units of purified
           kinase per ml.   Aliquots were analyzed for BrVdUDP
           formation by HPLC as reported elsewhere[5].   Chromatograms
           are shown for 0.03 hr (——), 0.67 hr (---), 2.5 hr
           (···), and 5.75 hr (—·—).

These results are consistent with anabolism studies with cell cultures. A close analog of BrVdUrd, (E)-5-(2-iodovinyl)-2'-deoxyuridine, was efficiently converted to its monophosphate derivative in cells infected with both HSV-1 and HSV-2. However, a diphosphate (or triphosphate) could only be detected with extracts of cells infected with HSV-1[6]. The inefficient phosphorylation of BrVdUMP by the dThd-dTMP kinase from HSV-2 correlates well with the observation that replication of HSV-2 is at least 100-fold less sensitive to BrVdUrd than is that of HSV-1[7] and suggests a cause-and-effect relationship. The sensitivities of the two viruses to acyclovir are about equal[7].

The characteristics of the metabolic changes imposed on the cell by HSV-1 versus HSV-2 appear to be different enough to affect the drug therapy used against these viruses. Specifically, the second step of BrVdUrd phosphorylation appears to be more selective than the first step and may help determine the effectiveness of the analog to inhibit virus replication. The apparent dependence of some pyrimidine derivatives on the virus-associated dThd-dTMP kinase for both phosphorylation steps may be a crucial difference between these compounds and the purine analog, acyclovir.

REFERENCES

1. Y.-C. Cheng and M. Ostrander, Deoxythymidine kinase induced in Hela TK⁻ cells by herpes simplex virus, Type I and Type II: purification and characterization, J. Biol. Chem. 251:2605 (1976).

2. J. A. Fyfe, P. M. Keller, P. A. Furman, R. L. Miller and G. B. Elion, Thymidine kinase from herpes simplex virus phosphorylates the new antiviral compound, 9-(2-hydroxyethoxymethyl)guanine, J. Biol. Chem. 253:8721 (1978).

3. M. S. Chen and W. H. Prusoff, Association of thymidylate kinase activity with pyrimidine deoxyribonucleoside kinase induced by herpes simplex virus, J. Biol. Chem. 253:1325 (1978).

4. W. H. Miller and R. L. Miller, Phosphorylation of acyclovir (acycloguanosine) monophosphate by GMP kinase, J. Biol. Chem. 255:7204 (1980).

5. J. A. Fyfe, Differential phosphorylation of (E)-5-(2-bromovinyl)-2'-deoxyuridine monophosphate by thymidylate kinases from herpes simplex viruses Type I and II and varicella zoster virus, Mol. Pharmacol. 21:432 (1982).

6. J. Descamps and E. DeClercq, Specific phosphorylation of (E)-5-(2-iodovinyl)-2'-deoxyuridine by herpes simplex virus-infected cells, J. Biol. Chem. 256:5973 (1981).

7. E. DeClercq, J. Descamps, G. Verhelst, R. T. Walker, A. S. Jones, P. F. Torrence and D. Shugar, Comparative efficacy of antiherpes drugs against different strains of herpes simplex virus, J. Infect. Dis, 141:563 (1980).

ADENOSINE DEAMINASE, PURINE NUCLEOSIDE PHOSPHORYLASE AND

5´-NUCLEOTIDASE ACTIVITIES IN INFECTIOUS MONONUCLEOSIS

Johannes Mejer*, Per Nygaard**, Jørgen Cohn***,
Ole Gadeberg*** and Viggo Faber***

*) Blood Serology Department, Bispebjerg Hospital
**) University Institute of Biological Chemistry B
***) Department of Infectious Diseases M, Rigshospita-
let. Copenhagen Denmark

The present study was undertaken to study 3 selected purine
pathway enzymes in infectious mononucleosis (IM). The enzymes concerned are adenosine deaminase (ADA), purine nucleoside phosphorylase (PNP) and 5´-nucleotidase (5´-N). The enzymes were selected
since the absence of these enzymes have been associated with clinically defective lymphoid function (1), and because abnormalities
in the activities of ADA and PNP have been observed in leukemia
(2,3)
    In 1974 Quagliata et al. (4) claimed that 5´-N may also be
usefull in the biochemical characterization of subgroups of patients with lymphoid leukemia. In the early phase of IM they found
a decrease of 5´-N.

MATERIALS AND METHODS

    Sixteen consecutively admitted patients with IM were included
(age range 16-24 years, 5 females and 11 males). The diagnosis was
established by clinical symptoms with sore throat, adenopathy and
fever together with heterophil antibodies and atypical lymphocytes
in peripheral blood. In one of the patients there was a negative
test for heterophil antibodies. Lymphocyte responsiveness to pokeweed mitogen was found to be depressed during the first week of
illness.
    Six patients with myeloid proliferation were investigated.
The six patients include one patient with myeloid proliferation
of uncertain origin, there were 7% immature cells in the peripheral blood, one patient with extramedullary hemopoiesis due to
bone marrow metastasis with 14% immature cells in the peripheral

249

blood, and 4 patients with acute myeloid leukemia with 8 to 89%
blasts in peripheral blood (age range 67-79 years, 4 females
and 2 males).

Twelve healthy donors were investigated as controls (age range
19-59 years, 5 females and 7 males).

In 9 patients with IM follow-up test blood samples were obtained 4 to 6 weeks later as a part of an out-patients follow-up.

Mononuclear cells from IM patients, patients with myeloid proliferation and controls were obtained by Ficoll-Isopaque gradient
separation (5) of heparinized peripheral blood.

ADA and PNP were measured spectrofotometrically in cell free
extracts as previously described (3). 5´-N was assayed radiochemically using AMP or CMP as substrate in cell free extracts (6). 5´-N
with AMP as substrate is expressed as that part of total AMP´ase
activity, which is inhibitable by 20 µM α,β-methylene-adenosine-di-
phosphate.

Protein was determined according to Lowry et al. (7).

STATISTIC

Mann-Whitney and Wilcoxon rank sum test.

Table 1. Comparison of purine enzyme values between normals and
patients with infectious mononucleosis or patients with
myeloid proliferation including 4 patients with acute
myeloid leukemia.
Values shown are median and range, nmoles/h/mg protein
at 37°. x:p<0.01. o:p<0.02.

Enzyme	IM N=16	Myeloid Proliferation N=6	Controls N=12
ADA	1574^{x} (528-3782)	1038^{x} (803-11169)	514 (337-794)
PNP	3418^{x} (2442-5366)	4270 (2715-5434)	5008 (3123-6880)
5´-N (AMP)	38^{x} (3-111)	31^{x} (0-85)	114 (35-346)
5´-N (CMP)	93° (49-278)	61 (42-80)	137 (94-458)

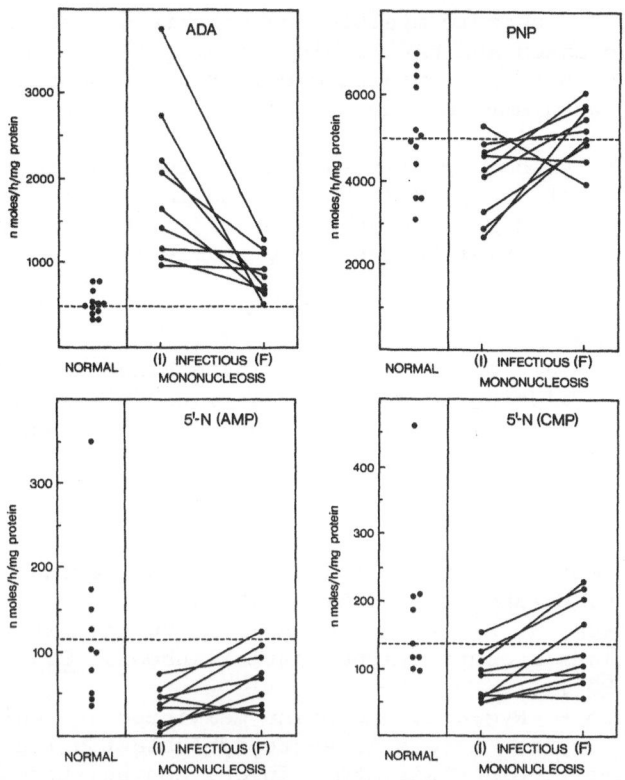

Fig. 1. Purine enzyme activities of mononuclear cells from patients
with infectious mononucleosis and from controls.
I, initial values within the first 3 days after admission.
F, follow-up 5+1 week after admission. Dotted line, normal
median values.

RESULTS AND DISCUSSION

IM is characterised by self limiting lymphocytosis largely
consisted of T-lymphocytes during the acute phase of the illness
(8).

As shown in table 1 elevated activity of ADA is present in
the early phase of the disease and both 5´-N (with both AMP and
CMP as substrate) and PNP are decreased. All enzyme activities return to practically normal values 4-6 weeks after the onset (Fig.
1). Comparison between initial values and follow-up values revealed
for all enzymes a significant difference (p<0.01).

The findings of the combination of elevated ADA and depressed
PNP and 5´-N activities in this T-cell proliferation is of special
interest because exactly the same abberation in enzyme activities
has been described in leukemic lymphoblasts that display T-cell
surface markers (2).

In acute and chronic myeloid leukemia it is well established that ADA is increased and PNP within normal control limits (3) and table 1. Further 5´-N is diminished compared with normal mononuclear cells (both with AMP and CMP as substrate) (table 1).

These data indicate that the biochemical heterogeneity within the spectrum of lymphoid proliferation is identical in the self limiting proliferation in IM (table 1) and in T-cell acute lymphoid leukemia (2). Myeloid proliferation differs from these lymphoid proliferations in the normal PNP activity (table 1).

ACKNOWLEDGMENT

Many thanks are due to Dr. Nis Nissen Finseninstitutet, Copenhagen, for permission to study patients under his care, and to Inge Carlsen for her technical assistance. The work was supported by the Danish Medical Research Council.

REFERENCES

1. I. H. Fox, Metabolic basis for disorders of purine nucleotide degradation. Metabolism. 30:616 (1981).
2. D. G. Poplack, J. Blatt, and G. Reaman, Purine pathway enzyme abnormalities in acute lymphoblastic leukemia. Cancer Res.. 41:4821 (1981).
3. J. Mejer and P. Nygaard, Adenosine deaminase and purine nucleoside phosphorylase levels in acute myeloblastic leukemia cells. Relationship to diagnosis and clinical course. Leukemia Res.. 3:211 (1979).
4. F. Quagliata, D. Faig, M. Conly, and R. Silber, Studies on the lymphocyte 5´-nucleotidase in chronic lymphocytic leukemia, infectious mononucleosis, normal subpopulations, and phytohemagglutinin-stimulated cells. Cancer Res.. 34:3197 (1974).
5. A. Bøyum, Separation of blood leukocytes, granulocytes and lymphocytes. Tissue Antigens. 4:269 (1974).
6. J. Mejer and P. Nygaard, Ageing and activities of purine metabolizing enzymes in leukocytes. In Inborn Errors of Immunity and Phagocytosis (F. Güttler, J. W. T. Seakins and R. A. Harkness, Eds.) p. 181 MTP Press.
7. O. H. Lowry, N. J. Rosebrough, A. L. Farr, and R. L. Randall, Protein measurement with the Folin phenol reagent. J. biol. Chem.. 193:265 (1951).
8. M. Papamichail, P. J. Sheldon, and E. J. Holborow, T and B-cell subpopulations in infectious mononucleosis. Clin. exp. Immunol.. 18:1 (1974).

T CELL ACTIVATION IN TYPHOID FEVER DETECTED BY INCREASED LEVELS OF ADENOSINE DEAMINASE

Michele Russo, Teresa Pizzella, Salvatore Nardiello, Flavio Fiorentino and Bruno Galanti

Clinic of Infectious Diseases, 1st Medical School Naples, Italy

INTRODUCTION

The demonstration of an inherited adenosine deaminase (ADA) deficiency in patients with combined immunodeficiency (Giblett et al., 1972) suggests a causal association between lymphocyte ADA levels and immune function. ADA activity is greatly increased in in vitro stimulated lymphocytes (Hovi et al., 1976) and in peripheral mononuclear cells from subjects developing an active immune response (Galanti et al., 1981). Recent data indicate that mechanisms of cell-mediated immunity play the major role in recovery from typhoid fever (Rajagopalan et al., 1982). The aim of this study is to assess the value of lymphocyte ADA assay in detecting T cell activation during typhoid fever.

MATERIALS AND METHODS

Patients and controls

We investigated 9 patients (all males, age range 14-28 years) with uncomplicated typhoid fever, diagnosed from blood culture. All patients were on therapy with chloramphenicol (2 g/die) at the time of the study. In 6 of them, samples were taken twice, the first within 7 days from the onset of fever and the second 7-10 days later. Eleven normal subjects (8 males, age range 22-36 years) from our hospital staff were controls.

Preparation of cell extracts

Peripheral mononuclear cells (PBMC) were isolated from heparinized blood by standard gradient centrifugation (Bøyum, 1968). T-enriched cell fractions were obtained by rosetting PBMC with 2-aminoethylisothiouronium bromide-treated sheep red blood cells (Madsen and Johnsen, 1979). Cell suspensions were adjusted to a concentration of 5×10^6 cells/ml and sonicated at 23 kc/sec. After centrifugation at 800 g for 10 min, supernatants were immediately frozen at $-20°C$.

ADA assay

ADA activity was measured on supernatants by a colorimetric method (Giusti, 1974), and expressed as nmol of adenosine deaminated/min/37°C/10^7 cells.

RESULTS

ADA levels were significantly increased in T cells from typhoid fever patients (Table 1). The increase was detected already during the first week of disease: ADA values increased further in 5 of the 6 re-assayed patients.

Table 1. Levels of adenosine deaminase (ADA) activity in peripheral mononuclear cells (PBMC) and T cells from typhoid fever patients. Mean ± standard deviation is reported.

| | Days of illness | No. of assays | ADA nmol/min/10^7 cells | |
			PBMC	T cells
Typhoid fever patients (9)	2-7	7	27.5 ± 3.3	32.9 ± 6.6[°°]
	8-20	8	36.1 ± 6.5[°]	41.4 ± 12.4[°°]
Normal controls (11)	–	11	25.7 ± 6.6	20.1 ± 4.3

[°]P < 0.005,   [°°]P < 0.001 versus controls

During the first week of disease, in spite of the high ADA values in T cells, enzymatic levels were not different from control values when assayed on total PBMC: this was due to a parallel reduction of ADA activity in non-T cells (data not shown).

DISCUSSION

This study confirms our previous reports of increased ADA levels in PBMC from typhoid fever patients (Galanti et al., 1978; Galanti et al., 1981). We have demonstrated here that high levels of ADA are present in T cells from these patients as early as during the first week of fever; in addition, ADA values tended to increase further in a more advanced stage of disease.

Activation of T cells has been recently documented by several in vitro techniques in uncomplicated, but not in complicated cases of typhoid fever, suggesting that mechanisms of cell-mediated immunity play the most important role in recovery from typhoid fever (Rajagopalan et al., 1982).

The high ADA levels in T cells during typhoid fever could be an expression of T cell activation, and could thus be a sensitive means for the evaluation and follow-up of cell-mediated immune responses in this disease.

REFERENCES

Bøyum, A., 1968, Separation of leucocytes from blood and bone marrow, Scand. J. clin. Lab. Invest., 21 (Suppl.97): 77.

Galanti, B., Russo, M., Nardiello, S., Fiorentino, F., 1978, Attività adenosina deaminasi dei linfociti in corso di febbre tifoide (dati preliminari), Boll.Soc.It.Biol.Sper., 54: 2609.

Galanti, B., Nardiello, S., Russo, M., and Fiorentino, F., 1981, Increased lymphocyte adenosine deaminase in typhoid fever, Scand. J. Infect. Dis., 13: 47.

Giblett, E.R., Anderson, J.E., Cohen, F., Pollara, B., Meuwissen, H.J., 1972, Adenosine deaminase deficiency in two patients with severely impaired cellular immunity, Lancet, 2: 1067.

Giusti, G., 1974, Adenosine deaminase, in: "Methods of Enzymatic Analysis", H.U. Bergmeyer, ed., Verlag Chemie, Weinheim.

Hovi, T., Smyth, J.F., Allison, A.C., and Williams, S.C., 1976, Role of adenosine deaminase in lymphocyte proliferation, Clin. exp. Immunol., 23: 395.

Madsen, M., and Johnsen, H.E., 1979, A methodological study of
        E-rosette formation using AET-treated sheep red blood
        cells, J. Immunol. Methods, 27: 61.
Rajagopalan, P., Kumar, R., and Malaviya, A.N., 1982, Immunological
        studies in typhoid fever. II. Cell-mediated immune
        responses and lymphocyte subpopulations in patients
        with typhoid fever, Clin. exp. Immunol., 47: 269.

OPTIMIZED MICRO-MEASUREMENT OF H.PRTase ACTIVITY

ON FIBROBLASTS AND HAIR FOLLICLES

P. Baltassat, B. Mousson, M.C. Rissoan, and M.T. Zabot

Laboratoire de Biochimie - Pr J. COTTE

Hôpital Debrousse - Lyon - France

INTRODUCTION

Detection of heterozygotes which are carriers of Lesh and Nyhan's syndrome, poses a difficult problem, which has up to the present, not been solved in a very satisfactory way in current practice. This had led us, for genetic counseling, to reconsider in detail the determination methods of H.PRTase activity in cultured cells and in hair follicles.

MATERIALS

Fibroblasts were obtained from skin biopsy specimens cultured with newborn calf serum (10 per cent v/v), antibiotics and antifungic agents.

Fibroblasts were grown in monolayer to confluence and then removed by treatment with trypsine, washed in NaCl 0.15 mol/1. Cells were resuspended in a volume of TRIS buffer 10 mmol/1 pH 7.4, calculated to obtain a protein concentration of between 1.0 and 2.5 mg/ml. Cells were then disrupted by ultrasonation (3 cycles of 10 sec each) centrifuged at 19 000 g for 40 min. The supernatant was used to measure enzyme activities, either immediately after its preparation or after being kept frozen at temperature of $- 80\ °C$.

Hair follicles (about 60) were carefully removed from the patients and examined to ascertain the presence or absence of a follicle sheath.

Reagents

Labelled substrates ($8 -^{14}C$ hypoxanthine and $8 -^{3}H$ adenine)

were purchased from Amersham UK. PRPP dimagnesium salt was purchased
from ICN (USA), thymidine triphosphate (TTP) from Sigma (USA),
chromatographic paper from Whatman (WG), XOMAT R film X-R1 for autoradiography from Kodak (France) and all other reagents from
Merck (WG).

H.PRTase in fibroblasts

Method
Incubation medium has been defined as follows : 8 $-^{14}$C hypoxanthine approximately 0.250 mmol (12 mCi/mmol), 8 $-^{3}$H adenine
approximately 0.150 mmol (0.800 Ci/mmol), PRPP 0.9 mmol, Mg Cl$_2$
5 mmol, TTP 5 mmol, TRIS 50 mmol pH 7.4 (final concentration/1),
in a final volume of 50 ul. The reaction was started by the addition of 25 ul of fibroblast extract. After a 50 min incubation
period at 37 °C, the reaction was stopped by 10 ul of EDTA 110
mmol/1.

The nucleotides formed is separated from the free bases
by a descending chromatography using DEAE cellulose paper strips
(1.5 cm x 17 cm - start at 5 cm), and Paladini - Leloir solvent
as described by CARTIER and HAMET (1). The radioactive spots were
located by autoradiography, cut out and radioactivity was measured
by liquid scintillation. Enzyme activity was expressed in ukat
(umol/sec)/kg protein.

A.PRTase activity was taken as a control value, and the
ratio H.PRTase/A.PRTase was calculated.

Results
The incubation conditions given above, have been studied
in detail :
- Optimum pH : 7.4 (plateau between 7.4 and 7.8)
- TTP concentration : TTP was used as an inhibitor of the 5'-nu-
cleotidase activity. In the presence of increasing concentrations
of TTP, H.PRTase activity rises rapidly to a peak at 5 mmol and
then decreases sharply after this concentration. H.PRTase activity
for a 7.5 mmol/1 TTP concentration is half what it is for a 5
mmol/1 TTP concentration.
- Apparent Km : + for hypoxanthine : it was measured in the supernatant of the crude extracts, without further purification ; Km
values for hypoxanthine have been found to range from 4.55 to
8.65 umol. (on 5 different normal strains) ; these values are similar to those measured in red blood cells.
            + for PRPP : the Km for PRPP is 0.150 mmol.
- Effect of incubation time : enzyme activity is linear up to 50
min, which is the optimum time for sensitivity (figure 1).
- Effect of protein concentration : enzyme activity is directly
proportional to the increase of protein content (0.1 to 0.9 mg/ml
final concentration).
- Precision : the variation coefficient of the overall procedure
is 0.063.

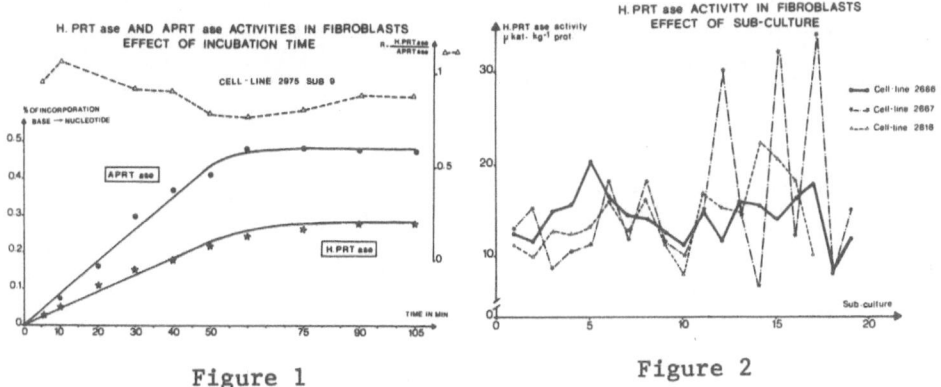

Figure 1   Figure 2

- **Effect of subculture** : H.PRTase activity has been measured in 3 cell strains from subculture 1 to subculture 19. Results are shown in figure 2. They indicate that the enzyme content is not influenced by the number of sub-cultures.
- **Normal values** : 34 different cell strains have been tested : the values obtained, range from 10 to 31 µkat/kg prot. (Mean : 17.8 µkat ±5.55). Three obligate heterozygotes for Lesh and Nyhan's syndrom have been tested : their respective value were 4.3, 3.6 and 5.2 µkat/kg prot.

H.PRTase in hair follicles

Method :
The method published by SILVERS et al (2) has been modified as follows : incubation medium consisted of an equal volume of TRIS 1 mol/1 pH 7.4, $MgCl_2$ 30 mmol/1, PRPP 6.25 mmol/1, TTP 4.5 mmol/1. Each hair follicles was immersed in 120 µl of this medium, disrupted by 3 cycles of freeze-thaw and then removed. 40 µl aliquots were transfered into test tubes, and the reaction was started by addition of 10 µl of 8 - ^{14}C hypoxanthine (final concentration 0.666 mmol/1) and 10 ul of 8 -^{3}H adenine (0.166 mmol/1).
After a 60 min incubation, the reaction was stopped by the addition of 10 µl of EDTA 220 mmol/1, and the nucleotides were separated as indicated above (see "Fibroblasts").

Results :
- Effect of incubation time : The reaction is linear until the 75th min of incubation (figure 3).
- TTP concentration : the reaction reaches a maximum for final TTP concentration of 0.75 mmol/1. The optimum H.PRTase/A.PRT ase ratio is obtained at this TTP concentration (figure 4).

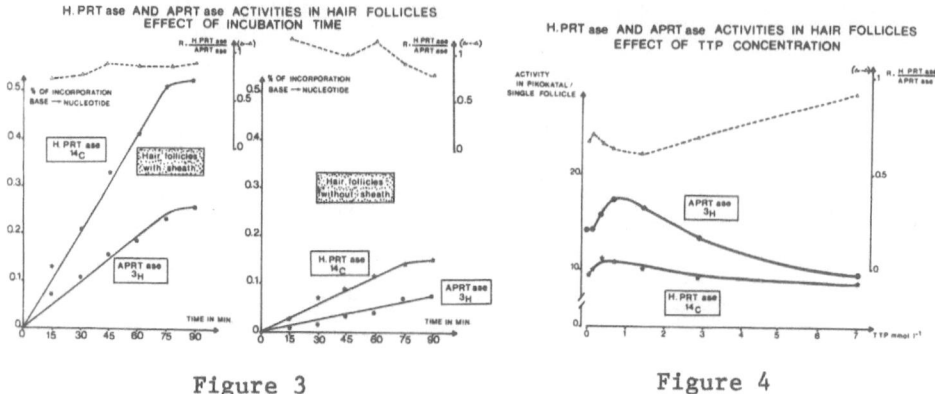

Figure 3   Figure 4

- Normal values : H.PRTase : greater than 0.5 pkat/single follicle. A.PRTase : greater than 0.1 pkat/single follicle. Ratio H.PRTase/A.PRTase : between 0.4 and 1.20.
- Effect of presence of sheaths : when the sheath is not visible, enzyme activities are unmeasurable in most follicles.

CONCLUSION

Fibroblasts : the procedure has been optimized to obtain satisfactory precision and to avoid the overlapping between normal and heterozygote values. The absence of subculture influence increases the reliability of the test.

Hair follicles : results are much more difficult to utilize. The quality of the hair must be carefully noted. At present, we are testing a modified procedure, in which the hair follicle, after a single freeze - thaw cycle, is kept immersed in the incubation medium throughout the incubation time. Sensitivity is greatly increased by this procedure and the active and inactive hair population are much more distinct, when the sheath is present. When the sheath is absent, A.PRTase is too low to provide reliable results.

REFERENCES

1 - CARTIER P., HAMET M., 1968, Les activités purine - phosphoribosyl transférasiques des globules rouges humains : technique de dosage. Clin. Chim. Acta, 20 : 205.

2 - SILVERS D.N. et all., 1972, Detection of the heterozygote in Lesh-Nyhan disease by hair-root analysis. N. Engl. J. Med., 286 : 390.

OPTIMAL CONDITIONS FOR PHOSPHORIBOSYLPYROPHOSPHATE

SYNTHESIS IN A CRUDE EXTRACT OF LIVER CELLS

Marcel Lalanne

Département de Biochimie, Faculté de Médecine,
Université Laval, Québec, P.Q., Canada G1K 7P4

We have recently shown that adenosine and some of its analogs decrease phosphoribosylpyrophosphate (PP-ribose-P) availability in rat liver cells (unpublished results). Tubercidin (7-deazaadenosine), 3'-deoxyadenosine, and 6-methylmercaptopurine ribonucleoside are the analogs which cause the greatest decrease of PP-ribose-P availability. They are three- to eight-times more potent than adenosine. This action could possibly result from the inhibition of PP-ribose-P synthesis by these compounds or one of their metabolites.

Our aim is to study the effects of the nucleotides of the various analogs on the synthesis of PP-ribose-P in a crude extract of rat liver cells. Although PP-ribose-P synthetase (EC 2.7.6.1) has already been purified from rat liver, and its physical and kinetic properties studied (1,2), we think that some properties of the enzyme in a crude extract could differ from those of the purified enzyme and be more representative of the kinetics of the enzyme in the cells. As a first step, we have characterized the kinetics of PP-ribose-P synthesis in a 100,000 g supernatant from an homogenate of liver cells isolated as described earlier (3).

The PP-ribose-P synthesized during the assays is measured simultaneously by the formation of radioactive AMP from [2-^{3}H]adenine in the presence of a dialyzed cat hemolysate, a crude source of adenine phosphoribosyltransferase (APRT, EC 2.4.2.7) in which no PP-ribose-P synthetase activity is detectable. Except if stated otherwise, the basic incubation medium contains 210 µM adenine, 680 µM dithiothreitol, 830 µM ribose-5-P, 3.0 mM MgCl$_2$, 3.3 mM CaCl$_2$, and 50 mM potassium phosphate buffer at pH 7.8.

First, we have determined if some ribonucleoside triphosphates,

Table 1.   PP-ribose-P synthesis from ribose-5-P (R5P) in a dialyzed
           supernatant

Nucleotide	PP-ribose-P formed, nmoles per h per mg protein		
1.67 mM	R5P alone	R5P + System I[a]	R5P + System II[b]
None	17.3	51.1	43.7
ATP	17.3	31.0	30.1
GTP	20.7	30.6	23.5
ITP	24.4	28.8	22.4
UTP	30.1	33.6	30.2
CTP	27.8	29.2	25.1

[a] Pyruvate kinase (10 U) and 2.5 mM phosphoenolpyruvate.
[b] Creatine kinase (2 U), 2.5 mM phosphocreatine, and 40 μg of bovine
serum albumin.

other than ATP, could act as pyrophosphoryl donors for PP-ribose-P
synthesis from ribose-5-P. With an undialyzed supernatant, 1.67 mM
ATP, GTP, ITP, UTP and CTP inhibit PP-ribose-P synthesis 25 to 55%
(data not shown). Using a dialyzed supernatant, PP-ribose-P is still
synthesized from ribose-5-P alone about half as rapidly as with an
undialyzed supernatant (Table 1). ATP has little effect on this synthesis, but GTP, ITP, UTP, and CTP increase it 25 to 120%. These
effects are variable from one extract to another.

Both AMP, a product of the reaction, and ADP, which is formed
from AMP by adenylate kinase present in the liver extract, inhibit the
purified rat liver PP-ribose-P synthetase (2). Furthermore, the activity of APRT in the cat hemolysate is reduced 35 and 75% by 1.67 mM
ADP and AMP, respectively (data not shown); however, it is not affected by the various nucleoside triphosphates. Thus, the addition
of an ATP regenerating system would tend to eliminate these inhibi-
tions by AMP and ADP. In fact, the stimulation of PP-ribose-P synthesis in the dialyzed extract by GTP, ITP, UTP, and CTP could be due to
ATP regeneration via the action of the nucleoside diphosphate kinase
present in the liver extract.

The addition of an ATP regenerating system (either phosphoenolpyruvate and pyruvate kinase, or phosphocreatine and creatine kinase)
to the incubation medium increases the amount of PP-ribose-P synthesized from ribose-5-P alone in the dialyzed extract (Table 1). Surprisingly, ATP and the other nucleoside triphosphates reduce PP-ribose-P
synthesis under these conditions. This effect results from the

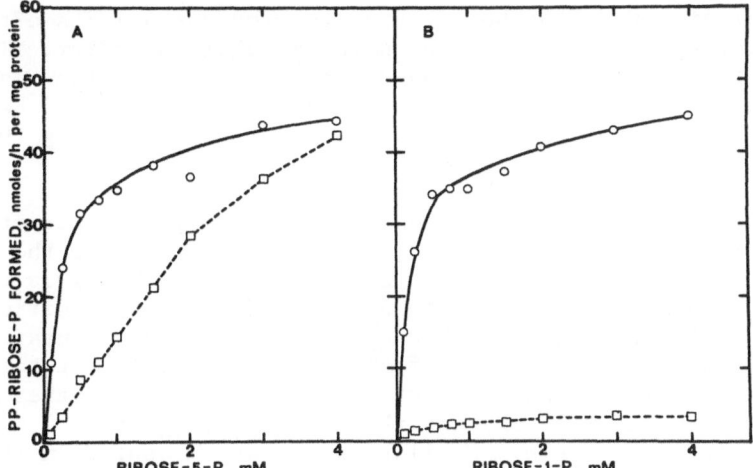

Fig. 1. PP-ribose-P synthesis in a dialyzed extract of liver cells
         at various concentrations of A) ribose-5-P, and B) ribose-1-P
         in the absence (dashed lines), or presence (solid lines) of
         phosphoenolpyruvate and pyruvate kinase.

chelation of Mg^{2+} by the various nucleotides (data not shown).
The addition of further $MgCl_2$ to the incubation medium increases
PP-ribose-P synthesis.

     As ATP and dATP are the only pyrophosphoryl donors of the puri-
fied rat liver PP-ribose-P synthetase (1), presumably PP-ribose-P
synthesis from ribose-5-P alone occurs because some residual ATP is
present in the incubation medium. The concentration of this residual
ATP is probably low, as the liver cell extract and cat hemolysate are
dialyzed 150 min against two changes of buffers prior to use. Nonetheless, this system appears saturated with ATP, because the addition
of equimolar amounts of ATP and $MgCl_2$ does not increase PP-ribose-P
synthesis. Thus, in a crude liver cell extract, PP-ribose-P synthesis can proceed maximally at a low concentration of ATP provided
that an ATP regenerating system is present in the incubation medium
(as in the cells). The residual ATP could arise from an impurity in
the commercial products used, such as ribose-5-P. It could also be
tightly bound to PP-ribose-P synthetase and be regenerated in situ
by an enzymatic system complexed with PP-ribose-P synthetase.

     The addition of phosphoenolpyruvate alone is almost sufficient
to obtain the maximal synthesis of PP-ribose-P from ribose-5-P, as
extracts of liver cells contain a significant amount of pyruvate
kinase. However, all the enzymes of the pentose phosphate, glycolytic,
and gluconeogenic pathways are also present in these extracts. Thus,

it is not certain that phosphoenolpyruvate produces all its effect by only regenerating the residual ATP in the incubation medium. The metabolism of ribose-5-P by the non-oxydative branch of the pentose phosphate pathway could be reduced, or some ribose-5-P could be generated in the presence of phosphoenolpyruvate, thus increasing the amount of ribose-5-P available to PP-ribose-P synthetase.

The rate of PP-ribose-P synthesis in the dialyzed supernatant increases almost linearly with ribose-5-P concentration in the absence of ATP and of an ATP regenerating system (Fig. 1A). There is nearly no PP-ribose-P synthesized from ribose-1-P under these conditions (Fig. 1B). In the presence of phosphoenolpyruvate and pyruvate kinase, but without any addition of ATP, both substrate saturation curves are hyperbolic. The apparent K_m of PP-ribose-P synthetase for ribose-5-P is 224 ± 57 μM, and the apparent V_{max} is 48.7 ± 5.6 nmoles per h per mg protein. Similar values are measured in the presence of ribose-1-P. However, it is not a substrate for the PP-ribose-P synthetase purified from rat liver (1). Ribose-1-P is probably converted first to ribose-5-P by a phosphoribomutase which is present in rat liver (4).

In the standard assay for PP-ribose-P synthesis, 2.5 mM phosphoenolpyruvate and 10 units of pyruvate kinase are added to the incubation medium described initially. Under these final conditions, PP-ribose-P synthesis is maximal between 50 and 70 mM phosphate (data not shown). The rate of synthesis is linear for more than 60 min. There is a linear relationship between the amount of PP-ribose-P synthesized and the amount of protein in the extract, up to 300 μg of protein.

This work was supported by the Medical Research Council of Canada.

REFERENCES

1. D.G. Roth, E. Shelton, and T.F. Deuel, Purification and properties of phosphoribosyl pyrophosphate synthetase from rat liver, J. Biol. Chem. 249: 291-296 (1974).
2. D.G. Roth and T.F. Deuel, Stability and regulation of phosphoribosyl pyrophosphate synthetase from rat liver, J. Biol. Chem. 249: 297-301 (1974).
3. M. Lalanne and F. Lafleur, Inosine synthesis and phosphoribosylpyrophosphate availability in rat liver cells in the presence of allopurinol, Can. J. Biochem. 58: 607-613 (1980).
4. A. Abrams and H. Klenow, On the action of surface forces on phosphoribomutase, Arch. Biochem. Biophys. 34: 285-292 (1951).

DETECTION OF HYPOXANTHINE GUANINE PHOSPHORIBOSYL TRANSFERASE

HETEROZYGOTES BY THIN LAYER CHROMATOGRAPHY AND AUTORADIOGRAPHY

Theodore Page, Bohdan Bakay, and William L. Nyhan

Department of Pediatrics
University of California, San Diego
La Jolla, California 92093

An almost complete deficiency of the enzyme hypoxanthine guanine phosphoribosyl transferase (HPRT) is known to be the cause of the Lesch-Nyhan syndrome (1,2). The gene for HPRT is located on the X-chromosome, so that heterozygous females show two populations of cells, one HPRT$^+$ and one HPRT$^-$, as predicted by the hypothesis of Lyon (3,4). Such mosaicism has been demonstrated in populations of cultured fibroblasts and in hair root follicles of heterozygotes (4,5).

The available techniques for the detection of these two population of cells in the heterozygote are of two types: those which measure the incorporation of hypoxanthine or guanine (or analogs of these purine bases) by cultured fibroblasts, and those which assay the activity of HPRT in individual hair roots. The tissue culture methods are quite time consuming, and it is always possible that only a single population of cells will be produced, which could give falsely positive results. The methods which measure the enzyme activity in hair roots are lacking in sensitivity and reliability, and may require highly specialized equipment.

The method described here utilizes a radiochemical assay with subsequent thin layer chromatography and autoradiography to measure the HPRT activity in individual hair roots collected from the possible heterozygote. This method is fast and sensitive, and requires no specialized equipment.

Statistical analysis has shown that heterozygotes can be reliably detected using 20 to 30 hair roots (6); 30 were used for this procedure. The hairs were carefully plucked from random areas of

the scalp so that a bulb of cells was visible on the tip of the
hair root. The hairs were inserted into Eppendorf microtubes containing 50 μl of lysis buffer, so that the sheath cells were
completely immersed. The lysis buffer consisted of 50 mM hydroxy-
ethyl piperazine ethylsulfonate (HEPES), 50 mM magnesium acetate,
20 mM magnesium hydroxide, and 5 mM phosphoric acid, at pH 7.2; the
buffer contained no monovalent cations. The tubes were frozen and
thawed three times in a methanol/dry ice bath. To these lysates
were added 5 μl of a 10 mM solution of phophoribosyl pyrophosphate
(PRPP), 5 μl of a 1 mM solution of α,β–methylene adenosine diphosphate (AOPCP), and 5 μl of a solution consisting of 1.09 mM ^{14}C
hypoxanthine and 0.81 mM ^{14}C adenine, so that adenine phosphoribosyl transferase (APRT) could be used as an internal standard.
Two positive and two negative controls were run in parallel with
the hair roots being tested. Normal hair roots were used for the
controls; the negative controls were prepared as above, but with 3
μl of 1.79 mM ^{14}C adenine in place of the adenine–hypoxanthine
solution. The tubes were incubated for three hours at 37° C in a
shaking water bath. After the incubation, 3 μl of each reaction
mixture was spotted on a polyethylimino cellulose thin layer chromatography sheet in three 1–μl portions; if the reaction products
were to be quantitated, 1 μl of a standard containing 10 mM adenosine monophosphate (AMP) and 10 mM inosine monophosphate (IMP) was
spotted with each reaction mixture. A filter paper wick was sta-
pled to the top of the sheet, and the sheet was washed overnight by
ascending chromatography in a 1:1 solution of methanol and water.
After the wash the wick was cut off and discarded, and the sheet
was thoroughly dried. The sheet was then developed sequentially in
four different buffers without drying in between as follows: 30
sec in 50 mM sodium formate, pH 4.5; 30 sec in 0.5 M sodium formate, pH 4.6; 2 min in 2 M sodium formate, pH 4.6; and then in 4 M
sodium formate, pH 4.7 until the solvent front reached the top of
the sheet. The developed TLC sheet was covered with X–ray film,
and exposed 24–48 hours. In the developed film, the reaction products were visible as dark spots. For quantitation of enzyme
activity, the standards were visualized under ultraviolet light;
the spots were cut from the sheet, and the reaction products were
quantitated by liquid scintillation spectrometry.

Chromatography of the hair roots of normal controls (Figure 1)
shows two radioactive products on autoradiography, corresponding to
IMP (R_f=0.78) and AMP (R_f=0.61), which indicated the prsence of
HPRT and APRT, respectively. In hair roots from individuals with
the Lesch–Nyhan syndrome, only the AMP spot was seen, indicating a
lack of HPRT activity. The hair roots of heterozygotes showed both
patterns, indicating the presence of both HPRT$^+$ and HPRT$^-$ hair
roots. The results of the screening of 10 individuals are shown in
Table 1. All of the viable hair roots of normal individuals were
HPRT$^+$. All of the hair roots from patients with the Lesch–Nyhan
syndrome were HPRT$^-$. The hair roots of heterozygotes showed HPRT$^+$
and HPRT$^-$ populations, as expected.

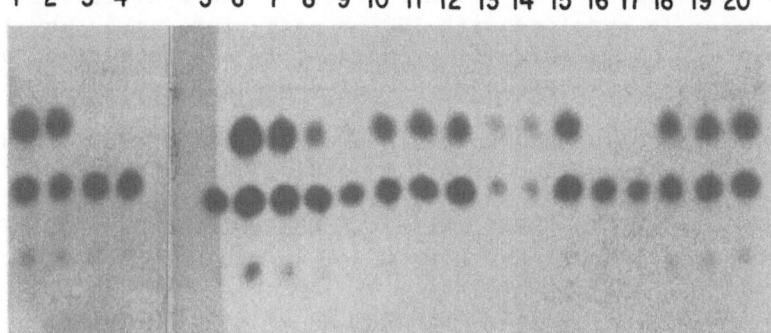

Figure 1. Autoradiography of the developed TLC sheet showing normal (1,2), Lesch–Nyhan (3,4), and heterozygote (5-20) patterns.

Table 1

Results of the Screening of 10 Individuals

Subject	HPRT$^+$	HPRT$^-$	Nonviable
Normal			
KK	30	0	0
JG	26	0	4
NC	27	0	3
Lesch–Nyhan			
CW	0	6	0
MJ	0	6	0
Heterozygote			
PG	18	12	0
LC	25	5	0
LR	19	9	2
JR	14	11	5
AW	16	13	1

For the development of this procedure, the reaction mixtures were analyzed by high pressure liquid chromatography. It was found that the yields of product in both reactions were more than doubled by the addition of the nucleotidase inhibitor AOPCP. In both reactions, a quantity of the corresponding nucleoside was formed equal to 1-3% of the main reaction product. The quantitiy of the product formed was linear, at least up to 8 hours. In the HPRT reaction it was found that 0.9% of the IMP produced had been converted to xanthosine monophosphate (XMP); the XMP occasionally produced a spot on autoradiography at R_f=0.53, but this in no way interfered with the interpretation of the results. In the absence of monovalent cations, less than 1% of the AMP produced was converted to IMP. If the reaction is supplied with 100 mM potassium, the physiological activator of AMP deaminase, approximately 12% conversion of AMP to IMP occurred. No other labeled biproducts were detected.

Quantitation of the reaction products after TLC showed that the AMP/IMP ratio varied from 1.6 to 0.6. The ratio increased with storage of the hairs at -20° C, presumably due to the relative instability of HPRT. HPRT$^-$ hairs showed a ratio of >50. The absolute activity in each case varied according to the number and condition of the sheath cells of the hair root. With hair roots which contained few cells, or roots which had been stored for long periods, it was found to be useful to increase the incubation time to 4–8 hours, spot 5–10 µl, or to increase the exposure time to 72 hours or more. If the reaction mixtures are frozen after spotting, the chromatography and autoradiography can be repeated until satisfactory results are achieved.

The advantages of this system over previous methods of heterozygote detection are sensitivity, reliability, simplicity, and the relatively short time in which it can be performed. The entire procedure is conveniently performed in 48 hours, although it can be shortened to 24 hours if scintillation counting is substituted for autoradiography. Tissue culture methods require 3–6 weeks, and there is always the danger of a false positive diagnosis resulting from the growth of only HPRT$^+$ cells from the explant. Other methods which assay HPRT activity in hair roots are not as sensitive or reliable. The use of AOPCP more than doubles the effective enzyme activity, and the measurement of APRT activity in the same assay makes the method highly reliable. Using this method, it was possible to detect heterozygotes whose hair roots did not contain enough enzyme activity to permit detection by other methods. Furthermore, since only a small amount of each reaction mixture is analyzed, any questionable results may be checked by repeating the analysis.

REFERENCES

1. Lesch,M., and Nyhan,W.L., 1964, Am. J. Med., 36:561.
2. Seegmiller,J.E., Rosenbloom,F.M., and Kelley,W.N., 1967, Science, 155:1682.
3. Lyon,M.F., 1972, Biol. Rev., 47:1.
4. Migeon,B.R., der Kaloustian,V.M., Nyhan,W.L., Yong,W.J., and Childs,B., 1968, Science, 160:425.
5. Goldstein,J.L., Marks,J.F., and Gartler,S.M., 1971, Proc. Nat. Acad. Sci., 68:1425.
6. Franke,U., Felsenstein,J., Gartler,S.M., Migeon,B.R., Dancis,J., Seegmiller,J.E., Bakay,B., and Nyhan,W.L., 1976, Amer. J. Hum. Genet., 28:123.

A RAPID AND RELIABLE METHOD FOR SEPARATION OF RIBONUCLEOTIDES, DEOXYNUCLEOTIDES, AND CYCLIC NUCLEOTIDES BY REVERSED PHASE HPLC

*H. Martínez-Valdez, M.W. Taylor and R.M. Kothari

Department of Biology, Indiana University, Bloomington
IN. 47405. * Department of Biochemistry, Sch. Med.
National University of México, 04510, México, D.F.

INTRODUCTION

We have recently reported a rapid method of separating ribo-nucleotides on a C_{18}-reversed phase column, using isocratic elution. The elution profiles were dependent on both pH and ionic strength (1). To our knowledge, none of the existing methods affords simultaneous detection of major ribo, deoxyribo and cyclic nucleotides using isocratic elution conditions. Accordingly, some protocols provide the separation of purine nucleotides alone (2), pyridine ones alone (3), exclusive separation of either ribo-nucleotides (4), or deoxynucleotides (5). Using an isocratic elution buffer and a reversed phase column, we now report the resolution of the deoxynucleotides and cyclic nucleotides from each other and from ribonucleotides. The deoxynucleotides and cyclic nucleotides elute, in general, later than the corresponding ribonucleotides, and depending on the analysis desired, their elution could be expedited using the same buffer fortified with 10% methanol (V/V). Combining periodate treatment to remove ribo-nucleotides, deoxynucleotides can be easily distinguished from ribonucleotides in biological samples. It is hoped that this protocol will be of use in the analysis of ribotides, deoxy-ribotides and cyclic ribotides from various cell types, cells under different pathological state and cells in different energy levels with varying metabolic conditions.

MATERIALS AND METHODS

Apparatus: A sample injector model U6K, solvent delivery systems M45, a variable wavelength detector model 440 (Waters Assoc., Milford, Mass.) and an automatic recorder Omniscribe

(Houston Inst.) were used. A prepacked reversed column (30x0.4 cm
I.D.) containing an octadecyl C_{18} chemically bonded (Waters Assoc.)
was utilized. A precolumn (5x0.4 cm) also packed with C_{18} material
was used to protect the main column. Main column was washed daily
with double distilled water, followed by methanol: water (30:70),
and preserved in the later in between use as previously described
(1).

    Periodate Oxidation: In principle, the method of Neu and
Heppel was followed (6). 10 mM stock solutions of each nucleotide
was prepared in distilled water and stored at -70 when not used.
Proportional volumes of ribonucleotides and deoxynucleotides were
mixed to a final concentration of 1.6 uM. Aliquots of 1.0 ml. of
this mixture were treated at ambient temperature for 10 min with
80 ul of 1 M sodium periodate after which 10 ul of 4 M cyclohexyl-
amine were added and the resultant mixture incubated at 45¤C for
90 min. After incubation, 10 ul of 0.5 M rhamnose were added to
remove the excess iodate ions. A 10 ul. sample was chromatographed at room temperature, at a flow rate of 1 ml/min, 1000 psi,
a chart speed of 1 cm/min and monitored at 254 nm at 1.0 AUFS.

RESULTS AND DISCUSSION

    From Fig. 1 (a,b,c, and d) it is clear that good separation
of each ribonucleotide and deoxynucleotide in all purine and pyrimidine series is achieved on a C_{18} column, using isocratic elution
with ammonium phosphate buffer (0.2M, pH 5.1). Fig. 2 (a,b,c, and
d) illustrates the effect of periodate treatment on the ribo-
nucleotide-deoxynucleotide mixture in each series. In the chromatography of ribonucleotides, the triphosphate always is eluted
first, closely followed by the diphosphate, with the monophosphate
being retained for a relatively longer period. A similar pattern
emerges with the deoxynucleotides. After periodate treatment, we
have noted a minor shift in the elution pattern.

    Cyclic nucleotides under our conditions have a long retention
time, and are reasonably separated from the ribonucleotides and
deoxynucleotides (c-CMP at 5.9, c-GMP at 36.4, and c-AMP at 112
minutes). However the time of elution of cAMP is longer than
normally desired with an HPLC system. It is evident from Table 1
and Fig. 3, that modification of the eluant by the addition of
differing amounts of methanol, expedites the elution without
affecting resolution. From Fig. 3 it is also clear that the
retention time of CTP, GTP, and ATP is influenced by the incorporation of methanol in the eluant.

    Five points emerge from the elution patterns: (i) the D-2
deoxyribose entity is retained in the column more tightly than
the D-ribose entity, probably due to the higher electronegativity

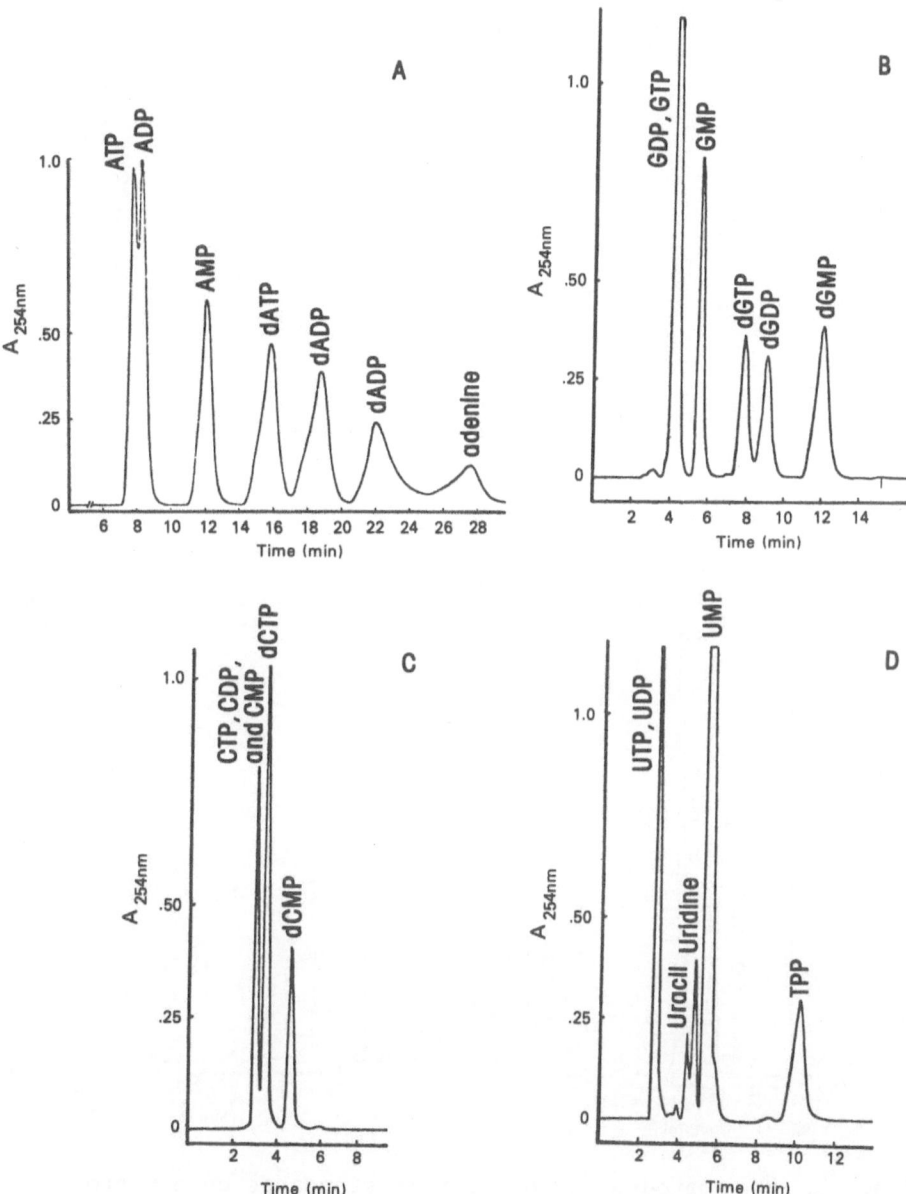

Fig. 1. Chromatogram of nucleotide standards on a micro
Bondapak C$_{18}$column before periodate oxidation.
A. Adenine series; B. Guanine series; C. Cytosine
series; D. Uracil plus thymine series.

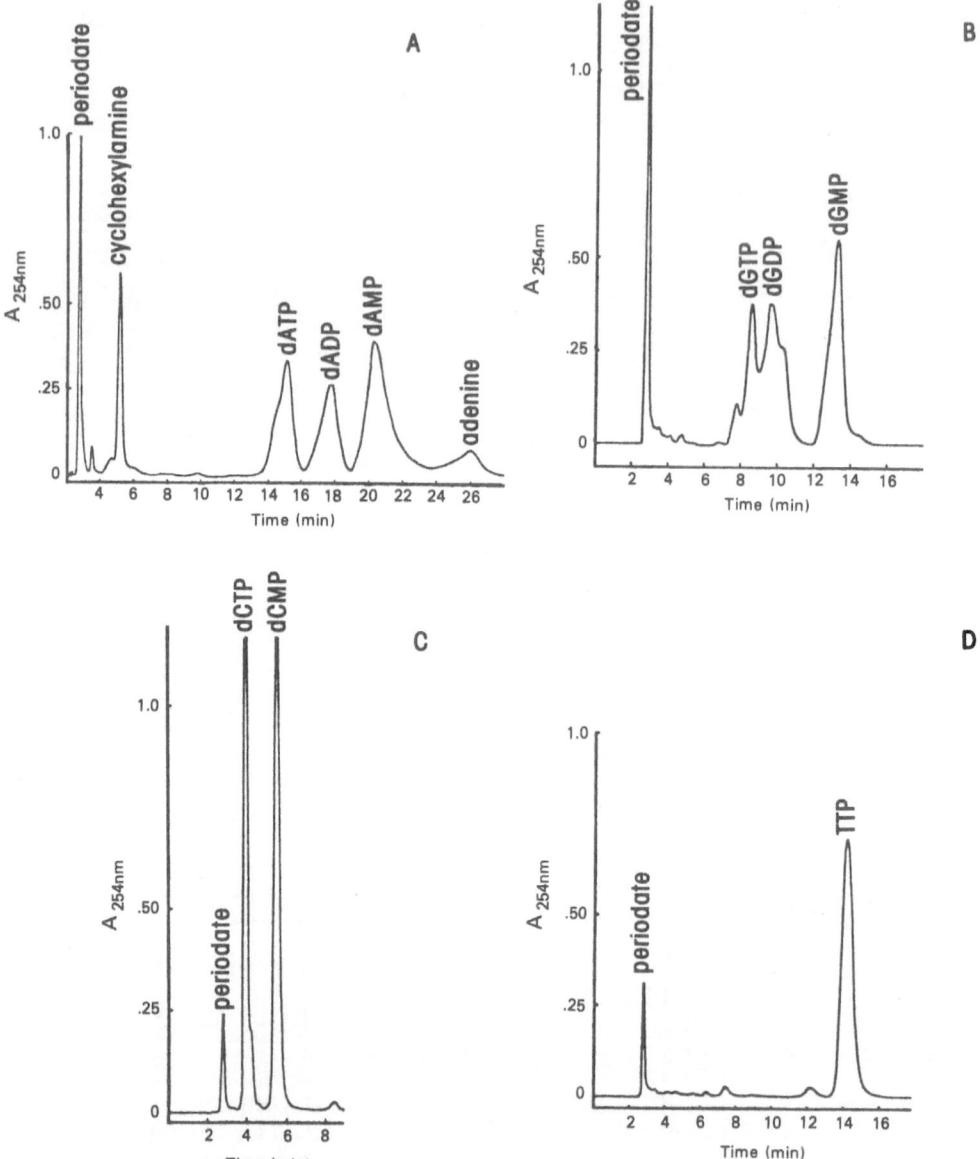

Fig. 2. Chromatogram of nucleotide standards on a micro Bondapak C_{18} column after periodate oxidation. A. Adenine series; B. Guanine series; C. Cytosine series; D. Thymidine triphosphate. Chromatography of series C and D was performed after silica cartridge fractionation (see Methods and Materials).

of the -OH ions, (ii) the purine entity is retained on the column
more tightly than the pyrimidine, probably due to its higher
charge, (iii) the elution pattern in each series is a function
ofthe degree of protonation of the PO_4 group, the entity with the
most PO_4 groups elutes earlier and the entity of the least elutes,
(iv) the degree of protonation is disproportionately masked by the
number of phosphate groups as is evident by the distance between
the elution of the tri-, di-, and monophosphates, and (v) cycliza-
tion of the monophosphates greately increases their retention on
the column (Table 1). Thus the affinity of the nucleotides for
the C_{18}-column is not simply a function of the charge on the PO_4
group alone. The recovery of nucleotides is quantitative, and the
sensitivity of this method allows us to measure samples at the
level of nMoles.

Recent efforts on the applicability of metal chelate affinity
chromatography to achieve the fractionation of AMP or GMP from
their respective deoxyhomologs were unsuccessful (8). Besides the
need for using gradient, on other procedures showed coelution or
overlapping of AMP, TMP, and UMP (9). The use of pH gradient (10),
molarity gradient (11), or pH and a molarity gradient (12) lead to
an incresed time and buffer for equilibration between analysis, and
the need for gradient programming, for optimal resolution. None of
these drawbacks are present in our procedure. It is convenient
since it uses an isocratic mode of elution, thus avoiding the
problems of gradient elution. It is simple for it does not need
prior enzymatic treatments done in other systems (13). Finally, it
is comprehensive for it affords clearcut and rapid separation of
ribo-, deoxyribo-, and Cyclic nucleotides.

TABLE 1

Elution of Cyclic Nucleotides

Nucleotide	Buffer + Methanol		
	Nil	3%	10%
c-CMP	5.9	5.7	3.85
c-GMP	36.4	21.2	8.55
c-AMP	112.4	53.5	17.35
ATP	7.8	--	3.6
GTP	4.4	--	3.2
CTP	3.05	--	2.8

The figures quoted in different columns are the retention time in
minutes.

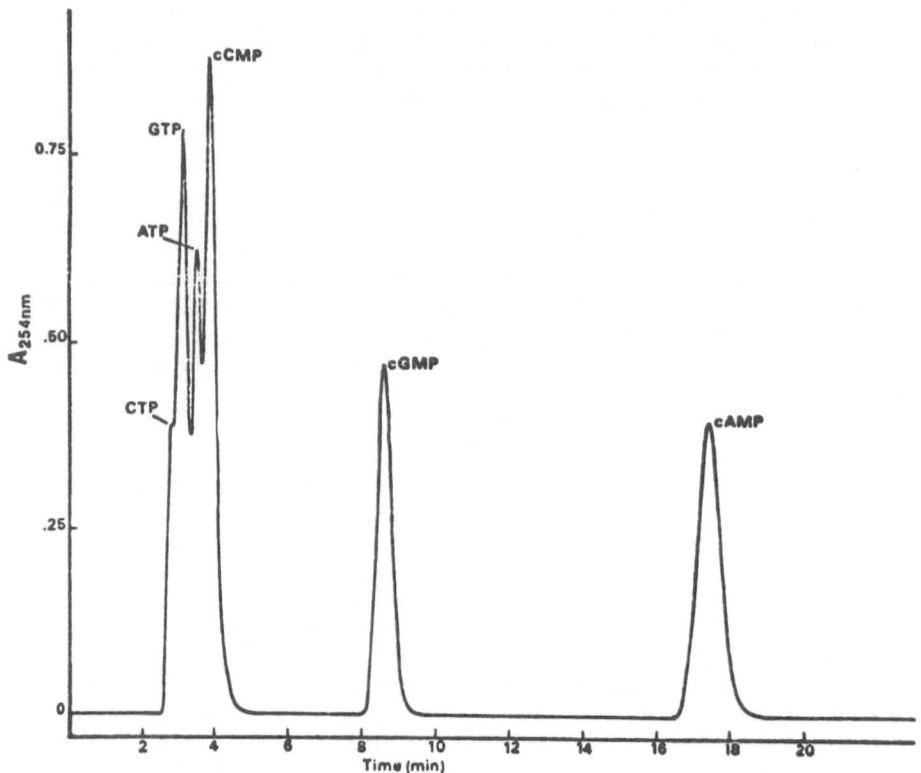

Fig. 3. Chromatogram of cyclic nucleotide standards on a
         micro Bondapak C_{18} column. The eluant was
         ammonium phosphate buffer (0.2 M, pH 5.1) con-
         taining 10% (v/v) methanol.

ACKNOWLEDGEMENTS

     This work was supported by grant GM18924 from U.S.P.H.S. The
authors wish to acknowledge the assistance of Mr.Garth E. Ziege
in the preparation of figures, as well as the typing work of miss
Alejandra Palomares.

REFERENCES

1.- Taylor, M.W., Hershey, H.V., Levine, R.A., Coy, K., and
    Olivelle, S. (1981) J. Chromatogr. 219, 133-139.

2.- Maybaum, J., Klein, F.K., and Sadee, W. (1980) J. Cromatogr.
    188, 149-158.

3.- Anderson, F.S. and Murphy, R.S. (1976) J. Chromatogr. 121,
    251-262.

4.- McKeag, M. and Brown, P.R. (1978) J. Chromatogr. 152, 253-254.

5.- Wakizaka, A., Kurosaka, K., and Okuhara, E., (1979) J. Chromatogr. 162, 319-326.

6.- Neu, H.C., and Heppel, L.A. (1964) J. Biol. Chem. 239, 2927-2934.

7.- Lothrop, C.P., Jr. and Uziel, M. (1980) Anal. Biochem 109, 160-166.

8.- Hubert, P. and Porath, J. (1981) J. Chromatogr. 206, 164-168.

9.- Hartwick, R.A. and Brown, P.R. (1975) J. Chromatogr. 112, 651-662.

10.- Pon, R.T. and Ogilvie, K.K. (1981) J. Chromatogr. 205, 202-205.

11.- Schweinsberg, P.D. and Loo T.L. (1980) J. Chromatogr. 181, 103-107.

12.- Floridi, A., Palmerini, C.A., and Fini, C. (1977) J. Chromatogr. 138, 203-212.

13.- Cohen, M.B., Maybaum, J., and Sadee W. (1980) J. Chromatogr. 198, 435-441.

TURNOVER AND DISTRIBUTION OF ^{14}C-URIC ACID IN PSORIATIC PATIENTS

Giuliano Mariani, Mario Tuoni, Roberto Gianfaldoni,
Nicola Molea, Luca Giubbolini, Lucio Fusani, Eneo Mian,
and Romano Bianchi

CNR Institute of Clinical Physiology, 5th Medical Pathology, and Institute of Dermatology of the University of
Pisa, Pisa (Italy)

INTRODUCTION

Psoriatic patients frequently present with altered serum urate
levels, a disturbance which is commonly attributed to some changes
in nucleoprotein metabolism directly linked to the pathological pro-
cess of the skin lesions; in fact, it is well known that an increased turnover rate of the epidermal cells (from the normal value of 27
days to 3-4 days only) is responsible for the psoriatic skin changes
(Fitzpatrick and Haynes, 1980). To date, no tracer turnover studies
with labelled uric acid have been reported in patients affected by
psoriasis. We present here the metabolic results obtained with the
aid of ^{14}C-uric acid in a group of psoriatic patients with various
degrees of severity of the disease. The aim of the study was to elucidate some pathophysiologic aspects of uric acid turnover in such
clinical conditions.

MATERIAL AND METHODS

Patients and Protocol of the Turnover Study

The metabolic investigation was carried out in a total of 18
subjects, including 5 healthy controls (age range 20-48 yrs.) and 13
psoriatic patients (age range 29-65 yrs.); 7 of the psoriatic patients
had a skin involvement corresponding to 15-30% of the body surface,
and 6 had a 50-100% skin surface involvement.

Uric acid-2-^{14}C (Amersham, sp. activity 51 mCi/mMole) was injected i.v. as a single bolus, and venous blood samples were taken at frequent intervals during the first 2 hours, then with a decreasing frequence algorithm until 96 hours p.i.; urines were cumulatively collected at 12 hour intervals.

Measurement of ^{14}C-Uric Acid Radioactivity

^{14}C-uric acid in the plasma and urine samples was purified by polyacrylamide gel chromatography, as previously described (Bianchi et al., 1978; 1979); P2 Biogel columns (Biorad Laboratories) were utilized to this purpose, with a 0.05 M phosphate elution buffer at pH 7. Fractions containing uric acid were pooled and counted in an automatic liquid scintillation counter. One ml samples were applied to each 0.9x28 cm column.

Data Analysis

The experimental data of plasma and urine radioactivities were analyzed by the non-compartmental approach (Rescigno and Gurpide, 1973), as described in details previously (Bianchi et al., 1979). The formulas utilized in this approach allow to determine the following parameters of uric acid kinetics: total metabolic clearance rate (MCR), mean residence time of the tracer, total distribution volume (TDV, plasma equivalent), fractional catabolic rate (FCR, relative to TDV), and clearance rate of ^{14}C-uric acid via the renal route (MCR$_r$). The total turnover rate (TR) and the total pool of exchangeable uric acid are then obtained as the product of, respectively MCR or TDV by the plasma urate concentration. The extrarenal disposal of uric acid (bacterial uricolysis in the gut, skin desquamation) is determined as the difference between the total metabolic clearance rate and the clearance rate of uric acid through the kidney route.

RESULTS

The metabolic results obtained in the control subjects and in the psoriatic patients are summarized in Table 1, and diagrammatically represented in Figure 1. In the average, psoriatic patients with 15-30% skin surface involvement were found to have total pool values significantly higher than both the controls and the 50-100% group; their serum urate levels and turnover rate values were also higher

Table 1. Metabolic results obtained in the patients studied (mean
         values ± 1 S.D.); all data normalized to sq.m. body surface

	Serum Urate mg/dl	Pool mg	MCR lt/day	TR mg/day	FCR %/day	MCR_r % MCR
15-30% (n = 7)	5.19 ± 1.26	710^1 ± 151	8.4 ± 2.2	416 ± 78	60.5 ± 15.4	53.7 ± 11.1
50-100% (n = 6)	3.98 ± 1.61	444 ± 234	14.3^2 ± 5.4	440 ± 161	110.4^3 ± 54.0	46.0 ± 25.8
Controls (n = 5)	4.47 ± 0.46	563 ± 37	8.5 ± 1.7	381 ± 78	67.3 ± 11.5	65.6 ± 7.0

(^{1}) Significantly different (P<0.05) from controls and 50-100% group;
(^{2}) Significantly different (P<0.025) from controls and 15-30% group;
(^{3}) Significantly different (P<0.05) from the 15-30% group.

than normal, though not significantly. Whereas, the 15-30% group had
virtually normal values of MCR, and slightly reduced renal clearance
rate values.

The patients with 50-100% skin surface involvement showed mean
MCR values significantly higher than both the controls and the 10-40%
group, and FCR significantly higher than the 15-30% patients; they
also had total turnover rate values slightly higher than both the control and the 15-30% groups, whereas their mean serum urate levels,
total pool, and renal clearance rates were lower than in the other
two groups; though without statistical significance. Two patients
with generalized psoriasis (erythrodermic stage) had the lowest values
of uric acid removal through the kidneys (respectively 30.2 and 8.6%
of total MCR), while their total turnover rates were very high (respectively 652 and 530 mg/day x sq.m.). Thus, removal of uric acid via
routes other than the kidneys accounted in these two patients for
about 70 to 90% of the total turnover rate (normal value: about 35%).

DISCUSSION

Some interesting and partly unexpected results as concerns the
uric acid kinetics were obtained in this study. In general, psoriatic
patients had much faster disappearance rates of tracer uric acid
than the control subjects had; this increase in tracer removal was

roughly proportional to the extent of skin surface involvement. For
instance, in one patient with 100% skin surface involvement plasma
concentration of ^{14}C-uric acid went down to nearly zero within about
60 hours after injection, as compared to more than 120 hours in control subjects. This metabolic behaviour is therefore reflected by
the high MCR, FCR, and TR values observed in the group of patients
with more severe disease involvement. Yet, these patients exhibited
serum urate levels and total pool values slightly lower than normal.
These findings are consistent with the assumption that in these patients the higher removal rate of uric acid depends on a direct loss
of uric acid from the body (as it may be due to extensive skin desquamation), whose extent is such to keep their serum urate levels
and total pool values lower than normal; in fact, the extrarenal disposal rate of uric acid is particularly high in these patients (and
especially so in those cases with erythrodermic disease), and it may
be assumed that most of this extrarenal quota is due to the skin desquamation processes.

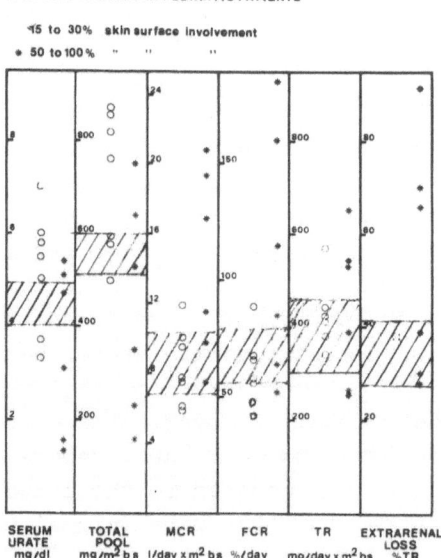

Fig. 1 - Diagrammatic representation of the metabolic results ob-
          tained in the individual psoriatic patients, respectively
          with 15-30% skin surface involvement and 50-100% involve-
          ment. The hatched areas indicate the ± 1 Standard Devia-
          tion range on the mean values observed in the control group.

On the contrary, the desquamation processes appear to be less pronounced in the patients with a 15-30% skin surface involvement. In
these patients the main effect of the increased cellular and nucleoprotein turnover is therefore able to induce an increase in both the
serum urate levels and the total pool values (besides the expected
increase in the total turnover rate), since the extrarenal loss due
to direct skin desquamation is not so important as to counterbalance
this primary metabolic effect.

REFERENCES

Bianchi, R., Vitali, C., Clerico, A., Pilo, A., Fusani, L., and Ma-
    riani, G., 1979, Uric acid metabolism in normal subjects and in
    gouty patients by chromatographic measurement of ^{14}C-uric acid
    in plasma and urine, Metabolism, 28: 1105.
Bianchi, R., Vitali, C., Clerico, A., Pilo, A., Fusani, L., Riente,
    L., and Mariani, G., 1978, A chromatographic method for the de-
    termination of uric acid turnover in man, J. Nucl. Med. All.
    Sci., 22: 37.
Fitzpatrick, T.B., and Haynes, H.A., 1980, Skin lesions of general
    medical significance, in: "Principles of Internal Medicine",
    K.J. Isselbacher, R.D. Adams, E. Braunwald, R.G. Petersdorf,
    and J.D. Wilson, eds., McGraw-Hill Kogakusha, Tokio.
Rescigno, A., and Gurpide, E., 1973, Estimation of average times of
    residence, recycle, and interconversion of blood-borne compounds
    using tracer methods, J. Clin. Endocrinol. Metab., 36: 276.

SERUM URIC ACID LEVELS IN PSORIASIS

J.T. Scott and M.A. Stodell

Kennedy Institute of Rheumatology

Bute Gardens
London W6 7DW, England

Psoriasis is commonly thought to be associated with hyper-uricaemia[1] and an association between the extent of psoriasis and serum uric acid has been observed[2]. These views have however been disputed[3]. The purpose of this study was to investigate the possible presence of hyperuricaemia in uncomplicated psoriasis.

PATIENTS AND METHODS

Three carefully age and sex matched groups were studied, each of 41 subjects.

1. Patients with psoriasis attending the Dermatology Clinic, Charing Cross Hospital, London. (21 men, 20 women; mean age 38 years, range 18-65 years).

2. Patients attending the same clinic with contact dermatitis (21 men, 20 women; mean age 41 years, range 18-64 years).

3. Hospital employees without skin disease (21 men, 20 women; mean age 38 years, range 18-65 years).

No subjects had any past or present evidence of inflammatory arthritis, or were taking drugs known to influence levels of serum uric acid.

Extent of psoriasis was estimated by the "rule of nines"[4]. Alcohol consumption was graded nil = 0, 2 glasses or spirit or 2 glasses of beer daily = 1, greater alcohol consumption = 2. Serum uric acid was measured by Technicon Autoanalyser and other biochemical variables by standard techniques.

283

RESULTS

The means and standard deviations of the serum uric acid levels in the three groups were as follows:-

Group 1 (psoriasis)	311 ± 85 µmol/l
Group 2 (dermatitis)	309 ± 45 µmol/l
Group 3 (hospital employees)	293 ± 66 µmol/l

There is no significant difference between results in the three groups. Nor were there significant differences with regard to mean alcohol consumption, alanine aminotransferase (ALT) and gamma glutamyl transpeptidase (GGT).

	Alcohol consumption	ALT	GGT
Group 1	0.8	23	13
Group 2	0.7	20	11
Group 3	0.7	21	11

In order to assess any possible association between extent of psoriasis and serum uric acid, another 31 patients with psoriasis (fulfilling inclusion criteria for the study) were included with the original 41 psoriatic patients. In these 72 patients no significant correlation could be observed between the extent of psoriasis and serum uric acid. (R = 0.2, p = 0.08).

CONCLUSIONS

1. Serum uric acid levels were not shown to be significantly raised in patients with uncomplicated psoriasis.

2. There was no significant association between extent of psoriasis and serum uric acid.

3. Patients with psoriasis, when compared to patients with dermatitis and healthy hospital employees, showed no differences with regard to alcohol consumption and liver enzymes.

REFERENCES

1. Kelley W M, Harris E D, Ruddy S, Sledge C B.
        Textbook of Rheumatology, W.B.Saunders Co. Philadelphia
        (1981) p497.

2.   Eisen A Z, Seegmiller J E. Uric acid metabolism in psoriasis.
         J. Clin. Invest. 40: 1486 (1961)

3.   Lambert J R, Wright V. Serum uric acid levels in psoriatic
         arthritis. Ann.Rheum.Dis. 36: 264 (1977)

4.   Evans E I, Purnell O J, Robinett P W, Batchelor A, Martin M.
         Fluid and electrolyte replacement in severe burns.
         Ann.Surg. 135: 804 (1952)

PURINE METABOLITES, URACIL AND cAMP DURING EXCHANGE TRANSFUSION

H. Manzke, H.G. Koke and K. Kruse

University Hospital for Children

Kiel, FRG

INTRODUCTION

The metabolism of newborn infants during their first week of life is characterized by an increased breakdown of nucleoproteins leading to transient hyperuricemia and hyperuricosuria which is particularly enhanced in infants with hemolytic diseases. Furthermore, there are physiologically considerable changes in the concentration of calcium in the blood of newborns.

AIMS OF THE STUDY

The purpose of the present study was threefold:
1) to investigate the influence of the exchange transfusion on the elimination of the oxypurines uric acid, xanthine, and hypoxanthine as well as of the pyrimidine uracil.
2) to determine the amount of adenine retained during and after the exchange transfusion with CPD-adenine blood and
3) to study the citrate-induced hypocalcemic stimulation of the neonatal parathyroid glands by measuring the cyclic AMP concentration in the blood and in the urine during and after exchange transfusion with CPD-adenine blood.

MATERIAL

The study was performed on five newborn infants with various causes of hyperbilirubinemia leading to exchange transfusion (E.T.) with CPD-adenine blood (200-240 ml/kg).
Pat. 1: B.R., f., premature, birth weight 780 g, idiopathic respiratory distress syndrome, hyperbilirubinemia, E.T. 2nd day.
Pat. 2: N.S., f., birth weight 4050 g, pyruvate kinase deficiency,

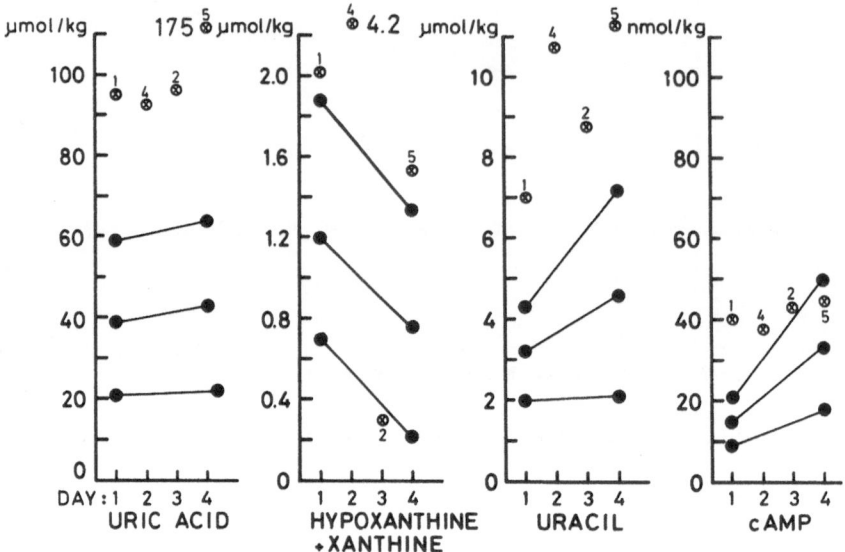

Fig.1:Urinary output of uric acid, hypoxanthine+xanthine, uracil
and cAMP in 18 normal newborns on their first and fourth day of
life. Mean and SD-range of the values. The 24-h. urinary excretions
of the same substances measured in 4 newborns during and after E.T.
are plotted. Numbers above the circles refer to the patient's number.

bilirubin 31 mg%, E.T. 4th day.
Pat. 3: R.D., m., premature, birth weight 2170 g, sepsis, E.T. 6th
day.
Pat. 4: I.G., m., birth weight 4460 g, immune thrombocytopenia,
hyperbilirubinemia (20.1 mg/dl), E.T. 2nd day.
Pat. 5: T.W., m., birth weight 3010 g, suspected sepsis, E.T. 4th
day.

BIOCHEMICAL METHODS

    Uric acid (uricase) and hypoxanthine-xanthine (xanthine oxi-
dase) by enzymatic methods. Hypoxanthine, xanthine, adenine, 2,8-
dihydroxyadenine by HPLC. Cyclic AMP by an assay kit (code TRK
432). Phosphate by Merckotest 3338 and calcium by an calcium analyzer (Corning Eel model 940). All analyses on serum and urine
samples were done in duplicate.

RESULTS AND COMMENTS

    There were only slight decreases in the plasma concentrations
of the oxypurines and uracil in the newborns during the E.T.s
despite great differences between their concentrations in the
donor blood (uric acid 264±80, hypoxanthine 23±13, xanthine

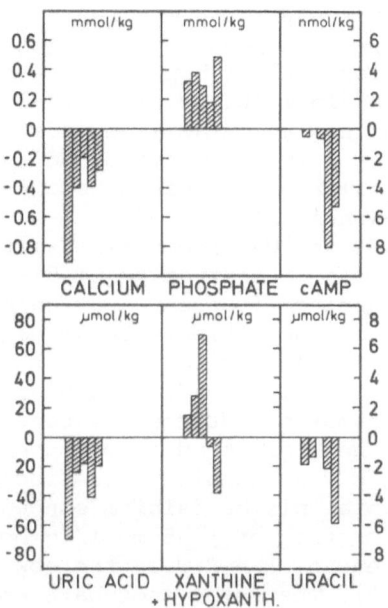

Fig.2: Net balance of calcium, phosphate, cAMP, uric acid, hypoxanthine, xanthine and uracil infused and removed per kg body-
weight in five newborns during E.T.s with CPD-adenine donor blood.

8.6±5.4, and uracil 3.0±1.4 µmol/l) and in the blood of the patients before the E.T.s. This indicates a rapid distribution of
the oxypurines and uracil among the body compartments during the
E.T.s. The mean values before and after E.T.s in the 5 newborns
were: uric acid 379±125 resp. 352±75, xanthine 10.7±4.9 resp.

11.0±5.5, hypoxanthine 15.0±7.5 resp. 12.7±4.4, and uracil 21.6±
11.2 resp.15.1±6.6 µmol/l. Twentyfour hours after E.T.s the mean
values in the plasma concentrations of uric acid (219±55 µmol/l),
xanthine (6.2±3.2 µmol/l), and uracil(10.5±5.0 µmol/l) were decreased
significantly, whereas the plasma concentration of hypoxanthine
($\bar{x}$=11.0±3.1 µmol/l) only fell slightly. The 24-h urinary excretion
of uric acid and uracil after the E.T.s was approximately twice
that measured in 18 normal newborn infants (Manzke et al.,1980),s.
fig.1. The 24-h excretion rate of hypoxanthine+xanthine was relatively high in 3 and low in 1 infant. Nevertheless, the effect of
the E.T.s on the oxypurines and uracil depends on its difference
between the concentrations in the donor blood and in the blood of
the newborns. The net balance of these substances, measured in
each individual, is shwon in fig. 2.
The donor blood (CPD-adenine 0.25) contained citrate-phosphate-
dextrose with adenine in a concentration of 81-199 µmol/l before
the E.T.s. The serum adenine disappeared at once after transfusion
and was not detectable in the patients serum samples as well as in
their urine samples collected during and after the E.T.s. Also,
there was no (nephrotoxic) 2,8-dihydroxyadenine detectable in all
the serum and urinary samples. Our results confirm the results of
Kreuger 1976 who found negligible adenine concentrations ranging
from 4-8 µmol/l during E.T.s. Adenine sideeffects were not observed. Although the total plasma calcium concentrations increased
under the i.v. administration of 1.35 mmol/kg body weight up to
3.2-4.5 mmol/l at the end of the E.T.s, the newborns had a negative
calcium balance (fig.2). The serum phosphate concentrations increased from $\bar{x}$=1.80±0.62 at the begin to $\bar{x}$=2.55±0.33 at the end of
the E.T.s. The urinary 24-h output of cAMP was markedly higher
than that measured in 18 newborns during their first four days of
life (fig.1). These findings suggest a stimulation of the parathyroid glands during E.T.s with CPD-adenine blood as also shwon by
Brown et al.,1978.

REFERENCES

Brown, D.R., Donavan, E.F., Tsang, R.C., Bobik, C.M., Chen, I.-W.,
and Johnson, J.R., 1978: Urinary phosphate and cyclic AMP excre-
        tion following citrate-induced hypocalcemic stimulation of
        the neonatal parathyroid glands, J.Pediatr. 93:842
Kreuger, A., 1976: Adenine metabolism during and after exchange
        transfusions in newborn infants with CPD-Adenine blood,
        Transfusion 16:249
Manzke, H., Spreter v. Kreudenstein, P., Dörner, K., and Kruse, K.,
1980, Quantitative measurements of the urinary excretion of crea-
        tinine, uric acid, hypoxanthine and xanthine, uracil,
        cyclic AMP and cyclic GMP in healthy newborn infants, Eur.
        J.Pediatr. 133:157

INCREASES IN METHYLATED NUCLEOSIDES DURING HUMAN PREGNANCY

Roger J. Simmonds, R.A. Harkness and S.B. Coade

Division of Perinatal Medicine, MRC Clinical Research
Centre,
Watford Road,
Harrow, HA1 3UJ, U.K.

In growing tissues there is an increased turnover of tRNA
and consequent increases in tRNA breakdown.   The methylated
bases and pseudouridine characteristic of tRNA are not 'salvaged'
unlike the majority of the constituent bases of tRNA.   There is
therefore an increased concentration of methylated purine nucleosides and pseudouridine in extracellular fluids in many cancer
patients (Mrochek et al., 1974). We have shown such an increase
in human pregnancy and in children.

A fetal extracellular fluid, amniotic fluid, has been studied
in late pregnancy.   Since amniotic fluid in late pregnancy is
largely derived from fetal urine, urinary excretion by newborn
infants has been compared to adults.   Plasma from women in late
pregnancy has also been compared to that from men.

Nucleosides were separated from urine, amniotic fluid and
plasma by chromatography on boronate columns with a high affinity
for vicinal glycols.   The concentrated nucleoside fraction was
then separated by high pressure liquid chromatography on 3 or 5
μm ODS (Hypersil) columns using a gradient of 1-30% .(v/v) methanol
in 0.004 mol/l KH_2PO_4 pH* 6.5 at 30^0C.   Components with correct
retention times and 254/280 nm absorbance ratios were estimated
by their absorbance.

The results in Table 1 show that concentrations in amniotic
fluid from infants born healthy and from those showing clinical
signs of intrapartum asphyxia are similar.   In contrast there
may be lower excretion by newborn infants who showed reduced

Table 1. Amniotic fluid concentrations of methylated
          nucleosides

	Concentrations in nmol/l			
	M^1I	M^1G	M^2G	M^2_2G
Normal mean	417	122	28	177
n = 7 SEM	83	28	7	26
Asphyxiated mean	425	119	22	269
n = 6 SEM	143	46	10	72

Table 2. Excretion of nucleosides by infants in the first month of life

	Ino	Ado	Guo	Psurd	M^1I	M^1G	M^2G	M_2^2G
				nmol/kg per h				
IUGR (n = 3)								
Mean	1.1	0.9	0.1	(191)	21.7	4.7	0.8	13.8
SEM	0.7	0.6	0.05	(n=1)	14.9	2.6	0.4	9.4
Asphyxiated (n = 6)								
Mean	8.0	0.1	1.2	366	40.1	9.9	1.8	30.2
SEM	2.0	0.04	0.4	33	9.0	2.0	0.6	6.0

Table 3. Excretion of nucleosides by 5 adults (3 male 2 female) for one week

	Ado D[†]	Ado N[†]	Psurd D[†]	Psurd N[†]	M^1I	M^1G	M^2G	M_2^2G
				nmol/kg per h				
Mean	0.9	1.0	116	99	12.7	5.3	2.0	12.9
SEM	0.2	0.2	5.4	4.9	0.8	0.5	0.2	0.7
	P 0.013*		P 0.0002*					

* Sign test   †D = Day   †N = Night

Table 4. Plasma concentrations of methylated nucleosides in pregnant women and in men and a non-pregnant woman.

	nmol/l				
	M^1I	M^1G	M^2G	M_2^2G	Ado
Pregnant (n = 4)					
Mean	27.8	7.0	7.3	36.5	210
SEM	3.0	4.6	2.0	13.5	43
Non-pregnant (n = 4)					
Mean	23	9.0	9.0	17.8	193
SEM	1.5	0.6	0.6	2.5	17

growth in utero (Table 2) compared to excretion by previously asphyxiated infants. Since fetal urine is the major source of amniotic fluid in late pregnancy it is probable that newborn children after asphyxia will show normal excretion because the amniotic fluid concentrations in infants showing intrapartum asphyxia are similar to normal (Table 1).

The excretion by newborn infants who showed signs of intra-partum asphyxia was greater than by adults, again consistent with some increased excretion associated with growth. Urinary excretion by 3 older children 12-17y was lower than that by the newborn infants after intrapartum asphyxia but higher than excretion by adults. In plasma pregnancy was not associated with increased concentrations of methylated nucleosides or adenosine. In contrast preliminary results suggest urinary excretion is higher during pregnancy. These findings are consistent with high renal clearances.

In conclusion urinary excretion of newborn and amniotic fluid concentrations of 1-methylinosine, 1-methylguanosine, N^2-methyl-guanosine and N^2-dimethylguanosine may provide an index of fetal tRNA turnover and thereby of growth.

REFERENCES

Mrochek, J.E., Dinsmore, S.R., and Waalkes, T.P., 1974, Analytic
     techniques in the separation and identification of specific
     purine and pyrimidine degradation products of tRNA:
     Application to urine samples from cancer patients. J. Nat.
     Cancer Inst., 53: 1553.

URINARY EXCRETION OF METHYLATED NUCLEOSIDES

IN DIFFERENT STATES OF FAILURE TO THRIVE

P. Clemens, Gisa Ziemer, J. Altenhoff, R. Grüttner,
Gesa Heller-Schöch[+] and G. Schöch[+]

Department of Pediatrics, University of Hamburg
Martinistrasse 52, D-2000 Hamburg 20, F.R.G. and
[+] Research Institute for Child Nutrition, Heinstück 11
D-4600 Dortmund, F.R.G.

INTRODUCTION

According to the cascade DNA→RNA→protein synthesis, disturbances of protein synthesis should be accompanied by alterations of RNA-metabolism. These are traceable by measuring "one way" urinary methylated RNA-catabolites.[1]
Therefore we followed the excretion of several methylated ribonucleosides in different states of failure to thrive caused by disturbances of nutrition or metabolism: anorexia nervosa, acute enteritis and hypothyreoidism. In all three states an effective treatment can change the metabolic situation towards anabolism.

PATIENTS AND METHOD

Anorexia nervosa: postpuberal female of 17 years. Before onset of therapy she had lost weight from 40 to 24 kg within six months. Acute enteritis: 2 months old boy. Before onset of therapy he had lost weight from 4700 to 4200 g within three days. Hypothyreoidism: 9 years old boy, somatic and psychointellectual developmental age 3-5 years, as yet untreated.
Details of the analysis method have been published[2]. We measured (in duplicate) creatinine and six methylated ribonucleosides: 1-methyladenosine (m^1A), 3-methyluridine (m^3U), 1-methylinosine (m^1I), 1-methylguanosine (m^1G), 2-methylguanosine (m^2G) and 2,2-dimethylguanosine (m_2^2G).

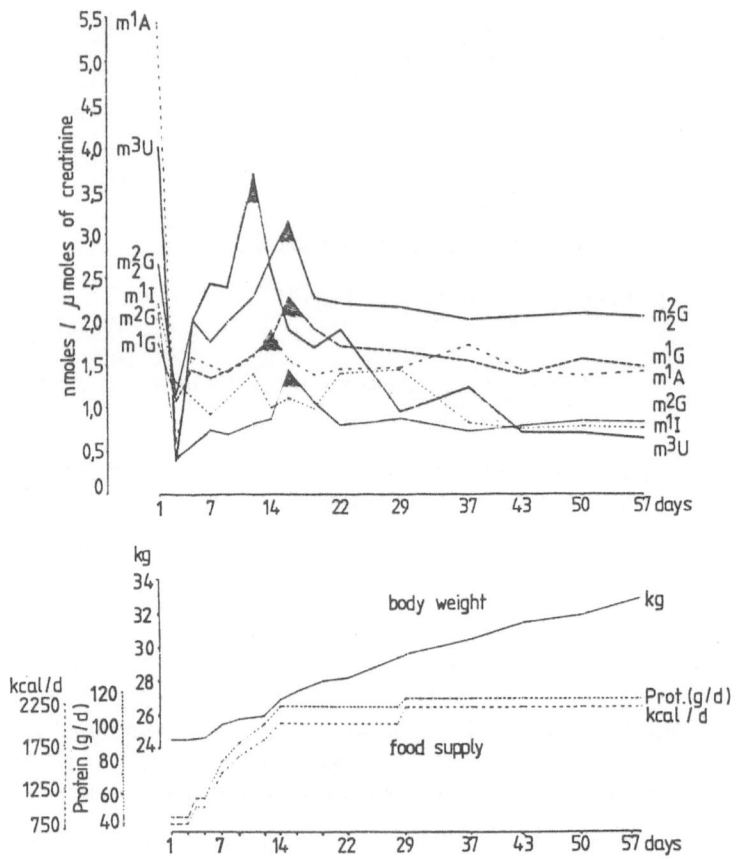

Fig. 1. Anorexia nervosa: the first 8 weeks of therapy.

RESULTS

Different phases of nucleoside excretion can be distinguished:
1. Before onset of therapy high values - in anorexia and enteritis.
2. After onset of therapy low values - in anorexia and enteritis.
3. After several days of therapy a short period of overshooting values - in all three cases.
4. Finally values stabilized on an intermediate level, as documented in anorexia and hypothyreoidism.

DISCUSSION

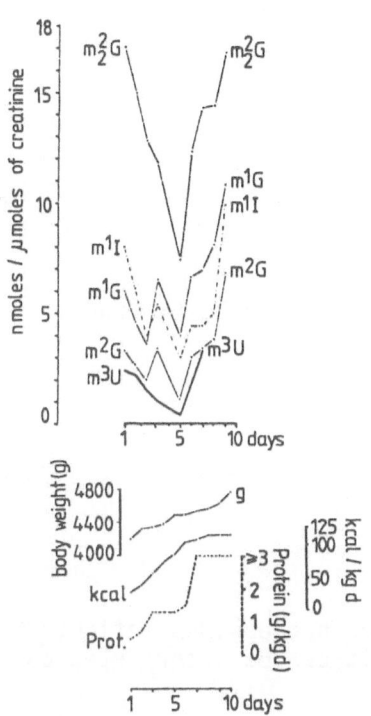

Fig. 2.: Acute enteritis: the first ten days of therapy.

1. Before onset of therapy our patients with anorexia and enteritis were in a clearly catabolic situation. Thus the observed high nucleoside excretion should reflect enhanced degradation of nucleic acids in train of tissue catabolism.
   In hypothyreoidism there are no initial high values as sign of catabolism. As in hypothyreoidism all metabolic processes are depressed there is no need for net catabolism.

2. At onset of feeding the nucleoside excretion decreases. This is interpreted as to reflect a situation in which gross catabolism has been stopped but net anabolism not yet started.

3. When normocaloric nutrition has been reached in anorexia and enteritis, a peak excretion can be observed. This obviously overshooting RNA turnover in an early anabolic state might tentatively be interpreted as to reflect an extremely active protein synthesis in the phase of regeneration of essential enzymes. In analogy the peak excretion in hypothyreoidism one week after administration of thyroxine can be interpreted as the first sign of an anabolic activation of metabolism. Normalisation of hormone values can be measured only later.
   Weight changes by themselves are no certain criterias for anabolism or catabolism as shown by rehydration in enteritis or disappearance of myxedema in hypothyreoidism.

4. In continuous anabolism the RNA-turnover is stabilized on an intermediate level as documented in anorexia and hypothyreoidism. The enteritis patient was dismissed before we could explore this stage.

The overall higher values in the baby with enteritis are due to the marked age dependency of the excretion of RNA catabolites.[3]

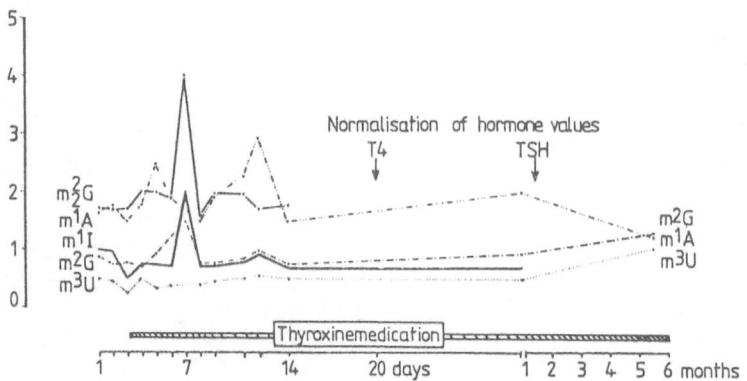

Fig. 3. Hypothyreoidism: the first six months of therapy.

SPECULATION

The excretion of "one way" methylated nucleosides reflects the activity of RNA turnover. Therefore it can be interpreted as a new indicator of catabolism-anabolism.

REFERENCES

1. G. Schöch and G. Heller-Schöch: Molekularbiologie und klinische Bedeutung des Stoffwechsels normaler und modifizierter Nucleo-basen, Helv. Paediatr. Acta (Suppl) 38:7 (1977).
2. G. Schöch, J. Thomale, H. Lorenz, H. Suberg, U. Karsten: A new method for the simultaneous analysis of unmodified and modified urinary nucleosides and nucleobases by high performance liquid chromatography. Clin. Chim. Acta 108:247 (1980).
3. G. Schöch, H. Lorenz, G. Heller-Schöch, H. Baisch, P. Clemens: Die Altersabhängigkeit der normalen und modifizierten Nucleo-basen im Urin als Ausdruck der Wachstumsgeschwindigkeit. Mschr Khlkde 129:29 (1981).

This study was supported by Deutsche Forschungsgemeinschaft.

REGIONAL VARIATION IN EXTRACELLULAR PURINE LEVELS IN VIVO

M. H. N. Tattersall, P. Slowiaczek, and A. De Fazio

Ludwig Institute for Cancer Research (Sydney Branch)
Blackburn Building,
University of Sydney, N.S.W. 2006, Australia

INTRODUCTION

In spite of many and varied roles for purine nucleosides, e.g. in immunocompetence, modulation of neurotransmission, blood vessel tone, modulation of antimetabolite action and platelet aggregation, there have been few reported measurements of purine concentrations in plasma and other tissues.

We have used an adaptation of the fluorimetric assay described by Gardiner[1] to measure purine levels in ultrafiltrates of human plasma (collection from different sites), cerebrospinal fluid and bone marrow aspirates. The assay technique allows the sequential determination of three major purine components, i.e. Hypoxanthine (Hx) plus Xanthine (X), Inosine (IR) and Adenosine (AR).

Collecting specimens into an inhibitor of adenosine deaminase (ADA), erythro-9-(2-hydroxy-3-nonyl) adenine (EHNA) markedly influences purine profiles and ensures more accurate quantitation of *in vivo* purine levels. We have further validated our adenosine measurements by the use of the riboside transport inhibitor dipyridamole (DP).

METHODS

All samples were collected into heparinized tubes containing EHNA (5–10µM final concentration) unless otherwise specified. On occasion, 1mM DP was used to completely inhibit riboside transport between intra- and extra-cellular compartments (data not shown). For comparative purposes, samples from different internal sites were collected virtually simultaneously from individual donors.

Purine metabolites were measured via fluorimetric detection of the rate of H_2O_2 production during sequential catabolism of uric acid[2]. H_2O_2 oxidizes nonfluorescent dichlorofluorescin to fluorescent dichlorofluorescein under conditions that allow for detection of 10pmoles of purine substrate. Deoxy- and ribo-nucleosides may be partitioned by boronate affinity chromatography[3] thus allowing their separate determination.

RESULTS

A)   Plasma

By using EHNA during blood collection, the amount of AR recovered always increased without substantially altering total purine levels. Very similar purine profiles were apparent when both EHNA and DP were used during blood collection indicating that results obtained using EHNA alone were not due to artefactual expansion of the extracellular AR pool due to AR transport from the red blood cells.

Blood clotting rapidly altered extracellular purine levels and where heparin was not present, extracellular levels rose rapidly due mainly to AR release. Even in the presence of heparin, cells must be removed from whole blood within minutes of venepuncture, otherwise elevated purine levels result.

Peripheral venous plasma from 18 normal volunteers (collected from the antecubital vein, using a tourniquet) contained AR $1.0 \pm 0.7\mu M$ (mean $\pm$ S.D.), IR $0.4 \pm 0.4\mu M$, Hx + X $1.3 + 1.1\mu M$, total purines $2.7 \pm 1.2\mu M$. No deoxynucleosides could be demonstrated in normal human plasma, however, elevated plasma deoxyadenosine has been demonstrated during clinical treatment with 2'-deoxycoformycin[4]. When 11 pairs of simultaneous peripheral venous and arterial plasma samples (collected by catheterisation of the femoral artery) were compared, no significant difference was found in total purines measured. However, the Hx + X fraction was significantly lower in arterial blood ($0.0025 < P < .005$). Some measurements of neonatal cord blood seem to confirm the suggestion that hypoxia tends to elevate extracellular purine levels, especially the Hx + X fraction (data not shown).

On average, there were found to be slightly higher AR levels in hepatic venous blood when compared with simultaneously drawn peripheral blood, however, the difference was not significant ($P > 0.1$). Purine levels (especially the AR fraction) in 3 patients with severe liver disease were very low (data not shown).

Table 1. Extracellular Purines in Man[a]

Source	No.	Purine (µMolar)			
		Hx + X	IR	AR	Total
Arterial	11	1.4+0.55[b]	0.2+0.19	1.2+0.76	2.8+0.98
Peripheral Venous	11	1.8+0.83	0.1+0.09	1.1+0.84	3.0+1.18
Hepatic Venous	6	1.5+1.98	0.2+0.12	2.2+0.82	3.9+1.55
Peripheral Venous	6	1.8+1.72	0.2+0.19	1.7+1.20	3.6+1.37
Bone Marrow	5	16.2+3.15	2.3+3.08	4.1+2.11	22.6+5.88
CSF	8	4.2+1.70	0.8+0.50	0.1+0.19	5.1+1.49

[a]*Samples collected into 5µM EHNA and heparin. Cells removed immediately by centrifugation. Purines were assayed in sample ultrafiltrates as described in reference[2].*

[b]*Mean ± Standard Deviation*

B)   Bone Marrow Aspirates

Bone marrow aspirate supernatants have an approximately 10-fold greater concentration of purines than do simultaneous peripheral venous samples (Table 1). AR comprises only 10 - 20% of bone marrow purines with Hx + X predominating.

C)   Cerebrospinal Fluid

Total purine levels in cerebrospinal fluid were always elevated when compared with simultaneous peripheral venous plasma (P < 0.05). AR was much lower, being detected in only 3 of 8 CSF specimens (P < 0.025), (Table 1). Since no significant metabolism of [14C-] AR to either IR and Hx or to phosphorylated products was detected in CSF, these results suggest that either AR in plasma does not readily cross into CSF, or that it is rapidly metabolised by the endothelial lining of the subarachnoid space.

DISCUSSION

We demonstrate the use of EHNA and rapid handling techniques to achieve more realistic quantitation of plasma purine levels *in vivo* and consequently report AR levels higher than in some previous reports[5]. A recent publication[6] using 2'-deoxycoformycin to similar effect, corroborates our reported AR levels.

Arterial plasma purines contain a significantly lower Hx + X fraction than simultaneously drawn peripheral venous plasma, lending support to reports that hypoxia leads to increased regional purine release[7]. AR levels were slightly higher in hepatic venous plasma, although no significant variation from peripheral venous samples was apparent. Plasma from bone marrow aspirates contained 10-fold higher purine concentrations than peripheral plasma and this is certainly an underestimate of *in vivo* bone marrow purines as samples are invariably haemodilute. Thus, the bone marrow may be a major contributor to plasma purines which have probably been derived through red cell nuclear degradation and white cell death in the marrow. The role of the liver in the supply of extra-cellular purines remains uncertain.

There appears to be a selective concentration of purines into CSF in favour of Hx and X, with AR levels being well below those of plasma. The relevance of such variations in regional purine levels to neurological and vascular function, and to the selective action of nucleic acid antimetabolites, remains to be determined.

REFERENCES

1.   D. Gardiner. A rapid and sensitive fluorimetric assay for adenosine, inosine and hypoxanthine. Anal. Biochem. 95:377. (1979).
2.   P. Slowiaczek, M. H. N. Tattersall. The determination of purine levels in human and mouse plasma. Anal. Biochem. (in press) (1982).
3.   E. H. Pfadenhauer, S. D. Tong. Determination of inosine and adenosine in human plasma using high performance liquid chromatography and boronate affinity gel. J. Chromatogr. 162:585 (1979).
4.   R. F. Kefford, R. M. Fox. Deoxycoformycin induced response in chronic lymphocytic leukemia: Deoxyadenosine toxicity in non-dividing lymphocytes. Brit. J. Haematol. 50:627 (1982).
5.   A. Leyva, J. Schornagel, H. M. Pinedo. High pressure liquid chromatography of plasma pyrimidines and purines and its application in cancer chemotherapy. Adv. Exp. Med. Biol. 122B:389 (1980).
6.   M. C. Capogrossi, M. R. Holdiness, Z. H. Israili. Determination of adenosine in normal human plasma and serum by high perform-ance liquid chromatography. J. Chromatogr. 227:168 (1982).
7.   O. D. Sangstad. Hypoxanthine as a measure of hypoxia. Pediat. Res. 9:158 (1975).

LYMPHOCYTE ADENOSINE DEAMINASE LEVELS IN ACTIVE AND IN INACTIVE

FORMS OF HBsAg-POSITIVE CHRONIC LIVER DISEASE

Salvatore Nardiello, Teresa Pizzella, Luciano Tarantino,
Michele Russo and Bruno Galanti

Clinic of Infectious Diseases, 1st Medical School
Naples, Italy

INTRODUCTION

Patients with congenital adenosine deaminase (ADA) deficiency
show an impairment of both cell-mediated and humoral immunity
(Giblett et al., 1972); exogenous ADA added to lymphocytes from
these patients partially restores their in vitro response to
mitogens (Polmar et al., 1975). On the other hand, ADA inhibitors
in association with deoxynucleosides inhibit lymphocyte proliferation
and T suppressor cell activity, while T helper cell function and
differentiation of B cells are unaffected (Gelfand et al., 1979).

Patients with chronic liver disease (CLD) frequently have a
reduced number of peripheral T cells (Colombo et al., 1977) and
their lymphomononuclear cells respond poorly to mitogens (Giustino
et al., 1972); moreover, abnormalities of T suppressor cell function
have been described in CLD patients (Chisari et al., 1981; Kashio
et al., 1981).

These defects of immune function prompted us to investigate
lymphocyte ADA levels in patients with HBsAg-positive CLD.

MATERIALS AND METHODS

Patients and controls

We investigated 30 patients with biopsy-proven CLD, all
HBsAg-positive by RIA (AUSRIA II, Abbott). Twenty-three patients
had chronic active hepatitis (CAH): 13 of them were untreated,

the remaining 10 had been on therapy with prednisone ± azathioprine
for 6-24 months and showed evidence of biochemical and clinical
remission. Seven patients had inactive cirrhosis (IC) evoluted
from a previously documented CAH. Forty-one normal subjects served
as controls.

Preparation of cell extracts

Peripheral mononuclear cells (PBMC) were isolated from
heparinized blood by standard gradient centrifugation (Bøyum, 1968).
Cell suspensions were adjusted to a concentration of 5×10^6 cells/ml
and sonicated at 23 kc/sec. After centrifugation at 800 g for 10
min, supernatants were immediately frozen at $-20°C$.

ADA assay

ADA activity was measured on supernatants by a colorimetric
method (Giusti, 1974), and expressed as nmol of adenosine deamin-
ated/min/37°C/10^7 cells.

E rosette assay

Percentages of T cells were evaluated by rosetting PBMC suspensions with 2-aminoethylisothiouronium bromide-treated sheep red
blood cells (Madsen and Johnsen, 1979).

RESULTS

Mean ADA values in CLD patients and controls are reported in
table 1. ADA levels were significantly decreased only in patients
with untreated CAH; they were within normal limits in patients
with CAH in remission and in those with inactive cirrhosis. T cell
counts were significantly lower in CLD patients compared with
control values (57.4 ± 12.1 versus 71.7 ± 6.4 , P <0.001) : no
differences were observed among the three subgroups of CLD patients.
No correlation was found between T cell counts and ADA levels
evaluated on the same PBMC preparations.

DISCUSSION

This study demonstrates that PBMC from patients with HBsAg-
positive CAH have lower ADA levels than normal controls, while
patients with CAH in remission under immunosuppressive treatment

Table 1. Levels of adenosine deaminase (ADA) activity in PBMC from patients with HBsAg-positive chronic liver disease. Mean ± standard deviation is reported.

	sGOT IU/1 (n.v. up to 18)	γ-globulins g/dl	ADA activity nmol/min/10^7 cells
CAH (23 patients)	72 ± 38	2.2 ± 0.6	24.7 ± 7.3
Untreated (13 patients)	84 ± 46	2.5 ± 0.7	22.1 ± 5.7 °
Treated (10 patients)	57 ± 28	1.9 ± 0.4	28.2 ± 7.9
Inactive cirrhosis (7 patients)	40 ± 21	2.0 ± 0.3	27.5 ± 6.3
Normal controls (41 subjects)	-	-	27.5 ± 6.6

° $P<0.01$ versus normal controls , $P<0.05$ versus treated CAH and inactive cirrhosis.

and those with inactive cirrhosis have normal ADA values. Reduced ADA activity has been reported in patients with solid malignancies (Uberti et al., 1976; Russo et al., 1981), while normal values have been observed in other conditions with acquired immune dysfunction (Sidi et al., 1980; Levinson et al., 1980). The mechanism underlying the decrease of ADA in CAH is unknown: it could be caused by a decreased activity in some lymphomononuclear cell population or by a change in the composition of peripheral mononuclear cells. However, our results suggest that different ADA levels could be related to the different functional state of immune cells in active and in inactive forms of chronic liver disease.

REFERENCES

Bøyum, A., 1968, Separation of leucocytes from blood and bone marrow, Scand. J. clin. Lab. Invest., 21 (Suppl.97): 77.

Chisari, F.V., Castle, K.L., Xavier, C., and Anderson, D.S., 1981,
    Functional properties of lymphocyte subpopulations in
    hepatitis B virus infection. I. Suppressor cell control
    of T lymphocyte responsiveness, J. Immunol., 126: 38.

Colombo, M., Vernace, S.J., and Paronetto, F., 1977, T and B
    lymphocytes in patients with chronic active hepatitis
    (CAH), Clin. exp. Immunol., 30: 4.

Gelfand, E.W., Lee, J.J., and Dosch, H.-M., 1979, Selective tox-
    icity of purine deoxynucleosides for human lymphocyte
    growth and function, Proc. Natl. Acad. Sci. USA, 76: 1998.

Giblett, E.R., Anderson, J.E., Cohen, F., Pollara, B., Meuwissen,
    H.J., 1972, Adenosine deaminase deficiency in two patients
    with severely impaired cellular immunity, Lancet, 2: 1067.

Giusti, G., 1974, Adenosine deaminase, in: "Methods of Enzymatic
    Analysis", H.U. Bergmeyer, ed., Verlag Chemie, Weinheim.

Giustino, V., Dudley, F.J., Sherlock, S., 1972, Thymus-dependent
    lymphocyte function in patients with hepatitis-associated
    antigen, Lancet, 2: 850.

Kashio, T., Hotta, R., and Kakumu, S., 1981, Lymphocyte suppressor
    cell activity in acute and chronic liver disease, Clin.
    exp. Immunol., 44: 459.

Levinson, D.J., Chalker, D., and Arnold, W.J.,1980, Reduced purine
    nucleoside phosphorylase activity in preparations of
    enriched T-lymphocytes from patients with sistemic lupus
    erythematosus, J. Lab. Clin. Med., 96: 562.

Madsen, M., and Johnsen, H.E., 1979, A methodological study of E-
    rosette formation using AET-treated sheep red blood cells,
    J. Immunol. Methods,, 27: 61.

Polmar, S.H., Wetzler, E.M., Stern, R.C., Hirschhorn, R., 1975,
    Restoration of in-vitro lymphocyte responses with exogenous
    adenosine deaminase in a patient with severe combined
    immunodeficiency, Lancet, 2: 743.

Russo, M., Giancane, R., Apice, G., and Galanti, B., 1981, Adenosine
    deaminase and purine nucleoside phosphorylase activities in
    peripheral lymphocytes from patients with solid tumours,
    Br. J. Cancer, 43: 196.

Sidi, Y., Boer, P., Pick, I., Pinkhas, J., and Sperling, O., 1980,
    Activity of adenosine deaminase and of purine nucleoside
    phoshorylase in peripheral lymphocytes from patients with
    acquired immunological disorders, Biomedicine, 33: 39.

Uberti, J., Johnson, R.M., Talley, R., and Lightbody, J.J., 1976,
    Decreased lymphocyte adenosine deaminase activity in tumor
    patients, Cancer Res., 36: 2046.

EFFECT OF PURINE RESTRICTION ON SERUM AND URINE URATE

IN NORMAL SUBJECTS

Wayne Stafford and Bryan T. Emmerson

University of Queensland Department of Medicine

Princess Alexandra Hospital, Brisbane 4102, Australia

We sought to determine in individual patients with gout the contribution which the diet was making to the serum urate concentration and the urine urate excretion. In order to interpret the results, we needed to know the findings in healthy subjects studied under similar conditions. Previous studies in 22 normal male subjects in USA (1) had shown an upper limit of urinary urate excretion on a purine-free diet of 575 mg (3.4 mmol) per 24 hours. The serum urate concentration and urine urate excretion reached a nadir after 5 days purine restriction, with a mean fall in the serum urate concentration of 1 mg/100 ml (0.06 mmol/l). Accordingly, we planned to compare the serum urate concentration and the mean 24 hour urinary urate excretion on a normal diet with values obtained after 5 days of a low purine diet. The resulting change would reflect the contribution by the purines in the normal Australian diet to the serum and the urine urate. The values during purine restriction would also provide information concerning the normal range for the 24 hour urinary urate excretion on a low purine food diet.

STUDY PLAN

The detailed plan of the study was set out clearly in writing so that all subjects would follow the same protocol. This plan (Fig. 1) began with two control days during which the person took the normal unrestricted diet, including their usual alcohol. During this time, two consecutive 24 hour urine collections were examined for volume, urate concentration and creatinine concentration. Urate was measured by an automated uricase method based on that of Liddle et alii (1959) (2) which gave a molar absorption of urate within 1% of published values. Blood was collected for urate and creatinine concentration. The subject then began 7 days of a low purine diet selected from a broad range of low purine foods and excluding all foods known to contain significant amounts of purines. After five days on the low purine diet, two further 24 hour urines were collected and examined for volume, urate and creatinine concentrations. On the seventh day of the diet, a further sample of blood was taken for urate and creatinine concentrations (Fig. 1).

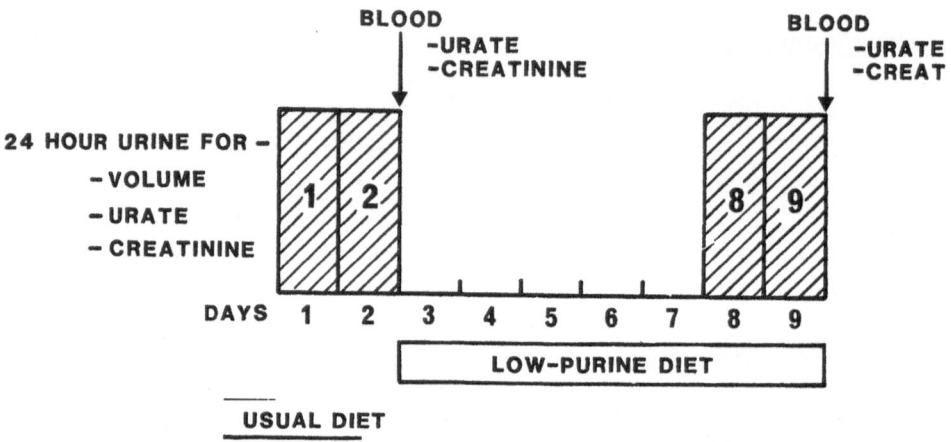

Fig. 1. Plan of study to determine changes in serum urate and mean urine urate excretion following purine restriction.

All subjects were volunteer students, 21 male medical students, one male dietetic student and 9 female medical students. The plan of study was carefully explained to them both verbally and in written form. Compliance was subsequently confirmed by questioning. All studies were carried out while the patient was pursuing his usual activities, except that strenuous physical exertion was avoided. No subject had a history of gout, renal disease or renal calculi and none had a family history of gout. Their usual alcohol consumption was documented and grouped as minimal, mild, moderate or heavy according to the average volume of alcohol consumed daily.

RESULTS

Initially the results were inspected for anomalies and results on consecutive days were then examined for differences. No significant differences were present between the values on the two consecutive days for both the normal diet and the low purine diet with regard to either urine volume, urate or creatinine excretion and urate or creatinine clearance.

A comparison between values during the normal diet and the low purine diet is recorded in Tables 1 and 2. Significant falls occurred in the serum urate and the urinary urate excretion with the

TABLE 1

COMPARISON BETWEEN NORMAL DIET AND LOW PURINE DIET IN 22 MALES
(Mean $\pm$ S.D.)

	Normal diet	Low-Purine diet	Significance (Wilcoxon)
S. urate (mmol/l)	0.42 $\pm$ 0.08	0.35 $\pm$ 0.05	p .001
U. urate excretion (mmol/24 hours)	5.1 $\pm$ 1.4	3.9 $\pm$ 0.7	.003
U. urate excretion (mmol/24 hrs/1.73m^2)	4.5 $\pm$ 1.3	3.5 $\pm$ 0.5	.003
S. creatinine (mmol/l)	0.098 $\pm$ 0.012	0.091 $\pm$ 0.008	.03
Cr. clearance/ (ml/min/1.73m^2)	127 $\pm$ 25	125 $\pm$ 19	NS

TABLE 2

COMPARISON BETWEEN NORMAL DIET AND LOW PURINE DIET IN 9 FEMALES
Mean $\pm$ SD)

	Normal diet	Low-Purine diet	Significance (Wilcoxon)
Serum urate (mmol/l)	0.27 $\pm$ 0.06	0.23 $\pm$ 0.05	.038
U. urate excretion (mmol/24 hours)	3.7 $\pm$ 0.64	2.8 $\pm$ 0.43	.008
U. urate excretion (mmol/24 hrs/1.73m^2)	4.0 $\pm$ 0.53	3.0 $\pm$ 0.35	.008
S. creatinine (mmol/l)	0.082 $\pm$ 0.01	0.073 $\pm$ 0.01	NS
Cr. clearance/ (ml/min/1.73m^2)	110 $\pm$ 15	115 $\pm$ 21	NS

institution of the low purine diet in both males and females. Since the urinary urate excretion on a low purine diet correlated significantly with body surface area in both males and females, urinary urate excretion was factored to standard surface area and comparisons were listed. There were no significant differences in either males or females between the normal diet and the low purine diet with regard either to urate clearance, creatinine clearance or urine volume. The serum creatinine was significantly higher in the males on the normal (meat containing) diet. [An increase in serum creatinine concentration has already been recorded after eating cooked meat, (3)]. The mean fall in the serum urate concentration with purine restriction in males was 0.07 mmol/l, which is comparable with the 0.06 mmol/l previously obtained by Seegmiller et alii, (1961), while the fall in the females was 0.04 mmol/l.

Significant differences between males and females were detected in relation to the following parameters:- Serum urate on a normal diet; Serum urate on a purine free diet; Serum creatinine on a normal diet; Serum creatinine on a purine free diet; Urinary urate excretion on a normal diet (this became non-significant when converted to standard surface area); Urinary urate excretion on a purine free diet; Urate clearance/1.73m^2 and creatinine clearance/1.73m^2 on a normal diet and urate clearance/1.73m^2 on a purine free diet. No significant differences were demonstrated between males and females in regard to the fall in serum urate with purine restriction or the fall in urinary urate excretion with purine restriction.

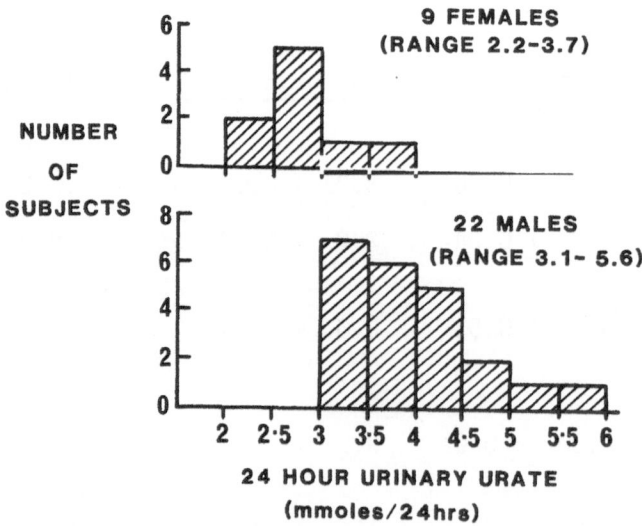

Fig. 2. Mean urinary urate excretion on a low purine diet in 9 females and 22 males.

The distribution of values for the mean 24 hour urinary urate excretion on a low purine diet for both males and females is shown in Fig. 2. This shows a skewed distribution in the males with a range of values between 3.1 and 5.6 mmol/24 hours, whereas in the female the distribution was more normal and the range narrower (2.2 - 3.7 mmol/24 hours). Since the urinary urate excretion on a low-purine diet correlated significantly with body surface area in both males and females, the 24 hour urinary excretion rates were corrected to standard surface area (Fig. 3). This narrowed the range and tended to normalize the distribution of the results. The greatest values obtained were 4.8 mmol/24 hours/1.73m^2 for males and 3.5 mmol/24 hours/1.73m^2 for females.

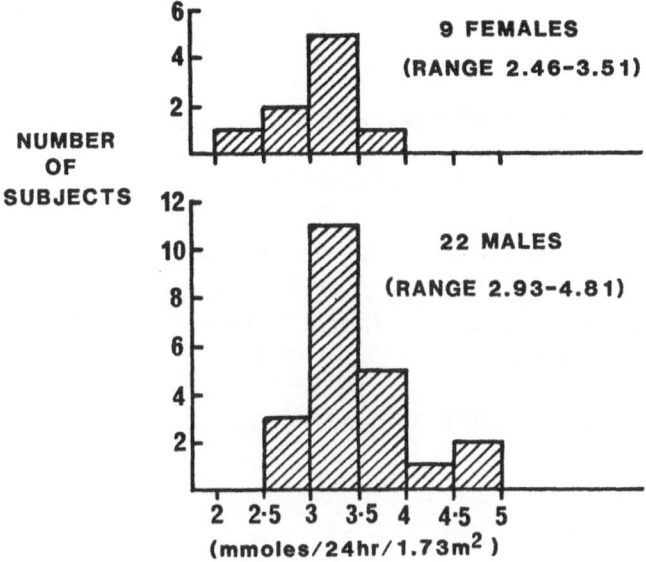

Fig. 3. Mean urinary urate excretion on a low purine diet in 9 females and 22 males when corrected to a standard surface area of 1.73m^2.

The degree of response to purine restriction was also related to the extent of usual alcohol consumption. Ten males whose usual alcohol consumption was minimal or, at most, less than 20 ml of ethanol daily were compared with another group whose usual ethanol consumption was consistently greater than 20 ml of ethanol daily. As shown in Table 3, there was a significantly greater fall in the serum urate concentration in the heavier alcohol consumers. Moreover, the serum urate concentrations were the same during the period of purine and alcohol restriction in the minimal alcohol drinkers as in the heavier alcohol drinkers. Thus, the fall in serum urate with purine and alcohol restriction was significantly greater in those with the usually greater alcohol consumption. The fact that the serum urate concentrations after purine and alcohol restriction were not significantly different in the two groups suggests that the difference was probably produced by the ethanol withdrawal. The fall in the urinary urate excretion on the low-purine no-alcohol diet was also greater in the heavier alcohol consumers but this difference did not reach significance. In the 6 females, all of whom took less than 20 ml of ethanol per day, the fall in the serum urate and the urinary urate excretion was not significantly different from that in males taking a comparable amount of alcohol. It would therefore appear that the regular consumption of alcohol results in significantly higher serum urate concentrations than would otherwise be present.

TABLE 3

COMPARISON OF RESPONSE TO PURINE RESTRICTION IN MALES
WITH USUAL LOW ALCOHOL CONSUMPTION AND THOSE WITH
HEAVIER ALCOHOL CONSUMPTION

	Usual Ethanol < 20 ml/day (Mean $\pm$ SEM) n = 10	Usual Ethanol > 20 ml/day (Mean $\pm$ SEM) n = 7	Difference (Mann-Whitney U Test) p
S. urate (mmol/1) (normal diet usual alcohol)	0.387 $\pm$ 0.017	0.477 $\pm$ 0.031	0.0275
S. urate (mmol/1) (low purine diet no alcohol)	0.352 $\pm$ 0.019	0.350 $\pm$ 0.010	NS
Fall in s. urate with purine and alcohol restriction	0.035 $\pm$ 0.012	0.127 $\pm$ 0.027	0.025

In order to determine the contribution of the low-purine food diet over that which would have been provided should a completely purine-free formula diet have been taken, the study was extended in one student of dietetics. After completing the 7 days of the low-purine diet, he then took a synthetic formula diet which was completely purine-free for another 5 days and collected urine in the last 2 days. This formula diet resulted in a further mean fall of urinary urate of only 0.26 mmol per day, thereby suggesting that the low-purine food diet which was taken during the study was not contributing greatly to the urinary urate excretion.

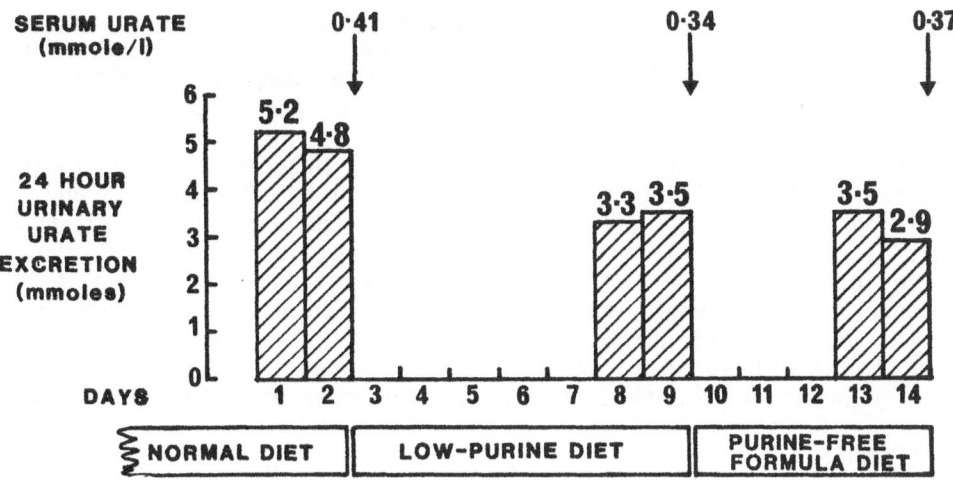

Fig. 4. Comparison between serum and urinary urate excretion on a normal diet, a low-purine food diet and a purine-free diet formula diet in one healthy dietitian.

DISCUSSION

This protocol was designed primarily to enable a comparison to be made between subjects with gout and normal subjects in relation to the contribution from their diet to their serum urate concentration and urinary urate excretion. Thus, gouty patients who consume unduly large amounts of dietary purines can be readily identified and the possible benefit of restriction of high purine foods considered. A surprising finding in the study, therefore, was the high mean 24 hour urinary urate excretions on the low purine diet, ranging up to 4.8 mmol/24 hours/$1.73m^2$, whereas the upper limit of the normal range had previously been regarded as up to 3.4 mmol/24 hours. Possible causes for this need to be considered. Subjects were studied as outpatients while undertaking their normal activities and they were able to choose their low purine diet at will. Thus, although the investigator was unable to detect lack of compliance by questioning the subjects being studied and the subjects appeared to be fully cooperative, purine restriction was unsupervised. Likewise, although only low-purine food was taken, it is possible, when a subject can choose a diet from a list of foods, that there may be some variation betwen subjects in the amount of purines consumed. Thus, large amounts of certain low-purine foods may make a significant contribution to purine consumption. Nonetheless, the difference between urate excreted on the low-purine diet and the urate excreted on a purine-free synthetic formula diet in one individual revealed a contribution of only 0.26 mmol (44 mg) of urate per 24 hour urine. Body size also appears to be an important factor determining total urinary urate excretion, so that correction to standard surface area should be undertaken. The usual pattern of alcohol consumption also has an effect. A high proportion of protein in a low-purine diet promotes an increase in urinary urate excretion (4) and even low-purine Australian diets tend to be high in protein. Moreover, all subjects studied were in the top 2% of the population intellectually, and a positive correlation has been established between a high IQ and gout (5), although its mechanism is conjectural. Urinary urate excretion may also vary with different populations and in different countries. It seems important therefore to obtain further information about the normal range of urinary urate on a low-purine diet in normal subjects from different populations before using this information to detect persons who are over-excretors of urate.

REFERENCES

1.   J.E. Seegmiller, A.I. Grayzel, L. Laster and L. Liddle, Uric acid production in gout. J Clin Invest 40:1304-1314, 1961.

2.   L. Liddle, J.E. Seegmiller, L. Laster, The enzymatic spectro-photometric method for determination of uric acid. J Lab Clin Med 54:903-913, 1959.

3.   F.K. Jacobsen, C.K. Christensen, C.E. Mogensen, F. Andreasen and N.S.C. Heilskov, Pronounced increase in serum creatinine concentration after eating cooked meat. Br Med J 1:1049-1050, 1979.

4.   W. Löffler, W. Gröbner and N. Zöllner, Nutrition and uric acid metabolism: Plasma level, turnover excretion. Fortschr Urol Nephrol 16:8-18, 1981.

5.   J.A. Sofaer and A.E.H. Emery, Genes for super-intelligence. J Med Genet 18:410-413, 1981.

ISOTOPE STUDIES OF URIC ACID METABOLISM DURING DIETARY PURINE ADMINISTRATION

Werner Löffler[1], Wolfgang Gröbner[1], Ramiro Medina[2] and Nepomuk Zöllner[1]

Medizinische Poliklinik der Universität München[1] and Lehrstuhl für Allgemeine Chemie und Biochemie der Technischen Universität München Freising-Weihenstephan[2]

Uric acid is excreted by the kidneys and intestinal tract, while other routes are negligible. By addition of purines to a purine-free or low purine diet changes in uric acid turnover and excretion occur, which cannot be investigated by metabolic balance methods as that part of the uric acid turnover secreted into the gut is completely destroyed by the intestinal flora (Sorensen, 1960). We therefore studied uric acid pool size and turnover in normal subjects both during a purine-free diet and during purine supplementation by stable isotope methods.

MATERIALS AND METHODS

Five healthy subjects (age 23 to 27 years; 2 females, 3 males) were studied on a purine-free, isoenergetic liquid formula diet with and without addition of 1 g adenosine-5´-monophosphate (AMP) and 1 g guanosine-5´-monophosphate (GMP) per 70 kg body weight. In each experiment the first ten days were allowed to reach steady state conditions. Isotopes were administered the following morning in the fasting state (labelled uric acid i.v.) and with the breakfast, respectively (labelled glycine). In three subjects uric acid pool size and turnover were estimated by the isotope dilution technique according to Benedict et al. (1949) using 15 N-uric acid. In two subjects endogenous uric acid

synthesis was measured after a single oral dose of
15 N-glycine (67 mg/kg body weight), simultaneously
pool size and turnover were estimated by use of 13 C-
uric acid. The isotopes administered were 1,3- 15 N_2-
uric acid, 95 percent enrichment; 2-13C-uric acid,
90 percent enrichment,15 N-glycine, 95 percent enrichment (Amersham).
Uric acid was isolated by anione exchange chromatography (Johnson and Emmerson, 1972) from freshly
collected 12-h-urine samples for 6 days, and from 5
additional 24-h-samples in the case of glycine incorporation.
For 15 N-measurements about 2 mg of pure uric acid were
degraded by a modified Kjeldahl procedure and molecular
nitrogen liberated for analysis of its isotope content
by Li O Br (Ross and Martin, 1970). For 13 C-analysis
uric acid was combusted according to Winkler and Schmidt
(1980). Uric acid concentrations were measured enzymatically (Zöllner, 1963).

RESULTS AND DISCUSSION

Results of the uric acid pool size and turnover studies
are shown in table 1. There was an increase in fractional turnover of the uric acid pool during purine supplementation in all subjects. This means that total body
uric acid clearance is enhanced in normal subjects
during dietary purine intake.

Table 1. Results of uric acid turnover studies by isotope dilution methods. I studies during purine-free, isoenergetic liquid formula diet; II studies during purine
supplementation. Subject 1-3 were studied using 15 N-uric
acid, Subject 4 and 5 were studied after injection of
13 C-uric acid.

Subject	A (mg)		k (1/day)		Renal uric acid excretion			
					UxV (mg/day)		$\frac{UxV}{AxK}x$ 100	
	I	II	I	II	I	II	I	II
1 female	569	736	0.28	1.21	237	575	51	64
4 female	805	1477	0.52	0.80	242	850	58	72
2 male	684	1592	0.66	0.81	277	726	79	62
3 male	562	1325	0.63	0.85	281	703	61	59
5 male	728	2116	0.63	0.88	330	967	72	52

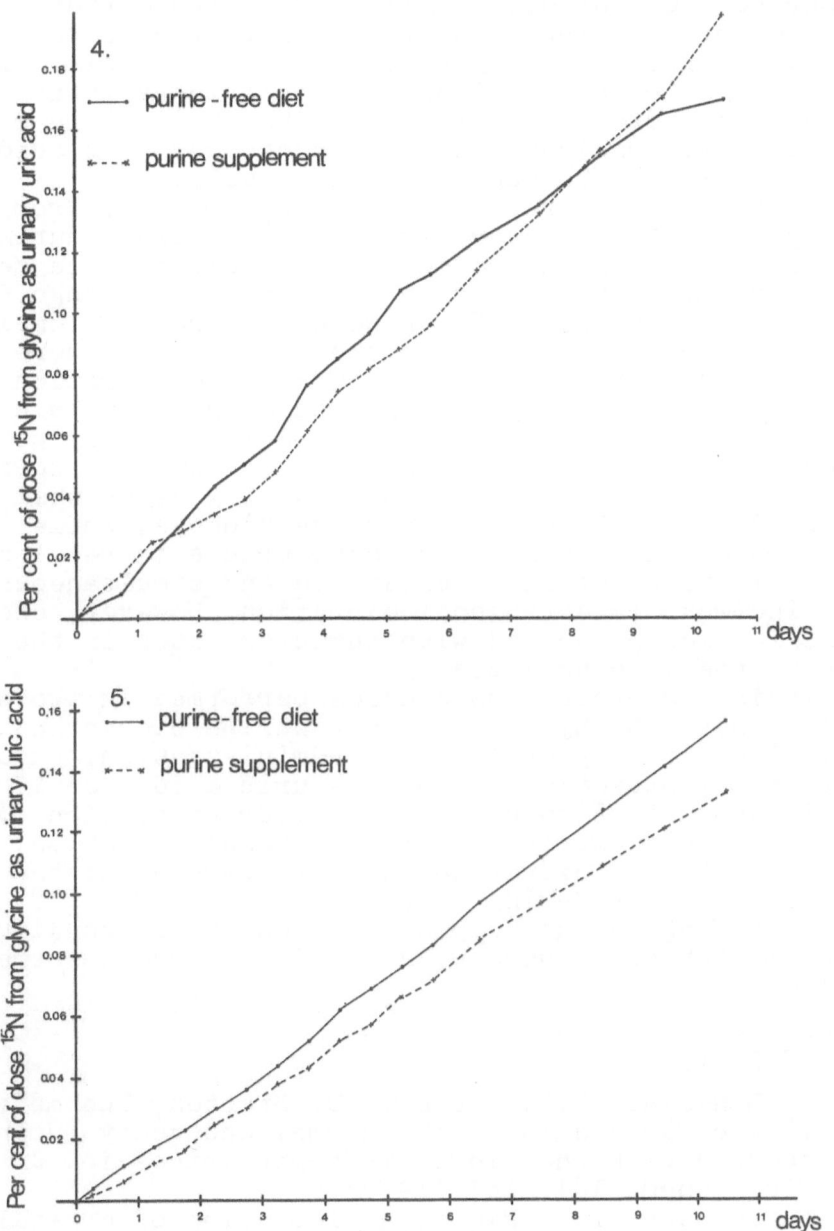

Figures 1 and 2. Cumulative excretion of 15 N as renal uric acid after a single oral dose of 15 N-glycine (67 mg/kg). 4 and 5 subject number.

Pool sizes in males were considerably smaller than previously reported by others from studies performed during
low purine diets or diets called purine-free (for summary see Löffler, 1980). As these diets were composed
of conventional food they were at best low purine diets,
also the preliminary periods for achieving a metabolic
steady state were too short (less than 7 days) in most
instances. We conclude that the endogenous uric acid
pool is smaller than suggested previously.
Fractional renal excretion of uric acid varied from 51
to 79 percent of turnover (average 64 percent) during
ingestion of the purine-free diet and did not change
during purine supplementation on an average (range 52
to 72, average 62 percent). However, comparing females
to males it was found that in the two females there was
an increase in fractional renal uric acid excretion of
13 and 14 percent, respectively,while in the male subjects this was constant or decreased (average 13 percent
decrease). These changes might be considered to correspond to the well known fact that females clear their
uric acid better than males via the kidneys, while in
normal males the increase in serum uric acid results in
a small increase of renal clearance and simultaneously
in an increase in extrarenal excretion. However, one
of the females presented with turnover rates in the same
range as the male subjects.
The glycine incorporation studies performed in two subjects support the hypothesis that purine biosynthesis
is not inhibited by oral purine administration. Cumulative renal excretion of 15 N as uric acid over 11
days is shown in Figure 1 and 2. After correction of
renal uric acid excretion for extrarenal excretion,
in Subject 4 15 N-excretion was 0.29 percent of the
dose given orally during purine-free diet and 0.28
percent during purine supplementation. Corresponding
values in Subject 5 were 0.22 and 0.26 percent,respec-
tively.

REFERENCES

1. J.D. Benedict, P.H. Forsham, D. Stetten, The metabolism of uric acid in the normal and gouty human
   studied with the aid of isotopic uric acid, J.
   Biol. Chem. 181, 183 (1949)
2. L.A. Johnson, B.T. Emmerson, Isolation of crystalline
   uric acid from urine, for urate pool and turnover
   measurements, Clin. Chim. Acta 41, 389 (1972)
3. W.   Löffler, Harnsäurepool and Harnsäureumsatz, in:
   "Hyperurikämie und Gicht", N. Zöllner, ed.,Springer,
   Berlin - Heidelberg - New York (1980)

4. J.   Ross, E. Martin, A rapid procedure for preparing
   gas samples for nitrogen-15 determination, The Analyst
   95, 817 (1970)
5. L.B. Sorensen, The elimination of uric acid in man
   studied by means of C 14-labbelled uric acid,
   Scand. J. Clin. Lab.Invest. 12, Suppl.54 (1960)
6. F.J. Winkler, H-L. Schmidt, Einsatzmöglichkeiten der
   13 C-Isotopen-Massenspektrometrie in der Lebens-
   mitteluntersuchung, Z. Lebensm. Unters.Forsch.
   171, 85 (1980)
7. N.   Zöllner, Eine einfache Modifikation der enzyma-
   tischen Harnsäurebestimmung, Z. klin. Chem. klin.
   Biochem. 1, 178 (1963)

INFLUENCE OF DIETARY PURINES ON ALLOPURINOL METABOLISM AND ALLOPURINOL INDUCED OROTICACIDURIA

S. Reiter, W. Löffler,
W. Gröbner and N. Zöllner

Medizinische Poliklinik
Universität München

The allopurinol induced oroticaciduria is markedly diminished by oral administration of purines[1]. The underlying mechanism is still unknown. We performed nutrition experiments to determine whether this effect could be due to an influence of dietary purines on either absorption or metabolism of allopurinol. A change of allopurinol metabolism by intravenous administration of adenine in the pig was previously reported[2].

METHODS

4 healthy male subjects received a purine-free synthetic diet for 2 weeks and 0.5-1g allopurinol daily. During the second week different purines were administered orally together with allopurinol. The urine was collected over 24 hours and urinary excretion of the following compounds was determined:
- Uric acid, hypoxanthine, xanthine, allopurinol, oxipurinol and allopurinol-1-riboside by a modification of the method of Webster et al.[3] on a Chrompack RP-8 column. Urine samples were preseparated on anion-exchange columns. For the determination of oxipurinol-7-riboside a preextraction of ribosides on Bio-Rad Affi-Gel 601 was necessary. For the determination of adenine the mobile phase contained 20% of methanol and was adjusted to pH 5.5 .
- Orotic acid and orotidine by an ion-pair method with a Beckman ion-pair kit. Urine samples were precleaned on Waters Sep-pak C18 cartridges.
All HPLC-determinations were done on a Hewlett-Packard 1084B system using isocratic conditions.

RESULTS AND DISCUSSION

The average increment of urinary oxipurine excretion
during oral administration of purines is shown in:

Table 1

subject, daily dose (day 8-14)	% of dose recovered as urinary oxipurines	% of increment of oxipurine excretion represented by hypoxanthine
1, hypoxanthine 1x7.35mmol	58	61
2, adenine 3x2.8 mmol	50	46
3, adenosine 3x2.9 mmol	42	46

In subject 2 the urinary excretion of adenine increased
from 16.4(+2.85)umol/24h during the first week to 175
(+68.8)umol/24h during the administration of adenine corresponding to 3.5% of the increment of purine excretion.

The total urinary recovery of allopurinol and its metabolites was not influenced by oral administration of purines
(data not shown).

Fig. 1 shows the urinary excretion of allopurinol and its
metabolites in subject 1(oxipurinol-7-riboside not determined). The administration of hypoxanthine as well as the
other purines tested(table 2) caused a marked decrease of
the excretion of allopurinol-1-riboside. This is most
likely due to the increased formation of hypoxanthine
(table 1) which can competitively inhibit the conversion
of allopurinol by PNP or HGPRT, allopurinol-1-riboside

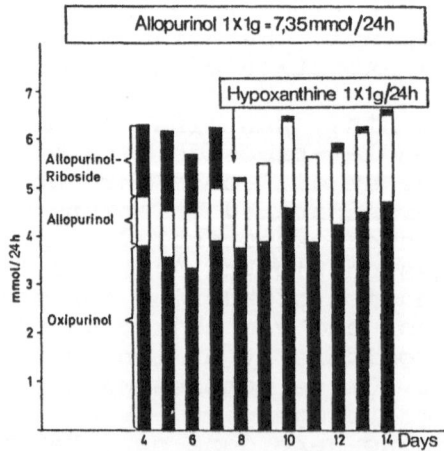

Fig. 1. Influence of dietary hypoxanthine(day 8-14) on the
urinary excretion of allopurinol and its metabolites.

Table 2. Reduction of the urinary excretion of allopurinol-
1-riboside by dietary purines.

subject, daily dose			allopurinol-1-riboside(% of total urinary excretion of allopurinol and its metabolites)	
			day 4-7	day 8-14
1, day 1-14:	allopurinol	1x7.35mmol	22.6(+3.5)	1.3(+0.8)
day 8-14:	hypoxanthine	1x7.35mmol		
2, day 1-14:	allopurinol	3x1.8mmol	24.6(+3.4)	1.8(+0.9)
day 8-14:	adenine	3x2.8mmol		
3, day 1-14:	allopurinol	3x1.9mmol	20.7(+2.6)	2.8(+1.1)
day 8-14:	adenosine	3x2.9mmol		

being a direct synthetic product of PNP and the final degradation product of allopurinol-1-ribotide as well.
In subject 4(fig. 2) the excretion of oxipurinol-7-ribo-
side was determined in addition and was found not to be
changed by dietary hypoxanthine(data not shown).

Table 3 shows the influence of dietary purines on the
allopurinol induced excretion of orotic acid and orotidine. The excretion of orotic acid and orotidine was reduced to a different degree by adenine as compared to
hypoxanthine and adenosine. This could be due to the fact
that nearly all the absorbed adenine was salvaged thus
consuming a large amount of PRPP, a substrate of PRPP.

Table 3. Reduction of urinary excretion of orotic acid
and orotidine by dietary purines.

subject, daily dose (day 8-14)		reduction of urinary excretion by %	
		orotic acid	orotidine
1, hypoxanthine	1x7.35mmol	59.6(+13.5)	33.7(+8.3)
2, adenine	3x2.8 mmol	42.3(+9)	51.6(+10)
3, adenosine	3x2.9 mmol	64.7(+11)	40.3(+3.2)

Comparing the excretion patterns of allopurinol-1-riboside
and orotic acid/orotidine in the urine over a period of
24 hours after allopurinol administration(subject 4,
fig. 2) we conclude from our data that allopurinol-1-ribo-
tide plays only a minor role in oroticaciduria(day 2,3).
Additional evidence provided the administration of hypoxanthine 5 hours after allopurinol(day 4): The excretion
of allopurinol-1-riboside was not influenced whereas the
excretion of orotic acid was reduced to the same extent
as on day 5 and 6. On these days hypoxanthine was given
together with allopurinol and inhibited the formation of
allopurinol-1-riboside completely.

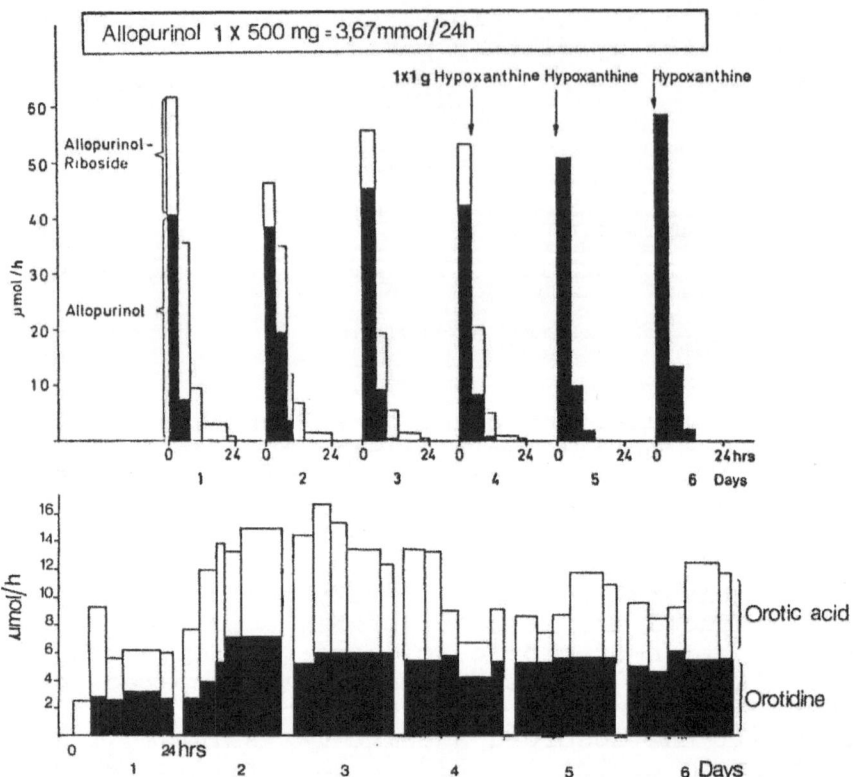

Fig.2. Influence of dietary hypoxanthine on the excretion
        patterns of allopurinol-1-riboside, allopurinol,
        orotic acid and orotidine.

These findings suggest that the reduction of allopurinol-
1-riboside formation by dietary purines is not the reason
for the simultaneous reduction of oroticaciduria.

REFERENCES

1. N. Zöllner and W. Gröbner. Dietary feedback regulation
   of purine and pyrimidine biosynthesis in man.
   Ciba Found. Symp. 48:165(1977)
2. J.S. Cameron, H.A. Simmonds, A. Cadenhead and D. Farebrother. Metabolism of intravenous adenine in the pig.
   Adv. Exp. Med. Biol. 76A:196(1977)
3. D.R. Webster, H.A. Simmonds, D.M.J. Barry and D.M.O.
   Becroft. Pyrimidine and Purine Metabolites in Ornithine Carbamoyl Transferase Deficiency.
   J. Inher. Metab. Dis. 4:27(1981)

THE EFFECT OF BEER INGESTION ON PLASMA AND URINE URIC ACID IN GOUT

AND NORMOURICAEMIC SUBJECTS

A.V. Rodgers, T. Gibson, H.A. Simmonds, and P. Toseland

Guy's Hospital, London, UK

INTRODUCTION

A relationship between hyperuricaemia, gout and excessive alcohol consumption is well established. The increase of serum lactate which follows alcohol ingestion may impair renal urate excretion and is claimed to be the principal mechanism.[1] Such an effect has been shown in studies using alcohol in amounts which have exceeded those commonly consumed by gout patients in our experience.[2] Furthermore, previous experiments have usually involved fasting subjects so that the role of lactate on urate clearance has not been clearly distinguishable from that of fasting alone.[3]

Gouty subjects in the United Kingdom include a large percentage of regular beer drinkers whose dietary habits are in other respects similar to those of a non-gouty population.[4] The aim of the present study was to examine the influence of alcohol on uric acid metabolism by simulating the drinking habits of gout patients without fasting.

METHODS

Five patients with primary gout and an equal number of normouricaemic volunteers were studied in a metabolic ward. Prior to admission they received a low purine, alcohol-free diet for one week and gout patients discontinued all medication apart from colchicine. In hospital they were maintained on an iso-caloric, constant purine diet for five days. On days 3 and 4 they drank 2.8 litres of beer between 12.00 - 16.00 h. containing 1.1mmol total purines and 53g ethanol. Days 1, 2 and 5 were control

327

periods when subjects drank an equivalent volume of diluted squash
(artificial fruit juice sweetened with glucose) over the same time
period. Plasma and urine uric acid were estimated at regular
intervals by a uricase method. Urine pH, ammonium and titratable
acid were measured on daily collections. Serum lactate and
alcohol were estimated by standard techniques. Excretion of hypoxanthine and xanthine was measured only in the normouricaemic
subjects.

RESULTS

    Essential details of participating subjects are seen in Table
1. On the days of beer consumption, blood alcohol levels achieved
similar levels in each group (Table 2). Both beer and squash
ingestion were associated with increases of serum lactate. There
were wide individual variations and the average value of the
controls was lower. A slight but consistent increase of plasma
urate occurred during beer ingestion. No such effect was observed
when squash was drunk. Table 2 illustrates the changes over days
2 and 3.

    Beer was associated with a fall of urine pH and a marked rise
of urine titratable acid in both groups but ammonium excretion
was unaltered.

    Urine uric acid excretion increased on days when beer was
drunk but this was slight in the gouty and more pronounced on the
first day of beer ingestion for both groups. Excretion of other
oxypurines altered very little throughout the study in the control
group. These were not measured in the gouty patients.

TABLE 1. Gout and normouricaemic groups

	Gout	Controls
No.	5	5
Mean age	46 (33–58)	38 (27–53)
Mean body wt (kg)	88 (60–122)	81 (74–92)

TABLE 2 Average blood alcohol, serum lactate, plasma uric acid levels, urate clearance

		DAY 2 SQUASH						DAY 3 BEER					
TIME		08.00	12.00	14.00	16.00	18.00	20.00	08.00	12.00	14.00	16.00	18.00	20.00
Blood alcohol (mmol/1)	Gout			–				0	0	9	28	33	15
	Controls			–				0	0	14	26	20	9
Serum lactate (mmol/1)	Gout	1.4	1.4	2.3	2.0	1.84	1.47	1.65	1.5	2.05	2.04	1.76	1.9
	Controls	0.82	0.63	1.1	0.97	1.8	1.05	0.87	1.1	1.1	1.24	1.27	1.41
Plasma urate (mmol/1)	Gout	0.37	0.37	0.36	0.37	0.37	0.37	0.36	0.37	0.4	0.39	0.39	0.37
	Controls	0.31	0.29	0.31	0.31	0.3	0.31	0.31	0.31	0.33	0.35	0.35	0.35
Urate clearance (ml/min)	Gout	–	5.7	7.5	9.8	6.7	6.6	4.8	5.2	4.4	7.7	8.2	6.4
	Controls	–	7.8	15.8	12.5	9.9	7.6	5.6	9.6	12.7	13.8	7.9	4.7

Urate clearance increased during the periods when squash and beer were being ingested (Table 2). Clearance values were consistently higher amongst the controls.

DISCUSSION

Although increases of serum lactate may reduce renal excretion of uric acid it is uncertain that this effect is a major mechanism of hyperuricaemia.[5] In a previous study, alcohol given in amounts associated with wine consumption, increased both urate and xanthine excretion.[6] These data and other investigations of hepatic production of uric acid,[7] suggest that ethanol may exert an alternative effect of increased uric acid synthesis.

In the current study, beer was drunk in amounts commonly ingested by patients with gout. The effect on plasma uric acid was definite but slight. Despite an increase of serum lactate and reduced urine pH with beer, urate clearance was accentuated. This was also seen with equal volumes of squash and was probably related to an increase of urine flow rates.[8] Urate clearance was consistently lower in the gouty patients, a characteristic displayed by the majority of those with primary gout.

The increase of uric acid excretion associated with beer may have been due to an effect of alcohol on urate synthesis but it is more likely that it reflected absorption of the purines contained in the beer. It has been estimated that beer contributes substantially to the total dietary purines of some patients with gout[4]. The current study indicates that amongst beer drinking gout patients, this may contribute more to the development of hyperuricaemia than any other alcohol related mechanism.

ACKNOWLEDGEMENTS

We are grateful to the Arthritis and Rheumatism Council of Great Britain for financial support and to Courage Breweries Ltd. for their kind assistance.

REFERENCES

1. D.S. Newcombe, Ethanol metabolism and uric acid. Metabolism 21: 1193 (1972)
2. C.S. Lieber, D.P. Jones, M.S. Losowsky, C.S. Davidson, Interrelation of uric acid and ethanol methabolism in man. J. Clin. Invest. 41: 1863 (1962)
3. M.J. Maclachlan and G.P. Rodnan, Effects of food, fast and alcohol on serum uric acid and acute attacks of gout. Am. J. Med. 42: 38 (1967)

4. T. Gibson, V. Rodgers, F. Court-Brown, H.A. Simmonds, Dietary intake of gout patients, Ann. Rheum. Dis. 40: 515 (1981)

5. T.S.F. Yu, J.H. Sirota, L. Berger, M. Halpern, A.B. Gutman, Effect of sodium lactate infusion on urate clearance in man, Proc. Soc. Exp. Biol. Med. 96: 809 (1957)

6. F. Delbarre, C. Auscher, H. Brouilhet, A. de Géry, Action de l'ethanol dans la goutte et sur le métabolisme de l'acide urique, Sem. Hôp. Paris, 43: 659 (1967)

7. J. Grunst, J. Dietze, M. Wicklmayr, Effect of metabolic changes on uric acid production of human liver in: "Urolithiasis Research", H. Fleisch, W.G. Robertson, L.H. Smith, W. Vahlinsiek, ed., Plenum Press, N. York (1976)

8. H.S. Diamond, R. Lazarus, D. Kaplan, D. Halbersam, Effect of urine flow rate on uric acid excretion in man, Arthritis Rheum. 15: 338 (1972)

INFLUENCE OF PURINES ON PYRIMIDINE METABOLISM IN RAT HEPATOCYTES

P. Banholzer, W. Gröbner, C. Barth, and N. Zöllner

Medizinische Poliklinik der Universität München

Pettenkoferstr. 8a, 8ooo München 2

INTRODUCTION

Nutrition experiments showed that dietary nucleic acids or purines diminish allopurinol induced orotic aciduria (Zöllner and Gröbner 1971, Zöllner and Gröbner 1977). Using lymphoblasts cultures we showed that this might be due to an inhibition of pyrimidine synthesis by purines (Banholzer et al. 1980, Banholzer et al. 1981). In this study the influence of purine bases and nucleosides as well as of oxipurinol on pyrimidine biosynthesis was investigated using rat hepatocytes in suspension. By measuring bicarbonate-^{14}C incorporation into orotic acid as well as the incorporation of orotic acid-^{14}C or uridine-^{14}C into nucleic acids it was possible to study the first and second part of pyrimidine biosynthesis and the salvage pathway separately.

METHODS

Liver cells were prepared as described previously (Barth et al. 1980). Hepatocytes in suspension containing approximately 10^6 cells per ml L-15 medium were used for each reaction. Viability was tested by trypan blue staining. Cells above 95 % viability were used.
After addition of 20 µCi NaHCO$_3$-^{14}C, 2 µCi orotic acid-6-^{14}C or 0,04 µCi uridine-2-^{14}C (isotopes had a specific activity of 50 mCi/mmol) 0,5 ml cell suspension was incubated with

shaking at 37°C for 60 min. in the presence of oxipurinol and/or different purines.
The incorporation of $NaHCO_3-^{14}C$ into orotic acid was measured according to the method of Wendler and Trembley (1980). The incorporation of orotic acid-6-^{14}C or uridine-^{14}C into nucleic acids was determined using the method of Castle et al. (1979).

RESULTS AND DISCUSSION

The incorporation of orotic acid into nucleic acids was markedly decreased by all added substances (fig. 1).

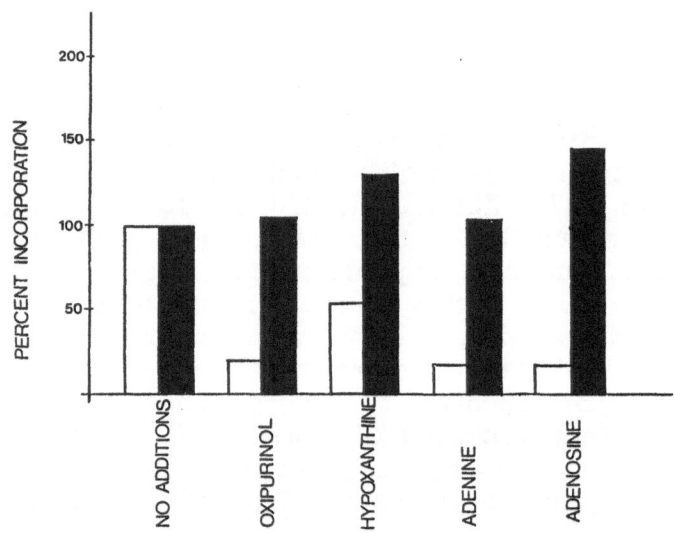

Fig. 1: Incorporation of orotic acid-6-^{14}C (white columns) or uridine-2-^{14}C(black columns) into nucleic acids with or without oxipurinol or purines. Results are shown in percent of controls. Controls contained no oxipurinol or purines.

This is in agreement with our results obtained in human cultured lymphoblasts. Oxipurinol inhibits orotidyl decarboxylase(ODC). Purines might inhibit the conversion of orotic acid to orotidine monophosphate by lowering the intracellular PRPP-concentration (Crandall et al. 1978).

Uridine incorporation is unchanged by oxipurinol. A slight increase is seen in the presence of purines which is probably due to an increase of ATP production (fig. 1). This increase also shows that the diminished orotic acid incorporation is not only a toxic effect.

Incorporation of $NaHCO_3-^{14}C$ into orotic acid is shown in fig. 2.

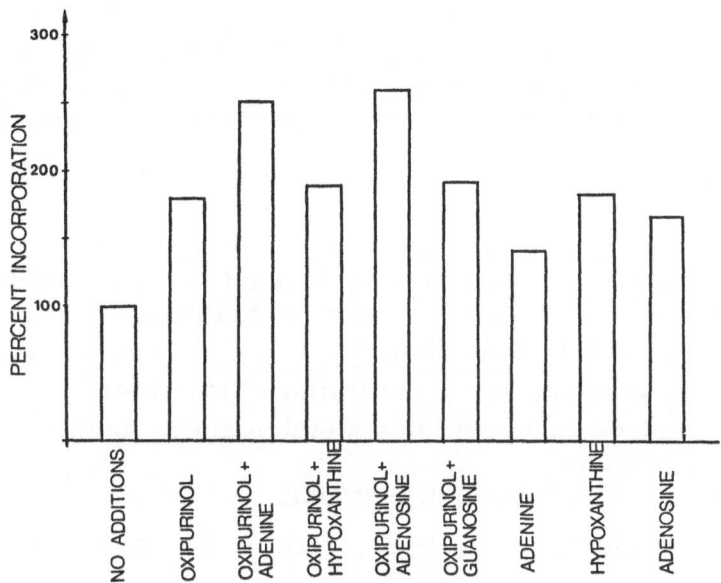

Fig. 2: Incorporation of $NaHCO_3-^{14}C$ into orotic acid of hepatocytes in suspension with or without oxipurinol and/or different purines in a concentration of 0,5 mM. Results are shown in percent of controls. Controls contained no oxipurinol or purines.

Oxipurinol leads to an increase of bicarbonate-^{14}C incorporation due to an inhibition of the second part of pyrimidine biosynthesis as shown above. This is in keeping with the orotic aciduria after administration of oxipurinol observed in vivo. It is also in agreement with our results obtained in cultured human lymphoblasts (Banholzer et al. 1980, Banholzer et al. 1981). Purines did not reduce bicarbonate incorporation into orotic acid as they did in human cultured lymphoblasts. Conversely they increased the amount of labelled orotic acid probably by inhibiting metabolisation of orotic acid. Crandall et al. (1978) found in keeping with our results an inhibition of the first part of pyrimidine biosynthesis by adenosine in lymphatic tissue of rats but not in rat liver slices. Results indicate that the reduction of allopurinol induced orotic aciduria by dietary purines is not due to inhibition of pyrimidine biosynthesis in liver tissue.

ACKNOWLEDGEMENT

We want to thank Dr. Hilmar for preparing hepatocytes and Mrs. Gamulin-Vucov for excellent technical assistance. This work was supported by the Deutsche Forschungsgem einschaft.

REFERENCES

Banholzer, P., Gröbner, W., and Zöllner, N., 1980,
Der Einfluß von Purinen sowie Oxipurinol auf den Pyrimidin-
stoffwechsel in menschlichen Lymphozytenkulturen,
Verh. dtsch. Ges. inn. Med. 86:928 .

Banholzer, P., Gröbner, W., and Zöllner, N., 1981,
Influence of Purines on Pyrimidine Metabolism in Human
Lymphoblasts in Culture,
XII. Int. Congr. of Nutrition, Diego/USA.

Barth, C.A., Willershausen, B.S., Walter, B., and Weis, E.E.,
1980,
Morphology and Metabolism of Adult Rat Hepatocytes in Primary
Culture,
Hoppe Seyler's Z. Physiol. Chem. 361:1017.

Castle, T., Kraemer, W., Liu, D.S.H., and Richardson, A.,
1979,
Characterization of RNA Synthesis by Isolated Hepatocytes in
Suspension,
Arch. of. Biochem. and Biophys., 195: 423.

Crandall, D.E., Lovatt, C.J., and Tremblay, G.C., 1978,
Regulation of Pyrimidine Biosynthesis by Purine and Pyrimidine
Nucleosides in Slices of Rat Tissues,
Archives of Biochemistry and Biophysics, 188:194

Wendler, P.A., and Tremblay, G.C., 1980,
Quantitative Isolation of Radiolabeled Metabolites without
Chromatography:

Measurements of the Biosynthesis of Purines, Pyrimidines, and
Urea in Isolated Hepatocytes,
Analytical Biochemistry 108: 406.

Zöllner, N., and Gröbner, W., 1971,
Influence of oral ribonucleic acid on orotic aciduria due to
allopurinol administration,
Z. ges. exp. Med. 156:317.

Zöllner, N., and Gröbner, W., 1977,
Dietary feedback regulation of purine and pyrimidine biosynthesis
in man.
In: Purine and Pyrimidine Metabolism,
Ciba Foundation Symposium 48:165 (Elsevier/Excerpta Medica,
North-Holland).

EFFECTS OF GLYCEROL ON PURINE METABOLISM

IN RAT LIVER CELLS

Christine Des Rosiers, Marcel Lalanne, and
Joan Willemot

Département de Biochimie, Faculté de Médecine,
Université Laval, Québec, P.Q., Canada   G1K 7P4

Glycerol phosphorylation in liver cells produces a rapid depletion of inorganic phosphate (P_i) and adenine nucleotides (1-3). These effects are similar to those produced by fructose (3,4). In addition, fructose administration to mice or humans also causes an increase in purine synthesis de novo (5-7).

Although the mechanism of fructose-induced purine nucleotide catabolism has been well studied in liver cells (8,9), this is not the case for glycerol-induced catabolism. In the present work, we have characterized the adenine nucleotide catabolism induced by glycerol in isolated rat liver cells and have compared it with that induced by fructose. We have also studied the effects of glycerol and fructose on purine synthesis de novo and on phosphoribosylpyrophosphate (PP-ribose-P) availability in these cells. The various changes caused by glycerol have been correlated in order to further our understanding of the control of purine nucleotide synthesis de novo.

To study the catabolism of adenine nucleotides, the intracellular pool is prelabeled by incubating freshly isolated liver cells (10) with [8-^{14}C]adenine. The cells are then washed to eliminate any residual [^{14}C]adenine, resuspended in the suspension buffer, and incubated under various conditions.

A rapid decrease in radioactive adenine nucleotides occurs in cells incubated with 10 mM glycerol (Fig. 1A). This is mainly due to ATP degradation. Concomitantly, the [^{14}C]ATP/[^{14}C]ADP and [^{14}C]ATP/[^{14}C]AMP ratios are decreased. A small but significant catabolism is observed with 0.25 mM glycerol. The extent of the catabolism increases as the concentration of glycerol is increased.

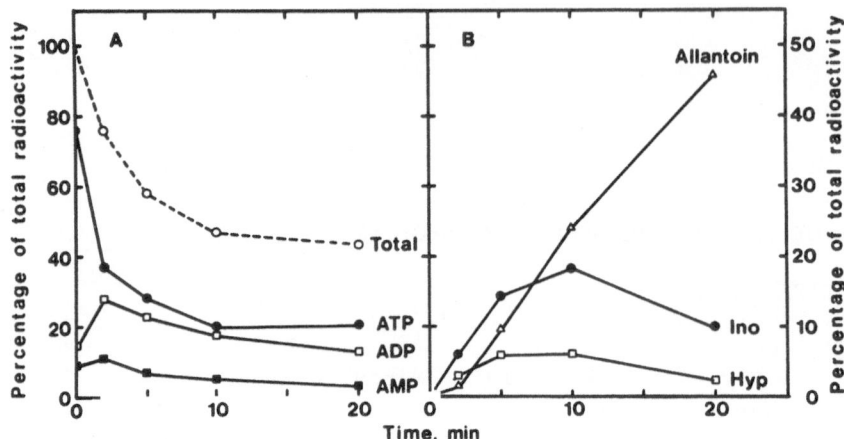

Fig. 1. A) Adenine nucleotide catabolism and B) formation of
         inosine (Ino), hypoxanthine (Hyp), and allantoin in
         liver cells incubated with 10 mM glycerol.

         The radioactivity lost from the adenine nucleotide pool is
recovered in various purine derivatives. There is a transient
increase in radioactive inosine and hypoxanthine, and a nearly
constant accumulation of radioactive allantoin (Fig. 1B). There is
also a transient accumulation of radioactive adenosine and IMP
(data not shown). However, the extent of that accumulation depends
upon the moment that the cells are exposed to glycerol. When glycerol
is added at the beginning of the incubation, about 8% of the total
radioactivity is found both in adenosine and in IMP. However, when
glycerol is added after a 15 min incubation, 17% of the total
radioactivity is found in IMP and only 3% in adenosine. All these
changes are similar to those induced by fructose under identical
conditions (data not shown).

         Initially, the rate of purine synthesis de novo, as measured
by the incorporation of [^{14}C]formate into purine (mainly adenine)
nucleotides (10), is either unchanged or decreased in cells incubated
with glycerol (Fig. 2A). This initial phase lasts longer as the
glycerol concentration is increased. Subsequently, there is a marked
stimulation of purine synthesis de novo with 1.25 and 2.5 mM
glycerol. The extent of the stimulation is greater at 2.5 mM
glycerol. This stimulatory phase is not observed during an 80 min
incubation with 5 or 10 mM glycerol. The PP-ribose-P availability,
as measured by the incorporation of [^{14}C]adenine into purine
nucleotides (10), is similarly modified in liver cells incubated
with glycerol, except that the increase occurs before that in purine
synthesis de novo (Fig. 2B).

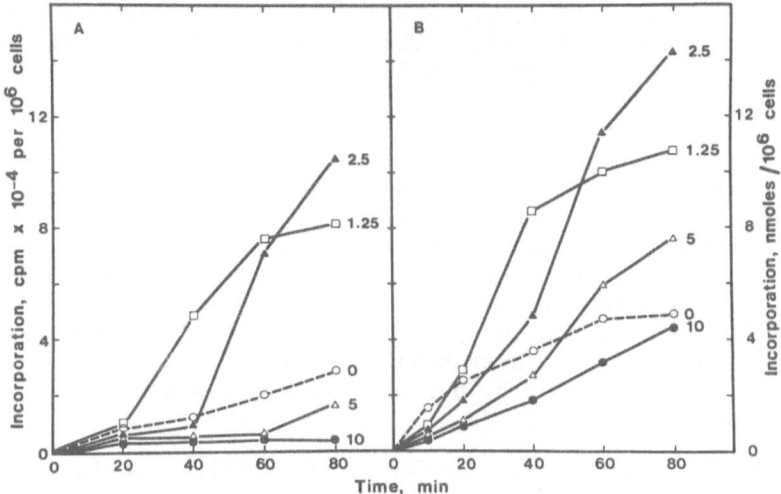

Fig. 2. A) Purine synthesis de novo and B) PP-ribose-P availability in liver cells incubated with various glycerol
concentrations.

These effects are similar to those produced by fructose under the
same conditions (data not shown). However, at identical concentrations, the period of latency or inhibition is shorter with fructose,
the extent of the stimulation at 1.25 and 2.5 mM fructose is smaller,
and an important stimulation occurs with 5 and 10 mM fructose.

The period of latency or inhibition is not explained by an
increased degradation of newly synthesized nucleotides in the
presence of glycerol. Only a small amount of [^{14}C]formate is
incorporated into purine nucleosides and bases during the entire
incubation period with 10 mM glycerol. Although a significant amount
of radioactivity from [^{14}C]adenine accumulates in inosine and
allantoin when cells are incubated with glycerol, nonetheless, the
effects of glycerol on PP-ribose-P availability are similar when
the incorporation of [^{14}C]adenine into total purine nucleotides,
nucleosides, and bases is considered.

Glycerol is rapidly phosphorylated to glycerol-3-P after its
uptake by the cells, and further metabolism is slow (data not shown).
Glycerol-3-P accumulates. As the initial concentration of glycerol
in the medium is increased, the amount of glycerol-3-P accumulating
is greater, and the period of accumulation lasts longer. Simultaneously, the intracellular P_i concentration decreases by a similar
amount in cells incubated 10 min with 1.25 to 10 mM glycerol

(data not shown). Thereafter, it returns to normal values, except
in the presence of 10 mM glycerol. The return to normal takes
longer at higher concentrations of glycerol.

The similarity between the adenine nucleotide catabolism induced
by glycerol and that induced by fructose indicates that the mechanisms of action of the two compounds are quite similar. The accompa-
nying changes in both the intracellular P_i concentration, and the
ATP/ADP and ATP/AMP ratios appear to control the rate of PP-ribose-P
synthesis in the cells. An increase in PP-ribose-P availability
always precedes the stimulation of purine synthesis de novo.
However, the lag between the two processes suggests that the rate of
purine synthesis de novo in these cells is controlled, not only by
the availability of PP-ribose-P, but also by another factor,
possibly the concentration of ATP.

This work was supported by the Canadian Arthritis Society.

REFERENCES

1. H.B. Burch, O.H. Lowry, L. Meinhardt, P. Max Jr., and K. Chyu,
      Effect of fructose, dihydroxyacetone, glycerol, and glucose
      on metabolites and related compounds in liver and kidney,
      J. Biol. Chem. 245: 2092-2102 (1970).
2. H.F. Woods and H.A. Krebs, The effect of glycerol and dihydroxy-
      acetone on hepatic adenine nucleotides, Biochem. J. 132: 55-60
      (1973).
3. M. Erecinska, M. Stubbs, Y. Miyata, C.M. Ditre, and D.F. Wilson,
      Regulation of cellular metabolism by intracellular phosphate,
      Biochim. Biophys. Acta 462: 20-35 (1977).
4. G. Van den Berghe, Metabolic effects of fructose in the liver,
      Curr. Top. Cell. Regul. 13: 97-135 (1978).
5. B.T. Emmerson, Effect of oral fructose on urate production,
      Ann. Rheum. Dis. 33: 276-280 (1974).
6. K.O. Raivio, M.A. Becker, L.J. Meyer, M.L. Greene, G. Nuki, and
      J.E. Seegmiller, Stimulation of human purine synthesis de novo
      by fructose infusion, Metabolism 24: 861-869 (1975).
7. M. Itakura, R.L. Sabina, P.W. Heald, and E.W. Holmes, Basis for
      the control of purine biosynthesis by purine ribonucleotides,
      J. Clin. Invest. 67: 994-1002 (1981).
8. G. Van den Berghe, M. Bronfman, R. Vanneste, and H.-G. Hers,
      The mechanism of adenosine triphosphate depletion in the
      liver after a load of fructose. A kinetic study of liver
      adenylate deaminase, Biochem. J. 162: 601-609 (1977).
9. G. Van den Berghe, F. Bontemps, and H.-G. Hers, Purine catabolism
      in isolated rat hepatocytes. Influence of coformycin,
      Biochem. J. 188: 913-920 (1980).
10. C. Des Rosiers, M. Lalanne, and J. Willemot, Purine synthesis
      de novo and its regulation in rat hepatocytes, Can. J. Biochem.
      58: 599-606 (1980).

PURINE RELEASE BY HUMAN ERYTHROCYTES

Alessandro Giacomello and Costantino Salerno

Institutes of Rheumatology and Biological Chemistry
University of Rome, and Center of Molecular Biology
National Research Council, 00100 Rome, Italy

The importance of blood and, in particular, of erythrocytes as a vehicle for transport of purines is well known[1,2]. Considerable quantities of purines enter and leave the nucleotide pools of red cells which take up adenine, guanine, hypoxanthine, and xanthine and convert them into nucleotides. No matter what purine is taken up by erythrocytes, hypoxanthine appears to be the main purine released in vivo. Human erythrocytes cannot synthesize purines de novo and are unable to convert hypoxanthine or guanine into adenine. Hypoxanthine release is mediated by prior conversion of the various purine nucleotides to IMP. In the present paper, some of the mechanisms which regulate the catabolic paths of this nucleotide are studied.

Human erythrocytes were prepared from blood freshly drawn in heparin and washed twice in 0.9% NaCl with removal of white cells by aspiration. In order to enrich erythrocytes with PRPP[3], the packed cells were preincubated in an equal volume of a medium containing 50 mM Tris-HCl (pH 7.4), 5 mM glucose, 0.12 mM Na phosphate, and the appropriate amount of NaCl to give isotonic solution and incubated up to 90 min at 37°C. The cells, washed twice in a medium not containing phosphate (50 mM Tris-HCl, 5 mM glucose, 12.8 mM NaCl, pH 7.4), were transferred in an equal volume of the last medium containing 5 μM ^{14}C hypoxanthine. The labeled purine base was incorporated within 2 min and IMP was the only radioactive substance present in detectable amount in the cells[3,4].

After a lag time which (like the intracellular PRPP concentration[3]) increased with increasing phosphate concentration and the time of preincubation in phosphate buffered saline, radioactive material, identified as hypoxanthine by paper chromatography[5], was released from erythrocytes in the medium (fig. 1). The lag phase was

not observed when an excess of hypoxanthine was added to PRPP-enriched erythrocytes in order to consume all the available intracellular
PRPP. These results suggest that the observed lag phase can be attributed, at least in part, to the presence of PRPP which causes HGPRT
-catalyzed reutilization[6] of hypoxanthine formed from IMP degradation.

When PRPP-enriched erythrocytes were transferred in buffered
saline containing not only 5 µM ^{14}C hypoxanthine but also 8 mM formycin B, the lag time increased and detectable amount of ^{14}C inosine
was present within the cells[3,5] (about 10% of the intracellular [14]C
IMP which was always the main labeled compound present within erythrocytes). After the lag time, the rate of hypoxanthine release was
about the same of that observed in the absence of formycin B (fig. 2).
Since, at the concentration employed, formycin B is known to inhibit
purine nucleoside phosphorylase in intact human erythrocytes[7], these
results confirm[2] that the cells sequentially degrade the intracellular IMP to inosine and hypoxanthine and suggest that the phosphory-
lase-catalyzed formation of hypoxanthine from its nucleoside is not
the rate limiting step in this catabolic path.

The addition of 0.03 ml of isotonic NaCl containing 1.4 mM
guanine or 1.6 mM adenine to 1.2 ml of a suspension of ^{14}C IMP-enriched erythrocytes in buffered saline led to an increase in the rate
of release of radioactive material identified as hypoxanthine by
paper chromatography[5]. The rate of guanine or adenine-induced hypoxanthine release was unaffected by the presence of formycin B up to
8 mM in the suspending medium. These results suggest that the presence of guanine or adenine in the incubation medium lead to an IMP
degradation path different from that occurring in the absence of
purine bases.

Using erythrocytes enriched with cold IMP and adding ^{14}C guanine or ^{14}C adenine, it has been possible to demonstrate that,
during the guanine or adenine-induced hypoxanthine release, guanine
or adenine are taken up by the cells as GMP or AMP respectively
(figs. 3,4). The labeled intracellular purine nucleotides were
identified by paper chromatography[5]. Erythrocytes were enriched
with cold IMP by adding buffered saline containing an excess of cold
hypoxanthine to an equal volume of packed erythrocytes preincubated
in phosphate buffered saline. Under the experimental conditions,
using ^{14}C hypoxanthine instead of cold hypoxanthine, it has been
possible to demonstrate that only one half of hypoxanthine was incorporated. Therefore, guanine or adenine were taken up by IMP-enriched erythrocytes as GMP or AMP respectively, even at intracellular
concentration of PRPP insufficient to allow a further uptake of hypoxanthine. Although more complex pathways cannot be excluded, the simplest mechanism capable of explaining GMP or AMP synthesis and guanine or adenine-induced hypoxanthine release from IMP-enriched erythrocytes is the transfer of the phosphoribosyl moiety of IMP to gua-

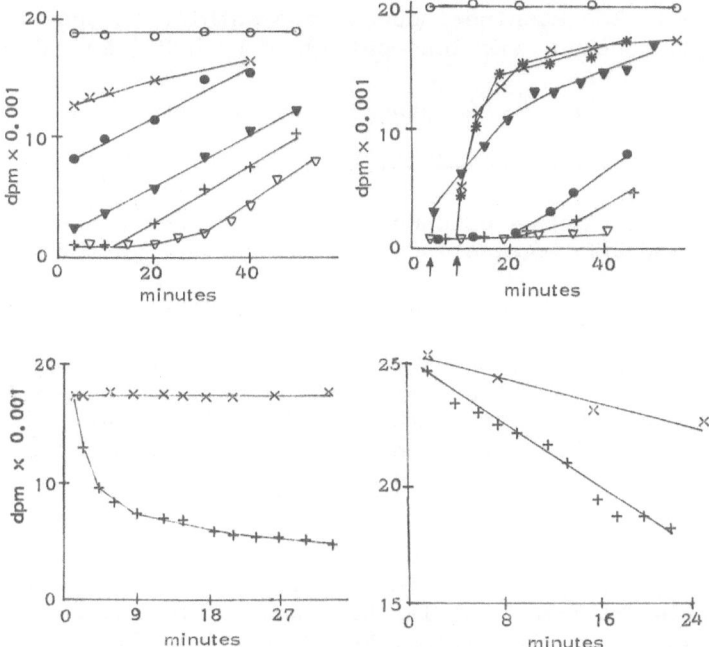

Fig. 1. (left) Effect of phosphate concentration (×3 mM, +9 mM, ∇ 12 mM) with a preincubation time of 40 min and of the pre-incubation time (o none, ● 15 min, ▼ 30 min) in 9 mM phosphate buffered saline on the release of radioactive material from 14C IMP-enriched human erythrocytes. Other experimental con-ditions are as described in the text. Abscissa: time of in-cubation in the medium containing 14C hypoxanthine.

Fig. 2. (right) Effect of adenine, guanine, and formycin B on the re-lease of radioactive material by 14C IMP-enriched erythrocy-tes. Preincubation time in 9 mM phosphate buffered saline: 0 (o), 60 (×,✳,●,+), 90 min (▼,∇). Formycin B concentration in the suspending medium: 0 (o,▼,●,✳,∇), 8 mM (×,+). Arrows: time of addition of adenine or guanine. In × and ✳ guanine and in ▼ adenine is present in the medium. Abscissa: time of incubation in the medium containing 14C hypoxanthine.

Figs. 3 (left) and 4 (right). 14C guanine (fig. 3) and 14C adenine (fig. 4) uptake by IMP-enriched erythrocytes. The cells were preincubated for 50 min (fig. 3) or for 30 min (fig. 4) in buffered saline containing 0 (×) or 9 mM phosphate (+), washed twice in buffered saline not containing phosphate and transferred in an equal volume of the last medium containing 12 μM (fig. 3) or 14 μM (fig. 4) hypoxanthine. After 4 (fig. 3) or 2 min (fig. 4) of incubation, 14C guanine (4 μM) or 14C adenine (8 μM) were added (figs. 3 and 4 respectively). Ordinate: dpm in the suspending medium. Abscissa: time of incubation in the medium containing the labeled purine base.

nine or adenine. The adenine-induced hypoxanthine release could be mediated by the coupled reactions catalyzed by HGPRT and APRT.

$$IMP + PP \xrightarrow{HGPRT} hypoxanthine + PRPP$$

$$PRPP + adenine \xrightarrow{APRT} AMP + PP$$

The HGPRT-catalyzed IMP pyrophosphorolysis has been recently studied[6]. GMP synthesis and IMP degradation could occur through the following reaction:

$$guanine + IMP \longleftrightarrow GMP + hypoxanthine$$

It has been previously demonstrated that purified human erythrocyte HGPRT catalyzes this reaction[8] whose equilibrium favors hypoxanthine and GMP formation. According to these hypothesis, HGPRT play an important role in IMP degradation and in purine interconversion.

REFERENCES

1. A. W. Murray, D. C. Elliot, and M. R. Atkinson, Nucleotide Biosynthesis from preformed purines in mammalian cells, in: "Nucleic
   Acid Research and Molecular Biology," J. N. Davidson and W. E.
   Cohn, eds., Academic Press, New York (1970).
2. A. W. Murray, The biological significance of purine salvage,
   Ann. Rev. Biochem. 40:811 (1974).
3. A. Giacomello and C. Salerno, Hypoxanthine uptake by human erythrocytes, Febs letters 107:203 (1979).
4. C. Salerno and A. Giacomello, Hypoxanthine transport in human
   erythrocytes, in: "Purine Metabolism in Man-III, Part B,"
   A. Rapado, R. W. E. Watts, and C. H. M. M. DeBruyn, eds.,
   Plenum Pub. Corporation, New York (1980).
5 E. Gerlach, R. H. Dreisbach, and B. Deuticke, Paper chromatographic separation of nucleotides, nucleosides, purines, and
   pyrimidines, J. Chromatog. 18:81 (1965).
6. A. Giacomello and C. Salerno, Human HGPRT - Steady state kinetics
   of the forward and reverse reactions, J. Biol. Chem. 253:
   6038 (1978).
7. M. R. Sheen, B. K. Kim, and R. E. Jr. Parks, Inhibition by the
   inosine analog formycin B of the isolated enzyme and of nucleoside metabolism in intact erythrocytes and sarcoma 180
   cells, Mol. Pharmacol. 4:293 (1968).
8. C. Salerno and A. Giacomello, Human HGPRT - IMP GMP exchange:
   stoichiometry and steady state kinetics of the reaction,
   J. Biol. Chem. 254:10232 (1979).

USE OF ADENINE IN BLOOD BANKING

Grant R. Bartlett

Laboratory for Comparative Biochemistry
4620 Santa Fe Street
San Diego, CA   92109   USA

The use of an acid-citrate-glucose mixture as a preservative/anticoagulant for the cold storage of human blood for purposes of transfusion developed out of research carried out during World War II in the United States and England. With this preservative it was possible to hold blood for three weeks with at least 70% post-transfusion survival of the red cells, longer storage leading to rapidly decreasing viability. The so-called ACD, and CPD (a slight variant), preservatives have been in use since the mid 1940's until last year when supplementation with adenine has allowed extension of the storage outdating time from 3 to 5 weeks. The purpose of the present communication is to summarize key events leading to this use of adenine.[1]

Studies in the late 1950's in the Laboratory of Finch at the Univ. of Washington showed that addition of adenosine to ACD could decrease the rate of breakdown of red cell organic phosphate and prolong cell survival.[2] Other studies by the same group found that the human red cell had high activities of the enzymes adenosine deaminase and nucleoside phosphorylase[3,4] and it was concluded that adenosine was converted by deaminase to inosine which was then split by phosphorylase to hypoxanthine and ribose phosphate, the latter being metabolized to provide energy to the cell. It seemed therefore that inosine should act the same as adenosine whose use appeared to be ruled out by its hypotensive activity. However subsequent studies by the Finch group and others failed to find an extension of red cell storage life with inosine.[5]

During this same period studies in my Laboratory at the Scripps Clinic in La Jolla and in that of Nakao in Tokyo greatly extended knowledge about the nature and amounts of organic phosphates in

fresh and stored human red cells and about effects of added inosine,
adenine and adenosine. Nakao et al[6] showed that additions of
adenine and inosine but not inosine alone, could delay breakdown of
ATP in stored red cells, and improve their survival. Shafer and
Bartlett[7] showed that long-stored human red cells, depleted of most
of their phosphate metabolites, including ATP, could resynthesize
ATP to the concentration of the fresh cell, by a brief incubation
at 37° with adenosine or with inosine plus adenine, but not with
inosine or adenine alone. It was suggested that part of the ribose
monophosphate formed from inosine by nucleoside phosphorylase was
converted to a pool of PRPP which reacted with the free adenine,
catalysed by adenine phosphoribosyltransferase (APRT), to form AMP
while much of the rest went to lactate providing high energy phosphate bonds needed to raise AMP to ATP. It was further proposed
in the case of adenosine that part of the nucleoside was converted
directly to AMP by a kinase and the rest metabolized to supply
phosphate for conversion of AMP to ATP.

It was inferred from the studies cited that there was insufficient PRPP in outdated stored human red cells for reaction with
adenine to form AMP. We had also incubated freshly drawn human
blood with ^{14}C-adenine and found only traces of radioactivity incorporated into adenine mononucleotides, indicating that the supply
of PRPP was limiting. It was a surprise then when Simon, working
in Finch's Laboratory, reported[8] that addition of a small amount of
adenine at the outset of storage of human blood could extend the
outdating time to 5 to 6 weeks and that the improved red cell survival was associated with a higher level of ATP. This important
discovery was confirmed in due course in several other laboratories.[9] The mystery about how adenine could be converted to adenylate was solved when we demonstrated a shift in the metabolism of
glucose, early in storage, into the pentose shunt with production
of a sizeable pool of PRPP.[10]

A Committee of the National Academy of Sciences/National Research Council, with input from various interested government agen-
cies and academic researchers decided that use of adenine in blood
banking looked promising and encouraged studies which still needed
to be done on the effects of adenine on blood typing and on recovery
of clotting factors and plasma proteins.[11] These studies were
carried out in several labs and no deleterious effects of adenine
were found.

The NRC Committee and its consultants had somehow neglected to
consider matters related to the pharmacology of adenine. However
my Laboratory elected to carry out a detailed study of the metabolism of adenine given intravenously and chose the rabbit as the
principal model. Adenine along with the ^{14}C-labeled derivative was
administered IV and the distribution of radioactivity determined at
various times thereafter in blood, urine and tissues.[12] Infused

adenine was quickly cleared from the blood and rapidly metabolized
so that none was left after two hours. Detailed studies were done
on tissue and urine metabolites 4 hours after a dose of 35mg/kg.
Approximately 10% of the adenine was eliminated unchanged in the
urine. Another 10% appeared in the urine as approximately equal
parts of two metabolites, 8-oxyadenine and 2,8-dioxyadenine. Most
of the rest of the radioactivity was in adenine mononucleotides of
different tissues, including red cells. Two initial metabolic reactions for adenine could be assumed: one with tissue PRPP, catalysed by APRT, to give adenylate and the other with xanthine oxidase
to form the oxyadenine metabolites. Adenine was then given IV to
rabbits in doses ranging from 2 to 200 mg per kg.[13] Surprisingly,
the same fraction of administered adenine was converted to oxyadenine derivatives no matter what the dose. With a dose of adenine
of 50 mg/kg and above insoluble particles of 2,8-dioxyadenine were
found in urine and bladder. With doses of 100 mg/kg and greater
considerable 2,8-dioxyadenine was found to be deposited in the
tubules of the kidney.

I carried out two experiments on myself with infusion IV each
time of 10 mg/kg of adenine.[14] Free adenine and two oxyadenine
metabolites were excreted in the urine to approximately the same
extent as had been found with the rabbit.

Our results on the in vivo metabolism of adenine led to a number of studies by several investigators, mostly in pharmaceutical
concerns, to determine the extent of excretion of the oxyadenine
metabolites, and their effects on kidney function, in a variety of
animal models given IV adenine. Included also were a few studies
on man. Although there were some interesting comparative differences the overall picture was essentially the same. The studies,
many of which were unpublished, have been brought together in an
important review by Warner.[15] It was tentatively concluded by some
of us involved that IV adenine at 50 mg/kg or greater to man or
other mammals is potentially dangerous as a consequence of possible
precipitation of 2,8-dioxyadenine in the urinary tract. It was also
concluded that a dose of up to 20 mg/kg, even when repeated daily,
was safe.

After deliberation by interested parties, it was decided to set
in motion the final countdown leading to approval by the Bureau of
Biologics of the FDA.[9,11] However concern about possible effects
of the 2,8-DOA metabolite led to an over-cautious recommendation of
17.5 mg of adenine per bag and this is the formula adopted in the
ruling published in the Federal Register of August 4, 1978. Contin-
uing tests have shown that this amount of adenine, marginal for the
preservation of red cells in whole blood for 5 weeks is too little
for packed red cells. Moreover it was belatedly realized that there
was not enough glucose to sustain the red cells for this period at
high hematocrit. Developments in the art of blood banking have led

to rapidly increasing use of all components with separation of
plasma and platelets at the time of collection and holding of the
red cells at a hematocrit of 80 or higher. Separation of the plasma
also removes a large fraction of the preservatives added to the
collection bag. Insiders now recommend 0.5mM adenine in the
collected whole blood (ca 35 mg per bag) with an increase of 50%
in the glucose. Such a formula for the preservative, called CPD-A2,
is now being processed for licensing by suppliers of blood bags.

More than ten million units (pints) of blood are currently
being collected in the United States annually. Almost all of this
blood is mixed with a preservative containing adenine. Following
our lead, use of an adenine preservative is spreading into other
countries. A substantial amount of the adenine in the preservative
will, on the average, be transfused into the recipient. Little is
now understood about the control of trace pools of adenine in the
body or about the effects which might be produced by their alteration. Neither the physician nor the patient will have any say as
to whether or not adenine is to be received. It is essential
therefore that both experts on purine metabolism in man as well
as professionals involved in blood banking carefully monitor this
new development.

REFERENCES

1. G.R. Bartlett, in: "The Human Red Cell in Vitro," T.J. Green-
   walt and G.A. Jamieson, eds., Grune & Stratton, Inc., New
   York (1974) p. 5.
2. D.M. Donohue, C.A. Finch and B.W. Gabrio, J. Clin. Invest. 35:
   562 (1956).
3. F.M. Huennekens, E. Nurk and B.W. Gabrio, J. Biol. Chem. 221:
   971 (1956).
4. B.W. Gabrio, C.A. Finch and F.M. Huennekens, Blood 11:103
   (1956).
5. R.D. Lange, W.H. Crosby, D.M. Donohue, et al, J. Clin. Invest.
   37:1485 (1958).
6. K. Nakao, T. Wada, T. Kamiyama, et al, Nature 194:877 (1962).
7. A.W. Shafer and G.R. Bartlett, J. Clin. Invest. 41:690 (1962).
8. E.R. Simon, Blood 20:485 (1962).
9. E.R. Simon, Transfusion 17:317 (1977).
10. G.R. Bartlett, in: "Erythrocytes, Thrombocytes, Leukocytes,"
    E. Gerlach, K. Moser, E. Deutsch, et al, eds., Thieme,
    Stuttgart (1973) p. 139.
11. S.N. Swisher, Transfusion 17:309 (1977).
12. G.R. Bartlett, Transfusion 17:351 (1977).
13. G.R. Bartlett, Transfusion 17:358 (1977).
14. G.R. Bartlett, Transfusion 17:367 (1977).
15. W.L. Warner, Transfusion 17:326 (1977).

SUBSTANCE RATES OF DIFFERENT STEPS OF PURINE METABOLISM IN NORMAL AND PRESERVED RED BLOOD CELLS (RBC) STUDIED IN EXPERIMENTS SIMULATING IN VIVO CONDITIONS

Carl-Henric de Verdier, Frank Niklasson and
Claes F. Högman

Department of Clinical Chemistry and Blood Center
University Hospital, S-750 14 Uppsala, Sweden

INTRODUCTION

In RBC the energy provided by glycolysis is transferred by
ATP to energy requiring processes as ion pumping, chemical synthesis etc by means of ATP. That explains the use of this substance as a marker for cell viability after preservation of
erythrocytes and also the addition of precursors – adenine, adenosine, inosine (for the ribose phosphate part) etc – to preservation media. The physiological role of RBC in the transport of
purine compounds between the organs of the body has been discussed
for years. In order to answer questions hidden within these fields
of exploration more firm data describing the substance rates of the
specific steps of purine metabolism in RBC are needed.

METHODS

The dialysis system for incubation of RBC, briefly described
in earlier publications from these laboratories [1-2] was applied
for the investigation of purine metabolism because it allowed continuous and calculable addition and removal of the substances at
low – close to physiological – concentrations. The dialysis
tubing Cuprophan [R], (thickness 20 μm; width 4 mm) was kindly supplied by Enka AG, P.O.B. 200916, D-5600 Wuppertal 2.

About 900 mm of the tubing was immersed into the RBC suspension. A tubing pump delivered the basic medium with some additi-
ves to the affluent end of the dialysis tubing. The volume rate
of the pump was measured and kept about 45 ml·h^{-1}. The volume of
the suspension was at the start of the experiment about 85 ml. The
experiment was carried out in a 100 ml glass beaker equipped with
a magnetic stirrer and placed in a water bath thermostated at 37°C.

RESULTS AND DISCUSSION

Adenine and Adenosine

Additon of adenine in high concentrations (50 $\mu mol \cdot l^{-1}$) to
RBC partly deprived of adenylates resulted under physiological
conditions in a net synthesis of adenylates at substance rates of
51, 26 and 33 $\mu mol \cdot h^{-1}$ (3 experiments). Higher phosphate concentration and simultaneous addition of inosine increased the rate of
synthesis two to three times.
        Addition of adenosine to comparatively low concentrations
resulted in rapid synthesis of adenylates. The proportion of adenosine deaminated to inosine increased at concentrations above
the physiological level as shown in Table 1.

Table 1. Metabolism of adenosine added to RBC in the dialysis
         incubation system. Substance rates calculated per litre
         RBC

Adenosine in susp. $\mu mol \cdot l^{-1}$	<0.01	0.04	0.75	1.5
Adenylate, net synthesis, $\mu mol \cdot h^{-1}$	50	250	240	340
Inosine, release, $\mu mol \cdot h^{-1}$	0	30	140	250
Hypoxanthine, release, $\mu mol \cdot h^{-1}$	60	430	420	750

Guanosine

        Addition of guanosine to RBC suspension resulted in the syn-
thesis of guanylates with an energy charge that was close to the
adenylate energy charge. The sum of adenylates + guanylates
reached a higher value than the substance concentration of the
adenylates alone, probably due to the fact that GTP exerts less
product inhibition than ATP at the phosphofructokinase step. Supplementation of blood preservation media with guanosine results in
higher $2,3-P_2$-glycerate concentration[2]. Our finding that GTP
loaded RBC exported hypoxanthine but no guanine indicates that
the catabolism passes the guanylate reductase steps and generates
oxidized nicotinamide cofactors which are known to stimulate the
synthesis of $2,3-P_2$-glycerate.

Liquid Preserved RBC

        RBC stored at +4 ^{o}C in CPD-adenine or SAG-mannitol for
different periods of time have,after neutralization with isotonic
Na-bicarbonate and washing,been studied in the dialysis incubator.
Figure 1 illustrates the data obtained for some storage units
when the RBC were incubated in a physiological medium for 4 h
with a supplementation of adenine in order to prevent purine depletion. Adenylate energy charge is shown to the left and total
adenylates to the right with identical symbols for the units in
the two diagrams. The letters C and S are used to indicate the

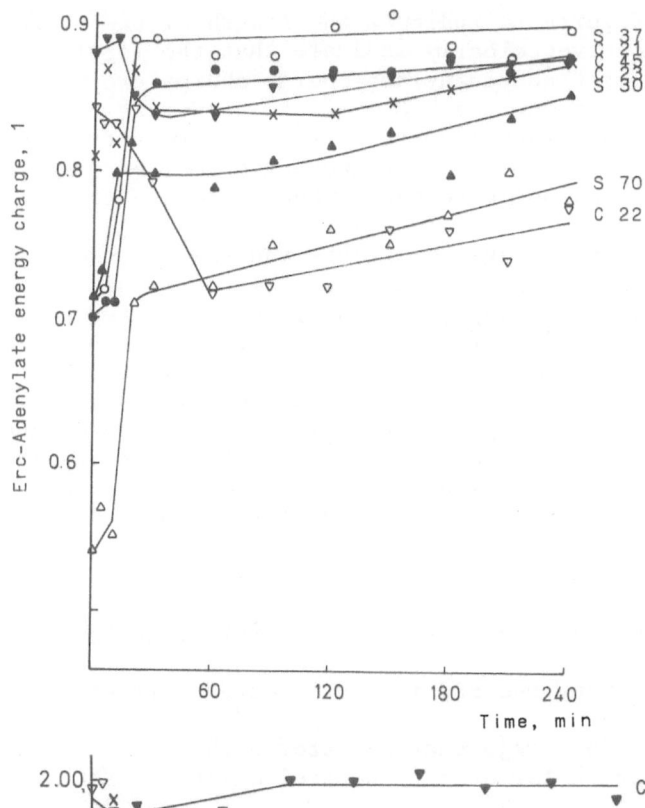

Figure 1

Adenylate energy
charge (left) and
total adenylates
(below) in RBC
from preservation
units during dia-
lysis incubation
experiments.

Storage media
C = CPP-adenine
S = SAG-manitol
Ref.
C.F. Högman et al.
Vox Sang 41:274
(1981)

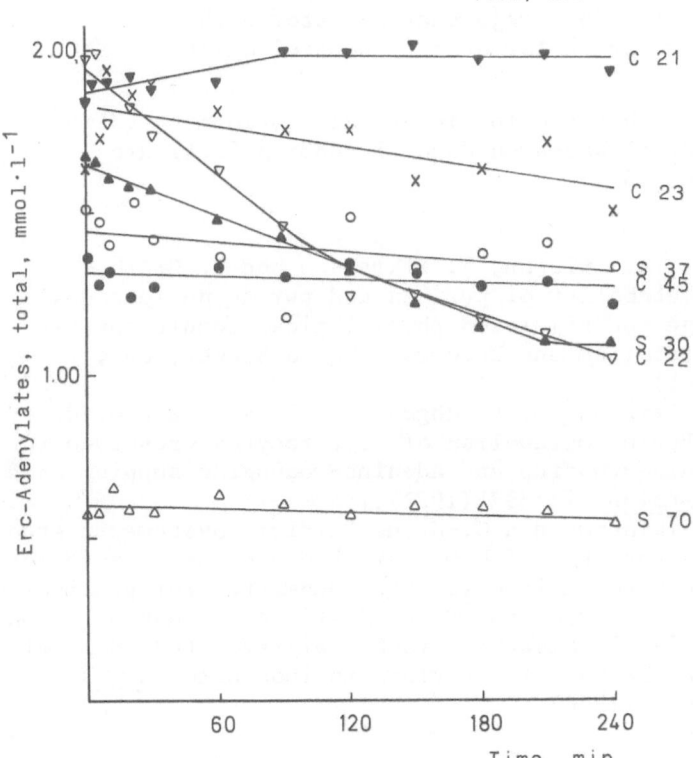

storage media and the figures to indicate the length of the storage
expressed in days. The curves clearly indicate that the initial
values for total adenylates and adenylate energy charge are not in
all cases representative for how the RBC will behave during the
incubation. Analysis of other components as K^+, lactate and hypoxanthine strengthens the conclusion that incubation may be used to
evaluate a membrane damage, not detectable when a storage cold
specimen is investigated.

With exception of the storage conditions and the kinetics of
the RBC eliminatory function of the spleen of the recipient, the
viability of transfused RBC is dependent on a) genetic effects on
metabolism or membrane function of the RBC b) dietary or drug
effects on the RBC at the time of the blood donation. In order to
investigate the reason for the decline of total adenylates and
primarily also of adenylate energy charge during incubation of RBC
in unit C 22 a new unit from the same donor was stored and specimens were taken after 1, 19 and 41 days. Normal values were then
obtained indicating a non-genetic explanation for the poor quality
of the RBC at the first storage occasion.

SUMMARY

A dialysis incubation system simulating the conditions in the circulation can be used:
1. To draw conclusions about the RBC in transporting purines
   between organs.
2. To improve solutions for "rejuvenating" stored RBC.
3. To design better systems for quality control of stored RBC.

ACKNOWLEDGEMENT
Grants from the Swedish Board for Technical Development (81-5490)
and the Swedish Medical Research Council (B83-3X-6524) are
gratefully acknowledged.

REFERENCES

1. C.-H. de Verdier, Å. Ericson, F. Niklasson and T. Groth,
   Erythrocyte metabolism of purines and purine nucleosides
   during storage and simulated physiological conditions, in
   "Red Cell Metabolism and Function," G. J Brewer, ed.,
   New York (1981).
2. O. Åkerblom, Å. Ericson, C. F. Högman, D. Strauss and C.-H.
   de Verdier, Purine metabolism of erythrocytes preserved in
   adenine, adenine-inosine and adenine-guanosine supplemented
   media. Transfusion, 21:397 (1981).
3. Å. Ericson, F. Niklasson and C.-H. de Verdier, Systematic study
   of nucleotide analysis of human erythrocytes using anionic
   exchanges and HPLC. Clin Chim Acta. Submitted for publication.
4. J. F. Henderson, G. Zombor and P. W. Burridge, Guanosine triphosphate catabolism in human and rabbit erythrocytes: Role of
   reductive deamination of guanylate to inosinate. Can J
   Biochem. 56:474 (1978).

BIOCHEMICAL STUDY OF A CASE OF HEMOLYTIC ANEMIA WITH INCREASED

(85-fold) RED CELL ADENOSINE DEAMINASE

J.L. Pérignon, M. Hamet, H.A. Buc, P. Cartier
and M. Derycke

Laboratoire de Biochimie, INSERM U 75, CHU Necker-
Enfants Malades, 156. rue de Vaugirard 75015 Paris

INTRODUCTION

Hereditary hemolytic anemia with increased red cell adenosine deaminase (ADA) activity has been reported in two families (1, 2). We present the clinical and biochemical study of a third case of this disease, with emphasis on metabolic disturbances.

MATERIAL AND METHODS

Assay of adenosine deaminase and of other enzymes of purine and pyrimidine metabolism were performed by previously described radioisotopic assays (3) . Electrophoresis of RBC ADA was performed on hemolysates after Spencer et al. (4). The enzymes of glucose and glutathione metabolism were assayed after Beutler (5). Erythrocyte glucose consumption was assayed according to Cartier et al. (6). Adenosine deaminase was partially purified from human red blood cells ; the specific activity of the preparation was 1.2 μmol.min^{-1}.mg protein^{-1}. Rabbit antiserum to human adenosine deaminase was raised by immunization of white male rabbits with this preparation.

Human RBC were incubated with 120 μM [^{14}C-8] adenosine in Gardner's medium at 37° during 30 minutes ; after extraction in 0.6 M PCA, and neutralization with amine/freon, the nucleotides were chromatographied by HPLC (7) with continuous counting in a Flow One radioactive detector. RBC were also incubated with 830 μM [^{14}C-8] adenine in an isotonic high phosphate (30 mM) medium at 37° during 60 minutes ; after centrifugation, 25 μl of the supernatant were chromatographied on Whatman CM 82 paper in distilled water and the spots counted in a Tricarb liquid scintillator.

CASE REPORT

    The propositus is a ten-year-old male Caucasian child born in
1971 in Algeria, where he lives. Since the age of six months, he
presented about ten episodes of anemia and jaundice, treated by
blood transfusions. A detailed study will be published elsewhere (8) .

RESULTS

    The patient's RBC ADA activity was 43 000 $nmol.min^{-1}.ml$ RBC^{-1}
(normal values : 495 $\pm$ 60). There was an about 3-fold increase of
red cell pyrimidine 5'-nucleotidase and orotate phosphoribosyltransferase, whereas other enzymes of purine and pyrimidine metabolism (inosine phosphorylase, adenosine kinase, adenine phosphoribosyltransferase, hypoxanthine-guanine-phosphoribosyltransferase,
phosphoribosylpyrophosphate synthetase) were normal or slightly
elevated. There was a 6-fold increase of pyruvate kinase activity
relatively to comparably reticulocyte-rich blood, and a 1.5 to 3-
fold increase of the other enzymatic activities of glucose and glutathione metabolism. Plasma ADA was much elevated (30.5 $\mu mol.min^{-1}$.
ml^{-1} ; normal value : 5.1 $\pm$ 2.5), probably reflecting intravascular
hemolysis. ADA activity in lymphocytes (2.13 $nmol.min^{-1}.10^{-6}$ cells ;
normal : 1.93 $\pm$ 0.61) and in fibroblasts (26 $nmol.min^{-1}.mg$ protein
$^{-1}$; normal range : 14-118) was normal, whereas the small increase
of activity in platelets (59.5 $nmol.min^{-1}.10^{-11}$ cells ; control
26.7) and in the liver (8.4 $\mu mol.min^{-1}.mg$ $protein^{-1}$; normal :
3.5 $\pm$ 0.3) is probably due to contamination by RBC.

    The patient's RBC ATP (1180 nmol/ml) and the content of total
adenine nucleotides (ATP + ADP + AMP) were respectively 64 % and
65 % of comparably reticulocyte-rich blood. Glucose consumption
and lactate production were normal.

    Kinetic studies on the RBC ADA of the patient as well as of
normal control subjects were performed on crude hemolysates. The
Km of the patient's ADA for adenosine (29 $\mu M/l$) and deoxyadenosine
(20 $\mu M/l$) were normal ; the enzyme showed a normal optimum pH
(7.4), and a normal electrophoretic pattern. Heat stability was
normal at 56°C (75 % inhibition after 60 minutes) and 68°C (97 %
inhibition after 30 minutes) ; the enzyme was completely inhibited
by erythro-9-(2-hydroxy-3-nonyl)-adenine (200 $\mu M/l$) and deoxycoformycin (20 $\mu M/l$). The immunological studies confirmed that patient's
ADA was intrinsically normal : the curves of neutralization showed
that, at equivalence, 1 μl of antiserum neutralized 3.9 $nmol.min^{-1}$
ADA activity of normal hemolysate, 1.9 $nmol.min^{-1}$ of patient's hemolysate and 1.9 $nmol.min^{-1}$ of purified ADA. This indicates that
patient's ADA is immunologically normal, and that its molecular
activity is normal.

    When RBC were incubated with radioactive adenosine (Table I)

Table I   INCORPORATION OF [14c] ADENOSINE IN RED CELL NUCLEOTIDES

Percent of radioactivity

	ADP	ATP	IMP	GTP
PROPOSITUS	0.2	3.8	95.8	0.2
	0.4	4.0	95.3	0.2
CONTROL (mean ± SD) (n = 5)	21 ± 0.4	21 ± 4.7	73.2 ± 4.3	n.d.

RBC were incubated at 37°C during 30 minutes with 120 µmol/l
[14c] adenosine. The nucleotides were analysed as described under
materials and methods. The results are expressed as percent of
total radioactivity in nucleotides. In all cases, no radioactivity
was detected in AMP. The total dpm in nucleotides were 151 500
and 179 400 in the patient, and 138 200 ± 14 000 in the controls.
n.d. : not detectable.

nearly all the radioactivity was found in IMP, indicating extensive adenosine deamination. Table II shows that when erythrocytes
are incubated with radioactive adenine, which is incorporated in
adenylic nucleotides, an abnormal amount of hypoxanthine is released in the incubation medium, indicating an excessive AMP degradation.

DISCUSSION

We report on the first case of hemolytic anemia with increased RBC ADA activity detected in childhood. RBC ADA activity (x 85)
is higher than in the patients of Valentine et al. : x 40-70 (1)
or Miwa et al. : x 40 (2). However, the level of RBC ATP is less

Table II RADIOACTIVITY IN PURINE BASES AND NUCLEOSIDES IN THE
INCUBATION MEDIUM OF RBC INCUBATED WITH [14c] ADENINE

	ADENINE	ADENOSINE	HYPOXANTHINE	INOSINE
PROPOSITUS	92.2	1.1	6.4	0.2
CONTROLS (mean ± SD) (n = 21)	97.7 ± 1.2	0.6 ± 0.3	0.8 ± 0.4	0.9 ± 0.7

RBC were incubated at 37°C during 60 minutes with 830 µmol/l [14c]
adenine. Bases and nucleosides were analysed as described under
materials and methods. The results are expressed as percent of total radioactivity in bases and nucleosides.

depressed in our patient (64 % of comparably reticulocyte-rich controls, vs <50 % in other cases (1, 2). The dyserythropoïesis documented in this patient (8) might account for the severity of the anemia. Like in the previous cases, the excess of RBC ADA activity is due to an excess of a catalytically and immunologically normal enzyme, and the marked elevation of ADA activity is found only in erythrocytes.

The low RBC ATP concentration is the most probable explanation for the hemolysis (1). "In vitro" studies of adenosine metabolism by intact patient's RBC showed as expected, a markedly decreased ATP synthesis from adenosine ; moreover, metabolic studies of adenylic nucleotides labelled with radioactive adenine indicate that AMP degradation (probably by hydrolysis of the phosphate ester followed by deamination of adenosine) is abnormally elevated in the patient's erythrocytes. Thus, the low RBC ATP concentration appears to be secondary to both a diminished synthesis of AMP from adenosine and an excessive catabolism of AMP.

REFERENCES

1. W.N. Valentine, D.E. Paglia, A.P. Tartaglia and F. Gilsanz, Hereditary hemolytic anemia with increased red cell adenosine deaminase (45-to 70-fold) and decreased adenosine triphosphate, Science 195:783 (1977).
2. S. Miwa, H. Fujii, N. Matsumoto et al., A case of red-cell adenosine deaminase overproduction associated with hereditary hemolytic anemia found in Japan, Am J Hematol 5:107 (1978).
3. J.L. Pérignon, M. Hamet, H.A. Buc, P.H. Cartier, M. Derycke and A.M. Houllier, Biochemical study of a case of hemolytic anemia with increased (85-fold) red cell adenosine deaminase, "in press"
4. N. Spencer, D.A. Hopkinson and H. Harris, Adenosine deaminase polymorphism in man, Ann Hum Genet 32:9 (1968).
5. E. Beutler, Red cell Metabolism. A manual of Biochemical methods. 2nd ed. New York : Grune & Stratton, 1975.
6. P. Cartier, J.P. Leroux and H. Temkine, Techniques de dosage des intermédiaires de la glycolyse dans les tissus. Ann Biol Clin (Paris) 25:791 (1967).
7. S.S. Matsumoto, K.O. Raivio and J.E. Seegmiller, Adenine nucleotide degradation during energy depletion in human lymphoblasts ; adenosine accumulation and adenylate energy charge correlation. J Biol Chem 254:8956 (1979).
8. M. Derycke, J.L. Pérignon, S. Bellucci, H. Buc and G. Schaison, Un cas d'anémie hémolytique congénitale par hyperactivité de l'adénosine désaminase. Actualités Hématologiques (Paris)(in press).

IMPAIRED METABOLISM OF DEOXYADENOSINE IN UREMIC ERYTHROCYTES

William P. Wiesmann and Horace K. Webster

Dept. of Nephrology and Hematology, Walter Reed Army
Institute of Research, Washington, D.C. 20012

INTRODUCTION

The metabolism of adenylates and guanylates by mammalian cells
is tightly controlled through a series of enzymatic reactions (7,
9). Adenosine (deoxyadenosine) is metabolized by erythrocytes
(RBC) to either adenylates via adenylate kinase (AK) or to inosine
by adenosine deaminase (ADA). Inosine, guanosine and their deoxy
forms are converted to the purine base hypoxanthine by purine nucleoside phosphorylase (PNP).

Congenital defects and chemical inhibitors of these enzymes
result in the accumulation of purine nucleosides and nucleotides
which have been directly linked to impaired lymphocyte function
and immunodeficiency syndromes (4,5). Since the uremic state is
complicated by an increased susceptability to infection largely
the result of acquired lymphocyte abnormalities, we have studied
the ability of uremic erythrocytes (RBC) to metabolize in vitro
radiolabelled adenosine and deoxyadenosine utilizying a combind
UV - radioactive high performance liquid chromatographic technique (HPLC) (1,2,3).

These data demonstrate a marked inability of uremic RBC to
metabolize adenosine and deoxyadenosine and may establish a partial basis for impaired lymphocyte function seen in this condition.

METHODS

Heparinized blood obtained from stable uremic patients on
chronic hemodialysis was obtained just prior to dialysis. Patients with a recent history of transfusions, antihypertensive or

immunosuppressive therapy were excluded from these studies. RBC
were incubated at 37° with 10-15 µCi of [^{14}C]-adenosine or [^{14}C]-
deoxyadenosine for one minute and 90 minutes. Purine nucleosides,
nucleotides and bases were then extracted with 1 M PCA and neutralized with KOH. Plasma obtained from control and uremic patients
was extracted with 0.6 Molar PCA on ice for 10 minutes.

    PCA extracts were analyzed by HPLC using described methods
(6). Purine nucleosides and bases were separated with a reversed-
phase column. Purine nucleotides were separated by anion-exchange
HPLC. Simultaneous UV monitoring and radioactivity detection were
performed with an on-line radioactivity flow detector.

RESULTS

Metabolism of ^{14}C-Adenosine by Control and Uremic RBC

    Inhibition of ADA can lead to accumulation of adenosine
(deoxyadenosine) which results in increased synthesis of ATP
(deoxy ATP) (7,9,10). Inosine, guanosine and their deoxy forms
are converted to the base hypoxanthine by PNP. Impairment of
this enzyme will lead to accumulation of these compounds (7,9,10).

    Table 1 represents the total concentration and radioactive
counts for adenosine, inosine, hypoxanthine and ATP in control
and uremic RBC which were incubated with [^{14}C]-adenosine. The
concentration and total radioactivity of adenosine was unchanged
in control and uremic blood. Likewise the amount and radioactive
labelling of inosine was unchanged in uremic RBC. In contrast,
there was a small but significant increase (14%, p<.05) in radioactive labelling and specific activity of hypoxanthine in control

Table 1. Metabolism of ^{14}C-Adenosine by Control and Uremic RBC

	AR	IR	HYP	ATP
CONTROL RBC				
(MEAN + SEM OF 3 CONTROLS)				
NMOLES[a]	63 ± 3	19 ± 2	49 ± 3	2720 ± 151
RADIOACTIVITY[b]	3583 ± 210	1788 ± 125	2993 ± 99*	656 ± 16
SPECIFIC ACTIVITY[c]	57	94	61	.25
UREMIC RBC				
(MEAN + SEM OF 5 PATIENTS)				
NMOLES	65 ± 3	24 ± 3	71 ± 5	4350 ± 120*
RADIOACTIVITY	3574 ± 175	2089 ± 100	2626 ± 100	678 ± 25
SPECIFIC ACTIVITY	55	87	36	.15

* INDICATES SIGNIFICANT DIFFERENCES AT P < 0.05
[a] NANOMOLES PER ML RBC
[b] INTEGRATED RADIOACTIVE COUNTS (AREA) PER CHROMATOGRAPHY PEAK
[c] SPECIFIC ACTIVITY = RADIOACTIVITY PER NMOLE

RBC verses uremic RBC. After 90 minutes of incubation ATP concentration and radioactivity were determined by anion-exchange HPLC. The ATP content was significantly greater in uremic RBC than in control RBC. Despite the increased ATP concentration in uremic RBC, incorporation of [14C]-adenosine into ATP was not increased. This resulted in a lower specific activity of ATP and suggests that the elevated levels of ATP are not due to increased synthesis from adenosine.

Metabolism of ^{14}C-deoxyadenosine by control and uremic RBC

Table 2 shows a significant impairment in uremic RBC for metabolism of [14C]-deoxyadenosine to [14C]-deoxyinosine and [14C]-deoxyinosine to [14C]-hypoxanthine. Approximately 7 times more [14C]-deoxyadenosine and 5 time more [14C]-deoxyinosine remain unmetabolized in uremic RBC.

Alterations in ADA should result in accumulation of deoxyadenosine which is metabolized to deoxy AMP and finally deoxy ATP (5). Anion-exchange HPLC of uremic whole blood extracts did not demonstrate evidence of deoxy ATP or deoxy GTP accumulation despite optimal conditions for the separation of these compounds.

Alterations in PNP in vivo results in the accumulation of deoxyinosine, deoxyguanosine and deoxyGTP (5). Plasma samples from the same patients were extracted and analyzed by reversed-phase HPLC. In four uremic patients studied, small but consistent peaks were observed co-eluting with deoxyinosine standards. Exogenous spiking with deoxyinosine was additive with the peak and the UV absorbance ratios (280/254 NM) were identical. Addition of purified PNP to the plasma abolished the peak. While these

Table 2. Metabolism of ^{14}C-Deoxyadenosine by Control and Uremic RBC

	DAR	DIR	HYP	DATP
CONTROL RBC (MEAN ± SEM OF 3 CONTROLS)				
NMOLES[a]	d	d	29 ± 5	d
RADIOACTIVITY[b]	414*± 25	114*± 5	3139 ± 175	1005 ± 100
SPECIFIC ACTIVITY[c]	---	---	108	---
UREMIC RBC (MEAN + SEM OF 6 PATIENTS)				
NMOLES	5 ± 1	24 ± 4	47 ± 5	d
RADIOACTIVITY	3009*± 125	1068*± 99	2451 ± 125	815
SPECIFIC ACTIVITY	600	44	52	---

* INDICATES SIGNIFICANT DIFFERENCES AT P < 0.05.
a NANOMOLES PER ML RBC.
b INTEGRATED RADIOACTIVE COUNTS (AREA) PER CHROMATORAPHY PEAK.
c SPECIFIC ACTIVITY = RADIOACTIVITY PER NMOLE.
d UNDETECTABLE.

data are not conclusive, they suggest that deoxyinosine is accumulating in uremic plasma. Attempts to identify deoxyadenosine and deoxyguanosine in plasma extracts have been hampered by the large amounts of UV absorbing material in uremic plasma which overlap these compounds. Deoxyadenosine, deoxyinosine and deoxyguanosine were not detected in control plasma.

DISCUSSION

These data demonstrate a marked abnormality in the ability of intact uremic RBC to metabolize adenosine (deoxyadenosine) to inosine (deoxyinosine) and hypoxanthine. The equivalent labelling by radioactive adenosine and deoxyadenosine of ATP pools suggests that the adenylate kinase pathway in uremic RBC is normal despite markedly elevated ATP levels. The latter may reflect decreased utilization of ATP in uremic RBC or an effect of high inorganic phosphate on ATP turnover (8).

We have tentatively identified nanomolar accumulations of deoxyinosine in uremic plasma. This compound is directly toxic to lymphocyte activation. Toxicity could result from "trapping" of deoxyinosine in lymphoid tissue by specific lymphoid deoxy kinases where it is phosphorylated to deoxynucleotides with resultant inhibition of DNA synthesis (9). Investigations are currently underway to examine the effect deoxynucleosides on deoxynucleotide formation in lymphocytes from uremic patients.

REFERENCES

1. Wilson W, Kirpatrick C, (1965). Ann Intern Med. 62:1-15.
2. Daniels J, Sakai H, (1971). Clin. Exp. Immunol. 8:213-218.
3. Goldblum S, Reed W, (1980). Ann. Int. Med. 93:597-613.
4. Carson D, Seegmiller J, (1976). J. Clin. Invest. 57:274-282.
5. Wilson J, Mitchel B. Daddona P, Kelley W, (1979). J. Clin. Invest. 64:1475-1484.
6. Webster H, Whaun, J, (1981). J. Chrom. 209:283-292.
7. Snyder F, Henderson J, (1973). J. Biol. Chem. 248:5899-5904.
8. Lichtman M, Miller D, (1970). J. Lab. Clin. Med. 76:267-279.
9. Carson D, Kaye J, Seegmiller J, (1977). PNAS (USA). 74:5677-5681.
10. Wortmann R, Mitchell B, Edwards N, Fox I, (1979). PNAS (USA) 76:2434-2437.

NUCLEOTIDE LEVELS AND METABOLISM OF ADENOSINE AND DEOXYADENOSINE

IN INTACT ERYTHROCYTES DEFICIENT IN ADENOSINE DEAMINASE

D. R. Webster, H. A. Simmonds, D. Perrett,
and R. J. Levinsky

Purine Laboratory, Guy's Hospital, Medical Unit, St.
Bartholomew's Hospital, Gt. Ormond Street Hospital,
London, U.K.

We previously reported severe ATP depletion in the erythrocytes of
an adenosine deaminase (ADA : EC 3.5.4.4) deficient child and also
found raised dAMP levels[1] in addition to the raised dATP and dADP
levels noted by others.[2]   Post-marrow graft ATP levels returned
to normal and the three deoxyadenosine nucleotides rapidly
disappeared.   No ATP depletion had been noted in earlier reports.[2]
Similar findings have now been obtained in the erythrocytes of two
further untreated ADA deficient children.   Enzyme replacement
therapy[3] also resulted in a rapid return of nucleotide levels toward
normal.

This paper reports in vitro studies to investigate mechanisms for
the formation and degradation of dATP in ADA deficient erythrocytes
from these two children.

PATIENTS AND METHODS

Clinical details of the two children (SY and KA) are given elsewhere (Levinsky et al - this symposium)

Nucleoside levels in erythrocytes were investigated before and
after exchange transfusions by high pressure liquid chromatography
(HPLC).[1,3]   Studies using intact erythrocytes were all carried out
prior to enzyme replacement therapy {8-^{14}C} adenosine (Amersham UK)
and {8-^{14}C} deoxyadenosine (New England Nuclear, Boston) were used.
One set of conditions - 40 min at 18mM P_i) was selected because of
the limited number of cells.[3]   Unlabelled adenosine at higher
concentrations (122 and 244µM) for 2h was also employed.

RESULTS

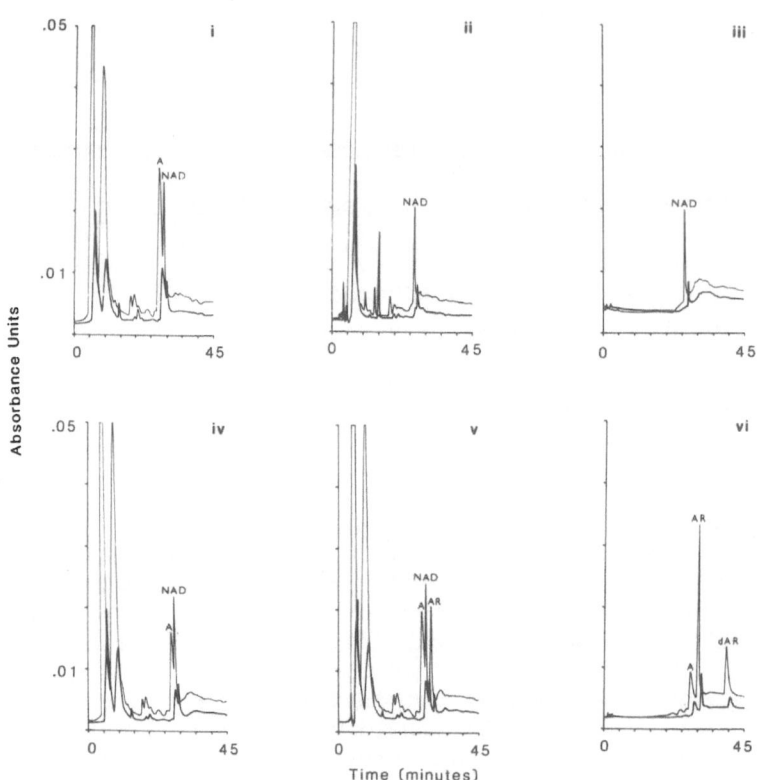

Figure 1. HPLC profiles of nucleosides and base levels in erythrocyte (RBC) extracts prior to treatment in SY (i,ii,iv) compared
with control (ii) and appropriate standards (iii, vi).   Note the
double frontal peaks corresponding to ATP and dATP in extracts from
SY compared with the single ATP peak in controls.   Two major
peaks with retention times corresponding to adenine (A Fig. 1,
vi) and NAD[+] (Fig. 1, iii) were also found in these extracts from
SY and KA (not shown) as compared with a single NAD[+] peak in
controls (Fig. 1, ii).   In separate experiments it was established
that the adenine corresponded to deoxyadenosine (dAR) degraded
during the TCA extraction step.[3]   dAR levels pretreatment in SY
equalled 243µmol/l packed RBCs; in KA 176µmol/l packed RBCs.   No
dAR was ever detected in the plasma[3] or in the RBCs post transfusion
from either SY or KA.   Following incubation with unlabelled
adenosine (AR) at either 122 or 244µmol/l, adenosine and deoxyadenosine nucleotide levels both increased, while intracellular
dAR levels decreased (Fig. 1, iv, v).   Significant amounts of AR
(~115µmol/l packed cells) were also found at 244µM AR, indicating
an upper limit to the RBCs ability to phosphorylate AR (Fig. 1,v)

Table 1. Metabolism of adenosine and deoxyadenosine in ADA deficient erythrocytes.

Substrate	Adenosine (AR)		Deoxyadenosine (dAR)			
Subject	Control	Patient	Control		Patient	
Concentration:	50µM	50µM	10µM	10µM	50µM	100µM
% counts in:						
Nucleotides	ATP 47.9	80.6	dATP 3.0	44.0	19.5	20.8
	ADP 1.0	10.0	dADP 0.5	4.4	2.3	2.6
	AMP 0.1	1.0	dAMP –	–	0.6	0.6
Deamination products	HR) H) 8.1	–	dHR) H) 4.0	–	–	–
	IMP 41.6	2.3	IMP 88.4	0.8	0.7	0.8
Unmetabolised substrate	AR 0.4	3.6	dAR 1.8	50.6	76.3	73.7

No counts from the $\{8\text{-}^{14}C\}$ radiolabelled AR or dAR were found in any deamination products at any substrate concentration using the erythrocytes from KA (Table 1) or SY (not shown) confirming the completeness of the ADA deficiency in these intact cells. At concentrations from 10 to 100µmol/l AR was completely metabolised - predominantly to ATP. Approximately 50-75% of the dAR, by contrast, remained unmetabolised at all concentrations in this range (Table 1); counts were also found in the three deoxyadenosine nucleotides in the approximate ratio 10:1:0.1).

DISCUSSION

The studies reported here confirm that dATP accumulation is assoc-iated with a corresponding depletion in ATP in erythrocytes of SCID patients with inherited ADA deficiency. The abnormally rapid fall in erythrocyte deoxynucleotide levels with treatment cannot be accounted for solely by dilution with transfused erythrocytes. This factor plus the finding of unmetabolised deoxyadenosine in the pretreatment erythrocytes of both children before but not after therapy stimulated the in vitro studies.

The studies with AR and dAR in these ADA deficient red cells have given results identical with previous studies in normal erythrocytes where ADA deficiency was simulated using the inhibitor erythro-9 (2·hydroxy-3-nonyl) adenine : EHNA[1]. They confirm that AR and, to a lesser extent, dAR, can be converted by the ADA deficient erythro-cyte to the three corresponding nucleotides in roughly the same proportion (10:1.0:0.1) presumably by the same nucleotide kinases. It would thus appear that the initial step is catalysed by adenine kinase (AK) for both AR and dAR. The studies also suggest that

rapid intracellular accumulation of dAR occurs in the ADA-deficient
red cell and may be necessary to sustain the high dATP levels.
Contact with ADA competent transfused red cells must enable the
latter to pick up and metabolise the deoxyadenosine, producing a
concomitant rapid fall in deoxyadenosine nucleotide levels.

These findings in inherited ADA deficiency could be explained if the
erythrocyte is normally dependent on a supply of adenosine (AR)
from the S-methylation pathway for the maintenance of its ATP pools.[2]
The erythrocyte could act as an effective scavenger (either by cell-
contact, or avid uptake of AR secreted as waste) which would
simultaneously serve to maintain its own ATP levels.   dAR is known
to be a suicidal inactivator of the enzyme responsible for AR
production by this pathway.[4]   Consequently dAR accumulating in vivo
could significantly inhibit AR production via this route.   Under
such circumstances the erythrocyte would have an enhanced capacity
for the uptake and phosphorylation of dAR - thereby producing the
raised dATP levels found not only in the erythrocytes of ADA deficient children, but also in patients treated with ADA inhibitors.[3]

Erythrocytes from a child with purine nucleoside phosphorylase (PNP)
deficiency (Watson et al - this symposium) and high levels of inosine
and guanosine, deoxyinosine and deoxyguanosine in plasma and urine
do not show intracellular accumulation of any of these nucleosides.
The finding of high intracellular levels of dAR in the ADA deficient
erythrocyte is thus in accord with a unique transport system for
adenosine-type compounds associated with both adenosine kinase (AK)
and ADA in the human red cell.

These studies have demonstrated that three deoxyadenosine nucleotides
may be formed in red cells of ADA deficient patients - at a high
ratio of dATP to the other two - and that they are derived initially
from dAR, and not vice versa.   They indicate that AK is responsible
for this conversion in the human erythrocyte and also suggest a
mechanism for the associated ATP depletion in inherited ADA
deficiency.

REFERENCES

1. D. Perrett, A. Sahota, H.A. Simmonds, and K.Hugh-Jones. Deoxyadenosine metabolism in the erythrocytes of children with severe,
combined immunodeficiency.   Bioscience Reports 1: 933 (1981)
2. Various authors reviewed in: Enzyme Defects and Immune Dysfunction
Ciba Symposium 68 (new series) Exerpta Medica, Amsterdam (1979).
3. H.A.Simmonds, R.J.Levinsky, D.Perrett and D.R.Webster. Reciprocal
relationship between erythrocyte ATP and deoxy-ATP levels in inherited ADA deficiency.   Biochem.Pharmacol 31: 947 (1982).
4. M.S.Hershfield, N.M.Kredich, D.R.Ownby, H.Ownby and R. Buckley.
In vivo inactivation of erythrocyte S-adenosylhomocysteine hydrolase by 2'-deoxyadenosine in adenosine deaminase-deficient patients.
J.Clin.Invest 63: 807 (1979).

OROTATE UPTAKE AND METABOLISM BY HUMAN ERYTHROCYTES

Peter Berman and Eric Harley

Department of Chemical Pathology
University of Cape Town
Cape Town, 8001
Republic of South Africa

The uptake of the pyrimidine nucleotide precursor, orotate, by non-nucleated human erythrocytes is not immediately explicable, (1). During the course of an investigation into the origin of the increased erythrocyte nucleotides seen in hereditary erythrocyte pyrimidine 5'nucleotidase deficiency, rapid uptake of ^{14}C orotate and its conversion to uridine nucleotides by normal red cells was observed (Figure 1(a) A). Moreover, free uridine was present within the cells and accumulated in significant quantities in the medium (Figure 2).

By progressively increasing the concentration of unlabelled orotate, the radio-activity associated with each uridine species showed a steady decline. This was in contrast to the intracellular ^{14}C orotate, which increased as the total orotate concentration rose (Figure 1). This indicates a high capacity, non-saturable transport system for orotate, with the rate limiting step being the conversion of orotate to UMP. From the specific activity of the orotate, an apparent Km and Vmax for the conversion of orotate to uridine appearing in the medium can be estimated. This gives a Km of about 30 µmol/litre and a Vmax of 25 pmol uridine formed per minute per 10^{10} red cells.

Availibility of released uridine for utilization by nucleated cells was demonstrated in co-culture experiments. Epstein Barr Virus transformed human lymphoblasts were incubated in quadruplicate in medium containing 5 µCi ^{3}H uridine and 0.5 µCi ^{14}C orotate. At 90 minute intervals, a suspension of washed erythrocytes was added to give a final cell density comparable to that in whole blood. Following incubation, the lymphoblasts were isolated using Ficoll Hypaque, and the ^{3}H and ^{14}C radioactivity in the washed acid soluble material was determined. An analogous experiment was performed using cultured human

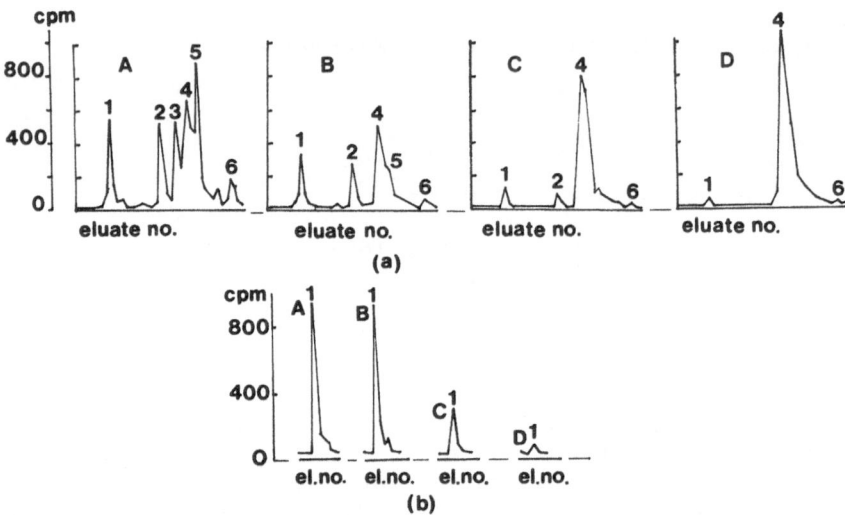

Figure 1. Washed erythrocytes incubated with 0.5 µCi ^{14}C orotate
            for 4 hours in tissue culture medium at a cell
            density comparable to whole blood. Unlabelled orotate
            was present at the following concentrations: A, o; B,
            10; C, 100; D, 1000 µmol/litre. Perchloric acid extracts
            of the cells (a) and medium (b) were prepared, and sub-
            jected to anion exchange HPLC on a Lichrosorb AN 10 anion
            exchange column under the following conditions: water/-
            ammonium phosphate 0.8 molar pH 4.5, gradient from 3 to
            100% with a delay of 1 minute, sweep time of 10 minutes
            and a flow rate of 2 ml/minute.   Elution positions of
            orotate and uridine species were determined by UV absorp-
            tion of simultaneously injected markers; fractions were
            collected and radio-activity plotted against eluate num-
            ber for the neutralized cell (a) and medium (b) extracts.
            In the latter case only the labelled uridine is depicted.
            The identity of the markers is as follows: 1, uridine; 2,
            UMP; 3, UDP-glucose; 4, orotate; 5, UDP; 6, UTP.

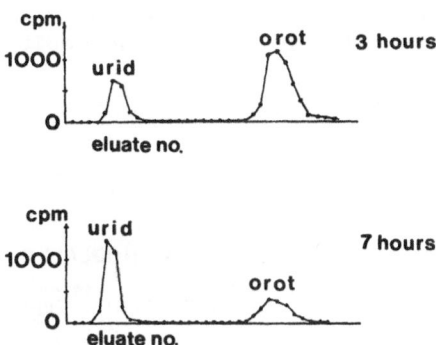

Figure 2. Anion change HPLC of the medium extract following incubation of red cells with 0.6 μCi ^{14}C orotate for the times
indicated. This shows the progressive disappearance of
orotate and accumulation of uridine in the medium.

fibroblasts. In this case the fibroblasts were separated from the red
cells by thorough washing, prior to trypsinization and acid precipitation. Using either lymphoblasts or fibroblasts, incorporation of ^{3}H
uridine into lymphoblasts proceeds irrespective of the presence of red
cells, whereas incorporation of orotate is dependent on the duration of
co-incubation with erythrocytes (Figure 3).

To confirm the mode of entry of ^{14}C orotate into nucleated cells,
the co-culture experiment in fibroblasts was repeated in the presence
of the uridine uptake inhibitor nitro — benzyl mercapto inosine (NBMI).
Flasks containing fibroblasts and ^{14}C orotate were prepated in triplicate (Figure 4). Flask A, the control, showed no ^{14}C uridine in the
medium, and low incorporation of label into acid precipitable material.
Flask B, which contained added erythrocytes, showed significant conversion of orotate to uridine in the medium and a high uptake of label

into acid precipitable material. In flask C containing erythrocytes
and NBMI in addition to fibroblasts, the medium showed the same distribution of label as the flask without inhibitor (B), but uptake of label
into acid precipitable material returned to the low level seen in the
control (A). Thus, NBMI blocks the erythrocyte mediated uptake of ^{14}C
orotate into nucleic acids of fibroblasts, but does not interfere with
the conversion of ^{14}C orotate to uridine in the medium.

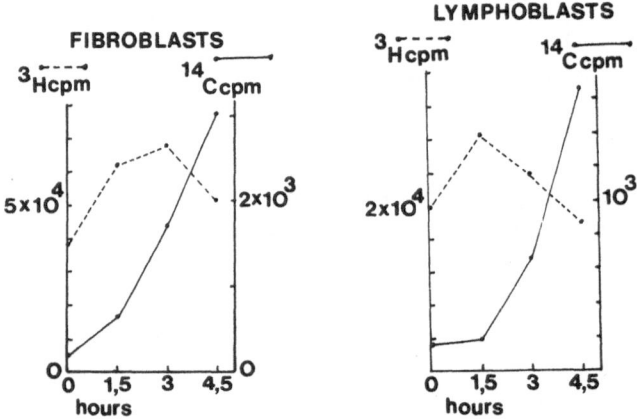

Figure 3. Uptake of radioactivity from ^{3}H uridine and ^{14}C orotate
          into the acid precipitable matrial of nucleated cells
          plotted against the duration of erythrocyte presence.

    The findings described above are sumarized in Figure 5. Erythro-
cytes are able to take up orotate from the medium, metabolize it to
uridine and secrete this uridine back into the medium when it becomes
available for nucleic acid synthesis in nucleated cells.

    A possible physiological role for this pathway would be the conver
sion of orotate derived from the diet or endogenous sources to a form
capable of utilisation by peripheral tissues. Experiments are under
way to determine whether the transport system for orotate across liver
cells function primarily to export orotate for metabolism by erythrocytes and distribution as uridine to peripheral tissues.

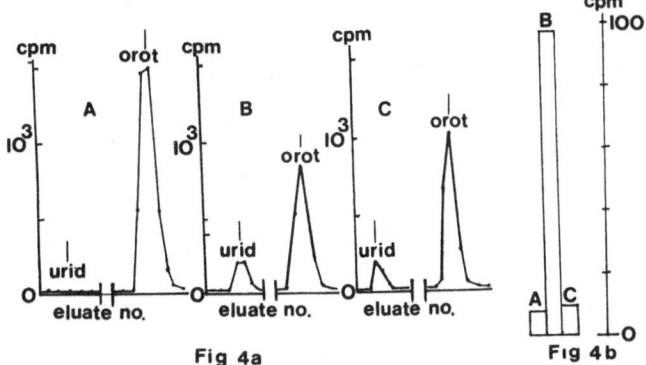

Fig 4a   **Fig 4b**

Figure 4. Distribution of label in the medium (a) and in acid precipitable material (b) of fibroblasts incubated with [14]C
orotate and A – no additions; B – erythrocytes; C – erythrocytes and 1 μmol/litre NBMI.

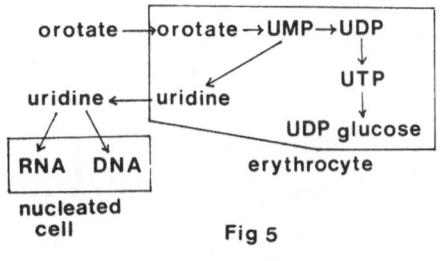

Fig 5

Figure 5.

REFERENCES

1.   E.H. Harley, A. Heaton and W. Wicomb, Pyrimidine metabolism in
        hereditary erythrocyte pyrimidine 5'nucleotidase deficiency,
        Metabolism 27: 1743 (1978).

IMP DEHYDROGENASE MUTANTS: CELL CULTURE MODEL FOR HYPERURICEMIA

Buddy Ullman

University of Kentucky

Lexington, Kentucky

INTRODUCTION

Primary gout with associated hyperuricemia is a genetically heterogeneous disorder characterized by hyperuricemia due to increased rates of purine synthesis or enhanced cellular turnover. The genetic lesions which lead to enhanced rates of purine synthesis are mostly undefined. Several groups have reported that hyperuricemia can be associated with different types of kinetic alterations in the enzyme phosphoribosylpyrophosphate (PRPP) synthetase (1-3) or a deficiency in hypoxanthine-guanine phosphoribosyltransferase (4-5). The purine metabolic pathway is subject to a variety of control mechanisms including substrate availability, feedback inhibition of enzyme activity, compartmentalization of substrates and effectors, and regulation of enzyme levels. This biosynthetic pathway has two regulatory domains. The first occurs before the IMP branchpoint in which both PRPP synthetase and PRPP-glutamine amidotransferase are subject to allosteric inhibition by nucleotide effectors. The second regulatory domain occurs subsequent to IMP synthesis, where both adenylosuccinate synthetase and IMP dehydrogenase, the penultimate enzymes in AMP and GMP synthesis, respectively, are also regulated by ribonucleotide effectors. The genetic and metabolic heterogeneity of overproduction hyperuricemia suggests that defects in any one of these enzymes might lead to excessive rate of purine synthesis.

Recently, Ullman et al. have characterized an adenylosuccinate synthetase-deficient murine cell line (6) which overproduces purines and excretes massive amounts of inosine into the culture medium (7). This genetic cell culture model for overproduction hyperuricemia was substantiated by Willis and

Seegmiller who pharmacologically simulated adenylosuccinate
synthetase deficiency in human cells with resulting purine
overproduction and overexcretion (8). These workers indicated
that increased rates of purine synthesis and excretion could also
be induced pharmacologically by inhibition of IMP dehydrogenase.

RESULTS AND DISCUSSION

This manuscript describes two murine T-cell lymphoma clones,
MYCO-1A and MYCO-1A-20, with genetically altered IMP dehydrogenase
activities. These clones were isolated by virtue of their
resistance to mycophenolic acid, a potent inhibitor of cellular
DNA synthesis (9) and of both IMP dehydrogenase and GMP synthetase
enzyme activities (10). The MYCO-1A and MYCO-1A-20 clones were
isolated from semi-solid agarose containing 1 μM and 20 μM
mycophenolic acid, respectively. In comparative growth rate
experiments, the MYCO-1A and MYCO-1A-20 cell lines were 3- and
50-fold less sensitive than wild-type cells to the growth

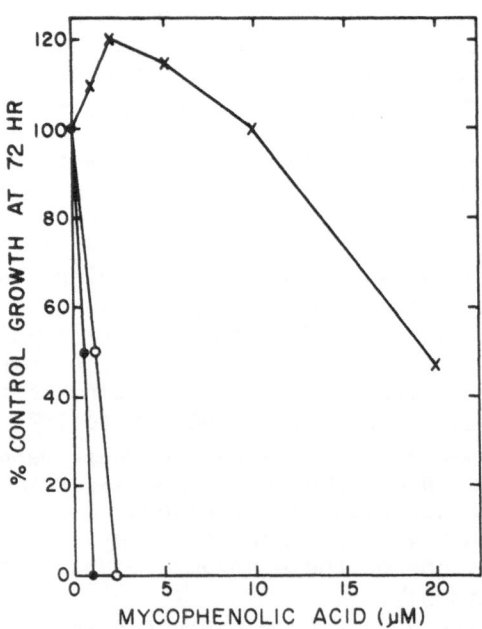

Figure 1. Growth Sensitivity of Wild-Type and Mutant Cells to
Mycophenolic Acid. The sensitivities of wild-type (o⎯o),
MYCO-1A (o⎯o), and MYCO-1A-20 (x---x) cells to growth
inhibition by mycophenolic acid were compared.

inhibitory effects of mycophenolic acid, Figure 1. These cell
lines were also cross-resistant to ribavarin, which can inhibit
IMP dehydrogenases as the monophosphate (11).

Since mycophenolic acid is an inhibitor of the penultimate
enzyme in guanyl nucleotide synthesis (10), the effects of
mycophenolic acid on the guanine nucleotide levels in wild-type
and mutant cells were examined, Figure 2. Whereas, a profound
depletion of GTP occurred in wild-type cells incubated 4 hr with 1
μM mycophenolic acid, a similar GTP depletion was seen only at 5
μM mycophenolic acid in MYCO-1A cells (Figure 2). No depletion of
intracellular GTP was observed in MYCO-1A-20 cells incubated with
exogenous mycophenolic acid at concentrations as high as 25 μM.
GTP depletion in wild-type and MYCO-1A cells resulted in a
concomitant increment in cellular pyrimidine ribonucleotides,
while ATP levels appeared to be unaffected (Figure 2).

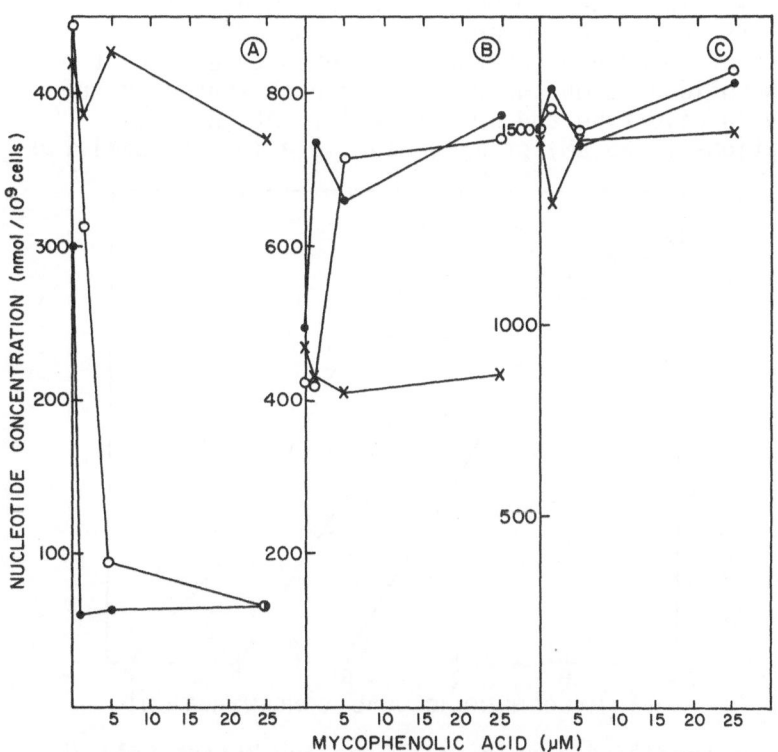

Figure 2. Ribonucleoside Triphosphate Concentrations in Wild-Type
and Mutant Cells Before and After Incubation with Mycophenolic
Acid. Wild-type (o___o), MYCO-1A (o___o), and MYCO-1A-20
(x----x) cells at a density of 10^6 cells/ml were incubated for 4
hr with either 0, 1, 5, or 25 μM mycophenolic acid. Nucleotides
were resolved by high performance liquid chromatography.

Since mycophenolic acid is known to inhibit mammalian IMP dehydrogenase (10) with a consequent depletion of guanine containing nucleotides (11), the IMP dehydrogenase activities in extracts of wild-type, MYCO-1A, and MYCO-1A-20 cell were compared in the presence of increasing mycophenolic acid concentrations. Figure 3 indicates the concentrations of mycophenolic acid required to inhibit IMP dehydrogenase in wild-type and mutant cell extracts. Clearly, the IMP dehydrogenase activities from MYCO-1A and MYCO-1A-20 cells are considerably less sensitive to mycophenolic acid inhibition than the wild-type activity.

Since, the sensitivities of the IMP dehydrogenase activities to mycophenolic acid inhibition were altered in mutant cell extracts, we compared several other kinetic parameters of the IMP dehydrogenase activities from wild-type and mutant cells. Table 1 indicates that for the IMP dehydrogenase activities in both mutant lines the Vmax and apparent K_m's for IMP are increased considerably over those found for wild-type enzyme activity. The apparent K_m values for NAD were also determined, Table 1.

Reports of purine overproduction induced by pharmacologic inhibitors of IMP dehydrogenase (8) led us to examine the effects of mycophenolic acid on the rates of purine synthesis and excretion in wild-type and mutant cells. At the same concentrations (1-25 μM) of mycophenolic acid that depleted

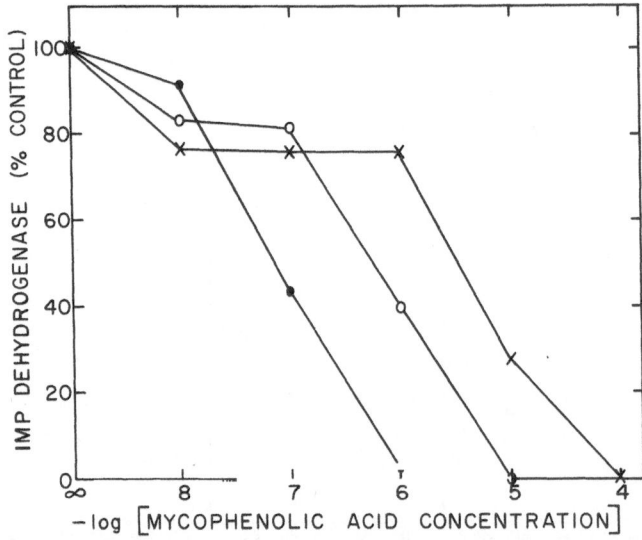

Figure 3. Sensitivities of Wild-Type and Mutant Cell IMP Dehydrogenase Activities to Inhibition by Mycophenolic Acid. Extracts from wild-type (o___o), MYCO-1A (o___o), and MYCO-1A-20 (x---x) cells were assayed for IMP dehydrogenase activity in the presence of varying mycophenolic acid concentrations. The enzymatic activity was assayed with 10 μM [14C]-IMP (25 mCi/mmol) and 0.3 mM NAD (13).

Table 1
IMP Dehydrogenase Kinetic Parameter

Cell Line	Vmax (pmol/m/µg protein)	K_m(IMP) 0(µM)	K_m(NAD) (µM)
Wild–Type	1.7	30	45
MYCO–1A	~25	~400	12
MYCO–1A–20	~25	~400	50

IMP dehydrogenase activities were assayed as described by Ullman (13).

wild–type cells of intracellular GMP and GTP, both overproduction and overexcretion of purines into the culture medium were observed, Figure 4. Mycophenolic acid at 1 µM in wild–type cells caused a large increase in purine excretion into the culture medium and increased the rate of total de novo synthesis two–fold as compared to unperturbed wild–type cells. The excreted purine

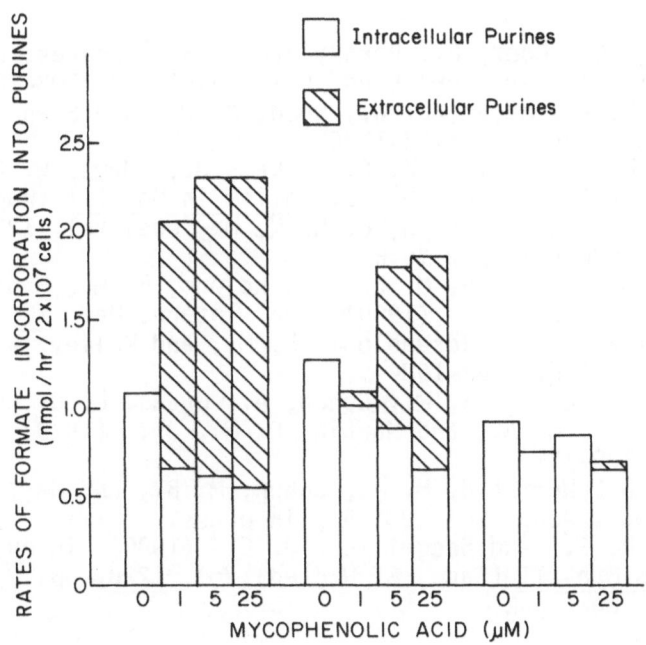

Figure 4.   Effect of Mycophenolic Acid on the Rates of Purine Synthesis and Excretion in Wild–Type and Mutant Cells.   The effects of either a 1 µM, 5 µM. or 25 µM mycophenolic acid incubation for 4 hr on the rates of purine synthesis were performed as described previously (7).

has been identified as inosine. A similar increment in purine
excretion and total synthesis is observed with MYCO–1A cells
incubated with 5 μM, but not 1 μM, mycophenolic acid. The
MYCO–1A–20 cell line is refractory to the mycophenolic acid
induced purine overproduction and excretion at concentrations as
high as 25 μM.

SUMMARY

These studies with wild–type and mutant cells defective in
IMP dehydrogenase and the previous data with the adenylosuccinate
synthetase–deficient cell line suggest that among the clinical
population with dominantly inherited hyperuricemia, patients with
partial deficiencies in these enzymes exist. It is hoped that
these pharmacogenetic cell culture models for overproduction
hyperuricemia will lead to the initiation of a search for
hyperuricemia patients with either of these deficiencies. If such
patients are found it may be possible to design chemotherapeutic
regimens by which effectors (inhibitors) of purine synthesis might
ameliorate the overproduction of purines by the de novo pathway.

REFERENCES

1. Sperling, O., Boer, P., Persky–Brosh, S., Kanarek, E., and de
Vries, A. (1972) Rev. Eur. Etud. Clin. Biol. 17, 703–706
2. Becker, M. A., Meyer, L. J., Wood, A. W., and Seegmiller, J.
E. (1973) Science 179, 1123–1126
3. Becker, M. A., Raivio, K. O., Bakay, B., Adams, W. B., and
Nyhan, W. L. (1980) in Purine Metabolism in Man–III (Rapado, E.,
Watts, R. W. E., and De Bruyn, C. H. M. M., eds) Vol. 122A, pp.
387–392, Plenum Press, New York
4. Kelley, W. N., Green, M. L., Rosenbloom, F. M., Henderson, J.
F., and Seegmiller, J. E. (1969) Ann. Intern. Med. 70, 155–206
5. Seegmiller, J. E., Rosenbloom, F. M., and Kelley, W. N.
(1967) Science 155, 1682–1684
6. Ullman, B., Clift, S. M., Cohen, A., Gudas, L. J., Levinson,
B. B., Wormsted, M. A., and Martin, D. M., Jr. (1979) J. Cell.
Physiol. 99, 139–152
7. Ullman, B., Wormsted, M. A., Cohen, M. B., and Martin, D. W.,
Jr. Proc. Natl. Acad. Sci. U.S.A., in press.
8. Willis, R. C., and Seegmiller, J. E. (1980) in Purine
Metabolism in Man–III (Sano, G. H., ed) Vol. 122B, pp. 237–242,
Plenum Press, New York

9. Franklin, T. J., and Cook, J. M. (1969) Biochem. J. $\underline{113}$, 515–524
10. Sweeney, M. J., Hoffman, D. H., and Esterman, M. A. (1977) Cancer Res. $\underline{32}$, 1803–1809
11. Robins, R. K., and Simon, L. N. (1973) Proc. Natl. Acad. Sci. U.S.A. $\underline{70}$, 1174–1178
12. Cohen, M. B., Maybaum, J., and Sadee, W. (1981) J. Biol. Chem. 256:8713–8717
13. Ullman, manuscript submitted.

BASIC MOLECULAR DEFECT IN ADA SCID

Elly Herbschleb-Voogt[1], Jan-Willem Scholten[1], Jaak
M. Vossen[2], Peter L. Pearson[1] and P. Meera Khan[1]

[1]Department of Human Genetics, State University Leiden, Wassenaarseweg 72, 2333 AL LEIDEN, The Netherlands
and [2]Department of Pediatrics, State University Leiden

INTRODUCTION

Severe deficiency of adenosine deaminase (ADA) is known to be
associated with an autosomal recessive form of severe combined immunodeficiency disease (SCID) in man[1]. Different theories have been
proposed to explain the basic defect in ADA$^-$SCID. They include: a
mutation in the structural gene for ADA or chromosome 20[1-3], a mutation in a gene regulating the expression of ADA[4], the occurrence of
an ADA specific inhibitor[5] and a defect in the post-translational
modification of ADA[6], implying a defect in the adenosine deaminase
complexing protein (ADCP) molecule, whose structural gene is on human
chromosome 2, or in its complex formation with ADA[7,8].
Many investigators studied the nature of basic defect by investigating the tissue or cell culture material derived from the patients using enzymological and/or immunological methods[9-14]. In the
present study, a small family with a patient, who died of ADA$^-$SCID
disease without any material being collected for further direct studies was investigated. Biochemical and immunochemical investigations
using somatic cell hybrids originating from a fusion between lymphocytes of the heterozygote father and Chinese hamster cells were performed to establish the basic molecular defect.

MATERIALS AND METHODS

The case of ADA$^-$SCID (A.Y.) and the methods concerning the peripheral blood cells, hybrid (a3/Y) cells, control fibroblasts and the
biochemical and chromosomal analysis of the hybrid clones have been
described previously[7]. The radioimmunoassay (RIA) procedure was adapted from Daddona et al.[10,15]. For these experiments pure human red
cell ADA, isolated by a combination of previously described

381

methods[16,17], was labelled with ^{125}I with the lactoperoxidase procedure[18]. Antibodies agains this pure ADA were raised in a rabbit[15].

RESULTS AND DISCUSSION

The quantities of ADA-cross-reacting material (CRM) in the red cell lysates of the parents and a sister of the proposita were found to be more or less proportional to the observed enzymatic activities, which were about half of the normal average (table 1). This might be an indication, that no detectable ADA protein produced by the mutant allele was present in their cells. It should, however, be noted, that red blood cells are not the most suitable material to study the natu re of possibly unstable mutant proteins, since they can not synthesize new protein.

A summary of the segregation patterns of the enzyme markers of chromosome 2 and 20 in the a3/Y hybrid clones is given in table 2. Quantities of ADA-CRM were determined by performing the RIA under various conditions (table 2). Under these conditions no measurable levels of ADA-CRM were found in the hybrids identified to contain the genetic information for the ADA SCID disease in absence of the normal human ADA gene (table 2).

The following conclusions on the causation of the ADA deficiency in this case can be drawn from the results in table 2, together with the previously reported marker analysis in these clones and 27 subclones derived from 4 of them[7]: a) If a specific inhibitor is involved, more or less normal amounts of ADA-CRM are expected in the affected. Therefore, inhibitor hypothesis can be ruled out in the present case. b) Since the presence or absence of human ADCP in the hybrid cells does not seem to influence the ADA expression (a3/Y14), the ADCP can not be implicated. c) A mutation in the structural gene such as a large deletion, a frame shift or a stopcodon will usually lead to a complete cessation or to the formation of an incomplete, immunologically unrecognisable gene product, while less radical changes such as point mutations might give rise to an extremely labile protein. In such a case, loss of CRM might easily occur for example during experimental handling. Furthermore, a mutation in a regulatory gene for ADA or in the flanking DNA sequences of the ADA structural gene might result in the extinction of the ADA gene. It is, therefo-

Table 1 The ADA activity and ADA-CRM in the hemolysates.

sample	ADA (IU/gHb)	ADA (ngCRM/mgHb)	Abs.Spec.Act. (IU/mgCRM)
normal adults (n=14)	0.95 +0.23 (0.63 - 1.38)	4.27 + 1.28 (2.64 - 7.82)	227 + 36 (177 - 304)
father Y	0.41	2.54	161
mother Y	0.46	2.72	169
sister Y	0.58	2.70	217

Table 2. Chromosomes, markers and ADA-CRM in the hybrids and a Chinese hamster line.

Cell line	Medium	ADA	ITPA	chr20 (intact)	MDH	IDH	ADCP	chr2 (intact)	unabs. exp.1	unabs. exp.2	abs.[3] exp.1	abs.[3] exp.2
		Chromosome 20			Chromosome 2				Anti-human ADA antiserum			
									unabsorbed		absorbed[3]	
a3/Y 1	NCS[2]	+	+	+			−	−	30.89	49.13	25.13	63.80
	HS[2]									42.02		31.50
a3/Y 3	NCS	+	+	+	−	−	−	−	42.58	41.34	39.19	
	HS									48.61		
a3/Y 4	NCS	−	−	−	−	−	−	−		4.10		∨
	HS								2.73	2.11		∨
a3/Y 6	NCS	−	−	−	+	−	−	−		3.70		
	HS								3.81	2.98	1.50	
a3/Y11	NCS	−	+	+	−	−	−	−		2.27		∨
	HS								2.98		2.11	∨
a3/Y13	NCS	−	+	+	−	−	−	−		3.21		∨
	HS								2.25		1.10	∨
a3/Y14[1]	NCS	−	+	+	+	+	+	−	2.19	3.08	∨	∨
	HS								4.70	3.57	1.75	1.45
a3	NCS	−	−	−	−	−	−	−	2.09		1.03	∨
	HS									2.33		

Legends: NCS: newborn calf serum, HS: donor horse serum, blank: not tested, ∨: means below the level of detection.

Footnotes: 1) Subcloning of this clone yielded 7 ADCP+ and 7 ADCP⁻ clones. The presence or absence of ADCP did not influence the ADA expression. 2) The hybrid cells were cultured on NCS as well as on HS, because of the possible uptake of calf ADA from the medium by fibroblasts[14]. 3) Since we found that our anti-ADA exhibits marked cross-reactivity towards Chinese hamster ADA[15], we used both crude anti-ADA and anti-ADA absorbed by a Chinese hamster extract bound to Sepharose 4B.

re, hard at present to distinguish between the various possibilities discussed under item c. Future studies at DNA level, using appropriate probes carrying the ADA gene sequences may provide further information on this unresolved problem not only in this particular case but also in several other ADA-CRM negative ADA SCID cases[10-14].

The RIA for ADA in the somatic cell hybrids has certain advantages. Their ADA protein is freshly synthesized. A quantification of ADA-CRM is not hampered by the presence of ADA-L[11,12]. Furthermore, the hybrid cell system seems to be a suitable tool to dissect and investigate separately the two different mutant alleles in a genetic compound, if involved in the causation of ADA deficiency in a given patient.

ACKNOWLEDGEMENTS

The help and advice of Dr. J. de Koning, Drs. E. Klasen and Drs. J. ten Kate for the radiolabelling and the RIA and the secretarial help of Corrie Bocxe are greatfully acknowledged.

REFERENCES

1. ER Giblett,JE Anderson,F Cohen,B Pollara,HJ Meuwissen, Lancet II:1067 (1972)
2. JA Tischfield,RP Creagan,EA Nichols,FH Ruddle, Hum Hered 24:1 (1974)
3. R Hirschhorn,N Beratis,F Rosen, Proc Natl Acad Sci USA 73:213 (1976)
4. MJ Siciliano,MR Bordelon,PO Kohler, Proc Natl Acad Sci USA 75:936 (1978)
5. PP Trotta,EM Smithwick,ME Balis, Proc Natl Acad Sci USA 73:104 (1976)
6. MB van der Weyden,WN Kelley, J Clin Invest 53:81a (1974)
7. E Herbschleb-Voogt,PL Pearson,JM Vossen,P Meera Khan, Hum Genet 59:317 (1981)
8. E Herbschleb-Voogt,KH Grzeschik,PL Pearson,P Meera Khan, Hum Genet 59:317 (1981)
9. PE Daddona,WN Kelley, Mol Cell Biochem 29:91 (1980)
10. PE Daddona,MA Frohman,WN Kelley, J Clin Invest 64:798 (1979)
11. PE Daddona,MA Frohman,WN Kelley, J Biol Chem 255:5681 (1980)
12. DA Wiginton,JJ Hutton, J Biol Chem 257:3211 (1982)
13. WP Schrader,AR Stacy,B Pollara, In:Pollara B et al(eds) Inborn errors of specific immunity. Academic Press,New York pp443 (1979)
14. DA Carson,R Goldblum,JE Seegmiller, J Immunol 118:270 (1977)
15. E Herbschleb-Voogt,JW Scholten,P Meera Khan, to be submitted for publication to Hum Genet.
16. WP Schrader,AR Stacy,B Pollara, J Biol Chem 251:4026 (1976)
17. CA Rossi,A Lucacchini,V Montali,G Ronca, Int J Pept Prot Res 7:81 (1975)
18. BE Chechik,A Madapallimattam,E Gelfand, J Natl Cancer Inst 62:465 (1979)

GENETIC MECHANISM(S) RESPONSIBLE FOR A DEFICIENCY OF ADENINE

PHOSPHORIBOSYLTRANSFERASE IN MAN

James M. Wilson, Peter E. Daddona, H. Anne Simmonds
and William N. Kelley
Departments of Internal Medicine and Biological
Chemistry, University of Michigan, Ann Arbor, Michigan
and Clinical Science Laboratories, Guy's Hospital
London, England

Adenine phosphoribosyltransferase (APRT) is a relatively non-abundant soluble enzyme which in man is coded for by a single structural gene on chromosome 16 (1). A partial deficiency of APRT in man was first described in 1968 (2). These subjects were asymptomatic and were shown to be heterozygous for the enzyme defect. A complete deficiency of APRT activity has now been described in several patients with renal calculi composed of 2,8-dihydroxyadenine who are homozygous for the enzyme defect (reviewed in ref. 3). The genetic mechanisms responsible for an inherited deficiency of APRT however have remained undefined.

In an attempt to further elucidate the mechanisms responsible for a deficiency of APRT in man, we have studied the catalytic, immunochemical, and electrophoretic properties of APRT in hemolysates from 30 patients with a deficiency of this enzyme in six unrelated families.

The level of enzyme activity and immunoreactive protein and the calculated absolute specific activity of APRT in hemolysates from control subjects and from the APRT-deficient patients are summarized in Table I. APRT enzyme activity and immunoreactive protein was markedly diminished to less than 1% of control in each of the four homozygous deficient patients studied. Patients who were heterozygous for a deficiency of APRT exhibited specific activities that were uniformly decreased to approximately 25% of the average normal value. However, the level of APRT immunoreactive protein in hemolysates of heterozygotes ranged from 22% to 112% of control. APRT immunoreactive protein was markedly decreased to 25% of control in heterozygotes from the B. family and F.family, one heterozygous parent from the R. family, and most of the heterozygotes from

TABLE 1
Characterization of erythrocyte adenine
phosphoribosyltransferase

Patient	Specific activity		CRM level		Absolute specific activity
	milliunits/mg	% control	ng CRM/mg	% control	milliunits/µg CRM
Controls[a]	0.46 ± 0.07	100	19.3 ± 4.5	100	24 ± 4
H. family					
Normal					
S.F.	0.43	93	21.5 ± 4.8 (3)[b]	111	20
Heterozygote					
M.S.	0.11	24	4.2 ± 0.4 (3)	22	26
A.S.	0.13	28	5.5 ± 0.8 (4)	28	24
R.S.	0.13	28	6.8 ± 0.8 (3)	35	19
E.A.F.	0.11	24	4.8 ± 1.0 (3)	25	23
A.G.	0.12	26	5.4 ± 0.7 (3)	28	22
A.D.H.	0.11	24	7.6 ± 1.1 (6)	39	14
D.S.	0.13	28	5.1 ± 1.0 (5)	26	25
J.C.S.	0.11	24	4.5 ± 1.0 (4)	23	24
M.F.	0.10	22	5.4 ± 0.8 (3)	28	19
J.F.	0.12	26	5.5 ± 0.4 (3)	28	22
E.R.	0.13	28	5.6 ± 1.3 (3)	29	23
K.R.	0.12	26	5.6 ± 0.4 (3)	29	21
G.V.P.	0.15	33	9.1 ± 0.2 (3)	47	16
L.R.	0.13	28	5.1 ± 1.4 (3)	26	25
J.E.S.	0.10	22	10.5 ± 1.4 (8)	54	9
L.S.	0.11	24	16.4 ± 2.5 (8)	85	7
C.D.H.	0.16	35	11.0 ± 0.6 (4)	57	15
Homozygote deficient					
F.D.H.	<0.0002	<0.04	0.08 (2)	0.4	N.D.[c]
B.D.H.	0.0044	1.0	0.14 ± 0.02 (3)	0.7	N.D.
R. family					
Heterozygote					
MR.	0.11	24	11.3 ± 2.5 (11)	59	10
F.R.	0.10	22	6.7 ± 1.2 (7)	35	15
Homozygote deficient					
S.R.	<0.0002	<0.04	0.08 (2)	0.4	N.D.
B. family					
Heterozygote					
N.B.	0.11	24	6.5 ± 0.7 (3)	34	17
Homozygote deficient					
S.B.	<0.0002	<0.04	0.18 ± 0.01 (3)	0.9	N.D.
F. family					
Heterozygote					
E.F.	0.11	24	5.0 ± 1.0 (6)	26	22
A.F.	0.09	20	4.4 ± 0.3 (4)	23	20
F.R.	0.12	26	5.6 ± 0.8 (5)	29	21
L. family					
Heterozygote					
L.L.	0.10	22	11.8 ± 1.2 (3)	61	8
D. family					
Heterozygote					
M.D.	0.13	28	18.6 ± 3.0 (4)	96	7
B.D.	0.13	28	21.6 ± 4.2 (7)	112	6

[a] Hemolysates from 12 normal adults were examined.
[b] Values in parentheses indicate the number of determinations.
[c] Not determined because of the low values for specific activity and CRM.

the H. family. The absolute specific activity of APRT from most of these patients was within the normal range. The single heterozygote from the L. family and the other parent from the R. family, in contrast, had only a moderate reduction in immunoreactive protein and a decreased absolute specific activity. The two heterozygotes from the D. family exhibited normal levels of immunoreactive protein and markedly reduced absolute specific activities.

The isoelectric properties of normal and mutant forms of APRT were studied using the recently described protein blot technique (5). Hemolysate samples were enriched for APRT activity approximately 20-fold by CM-Sephadex batch fractionation, and were focused in polyacrylamide slab gels under denaturing conditions. The

proteins were transferred electrophoretically from the gel to nitrocellulose paper and APRT was specifically located in situ by the
sequential binding of APRT antibody and ^{125}I-labeled protein A.
Autoradiographs of two protein blots are presented in figure 1.

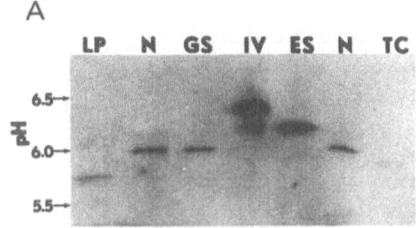
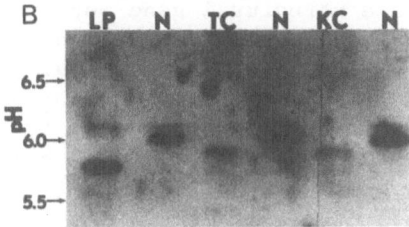

Fig. 1.   Denaturing isoelectric focusing of APRT.   Hemolysate samples enriched for enzyme activity were focused and analyzed for
immunoreactive material using the protein blot technique.   Panel A:
normal, Lanes A and C; L. family, Lane B (L.L.); H. family, Lanes
D (A.S.), E (F.D.H.), F (C.D.H.), G (J.E.S.), and H (L.S.); B.
family, Lanes I (N.B.) and J. (S.B.); and D. family, Lanes K (M.D.)
and L (B.D.).   Panel B: Lane A, F.R.; Lane B, S.R.; Lane C, M.R.;
and Lane D, normal.   Figures were taken from reference 4.

        Normal hemolysate exhibited a single isoelectric form of APRT
(Fig. 1A, lane A) that comigrated with the purified enzyme (data
not shown).   The specificity of the immunochemical detection was
demonstrated by the absence of detectable protein in hemolysates
from the CRM⁻homozygous deficient patients F.D.H. (Fig 1A, lane E)
and S.B. (Fig 1A, lane J).   Heterozygotes from the L. family (Fig
1A, lane B), H. family (Fig 1A, lanes D,F,G, and H), B. family
(Fig 1A, lane I), and D. family (Fig 1A, lanes K and L) demonstrated
a single form of the enzyme with an apparently normal isoelectric
point.

More complex patterns of immunoreactive protein were observed in hemolysates from the R. family. Patient F.R. (the father of the homozygote S.R.) exhibited a single isoelectric form of the APRT subunit which was indistinguishable from normal (Fig 1B, lane A). In contrast, patient M.R. (mother of S.R.) had two distinct protein bands (Fig 1B, lane C): an acidic (open arrow) and an apparently normal (closed arrow head) subunit species.

The finding of 25% of normal APRT catalytic activity in heterozygotes, a value which seems inappropriately low for an autosomal recessive disorder, was first described in 1968 by Kelley et al. (2). These investigators argued that this finding is consistent with a structural gene mutation in one allele if the native enzyme exists as a dimer and the hybrid dimer is nonfunctional (2). We have since provided conclusive biochemical evidence that native human APRT is a dimer which is formed by the random aggregation of monomer subunits expressed from each allele (6). Patients that are heterozygous for a variant allele coding for a structurally altered APRT subunit will, therefore, express three populations of native enzyme molecules: 25% of a normal-normal dimer, 50% of a normal-mutant or hybrid dimer, and 25% of a mutant-mutant dimer. The dimer formed by the aggregation of mutant subunits is probably labile in vivo since each of the homozygous deficient patients had very low levels of APRT immunoreactive protein. If the hybrid dimer is inactive and/or unstable in vivo, the residual enzyme activity would be 25% of normal.

Direct evidence for the existence of a nonfunctional mutant-normal hybrid dimer in the heterozygotes of the D. family, L.family, and R. family was provided by the finding of proportionately greater levels of APRT immunoreactive protein than catalytic activity. The putative variant enzyme subunit from patient M.R. (R. family) is apparently more acidic than normal. The heterozygotes of the B. family and F. family, and most of the heterozygotes of the H. family had coordinately decreased levels of APRT catalytic activity and immunoreactive protein suggesting that these particular variant enzyme subunits form hybrid dimers with the normal subunit that are labile in vivo.

In conclusion, our studies have shown that a deficiency of APRT activity in man results from a variety of different structural gene mutations that render the APRT molecule more labile in vivo and/or catalytically nonfunctional.

REFERENCES

1.   J.A. Tischfield and F.H. Ruddle, Assignment of the gene for adenine phosphoribosyltransferase to human chromosome 16 by mouse-human somatic cell hybridization, Proc. Natl. Acad. Sci. USA 71:45-49 (1974).

2.   W.N. Kelley, R.I. Levy, F.M. Rosenbloom, J.F. Henderson, and
     J.E. Seegmiller, Adenine phosphoribosyltransferase deficiency:
     a previously undescribed genetic defect in man, J. Clin. In-
     vest. 47:2281-2289 (1968).

3.   H.A. Simmonds and K.J. Van Acker in The Metabolic Basis of
     Inherited Diseases (J.B. Stanbury, J.B. Wyngaarden, and D.A.
     Fredrickson, eds) 5th ed, McGraw-Hill Book Co., London, in
     press.

4.   J.M. Wilson, P.E. Daddona, H.A. Simmonds, K.J. Van Acker,
     and W.N. Kelley, Human adenine phosphoribosyltransferase:
     immunochemical quantitation and protein blot analysis of mu-
     tant forms of the enzyme, J. Biol. Chem. 257:1508-1515 (1982).

5.   H. Towbin, T. Staehelin and J. Gordon, Electrophoretic trans-
     fer of proteins from polyacrylamide gels to nitrocellulose
     paper: procedure and some applications, Proc. Natl. Acad.
     Sci. USA 76:4350-4354 (1979).

6.   J.A. Holden, G.S. Meredith, and W.N. Kelley, Human adenine
     phosphoribosyltransferase: affinity purification, subunit
     structure, amino acid composition, and peptide mapping, J.
     Biol. Chem. 254:6951-6955 (1979).

CORRELATION BETWEEN A MUTANT APRT PROTEIN AND ALTERED DNA IN CHO

CELLS

Anne E. Simon and Milton W. Taylor

Program in Genetics
Department of Biology
Indiana University
Bloomington, Indiana 47405

INTRODUCTION

The frequency of mutation to 2,6-diaminopurine resistance occurs at a much higher than expected frequency for a diploid autosomal locus. Spontaneous 2,6-diaminopurine resistant mutants have been reported to occur at frequencies of 3×10^{-6} in CHO cells[1] and as high as 1×10^{-3} in mouse L-cells[2]. Such mutants appeared to arise in a single step; all were defective in the enzyme APRT and resistant to high concentrations (20-40 µg/ml) of 2,6-diaminopurine (DAP).

Others[3,4] have reported cells resistant to low concentrations of DAP or 8-azaadenine (AA) (4-8 µg/ml) that had the characteristics of heterozygotes (i.e., approximately 50% wild-type APRT activity) and that could give rise to mutants resistant to high concentrations of DAP or AA. Two distinct classes of aprt heterozygotes have been reported: those that give rise to fully resistant cells at a low frequency (10^{-6}–10^{-7} per cells plated) (class 1), and those that give rise to such mutants at a high frequency (10^{-3}–10^{-5}) (class 2).[4] We report here the analysis of one class 2 aprt heterozygote which gives rise to APRT⁻ cells spontaneously at very high frequencies (1×10^{-3} to 1×10^{-5}), and six APRT⁻ cell lines derived from this heterozygote.

MATERIALS AND METHODS

1. Cell culture conditions, APRT assay, APRT immunoprecipitation, and two-dimensional gel electrophoresis have been previously described.[5] D416 is a presumptive aprt

391

heterozygote resistant to 7 µg/ml DAP isolated from the wild-type
CHO pro⁻ after treatment with ethyl methane sulfonate.[4]

2. Gel blot hybridizations: 20 µg of purified mutant or
wild-type DNA were digested with one or more restriction enzymes,
electrophoresed through 0.8% or 1.5% agarose gels, and transferred
to nitrocellulose or DBM paper according to the method of Southern[6]
or Wald et al.[7] Hybridization was carried out initially under
aqueous conditions[8] and later with the dextran sulfate
modification,[7] using ³²P nick-translated aprt DNA[8] as a probe.
Blots were then washed and exposed to Kodak X-ray film using Dupont
lighting plus intensifying screens for 5 to 24 hr. at -70°C.

RESULTS AND DISCUSSION

APRT protein can be immunoprecipitated by antibody (raised in
rabbits against purified Syrian hamster liver APRT) and analyzed by
two-dimensional polyacrylamide gel electrophoresis. Gels showing
position of APRT immunoprecipitated from wild-type CHO pro⁻,
heterozygote D416, and one APRT⁻ mutant derived from D416 are shown
in Fig. 1. Wild-type APRT migrated to a single position
distinguishable from antibody. Heterozygote D416 produced two
types of APRT protein: one which co-migrated with wild-type, and a
second electrophoretically variant protein. All APRT⁻ cell lines
derived from D416 had lost the wild-type protein and retained the
mutant protein.

We next analyzed these cell lines at the DNA level using the
CHO aprt gene as a probe.[8] Genomic Southern blots[6] digested with
Hind III + Eco RI, Pvu II, or Taq I revealed no major differences
(Fig. 2). However, when DNA was digested with Msp I, differences
in restriction patterns specific to the aprt gene probe were
discernable (Fig. 2). Wild-type CHO DNA digested with Msp I and
probed with both the 1.5 kb Hind III-Pvu II and 1.8 kb Pvu II-Pvu
II fragments (see Fig. 3) showed two major bands at 3100 and 1150,
and two faint bands at 520 and 350 bp. D416, however, had a new
band at 1500 bp as well as the 1150 band, and D416d^r c26 (a mutant
derived from D416) had lost the band at 1150 and retained the band
at 1500 bp. Reprobing the DBM paper with the 1.5 kb Hind III-Pvu
II fragment or the 1.8 kb Pvu II-Pvu II fragment of the cloned CHO
aprt showed that the change in the Msp I site was within the 1.8 kb
fragment. The Msp I sites were mapped and are shown in Fig. 3.
Our data are consistant with the starred Msp I site being lost in
one chromosome in D416, whereas D416d^r c26 is homozygous for the
loss of the Msp I site. We believe this is a loss in the
restriction site rather than a change in methylation pattern, since
this site is also cleaved by the restriction enzyme Hpa II, which
recognizes the identical nucleotide sequence as Msp I but will not
cleave if the DNA is methylated at the site.

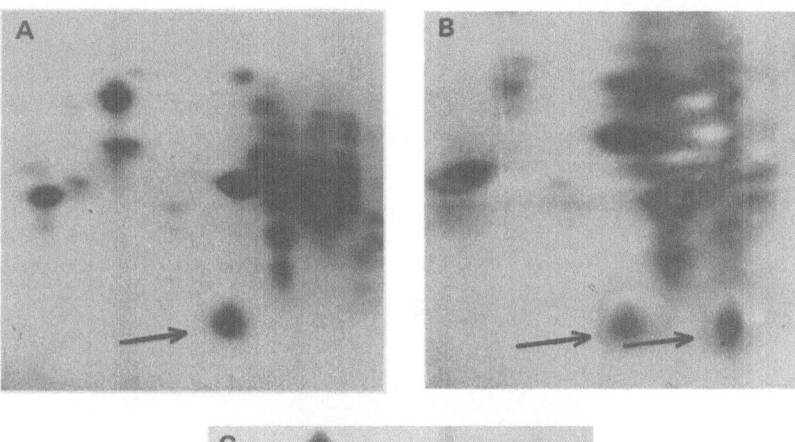

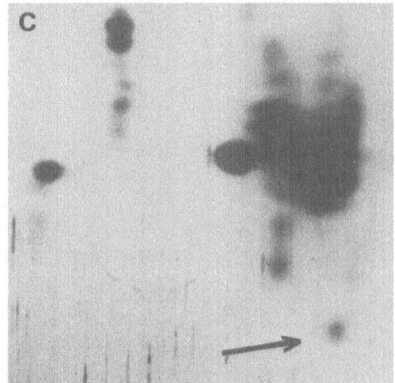

Fig. 1.   Sections of two dimensional gels showing APRT
          immunoprecipitated from (A) CHO pro⁻; (b) D416; (C)
          D416dʳc26. Gels were protein stained with silver nitrate.
          Arrow denotes APRT. All other darkly staining regions
          correspond to antibody.

        These results indicate that: (a) the electrophoretic variant
APRT protein, detected in both the heterozygote and its APRT⁻
derivatives, is due to a change in the DNA sequence; (b) this
change involves a small number of nucleotides and is either a small
deletion or a small insertion or a simple base-pair change in the
Msp I recognition sequence, CCGG; (c) the homozygotes have arisen

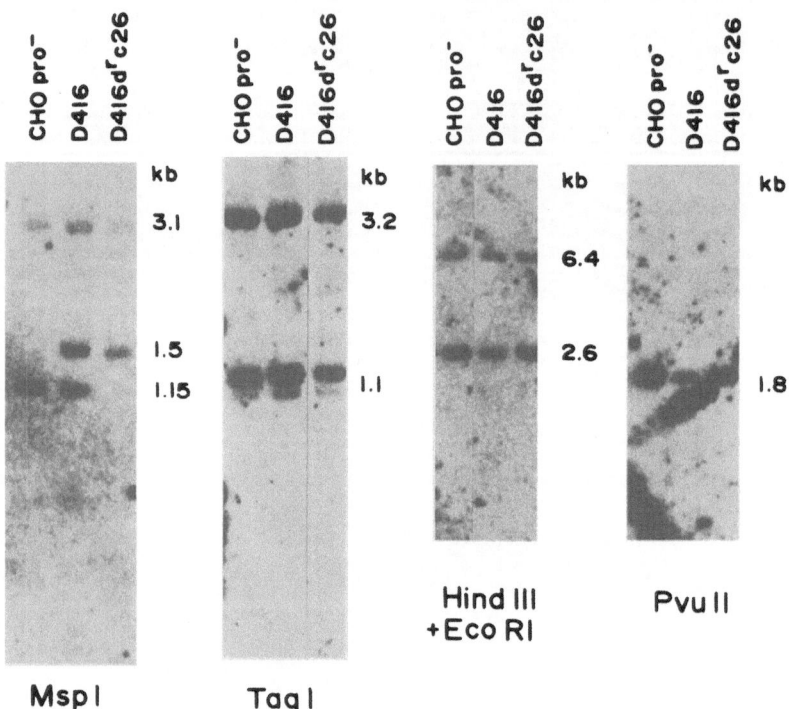

Fig. 2.   Genomic Southern Blots (see text).

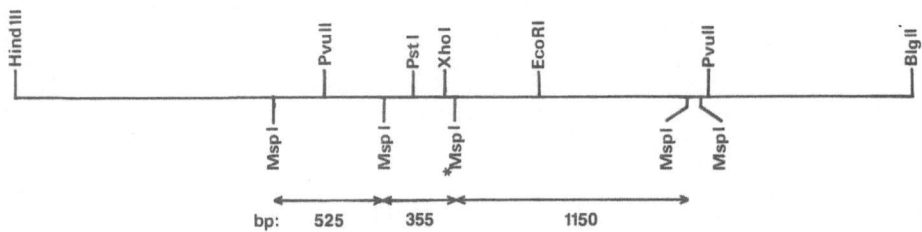

Fig. 3.   Map of the CHO APRT gene.   *Indicates altered Msp 1 site
in D416 and D416d^{r}c26.

as the result of a high frequency loss of the wild-type allele. This could be due to a large deletion or total loss of the chromosome and duplication of the mutant chromosome, or even possible mitotic recombination. We are currently attempting to measure the number of copies of the mutant allele in the CHO cell which would allow us to distinguish some of these possibilities.

REFERENCES

1. M. W. Taylor, J. H. Pipkorn, M. K. Tokito, and R. O. Pozzatti, Jr., Purine mutants of mammalian cell lines III: Control of purine biosynthesis in adenine phosphoribosyltransferase mutants of CHO cells. Somat. Cell Genet. 3:195-206 (1977).

2. J. A. Tischfield, J. J. Trill, U. I. Lee, K. Coy, and M. W. Taylor, Genetic instability at the adenine phosphoribosyltransferase locus in mouse L-cells. Mol. and Cell. Biol. 2:250-257 (1982).

3. L. H. Thompson, S. Fong, and K. Brookman, Validation of conditions for efficient detection of HPRT and APRT mutations in suspension-cultured Chinese hamster ovary cells. Mutation Res. 74:21-36 (1980).

4. W. E. C. Bradley and D. Letovanec, A high-frequency non-random mutational event at the adenine phosphoribosyltransferase (APRT) locus of sib selected variants heterozygous for APRT. Somat. Cell Genet. 8:51-66 (1982).

5. A. E. Simon, M. W. Taylor, W. E. C. Bradley, and L. H. Thompson, A model involving gene inactivation in the generation of autosomal recessive mutants in mammalian cells in culture. Mol. and Cell. Biol. (in press).

6. E. M. Southern, Detection of specific sequences among DNA fragments separated by gel electrophoresis. J. Mol. Biol. 98:503-517 (1975).

7. G. M. Wald, M. Stern, and G. Stark, Efficient transfer of large DNA fragments from agarose gels to diazobenzyloxmethyl-paper and rapid hybridization by using dextran sulfate. Proc. Natl. Acad. Sci. (USA) 76:3683-3687 (1979).

8. I. Lowy, A. Pellicer, J. F. Jackson, G. Sim, S. Silverstein, and R. Axel, Isolation of transforming DNA: Cloning the hamster APRT gene. Cell 22:817-823 (1980).

PURINE, THYMIDYLATE AND AMINO ACID REQUIREMENTS FOR HUMAN LYMPHOCYTE

TRANSFORMATION AND FRAGILE CHROMOSOME SITE EXPRESSION

Floyd F. Snyder, Carol A. Harasym and C. C. Lin

Departments of Medical Biochemistry and Paediatrics
Faculty of Medicine, University of Calgary
Calgary, Alberta T2N 4N1 Canada

Observations of heritable fragile site expression on human chromosomes being dependent on the type of tissue culture medium[1,2] have provided a basis for investigating the metabolic aspects of fragile site expression. Fragile sites are non-staining gaps of variable width usually involving both chromatids. The sites are inherited in a Mendelian dominant manner and except for fragile site Xq27, which is associated with X-linked mental retardation, the rest of the fragile sites have not been associated with any phenotypic abnormality. Restriction of folic acid was found to be required for the expression of all but one fragile site. The expression of heritable fragile site 10q25 requires bromodeoxyuridine (BUdR) in medium deficient in folic acid and thymidine[3]. Recently, the frequency of fragile site Xq27 was found to be increased by 5-fluorodeoxyuridine (FUdR)[4], a potent inhibitor of thymidylate synthetase.

In order to systematically investigate the mechanism(s) of fragile chromosome expression in folate deficient media we developed a basic culture medium. The medium is deficient in folic acid and a number of metabolites produced by folate mediated one-carbon transfer reactions (Fig. 1). These include the amino acids, serine, glycine and methionine; the purine source hypoxanthine; and the thymidylate source thymidine. Previous studies of fragile site expression have used leucocytes isolated in autologous plasma and cultured in medium supplemented with undialyzed serum. In order to eliminate serum sources of folate, amino acids and nucleotide precursors, we have used washed lymphocytes isolated from ficoll-hypaque banding and medium containing dialyzed human AB serum.

The transformation of peripheral blood lymphocytes stimulated with PHA was examined with single and multiple metabolite restriction

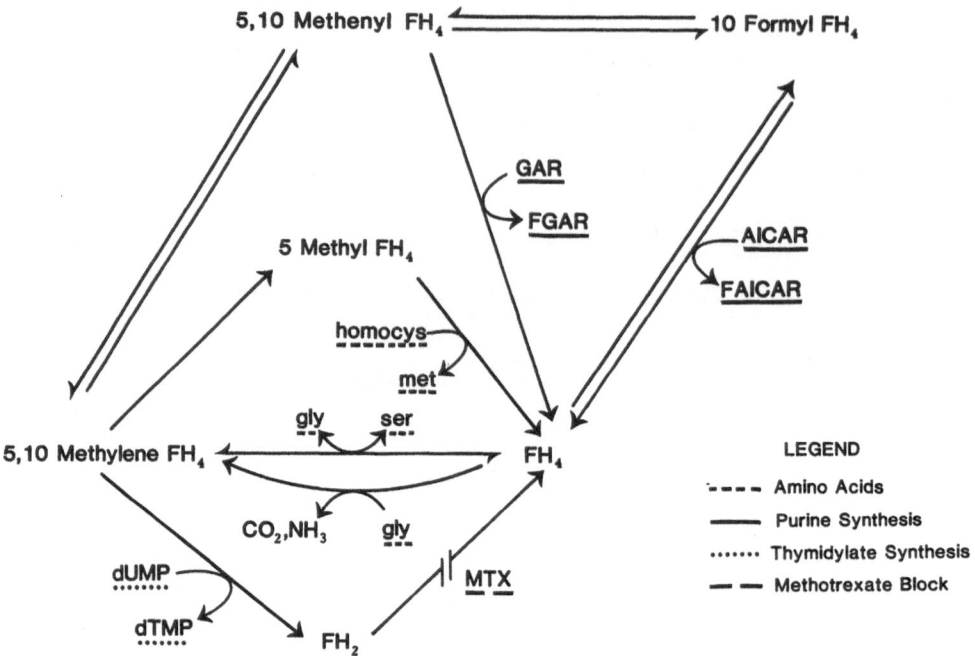

Fig. 1.   Folate Interconversions Involving 1-Carbon Transfer

by following [5-³H] uridine incorporation, 0.4 μM, 5Ci/mMol, into
RNA. Our aim was to develop a set of restrictive conditions which
are on the threshold of metabolite dependence but which support
lymphocyte transformation and thereby provide dividing cells suitable
for karyotypic analysis. The results for control subjects, subjects
with fragile site Xq27 and fragile site 10q25 are given in Table 1.
For all subjects examined folate was not required in the complete
medium. Transformation in folate deficient media showed a partial
requirement for serine and glycine and a more stringent requirement
for methionine and hypoxanthine. No requirement for thymidine could
be demonstrated in the absence of folate although a dependence on
thymidine could be demonstrated in the presence of methotrexate.
Common to the subjects having either fragile site Xq27 or 10q25 was
an increased dependence on glycine as compared to control subjects
(p < 0.05).

The set of restrictive conditions outlined in Table 1 was used
to examine the specificity of folate deficiency producing fragile
chromosome expression. Chromosomes were prepared by adding colcemid,
0.1 μg/ml, to lymphocyte cultures 1.5 h before harvesting. Cells
were treated with a hypotonic solution, 0.075 M KCl, for 20 min and

Table 1.   Lymphocyte Transformation in medium restricted in folate,
            amino acid and nucleotide precursors.
            PHA stimulated cultures were pulsed at 64h for 3 hours
            with [5-^{3}H] uridine and harvested

	Control Subjects	Fragile Site Xq27	Fragile Site 10q25
Restriction	Relative Rate of RNA synthesis		
None	100	100	100
Folate	93 ± 24	97 ± 24	106 ± 6
Folate, serine	68 ± 13	52 ± 34	44 ± 16
Folate, glycine	63 ± 9	40 ± 4	30 ± 3
Folate, methionine (7 µM)	50 ± 14	54 ± 20	60 ± 1
Folate, hypoxanthine	40 ± 11	23 ± 12	20 ± 5
Folate, thymidine	87 ± 21	90 ± 20	94 ± 10

fixed with methanol-acetic acid.   Air drying technique was used for
chromosome preparation and Q- and G-banding techniques were employed
for chromosome preparation.

The expression of fragile site Xq27 in folate deficient media
was investigated in 3 affected males with X-linked mental retardation.
The results indicate individual variation in fragile site Xq27 expression in response to media composition (Table 2).   Folate deficient media is not essential for fragile Xq27 expression but in
general expression is greatest in medium lacking folate and thymidine.
Deoxyuridine appears to further increase the frequency of expression.
Fluorodeoxyuridine, which inhibits thymidylate synthetase when phosphorylated also increased fragile site expression.   Restriction of
amino acids or the purine source, hypoxanthine, had no effect on
fragile site expression.   These observations suggest that thymidine
nucleotide depletion may be involved in the mechanism of fragile X
chromosome expression.

A fragile site 10q25 was also found in 60% of the cells from a
male with dermatitis herpetiformis and in 24% of his mothers cells
when cultured in folate and thymidine deficient medium containing
methotrexate and bromodeoxyuridine (Table 2).   This medium imposes
the use of bromodeoxyuridine for DNA synthesis in place of thymidine.
Individuals with fragile site 10q25 expressed on both homologs of
chromosome number 10 are rare.   This is the first case with as many
as 26% of the patients cells having both chromosomes number 10's
expressing fragile q25 sites.

These studies indicate a specific depletion of thymidine nucleotides appear to be involved in fragile chromosome expression, whereas
the restriction of amino acids or purines have no apparent affect.

Table 2. Metabolite restriction and frequency of fragile sites Xq27 in 3 affected males and 10q25 in male patient and mother

Medium (Folate Deficient) Restriction	Addition	% Cells with Fragile Sites		
A.	Fragile Site Xq27	A	B	C
	+ Folate	10	4	2
None		6	2	2
Glycine		0	2	2
Serine		2	0	0
Methionine (7 µM)		0	0	0
Hypoxanthine		0	0	0
TdR		6	8	16
TdR	+ UdR	22	4	8
TdR	+ BUdR + Mtx	2	0	0
TdR	+ FUdR	–	2	15
TdR	+ FUdR + UdR	–	3	27
B.	Fragile Site 10q25	subject		mother
TdR		0		–
TdR	+ BUdR	10		–
TdR	+ Mtx	0		–
TdR	+ Mtx + BUdR	60		24

TdR, thymidine; UdR, deoxyuridine; BUdR, bromodeoxyuridine; FUdR, 5-fluorodeoxyuridine; Mtx, methotrexate.

The expression may require bromodeoxyuridine as in the case of fragile site 10q25 or may also involve replacement of thymidine with deoxyuridine as indicated with studies on fragile site Xq27. Further studies of pyrimidine nucleotide metabolism and incorporation into DNA may clarify the molecular and hereditary bases of fragile chromosome site expression.

Acknowledgement: This work was supported by the Alberta Heritage Trust Fund, Provincial Cancer Hosptials Board grant H-105

REFERENCES

1. G.R. Sutherland, Am. J. Human Genetics, 31:125 (1979).
2. G.R. Sutherland, Science, 197:265 (1977).
3. G.R. Sutherland, E. Baker, and R.S. Seshadri, Am. J. Human Genetics 32:542 (1980).
4. T.W. Glover, Am. J. Human Genetics, 33:234 (1981).

GENETIC VARIABILITY OF PURINE NUCLEOSIDE PHOSPHORYLASE IN THE MOUSE

Floyd F. Snyder, Fred G. Biddle, Trevor Lukey and
Marcia J. Sparling

Departments of Medical Biochemistry and Paediatrics
Faculty of Medicine, University of Calgary
Calgary, Alberta   T2N 4N1   Canada

The inherited human deficiency of purine nucleoside phosphorylase is associated primarily with cellular immune dysfunction[1]. We
have begun to screen for quantitative activity variants of purine
nucleoside phosphorylase in Mus musculus in attempts to establish an
animal model for the enzyme deficiencies associated with immunodeficiency disease. A preliminary survey of feral and inbred mouse
strains revealed a 4-fold range in erythrocyte purine nucleoside phosphorylase activity. We have examined some of the biochemical and
genetic aspects of this variability.

We have screened 37 inbred mouse lines for erythrocyte purine
nucleoside phosphorylase activity. Most of these strains have
activity between 14 and 21 nmol/min/mg protein. Six of the strains
have significantly greater activity than the remainder, the range
being from 36 to 50 nmol/min/mg protein. We chose two strains, DBA/2J
and C57BL/6J, representative of the low and high activity strains
respectively, in which to study the biochemical and genetic features.
As shown in Table 1, erythrocytic purine nucleoside phosphorylase
activity is approximately 2.5-fold greater in C57BL/6J than DBA/2J
mice. Comparison of a second activity between these lines did not
show significant variation. Adenosine kinase, assayed with $[8-^{14}C]$
adenosine as previously described[2], was 0.66 ± 0.08 and 0.72 ± 0.21
nmol/min/mg protein for DBA/2J and C57BL/6J respectively. Mixing
experiments with lysates from DBA/2J and C57BL/6J gave the expected
additive contribution for purine nucleoside phosphorylase, ruling out
the possibility of endogenous activators or inhibitors.

In order to examine the basis of the quantitative variability,
several properties of purine nucleoside phosphorylase were compared

Table 1. Purine Nucleoside Phosphorylase activity in DBA/2J,
          C57BL/6J, F_1 and Backcross Offspring

Strain	nmol/min/mg protein
DBA/2J	17.2 ± 1.4 (20)
C57BL/6J	43.2 ± 2.2 (20)
(DBA/2J x C57BL/6J) F_1	32.1 ± 2.7 (20)
F_1 x DBA/2J a.	16.9 ± 1.8 (11)
b.	38.6 ± 3.2 (10)
F_1 x C57BL/6J a.	31.9 ± 3.0 (10)
b.	46.0 ± 3.5 (10)

Purine nucleoside phosphorylase was assayed in erythrocyte
lysates with 450 µM [8-^{14}C] inosine[3].

between DBA/2J and C57BL/6J. Thermal stability profiles showed only
minor differences between these strains. The activity from both
strains was stable in the presence of 100 mM phosphate at 60° C. In
the absence of phosphate, purine nucleoside phosphorylase was rapidly
inactivated, having half lives of 15 and 18 min for DBA/2J and
C57BL/6J respectively at 50° C. Kinetic studies of crude purine
nucleoside phosphorylase were also conducted with both inosine and
deoxyguanosine as nucleoside substrates. The results presented in
Table 2 were obtained from Hanes—Woolf transformation and indicate
a change in Vmax is the major kinetic difference between strains.
Thus the Vmax was 4-fold greater with inosine and 3-fold greater with
deoxyguanosine for C57BL/6J than for DBA/2J. We also examined the
starch gel electrophoretic profile for purine nucleoside phosphorylase in these two strains. DBA/2J and C57BL/6J have in common the
Np-1a band[4], and C57BL/6J also has a more cathodally migrating band
previously designated Np-2[5].

Table 2. Kinetic studies of erythrocyte purine nucleoside
          phosphorylase

Substrate/Strain		km (µM)	Vmax (nmol/min/mg)
Inosine:	DBA/2J	99	19
	C57BL/6J	136	79
Deoxyguanosine:	DBA/2J	52	8.9
	C57BL/6J	78	28

The inheritance of the quantitative and electrophoretic traits of purine nucleoside phosphorylase were further examined in crosses between DBA/2J and C57BL/6J. Offspring of crosses between DBA/2J and C57BL/6J (F_1) had intermediate activity characteristic of the additive contribution of both parents (Table 1). The backcross of the F_1 with DBA/2J gave two activity classes, one low and one intermediate. Backcrosses of the F_1 with C57BL/6J also gave two activity classes, one intermediate and one high (Table 1). Erythrocyte purine nucleoside phosphorylase segregated independent of sex in both backcrosses and shows an autosomal codominant mode of inheritance. We also tested for coincidence of the Np-2 electrophoretic band with activity in the backcross. Backcrosses of the F_1 with DBA/2J having intermediate activity were all found to have the Np-2 band, whereas none with low activity characteristic of the DBA/2J parent have the Np-2 band.

These studies have identified quantitative variation of purine nucleoside phosphorylase which is heritable and stable within inbred mouse lines. Associated with the quantitative variability are electrophoretic differences between DBA/2J and C57BL/6J purine nucleoside phosphorylase and minor differences in thermal stability and affinity for nucleoside substrates. Additional studies are underway to characterize the gene products of these strains and map the Np-2 locus.

Acknowledgements: Supported by the Medical Research Council of Canada

REFERENCES

1. E.R. Giblett, A.J. Ammann, R. Sandman, D.W. Wara, and L.K. Diamond, Lancet 2:1010 (1975).
2. T. Lukey and F.F. Snyder, Can. J. Biochem. 58:677 (1980).
3. F.F. Snyder, J. Mendelsohn, and J.E. Seegmiller, J. Clin. Invest. 58:654 (1976).
4. J.E. Womack, M.T. Davisson, E.M. Eicher, and D.A. Kendall, Biochem. Genet. 15:347 (1977).
5. T.A. Bremner, E. Premkumer-Reddy, K. Nayar, and R.E. Kouri, Biochem. Genet. 16:1143 (1978).

DEOXYCOFORMYCIN RESISTANT MAMMALIAN CELLS THAT OVERPRODUCE
ADENOSINE DEAMINASE

Patricia A. Hoffee and Stephen W. Hunt, III

Department of Microbiology, School of Medicine
University of Pittsburgh
Pittsburgh, PA 15261 USA

Adenosine deaminase (ADA) catalyzes the deamination of adenosine and deoxyadenosine to inosine and deoxyinosine, respectively. The study of ADA in mammalian cells is of particular importance because of (a) its indicated association with combined immunodeficiency disease in which patients with a deficiency of ADA activity exhibit a loss of both B and T cell function (1); (b) the occurrence of patients with hereditary hemolytic anemia who have a 40-70-fold increase in erythrocyte ADA levels (2); and (c) recent reports that in acute lymphoblastic leukemia high levels of ADA are found in T lymphoblast cells (3). It would be extremely useful, therefore, to have available a model cell culture system in which cells with elevated levels of ADA could be isolated and used to study the regulation and expression of ADA.

There are now known several cell culture systems in which variants selected as resistant to a particular inhibitor show elevated levels of the target enzyme(s), the most studied being that of methotrexate resistance which results in elevated levels of dihydrofolate reductase (4). To develop a similar system for ADA overproduction would require (a) a potent inhibitor of ADA activity and (b) growth conditions under which ADA activity was essential for cell survival. Deoxycoformycin (dCF), a specific inhibitor of ADA, has been described. This nucleoside analogue has a high affinity for ADA with a K_i of $10^{-10} - 10^{-11}M$ (5). Under normal growth conditions, dCF is not particularly toxic to rat hepatoma, CHO, or mouse L cells. If, however, one manipulates the growth conditions so that the nucleoside, adenosine, is the only carbon source available to the cells, deoxycoformycin becomes a potent inhibitor, presumably due to its inhibition of ADA.

Adenosine serves as a carbon source for mammalian cells by its
deamination to inosine in the presence of ADA, the cleavage of
inosine to hypoxanthine and ribose-1-P in the presence of purine
nucleoside phosphorylase, the conversion of ribose-1-P to ribose-
5-P in the presence of phosphopentomutase and then the entrance of
ribose-5-P into the glycolytic cycle. To insure that adenosine
utilization is mediated by ADA and to avoid the inhibitory effects
of adenosine on cell growth, a tubercidin resistant mutant (tub^r)
deficient in adenosine kinase activity was selected initially.
Tub^r mutants were selected in three different cell lines; rat
hepatoma, Chinese hamster ovary, and mouse L-cells. These tub^r
cell lines when placed in medium containing adenosine as a sole
carbon source, grew with a generation time of 18-24 hours.
Exposure of these cells to low levels of deoxycoformycin resulted
in death of the culture. Resistant clones appeared at a frequency
of about 10^{-6}. This frequency was enhanced 10-20 times by prior
treatment of the culture with the mutagen ethylmethane sulfonate
(6).

   Resistant cells (dCF^r) were assayed to determine ADA levels.
Initial isolates, selected as resistant to low levels of dCF
(0.04-0.4 µM), showed a 6-7-fold increase in ADA levels (Table 1).
Sequential isolation of dCF^r cells resistant to higher levels of
dCF resulted in cell lines with increasing levels of ADA. As seen
in Table 1, rat hepatoma cells resistant to 12 µM dCF have greater
than a 350-fold increase in the level of ADA activity. Similar
selections are currently underway with CHO and mouse L-cells. At
the present time we have available only the initial steps of
resistance. However, even at a level of 0.12 µM dCF, CHO cells
already have a 15-fold increase in ADA levels. All subsequent
studies described in this report deal exclusively with the rat
hepatoma cells.

   To determine the molecular basis for the increase in ADA in
the rat hepatoma cells, we compared the kinetic, physical and
immunological properties of ADA purified from parental rat
hepatoma cells, dCF^r hepatoma cells and rat liver. ADA was
purified using a method similar to that developed by Schrader
et al (7) for purification of human erythrocyte ADA. ADA in dCF
resistant cells is present as 2-4% of the total soluble cell
protein and can be purified to homogeneity by two passages over an
Ado-Sepharose affinity column. As shown in Table 2, the final
specific activities of ADA purified from rat liver and dCF
sensitive and resistant cells did not differ significantly.
Purified ADA from all sources migrated as a single band in SDS
polyacrylamide gels with an estimated molecular weight of 45,000.
The K_m values for adenosine were determined for the three enzymes
to be approximately 2.7×10^{-5}M. The K_i values for dCF were
determined to be $1.2-1.4 \times 10^{-10}$M. Determination of K_i for dCF
using the procedure for tight binding inhibitors (5) gave a
somewhat lower K_i of 0.8×10^{-10}M.

Table 1.   ADA Levels[a] in dCF Sensitive and Resistant Cells

dCF (μM)	Rat hepatoma	CHO	Mouse L
None	50[b]	85[b]	65[b]
0.04	-[c]	500	-
0.12	-	1240	356
0.4	334		
1.0	1,503		
2.0	4,090		
6.0	10,180		
12.0	17,950		

a)   ADA levels expressed as nmoles/min/mg protein.
b)   ADA level in sensitive parental cells.
c)   Not determined (-).

Table 2.   Properties of Purified ADA

Property	Sensitive cells	Resistant cells	Rat liver
Specific activity[a]	490	505	486
K_m (Ado)	28 μM	27 μM	26 μM
K_i (dCF)	0.14 nM	0.12 nM	0.12 nM
Mol. Wt.	45,000	45,000	45,000

a)   nmoles/min/mg protein x 10^{-3}.

Antibody made against purified rat liver ADA gave a single precipitin band with lines of identity between purified rat liver ADA and crude extracts from both dCF sensitive and resistant cells. In addition, purified enzyme from the three sources reacted with complete identity.

Quantitative immunoprecipitation experiments were performed to determine the relationship between enzyme activity and immuno-precipitable material for purified rat liver, N1S167 (sensitive parent) and 5-2 (resistant cell) ADA's. Figure 1 shows that the same amount of antibody precipitated the same amount of enzymatic activity for all three proteins. Thus, ADA's from rat liver and dCF sensitive hepatoma and resistant hepatoma cells are indistinguishable in terms of kinetic, physical and immunological properties. The increase in ADA activity in dCF[r] cells is clearly due to an increase in the number of molecules of a structurally normal enzyme.

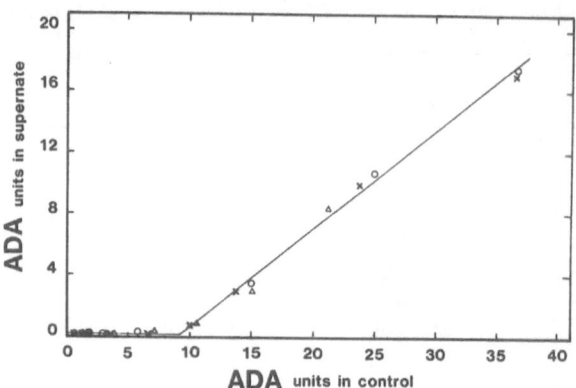

Figure 1. Immunotitration of purified ADA. Increasing amounts of ADA purified from rat liver (O), sensitive cells (Δ) or resistant cells (X) were incubated with a constant amount of immune or non-immune sera. The amount of ADA activity remaining in the supernatant fluid was determined.

We next determined if this increase was due to a change in the rate of synthesis or degradation of the ADA protein. Parental and dCF^r cells were pulse-labeled with ^{3}H-leucine and ^{14}C-leucine, respectively. Labeled cell extracts were mixed and the ratio of ^{3}H/^{14}C present in total soluble protein determined. ADA protein was immunoprecipitated from mixed cell extracts with purified IgG. The immunoprecipitates were run on SDS polyacrylamide gels, the gels sliced, and slices containing ADA assayed for radioactivity. Ratios of ^{3}H/^{14}C for total protein synthesis and for ADA protein were compared. Table 3 shows that the relative rate of ADA synthesis clearly paralleled the relative enzyme levels in 3 resistant cell lines with different levels of ADA activity. Degradation rates of ADA, on the other hand, did not differ significantly in the sensitive or resistant cells.

ADA mRNA activity was measured by _in vitro_ translation of (8) polysomal RNA in a rabbit reticulocyte system. The amount of ADA protein synthesized was determined by immunoprecipitation and gel electrophoresis. One cell line tested, 4-3-15, which has a 181-fold increase in ADA activity had a corresponding 171-fold increase in mRNA activity as compared to the dCF-sensitive parental cell line.

It would appear that the increase in ADA activity seen in dCF^r rat hepatoma cells is due to an increased synthesis of a

Table 3.   Rate of ADA Synthesis

Cell line	ADA Activity[a]	ADA Synthesis[a]
N1S167	1	1
5-2-3	58	69
4-3-5	90	92
4-3-15	181	185

a)   Relative rate.

structurally normal enzyme. This increased protein synthesis is associated with increased ADA mRNA activity. Is the increase in ADA mRNA activity a result of gene amplification?

In other examples of enzyme overproduction that have been a result of gene amplification, several chromosome anomalies have been noted. One of these is the appearance of homogeneously staining regions (HSRs) in banded chromosome preparations (4). We have examined our hepatoma cell lines for the presence of HSRs. Neither parental cell lines nor cell lines with less than a 50-fold increase in ADA activity contain HSRs. However, cell lines with greater than an 80-fold increase in ADA activity show one of several types of HSRs on their chromosomes. In fact, one line with a 200-fold increase in ADA activity appears to have a major part of a large chromosome as an HSR (Fig. 2). Confirmation that the HSR contains amplified ADA structural genes will require in situ hybridizations.

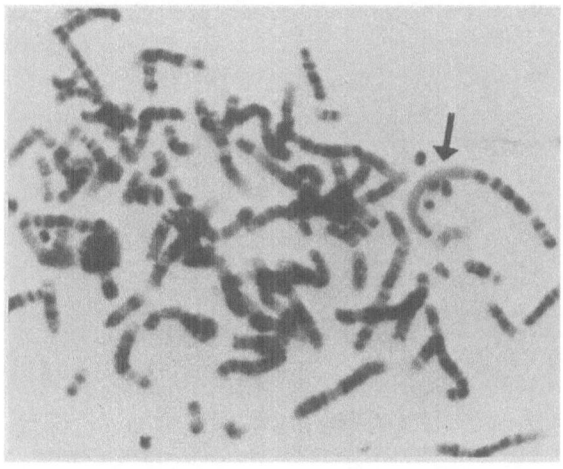

Figure 2.   G-Banded chromosome preparation from dCF[r] cell line with a 200-fold increase in ADA activity. Arrow marks an HSR.

In summary, we have developed a selection system that results in the isolation of mammalian cell lines with increased ADA levels, up to 350 times the normal level. The increase in ADA activity is due to an increase in ADA synthesis which is correlated to a corresponding increase in ADA mRNA activity. Additional studies are currently underway to determine gene copy number, and eventually to isolate and characterize the structural gene for ADA.

ACKNOWLEDGEMENT

This work was supported by U. S. Public Health Research Grant AM 12151.

REFERENCES

1. E. R. Giblett, J. E. Anderson, F. Cohen, B. Pollara, and H. J. Meuwissen, Lancet II:1067 (1972).
2. W. N. Valentine, D. E. Paglia, A. P. Tartaglia and F. Gilsanz, Science 195:783 (1977).
3. P. Daddona, J. Biol. Chem. 256:12496 (1981).
4. R. T. Schimke, "Gene Amplification", Cold Spring Harbor Symposium, Cold Spring Harbor, NY (1982).
5. R. P. Agarwal, T. Spector, and R. E. Parks, Jr., Biochem. Pharm. 26:359 (1977).
6. P. A. Hoffee and S. W. Hunt, III, Somatic Cell Gen. 8:465–477 (1982).
7. W. P. Schrader, A. R. Stacy and B. Pollara, J. Biol. Chem. 251:4026 (1976).
8. R. A. Padgett, G. M. Wahl, P. F. Coleman and G. R. Stark, J. Biol. Chem. 254:974 (1979).

CHARACTERIZATION AND USE OF CLONED SEQUENCES OF THE HYPOXANTHINE-GUANINE PHOSPHORIBOSYLTRANSFERASE GENE

A.C. Chinault, J. Brennand, D.S. Konecki, R.L. Nussbaum and C.T. Caskey

Howard Hughes Medical Institute Laboratories and Departments of Medicine and Biochemistry, Baylor College of Medicine, Houston, Texas 77030 U.S.A.

INTRODUCTION

The hypoxanthine-guanine phosphoribosyltransferase (HPRT) gene has been one of the most extensively used loci for the study of mutation in cultured mammalian cells. Such studies have been facilitated by the localization of the HPRT gene to the X-chromosome and the existence of powerful selection systems for the isolation of mutants and revertants. In addition, the fact that HPRT enzyme deficiency resulting from mutation of this gene in man leads to the X-linked recessive disorders of Lesch-Nyhan Syndrome and gouty arthritis makes it an excellent genetic locus for examination of the heterogeneity of spontaneous mutational events which occur in the human population. A large number of studies have focused on characterization of the gene product resulting from mutations at this locus by using immunological assays, kinetic and thermal sensitivity assays and peptide mapping (1). However, the detailed characterization of mutational events at the DNA level has not been possible due to the lack of appropriate molecular probes. Recently the availability of a mouse cell line with amplified HPRT genes has facilitated the cloning of HPRT cDNA recombinants (2). This report will briefly describe recent progress on the characterization of these cloned HPRT gene sequences and their use as probes to examine the genomic organization at the HPRT locus.

RESULTS AND DISCUSSION

NBR4 is a mouse neuroblastoma cell line which has been shown to have elevated levels of HPRT protein in vivo (3) and higher than

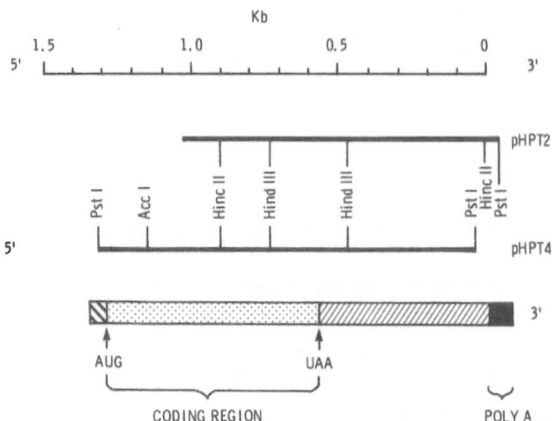

Figure 1. Partial restriction maps for the HPRT cDNA recombinants
pHPT2 and pHPT4 and a summary of the sequence organization derived
from nucleotide sequencing.

normal levels of mRNA coding for this enzyme as determined by in
vitro translation studies (4).   A sucrose gradient-enriched fraction
of mRNA from this cell line was used to construct cDNA libraries in
pBR322 by the G-C tailing procedure and cDNA recombinants complementary to HPRT mRNA were identified by differential hybridization and
a message selection/in vitro translation assay (2).   The initial
recombinants that were isolated were subsequently used as hybridization probes to identify additional plasmids carrying more extensive
HPRT sequences.   Based on the estimated length of HPRT mRNA (1550
bases) the plasmids isolated contain at least 85% of the full-length
sequence.

Characterization of cDNA sequences

    Partial restriction maps for two of the recombinant plasmids,
pHPT2 and pHPT4, are shown in Figure 1.   Nucleotide sequencing of
the mouse HPRT cDNA by the Maxam-Gilbert procedure has been completed.   Analysis of this information shows that the cloned sequences
contain about 100 nucleotides of 5'-noncoding sequence followed by
an open reading frame of 654 nucleotides and an additional 550
nucleotides of 3'-noncoding sequence preceding the polyA addition
site (Figure 1).   The assumption that the open reading frame represents the protein coding sequence leads to the prediction that mouse
HPRT contains 218 amino acids, corresponding to a molecular weight
of 24,500.   This is in reasonable agreement with the estimated size
of purified HPRT derived by SDS-polyacrylamide gel electrophoresis
(data not shown).   Comparison of the predicted amino acid composition for this mouse protein with published results for human (5) and

Table 1. Comparison of predicted amino acid composition for mouse
HPRT with those for human (5)and bovine (6) HPRT.

Amino acid	# Residues			Amino acid	# Residues		
	Mouse	Human	Bovine		Mouse	Human	Bovine
Ala	9	12	16	Met	7	5	5
Arg	11	12	9	Phe	9	9	8
Asx	28	31	28	Pro	9	14	10
Cys	4	5	6	Ser	14	14	16
Glx	12	16	19	Thr	11	12	12
Gly	15	19	23	Trp	0	0	1
His	5	5	6	Tyr	11	10	9
Ile	15	11	11	Val	20	18	17
Leu	21	20	19				
Lys	17	17	18	TOTALS	218	230	233

bovine (6) HPRT is shown in Table 1. There appears to be excellent
qualitative agreement between these results, although the total
number of amino acids is 12-15 fewer for the mouse. Attempts to
correlate the predicted tryptic digestion products with published
results (7) has been less successful; the reason for the discrepan-
cies is presently unclear.

Use of cloned probes for genomic studies

      The cloned mouse HPRT sequences have been shown to exhibit sig-
nificant cross-hybridization with mRNA and DNA from other mammalian
sources, including man (2). This homology has made it possible to
use these sequences as probes in Southern hybridization experiments
to examine genomic organization and DNA sequence polymorphisms.
Results from preliminary experiments are summarized in Figure 2. As
shown in lane G, a human genomic blot using DNA cut with the restriction endonuclease BamH I (using the cDNA insert from pHPT2 as
the hybridization probe) shows bands at 25, 17, 3.7 and 3.4 kb. In
addition it has been shown that normal humans may demonstrate two
extra bands at 20 and 12.5 kb which represnt restriction fragment
length polymorphisms (see R.L. Nussbaum et al., this volume).
Somatic cell hybrid lines have been used to examine the association
between these hybridizing bands and the human X chromosome. Somatic
cell hybrids between a human cell line bearing the 12.5 kb polymorphism and Chinese hamster V79 cells are analyzed in lanes A
through D. The cell line in lane A contains only a single human X,
as determined cytogenetically; DNA from a 6-thioguanine (6-TG) resistant derivative of this line, lacking the X, is analyzed in lane
B. Elimination of the human X leads to a diminution in intensity of
the 25 kb band (which subsequent work has shown to be a doublet) and
disappearance of the 12.5 and 3.7 kb bands. The remaining bands

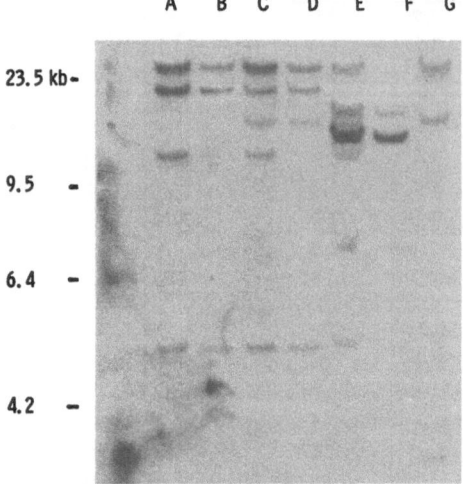

Figure 2. Southern hybridization analysis of DNA from hamster/human somatic cell hybrids (lanes A – D), mouse/human hybrids (lanes E – F) and normal human cells (lane G) using radiolabelled HPRT cDNA sequences as probes.

seen in lane B represent fragments from the hamster genome which hybridize to the probe. It should be noted that the 17 kb band seen in normal human DNA (lane G) is absent in these cell lines. In lanes C and D, DNA from a hamster/human hybrid cell line containing human X, 11, 16 and 21 (lane C) is compared with its 6TG-resistant derivative lacking the X (lane D). The 25, 12.5 and 3.7 kb bands are again shown to be X dependent, while the 17 kb band which is seen in these lines is not dependent on the presence of the human X and is, therefore, assumed to be a sequence from one of the human autosomes carried by the cell line. Similar comparisons of mouse/human hybrid lines are shown in lanes E and F. The hybrid used in lane E contains human chromosome 11 bearing translocated distal Xq (8); in lane F the 6-TG resistant derivative lacking the translocation chromosome is shown. The BamH I fragments of 25, 12.5 and 3.7 kb are shown to be carried by the human Xq. The remaining bands represent fragments derived from the mouse chromosomal background.

A number of general conclusions concerning the structure of the HPRT gene can be made from the results in Figure 2 where the BamH I fragment patterns of hamster and mouse, as well as human, can be compared. The observations that a number of BamH I fragments show hybridization (even though the probe sequence contains no site for this enzyme) and that the sum of the fragments is >40 kb indicate that the gene is very large and contains multiple introns. Although there is significant sequence homology between the coding sequences for the different species, the overall patterns are distinct, probably due to DNA sequence polymorphisms within the intron sequences. These conclusions are corroborated by preliminary studies on λ recombinants isolated from genomic libraries (unpublished observations), but further analysis will be required to establish the actual HPRT gene organization.

ACKNOWLEDGEMENTS

This work was supported by the Howard Hughes Medical Institute and by a Robert A. Welch Foundation grant (Q-533) to C.T.C. J.B. was supported by a Arthritis Foundation Fellowship.

REFERENCES

1. C.T. Caskey and G.D. Kruh, The HPRT locus, Cell 16:1 (1979).
2. J. Brennand, A.C. Chinault, D.S. Konecki, D.W. Melton and C.T. Caskey, Proc. Natl. Acad. Sci. USA 79:1950 (1982).
3. D.W. Melton, Somatic Cell Genet. 7:331 (1981).
4. D.W. Melton, D.S. Konecki, D.H. Ledbetter, J.F. Hejtmancik and C.T. Caskey, Proc. Natl. Acad. Sci. USA 78:6977 (1981).
5. A.S. Olsen and G. Milman, Biochem. 16:2501 (1977).
6. V.A. Paulus, R.G. Ingalls, B. Vasquez and A.L. Bieber, J. Biol. Chem. 255:2377 (1980).
7. M.R. Capecchi, R.A. Von der Haar, N.E. Capecchi and M.M. Sveda, Cell 12:371 (1977).
8. B.R. Migeon, T.R. Brown, J. Axelman and C.J. Migeon, Proc. Natl. Acad. Sci. USA 78:6339 (1981).

SOUTHERN ANALYSIS OF THE LESCH-NYHAN LOCUS IN MAN

R.L. Nussbaum, C.T. Caskey, F. Gilbert and W. Nyhan

Howard Hughes Medical Institute Laboratories and Departments of Medicine and Biochemistry, Baylor College of Medicine, Houston, Texas, Department of Pediatrics, University of Pennsylvania and Department of Pediatrics University of California, San Diego, USA

INTRODUCTION

The Lesch-Nyhan (LN) syndrome of mental retardation, hyperuricemia and self-mutilation is an X-linked disease of man resulting from a deficiency of the enzyme hypoxanthine-guanine phosphoribosyltransferase (HPRT) (1). It is known that a number of LN patients exhibit no enzyme activity and no crossreacting antigen (1,2); mutation in these patients may be due to substantial deletions or insertions detectable by Southern analysis. We have undertaken an initial survey of LN patients looking for major gene alterations detectable by Southern analysis; our findings constitute the body of this report.

METHODS

Cloned cDNA for HPRT (3) from a mouse neuroblastoma line was used as a probe against DNA from human fibroblasts and leukocytes digested with a variety of restriction enzymes and transferred to nitrocellulose by the method of Southern. The probe cross-hybridizes well with human genomic DNA (4). Fibroblasts were obtained from the Mutant Cell Repository, Camden, New Jersey, and from patients and their families.

RESULTS

Digestion of DNA from 15 patients with the LN syndrome using 7

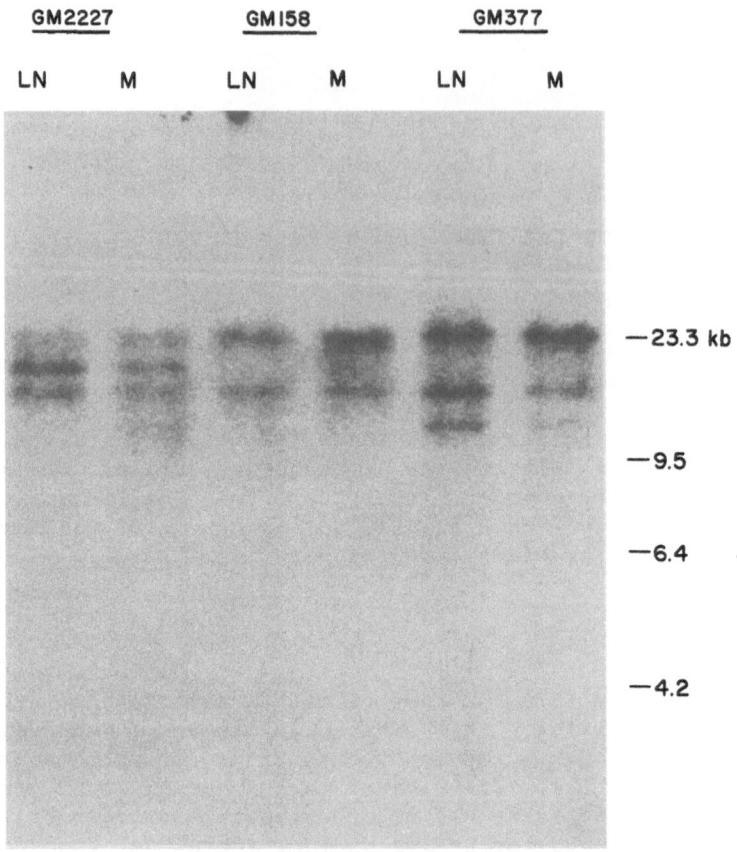

Figure 1. Southern analysis of Bam HI digested DNA from three LN
         patients (GM2227, GM158 and GM377) and their mothers (M).

different restriction enzymes all gave identical patterns with the
exception of Bam HI. In Figure 1 are shown the restriction patterns
of three Lesch-Nyhan patients and their mothers. The most common
pattern seen is represented by GM158 where 4 fragments of size 24,
17, 3.7 and 3.4 kb were seen; the small fragments are faint in this
study. In addition to these fragments, extra bands at 20 kb
(GM2226) and 12.5 kb (both GM377 and 2226's mother) were seen in
some patients. Control DNAs were also studied and their Bam HI
restriction patterns were found to include the common pattern as
well as the extra 12.5 and 20 kb fragments (Fig. 2). We hypothesized that the 12.5 and 20 kb fragments were restriction fragment
length polymorphisms not directly involved with the LN mutations in
the patients studied.

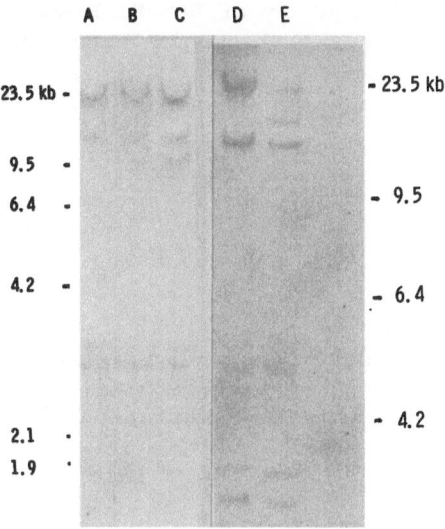

Figure 2. Southern analysis of Bam HI digested DNA from 5 normal
          females showing the 12.5 kb polymorphism (Lanes B and C)
          and the 20 kb polymorphism (Lane E). Simple Bam HI
          pattern is shown in Lanes A and D.

We proceeded to test if these polymorphic variants were X-linked
and had arisen at the HPRT locus. Somatic cell hybrids containing
an X chromosome from a normal individual bearing the 12.5 kb polymorphism were constructed and their Bam HI restriction pattern analyzed [See Figure 2, reference (4)]. The 12.5 kb polymorphism was
present in two hybrids containing human X and segregated with the X
when 6-thioguanine was used to select for cells that had spontaneously lost the human X chromosome. A third line bearing only
distal human Xq also contained the polymorphism and also showed
segregation of the polymorphism with Xq.

Further confirmation of the X-linkage of the 12.5 kb polymorphism was sought by studying the family of GM377. Precise concord-
ance was found between the presence of the mutation and the 12.5 kb
polymorphism. These studies have been extended to include 7 oppor-
tunities for segregation of the LN locus and the polymorphism; no
recombination has been seen.

The 20 kb polymorphism has also been studied and, in a small
family cluster (GM2226), appeared to segregate with the LN mutation.
Further studies are in progress.

Table 1.   Frequency of Bam HI Restriction
Polymorphisms in Human Populations

SOURCE	12.5 kb	20 kb	TOTAL
Normal	2	1	13
L-N	2	2	15
Other	2	1	10
Total	6	4	38

Table 1 shows the frequency of polymorphic variants for the restriction enzyme Bam HI at the HPRT locus. To date, approximately 25% of X chromosomes demonstrate a variation with this enzyme.

DISCUSSION

Our initial Southern analyses of patients with LN failed to show major deletions or insertions detectable as alterations in restriction pattern with a variety of enzymes. This negative result by no means proves that such mechanisms play no role in producing mutation at the HPRT locus since such studies would detect only large changes and would miss small deletions or insertions. The studies are further limited by the number of enzymes employed, the number of patients studied and the heterologous nature of the probe. However, at a first approximation, major alterations detectable by Southern analysis seem not to be responsible for the majority of Lesch-Nyhan mutations. Alterations in the Bam HI digestion pattern at the HPRT locus are present, however, and represent restriction fragment length polymorphisms at this locus. The frequency of the polymorphisms is high making them useful for gene mapping and linkage analysis in man.

REFERENCES

1.   W.N. Kelley and J.B. Wyngaarden, The Lesch-Nyhan syndrome, in "The Metabolic Basis of Inherited Disease," J.B. Stanbury, J.B. Wyngaarden, D.S. Fredrickson, eds., McGraw-Hill, New York (1978).
2.   C.T. Caskey and G.D. Kruh, The HPRT locus, Cell 16:1 (1979).
3.   J. Brennand, A.C. Chinault, D.S. Konecki, D.W. Melton and C.T. Caskey, Cloned cDNA sequences of the hypoxanthine/guanine phosphoribosyltransferase gene from a mouse neuroblastoma cell line found to have amplified genomic sequences, Proc. Natl. Acad. Sci. USA 79:1950 (1982).
4.   A.C. Chinault, J. Brennand, D.S. Konecki, R.L. Nussbaum and C.T. Caskey, Characterization and use of cloned sequences of the hypoxanthine-guanine phosphoribosyltransferase gene, this volume.

DE NOVO PURINE SYNTHESIS IN HUMAN LYMPHOCYTES

Peter B. Rowe, Eric McCairns, Dorit Sauer and Dale Fahey

Children's Medical Research Foundation, University of Sydney
P.O. Box 61, Camperdown, N.S.W., Australia, 2050

In previous studies on de novo purine synthesis in avian liver we established that, although the enzymes of the pathway did not constitute a macromolecular complex, there did appear to exist between them some particular relationship which permitted their partial copurification. This enabled us to define some of the properties of the pathway including aspects of its regulation.

Cell lysates of "resting" human peripheral blood lymphocytes incorporate 2-3 nmol glycine mg protein^{-1}h^{-1} into purine ribo-nucleotides[2], one-twentieth the activity observed in avian liver homogenates[1]. Lectin (phytohemagglutinin) activation of cultured lymphocytes results in a fifteen to twenty-fold increase in total activity and a four to five-fold increase in specific activity after 72 h[2], with equivalent changes in the activity of the first pathway enzyme amidophosphoribosyltransferase (EC 2.4.2.14).

Using modifications of the purification procedures used for the avian liver studies we have partially copurified the enzymes of the de novo synthetic pathway from lectin-transformed lymphocytes (Table I). Cells (1-2 x 10^9) were harvested by centrifugation at room temperature; the cell pellet was resuspended in a buffer consisting of 50 mM Hepes pH 7.4 containing 50 mM 2-mercaptoethanol, 5.0 mM MgCl$_2$, 25 mM KCl and 40% sucrose (w/v) at a cell density of 5 x 10^8 per ml, and disrupted by sonication at 0^{0}C. After centrifugation at 100,000 g for 30 min the supernatant solution was fractionated between 5 and 30% (w/v) polyethylene glycol 400 (PEG). The protein pellet was dissolved in a minimal volume of 50 mM Hepes pH 7.4 containing 50 mM 2-mercaptoethanol, 5.0 mM MgCl$_2$ 300 mM KCl

Table 1. Purification of enzymes of de novo synthesis from lectin
          transformed human peripheral blood lymphocytes

          1.83×10^8 lectin-transformed human lymphocytes were used
for this enzyme preparation. The values in parentheses are for
amidophosphoribosyltransferase.

Step	Specific Activity	Total Protein	Recovery	Purification
	$nmol \ mg^{-1} \ h^{-1}$	mg	%initial activity	-fold
1. Cell lysate	12 (25)	120	100 (100)	1
2. High speed super- natant solution	30 (59)	53	109 (103)	2.5 (2.4)
3. 5-30% PEG fraction	73 (152)	20	102 (102)	6.0 (6.0)
4. Column eluate pool	151 (833)	3.3	34 (92)	12.5 (33.3)
5. Sucrose density gradient	163 (1731)	1.4	16 (83)	13.6 (70.4)

and 1% (w/v) PEG 2000. This solution was chromatographed on a
controlled pore glass (CPG 10-350) column 98 x 1.5 cm equilibrated
in the same buffer. The peak purine synthetic activity fractions
which cochromatographed with amidophosphoribosyltransferase activity
in a position corresponding to a molecular weight of 150,000 were
pooled and the enzyme precipitated with 30% (w/v) PEG. The protein
pellet was redissolved in the same buffer and subjected to sucrose
density gradient centrifugation on a 12 ml 5-20% (w/v) linear
sucrose gradient in an SW 41 Rotor for 22 h at 40,000 rpm.

        Enzyme assays were performed as described previously[1] except
that the concentration of glycine was reduced to 0.1 mM and both
ammonia (as 40 mM ammonium chloride) and glutamine were used in
parallel assays for providing N-9 of the purine ring. The enzyme
preparation, in the presence of the appropriate cofactors GTP and
NAD synthesized both adenine and guanine nucleotides. The products
of the purine synthetic reactions were analysed by HPLC as previously
described[1].

The loss of purine synthetic activity on glass column chromatography and on density gradient centrifugation with very little loss of amidophosphoribosyltransferase activity implied a loss of or separation from the activity of GAR synthetase, the second enzyme of the pathway. This could not however be established due to the insensitivity of the assay for this enzyme.

FGAR amidotransferase (EC 6.3.5.3) was rate limiting at all stages of purification for reasons that are still not known. Certainly FGAR has not been identified in experiments involving purine-labelling in cultured lymphocytes. Ultimately all of the glycine was incorporated into the end product IMP. No purine nucleotide degradation products, either nucleosides or bases were detected.

Both glutamine and ammonia were effective donors for the N-9 of the purine ring with $S_{0.5}$ values of 2.8 mM and 0.2 mM respectively and the same $S_{0.5}$ of 0.125 mM for PP-ribose-P. Glutamine was essential for providing N-3 of the purine ring. In the presence of saturating levels of both of these substrates, the rates of glycine incorporation were not additive. Azaserine (4.0 mM) and GMP (I_{50} = 2.0 mM) inhibited both the glutamine and ammonia dependent synthesis of phosphoribosylamine. The observations suggest that a single or closely related active sites are involved. Furthermore glutamine and ammonia-dependent glycine incorporation was identically distributed in the active fractions from both the chromatographic columns and sucrose density gradients.

Both AMP and GMP inhibited purine synthesis at the level of formation of phosphoribosylamine irrespective of whether glutamine or ammonia was the N-donor. Detailed analysis of the AMP studies however was difficult because of the rapid enzymatic deamination of AMP with this enzyme preparation in the absence of GTP.

GMP inhibited both adenine and guanine nucleotide synthesis (Table II compare assays (ii) and (v); (iii) and (vi); (vii) and (ix)). The branch point regulation i.e. inhibition of adenylosuccinate synthetase (EC 6.3.4.4) and IMP dehydrogenase (EC 1.2.1.13) may be more apparent than real due to the extremely low rate of IMP synthesis secondary to proximal pathway inhibition. GTP apparently also inhibited IMP dehydrogenase as the synthesis of guanine nucleotides was significantly decreased in the assay for simultaneous adenine and guanine nucleotide synthesis (Table II compare assays (iii) and (vii)).

In the presence of GTP, deamination of AMP was virtually abolished although almost total conversion to ATP occurred during the course of a 30 min assay. Nevertheless AMP reduced the overall incorporation of glycine, the incorporation into IMP and into

Table II   The effect of purine nucleotides on the de novo synthesis
           of adenine and guanine nucleotides

       The assays were performed as described[1] with the addition
of 2.0 mM NAD or 2.0 mM GTP for the synthesis of guanine and adenine
nucleotides respectively. AMP and GMP were added at a concentration
of 2.0 mM.

Additions to standard assay		Glycine Incorporation nmol mg^{-1} h^{-1} into				
		Total	IMP	Intermediates[a]	Nucleotides Ade	Gua
(i)	None	88.2	46.2	42.0	0	0
(ii)	GTP	77.0	18.2	44.8	14.0	0
(iii)	NAD	84.0	29.4	43.4	0	11.2
(iv)	GTP + AMP	57.4	14.0	37.8	5.6	0
(v)	GTP + GMP	32.2	11.2	21.0	0	0
(vi)	NAD + GMP	29.4	8.4	21.0	0	0
(vii)	GTP + NAD	72.8	15.4	40.6	15.4	1.4
(viii)	GTP + NAD + AMP	58.8.	16.8	35.0	7.0	0
(ix)	GTP + NAD + GMP	30.8	9.8	21.0	0	0

[a] The intermediates were predominantly FGAR with traces of GAR and
FGAR polyphosphate.

adenine nucleotides (Table II compare assays (ii) and (iv)). A
similar AMP effect was observed in the assay for simultaneous
adenine and guanine nucleotide synthesis (Table II assay (viii)).

      Sucrose density gradient centrifugation of enzyme for step 3
of the purification procedure, the 5-30% PEG fraction, showed that
both purine synthetic and amidophosphoribosyltransferase activity
sedimented in a position corresponding to a protein of molecular

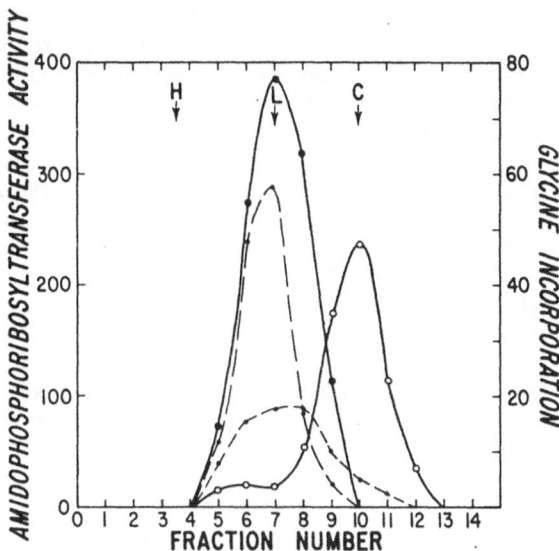

Fig. 1.   Sucrose density gradient profiles of the 5–30% PEG fraction
in the presence of (o) and absence of (●) 5.0 mM GMP in
the gradient medium.   Glycine incorporation into purines
(---) and amidophosphoribosyltransferase activity (—) are both
expressed as nmol h^{-1} fraction^{-1}.   In assays containing
GMP the PP-ribose-P concentration was increased to 5.0 mM
and $MgCl_2$ to 15 mM.   Protein markers (↓) were; hemoglobin
(H) M_r = 68,000, $S_{20,w}$ 4.11; lactate dehydrogenase (L),
M_r = 139,400, $S_{20,w}$ 7.35; catalase (C) M_r = 247,500, $S_{20,w}$
11.3

weight 150,000 with an $S_{20,w}$ of 7.25 (Fig.1).   While virtually all
of the amidophosphoribosyltransferase activity was recovered only
50% of the purine synthetic activity was detected in the gradient
fractions.   In the presence of 5.0 mM GMP in the gradient medium,
90% of the amidophosphoribosyltransferase activity was recovered in
the position of the catalase marker with an $S_{20,w}$ of 11.30 and an
approximate molecular weight of 250,000.   While all of the amidophosphoribosyltransferase activity was recovered, only 20% of the
purine synthetic activity was recovered, all in the lower molecular
weight region.   Synthesis, moreover, only proceeded as far as FGAR.
Analysis of the active fractions from the sucrose density gradients
on SDS-polyacrylamide gel electrophoresis demonstrated some thirty
distinct protein bands ranging in molecular weight from 30,000 to
150,000.

In summary, a de novo purine synthetic enzyme preparation was
isolated from human lymphocytes and the following pathway properties
established (i) ammonia could be as readily utilised as glutamine

for the synthesis of phosphoribosylamine but only glutamine could
provide N-3 of the purine ring (ii) in the presence of either GTP
or NAD, AMP or GMP were synthesized (iii) purine synthesis was
inhibited at the level of phosphoribosylamine synthesis by both
AMP and GMP irrespective of whether glutamine or ammonia was the
N-donor (iv) while the synthesis of AMP and GMP from IMP was self-
regulated, GTP also appeared to be an inhibitor of GMP synthesis
from IMP (v) amidophosphoribosyltransferase was isolated in a low
molecular weight form which was converted to a high molecular
weight form in the presence of GMP (vi) no evidence was obtained
for the existence of a   classical multienzyme complex.

ACKNOWLEDGEMENTS
     Thes studies were supported by grants from the National Health
and Medical Research Council of Australia and The University of
Sydney Cancer Research Fund.

REFERENCES
1.   P.B. Rowe, E. McCairns, G. Madsen, D. Sauer and H. Elliott,
         De novo purine synthesis in avian liver. J. Biol. Chem. 253:
         7711 (1978).
2.   P.B. Rowe, E. Tripp and G. Craig, Folate metabolism in lectin
         activated human peripheral blood lymphocytes in "Chemistry
         and Biology of Pteridines", R.L. Kisliuk and G.M. Brown, eds.
         Elsevier, New York (1979).

HUMAN PHOSPHORIBOSYLPYROPHOSPHATE (PRPP) SYNTHETASE REQUIREMENTS FOR SUBUNIT AGGREGATION

Michele J. Losman and Michael A. Becker

Department of Medicine
University of Chicago
Chicago, Illinois 60637 U.S.A.

The high energy sugar phosphate 5-phosphoribosyl 1-pyrophosphate (PRPP) is synthesized from MgATP and ribose-5-phosphate in a reaction catalyzed by PRPP synthetase (E.C. 2.7.6.1). In vitro studies of this enzyme,[1-6] as well as evaluations of PRPP production in intact cells, [7,8] indicate that PRPP synthetase activity is regulated in a complex fashion involving the interaction of substrates, activators, reaction products and diverse inhibitors. Enzyme activity is strictly dependent upon the presence of inorganic phosphate (Pi) and Mg^{2+} which serve both as activators and cofactors.[3] Among inhibitors of PRPP synthetase activity are purine, pyrimidine and pyridine nucleotide end-products of the pathways of PRPP utilization, the reaction products (PRPP and AMP), and 2,3-DPG.[3,6]

The quaternary structure of PRPP synthetase is similarly complex.[2,9,10] Human erythrocyte PRPP synthetase is composed of a single polypeptide subunit of molecular weight approximately 33,000 daltons.[9] The enzyme monomer is capable of reversible self-association under appropriate conditions of buffer composition and enzyme concentration to aggregated forms containing 2,4,8,16 and 32 subunits.[9] Monomeric and aggregated forms of PRPP synthetase have been isolated and characterized by means of sucrose density gradient ultracentrifugation and gel filtration chromatography,[2,9] and a relationship between quaternary structure and enzyme activity has been defined in vitro.[10] Under conditions in which association and dissociation of enzyme subunits is precluded, significant PRPP synthetase activity has been demonstrable only in the largest (16 and 32 subunit) aggregated forms.[10] Smaller forms of the enzyme are inactive or nearly inactive, with the apparent activity of these forms reflecting their reaggregation to the larger, active forms under the conditions of enzyme assay. The correlation between subunit aggregation and enzyme activation is very close, suggesting a potential mechanism for the regulation of enzyme activity.[10] Indeed, among the potent promotors of subunit association and dissociation are known activators and inhibitors of PRPP synthetase activity,[2][9] indicating a possible molecular basis for the action of these compounds on enzyme activity.[10] While subunit aggregation and disaggregation are enzyme concentration-dependent and effector-mediated processes[2][9] which may be important determinants of intracellular PRPP production, critical evidence that this is so remains to be obtained.

When purified enzyme preparations are subjected to sucrose gradient ultracentrifugation after incubation in a series of buffer solutions containing specific effectors of enzyme activity, peaks of enzyme activity corresponding to individual aggregated forms can be demonstrated.[10] The activity of PRPP synthetase in the 2 heaviest forms of the enzyme remains unaffected by dilution, while dilution diminishes the activity of smaller aggregated forms. The basis of this difference between aggregated forms lies in the inactivity of the smaller forms which show progressive reactivation during exposure to the components of the reaction mixture. The components of the reaction mixture essential for reactivation are MgATP, Mg^{2+} and Pi which in combination promote reactivation as a consequence of reaggregation.

As a prelude to studies comparing the capacity of normal and mutant superactive forms of PRPP synthetase to undergo in vitro subunit association and dissociation, we have examined the roles of MgATP, Mg^{2+} and Pi in promoting aggregation and activation of different forms of the normal enzyme. Previous studies have established that when purified PRPP synthetase is diluted to an enzyme concentration below about 1 μM in 1 mM Pi and 1 mM dithiothreitol (pH 7.4) and is incubated at 37°, the enzyme undergoes disaggregation with the majority of the enzyme assuming sedimentation characteristics corresponding to those of 8 and 4 subunit aggregates.[9] In preliminary studies, purified normal PRPP synthetase was diluted to 300-fold to an enzyme concentration of less than 0.5 μM in 1 mM Pi and 1 mM dithiothreitol and was incubated for 30 minutes at 37° with various combinations of MgATP, Mg^{2+} and phosphate prior to initiation of the reaction by completion of the reaction mixture with ribose-5-P and the previously omitted components. The enzyme assay was carried out at 26° in order to reduce the reaction velocity, and the course of PRPP production was followed by frequent sampling over 10 minutes of incubation.

When either MgATP or magnesium or both were absent during the preincubation period, delay in the expression of full enzyme activity was observed. Similarly when phosphate was omitted during preincubation (and replaced by 10 mM HEPES buffer), delay was apparent during the subsequent assay. When, however, preincubation was carried out in the presence of MgATP, Mg^{2+}, and as little as 1 mM Pi, no delay in the expression of full enzyme activity occurred.

Although all 3 effectors appear to be required to promote activa-ion during the preincubation period, the role of Pi is apparently a ermissive one, since on the basis of additional experiments,[9] the resence of Pi is required for dissociation as well as association of nzyme subunits. Moreover, Pi activation of PRPP synthetase cannot be xplained by an effect of this anion on subunit aggregation since the timulation of enzyme activity over the range from 1 to 32 mM Pi is naccompanied by a significant increase in subunit aggregation in re-ponse to MgATP and Mg^{2+}. Although Pi is a prime determinant of PRPP ynthetase activity,[7] the molecular basis of activation of the enzyme y Pi remains obscure.

Aggregates of PRPP synthetase containing 4,8,16 and 32 subunits ere isolated from sucrose density gradients run in 1 mM Pi and 1 mM lithiothreitol and studied in order to determine the concentrations of lgATP and Mg^{2+} required by each form for full reactivation (Figure 1). n the absence of Mg^{2+}, preincubation of the 4 subunit form with MgATP 20-1000 μM) failed to restore full enzyme activity. Similarly, pre-ncubation with Mg^{2+} (0.5-5 mM) in the absence of MgATP did not abol-sh the delay in enzyme activation. Only preincubation in the pres-nce of the combination of 5 mM Mg^{2+} and 1000 μM MgATP resulted in en-zyme activity proportional to the time of the subsequent incubation.

In contrast to the requirement for preincubation of the smaller ıggregates with MgATP and Mg^{2+} to restore full activity, the 16 and 32 subunit aggregates can express full activity without preincubation. Jevertheless, these two forms, previously both considered fully ac-tive,[10] could be distinguished by their responses to Mg^{2+} in the reac-tion mixture. Figure 2 summarizes the activation patterns of 4,16 and 32 subunit forms of PRPP synthetase. Activity is presented here as PRPP formed per minute of assay with 0.5 mM (left-hand panels) and 1.0 mM (right-hand panels) Mg^{2+} in the reaction mixture. Note that the 16 subunit form of PRPP synthetase, like the 4 subunit form, showed a de-lay in full activation when incubated with 0.5 mM Mg^{2+}. This brief delay occurred regardless of MgATP concentration but was abolished in the presence of 1 mM Mg^{2+}. The 32 subunit form was immediately active under all conditions tested, including Mg^{2+} concentrations as low as 150 μM. These findings suggest that the 32 subunit form of PRPP syn-thetase is the only fully active aggregate of the enzyme and that the 16 subunit forms is only partially active, probably requiring conver-sion to the larger form for full expression of activity.

We have recently examined the responsiveness to Mg^{2+} and MgATP of the subunit aggregates from a superactive form of PRPP synthetase with normal substrate and inhibitor binding properties but an increased maximal reaction velocity.[6] The distribution of forms of the mutant enzyme on sucrose gradient after the dilution and incubation proced-ures described above was indistinguishable from that of normal PRPP synthetase.[9] In several respects, however, the mutant enzyme behaved in an abnormal fashion. First, under conditions of suboptimal Mg^{2+}

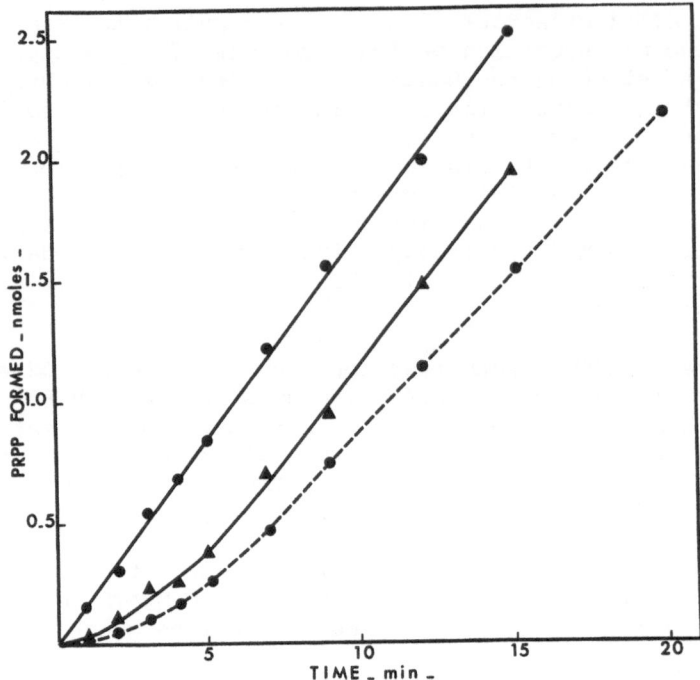

Figure 1. Effects of preincubation with Mg²⁺ and MgATP, singly or in combination, on the time course of PRPP synthesis catalyzed by the tetrameric form of PRPP synthetase. Highly purified PRPP synthetase was subjected to sucrose density gradient ultracentrifugation in 1 mM Pi, 1 mM dithiothreitol (pH 7.4) as previously described.[9] Enzyme sedimenting at 7.1s (corresponding to the 4 subunit aggregated form) was preincubated for 30 minutes at 37⁰ in 1 mM Pi, 1 mM dithiothreitol (pH 7.4) with the additions indicated below. PRPP synthesis was then measured at 26⁰ with final concentrations of 32 mM Pi, 1 mM MgATP, 5 mM Mg²⁺, and 500 µM ribose-5-P in the reaction mixture. Additions to the preincubation buffer were: ●----●, 5 mM Mg²⁺; ▲——▲, 1 mM MgATP; ●——●, 5 mM Mg²⁺ + 1 mM MgATP. Comparable results were obtained with enzyme sedimenting at 9.7s (8 subunit form).

and MgATP concentrations at 26⁰, mutant enzyme showed greater activity than normal enzyme even when activities of each (measured at 37⁰ under saturating substrate concentrations) were adjusted to be equal. Second, the tetrameric form of the mutant enzyme underwent full reactivation after preincubation with 100 µM MgATP and 1 mM Mg²⁺, in contrast to the requirement for 1000 µM MgATP and 5 mM Mg²⁺ for reactivation of the normal tetrameric form. Finally, full activation of the 16 subunit form of the mutant enzyme was found with 0.5 mM Mg²⁺ in the reaction mixture, in contrast to the delay noted under identical circumstances for the corresponding normal 16 subunit aggregate. Whether the apparent tendency of the mutant enzyme to reactivate more readily

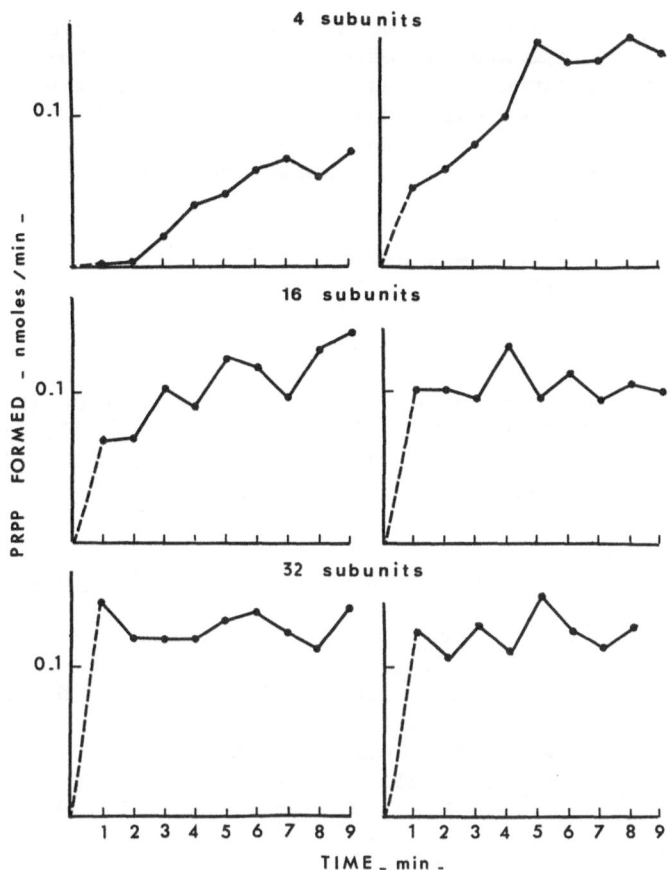

Figure 2. Activation of 4,16, and 32 subunit forms of PRPP synthetase during incubation in reaction mixtures containing 32 mM Pi, 100 µM MgATP, 200 µM ribose-5-P, and either 0.5 mM (left-hand panels) or 1.0 mM (right-hand panels) Mg^{2+}. Highly purified enzyme was subjected to sucrose density gradient ultracentrifugation in 1 mM Pi, 1 mM dith-iothreitol (pH 7.4) as previously described.[9] Forms of the enzyme sedimenting at 7.1s, 15.9s, and 22.1s (corresponding to 4,16, and 32 subunit aggregates, respectively) were equilibrated at 26° and PRPP synthesis over time was measured in the above reaction mixtures. Data are expressed as nmoles of PRPP generated per minute of incubation.

than the normal enzyme in any way contributes to in vivo superactivity remains to be determined.

In summary, our studies confirm previous suggestions that MgATP and Mg^{2+} directly promote subunit association in the presence of Pi which exerts a permissive effect on both aggregation and disaggregation.[2,9] In addition, we have found that subunit aggregates differ in their requirements for effectors to promote full activation and that only the 32 subunit aggregate of PRPP synthetase expresses full enzyme activity under all conditions examined. Finally, we have found preliminary evidence of an increased responsiveness to promotors of subunit aggregation in a mutant superactive form of PRPP synthetase.

ACKNOWLEDGEMENT: This work was supported in part by Grant AM-28554 from the National Institutes of Health

REFERENCES

1.   Wong, P.C.L., and Murray, A.W., Biochemistry 8:1608 (1969).
2.   Fox, I.H., and Kelley, W.N., J. Biol. Chem. 246:5739 (1971).
3.   Fox, I.H., and Kelley, W.N., J. Biol. Chem. 247:2166 (1972).
4.   Roth, D.G., Shelton, E., and Deuel, T.F., J. Biol. Chem. 249:291 (1974).
5.   Roth, D.G., and Deuel, T.F., J. Biol. Chem. 249:297 (1974).
6.   Becker, M.A., Kostel, P.J., and Meyer, L.J., J. Biol. Chem. 250: 6822 (1975).
7.   Hershko, A., Razin, A., and Mager, J., Biochim. Biophys. Acta 184:64 (1969).
8.   Bagnara, A.S., Letter, A.A., and Henderson, J.F., Biochim. Biophys. Acta 374:259 (1974).
9.   Becker, M.A., Meyer, L.J., Huisman, W.H., Lazar, C. and Adams, W.B., J. Biol. Chem. 252:3911 (1977).
10. Meyer, L.J. and Becker, M.A., J. Biol. Chem. 252:3919 (1977).

THE HORMONAL REGULATION OF PURINE BIOSYNTHESIS

M. Pizzichini°, A. Di Stefano°, G. Bruni°°, R. Leoncini°
and E. Marinello°

° Dept. of Biological Chemistry, Univ., Siena (Italy)
°° Dept. of Pharmacology, Univ., Siena (Italy)

INTRODUCTION

The hormonal regulation of purine biosynthesis is not well known and there are more indications taken from clinical data then based on experimental ground.

Feigelson and Feigelson,[1], have demonstrated that adenine biosynthesis is enhanced in rat liver, following cortisone administration in adrenalectomized rats.

The present researches were carried out in order to ascertain the effect of glucocorticoid hormones on all purine based biosynthesis in different rat organs.

MATERIALS AND METHODS

Albino male Wistar rats, weighing 200-250 g remained intact or were adrenalectomized through intraperitoneal way, under ether anesthesia.

Adrenalectomized animals were allowed water and 1% NaCl ad libitum and were kept five days at 25°C before any treatment.

The incorporation in vivo of formate-C^{14} into acid soluble adenine and the other purine bases (guanine and hypoxanthine) was taken as an index of the behaviour of purine biosynthesis in different organs (liver, kidney, heart, spleen).

Na-formate-C^{14} (54,8 mC/mmole) was administered through intraperitoneal way. The animals were killed one hour after formate ad-

ministration (10 μC/100 g b.w.), and six hours after cortisone acetate administration (5 mg/100 g b.w.), prepared as 1% suspension.
The organs were rapidly removed, chilled, washed with saline, homogenized in cold 5% trichloroacetic acid (TCA).

The nucleotide bases were obtained after acid hydrolisis, AgNO$_3$
precipitation, final elution with 1 N HCl,[2],.

The final mixtures of bases were separated on Dowex resin
(AG 50W-X8, 200-400 mesh, hydrogen form), eluting the bases with a
scalar gradient of HCl (1 N, 2 N, 3 N, 4 N, 6 N),[3],.

The single peaks thus obtained were also submitted to thin-
layer chromatography,[4], and eluted with HCl 0.1 N. The absorption
spectra of various bases were controlled with a G.25 Beckman recording spectrophotometer.

The specific radioactivity of the various bases, after purification, was checked on a Nuclear Chicago Delta Scintillation Counter.

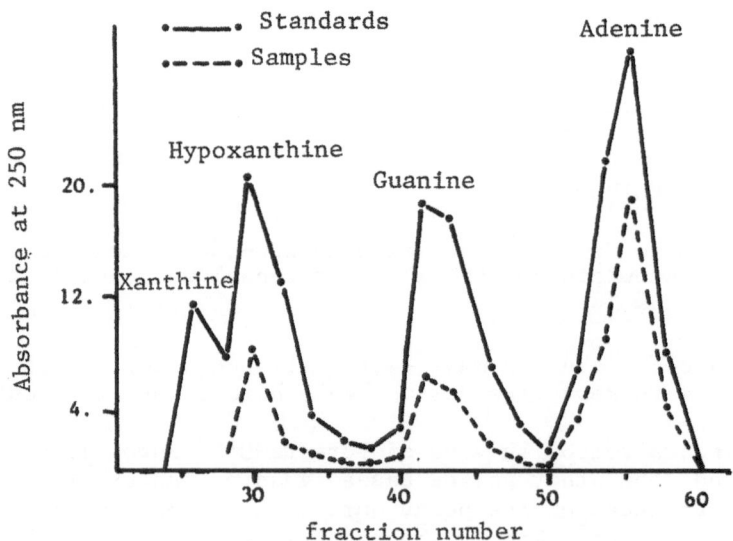

Figure 1 - Chromatographic Separation of Adenine, Hypoxanthine, Xan-
       thine and guanine on Dowex Resin (AG 50W-X8, 200-400mesh
       hydrogen form).

RESULTS

The separation of standard bases on Dowex resin, through scalar gradient of HCl is reported in Figure 1 .

The picture of the bases obtained from a liver thrichloroacetic extract is comparatively reported .

The specific activity of adenine, hypoxanthine, guanine, in the different organs, in normal and adrenalectomized rats, after formate-^{14}C, is shown in Table 1 .

The effect of cortisone on formate-^{14}C incorporation into acid soluble purines is shown in Table 2 .

DISCUSSION

Our results clearly indicate:
1) purine biosynthesis is decreased in all organs of adrenalectomized rats (liver, heart, kidney, spleen);
2) cortisone administration enhances purine biosynthesis only in the liver of adrenalectomized rats. No restoration of original activity is observed in kidney, heart and spleen;
3) the fact that purine biosynthesis is diminished in all organs considered, after adrenalectomy, lets us suppose that some adrenal hormones regulate this metabolic pathway;
4) cortisone enhances purine biosynthesis, after adrenalectomy, only at hepatic level, not in the other organs: it is therefore evi-

Table 1 - Formate-^{14}C Incorporation into acid soluble purines of different organs of normal (N) and adrenalectomized rat (A).

ORGANS		ADENINE dpm/μmole	Change %	GUANINE dpm/μmole	Change %	HYPOXANTHINE dpm/μmole	Change %
LIVER	N	1577		1955		2843	
	A	735	− 53	1421	− 27	1451	− 49
KIDNEY	N	13836		5492		12605	
	A	6207	− 55	3271	− 40	7939	− 37
HEART	N	3735		° ° °		4041	
	A	3344	− 11	° ° °		3421	− 15
SPLEEN	N	36264		17165		14974	
	A	14608	− 60	10329	− 40	7999	− 47

° ° ° The guanine levels were too low and the specific activity was not calculated

Table 2 – Effect of cortisone on formate-^{14}C incorporation into acid
soluble purines of different organs of adrenalectomized rat.

ORGANS	ADENINE dpm/μmole	Change %	GUANINE dpm/μmole	Change %	HYPOXANTHINE dpm/μmole	Change %
LIVER						
Control	735		1421		1451	
Treated	5003	+580	9457	+565	6219	+328
KIDNEY						
Control	6207		3271		7939	
Treated	4029	− 35	3405	+ 4	5853	− 26
HEART						
Control	3344		∘∘∘		3421	
Treated	3583	+ 7	∘∘∘		3274	− 4

∘∘∘ The guanine levels were too low and the specific activity was
not calculated

dent that, in such organs, purine biosynthesis is regulated by adrenal hormones (hydrocortisone, corticosterone), different from cortisone or that its incorporation by cortisone occurs at times and
doses different from the liver;

5) in the kidney, cortisone depress purine biosynthesis; in the
spleen, results till now obtained by us,[3], show a variable behavior,
but never an increase of purine biosynthesis after cortisone treatment.

REFERENCES

1. P.Feigelson and M.Feigelson –Studies on the mechanism of regulation by cortisone of the metabolism of liver purine and ribonucleic
acid, J. Biol. Chem. 238: 1073 (1963).
2. G.Schimdt –Chemical and enzymatic methods for the identification
and structural elucidation of nucleic acids and nucleotides, in:
"Methods in Enzymology" vol.III. S.P.Colowick and N.O.Kaplan, ed.,
Academic Press inc. New York (1957).
3. A.DiStefano, M.Pizzichini, G.Bruni and E.Marinello –Effetto del
cortisone sull'incorporazione di Formiato^{14}C nell'adenina di vari
organi di ratto, 27°Congresso Naz. SIB, Parma 14-16 ottobre (1981).
4. O.Sperling, G.Eilam, S.Persky-Brosh and A.DeVries – Simpler method for the determination od 5-phosphoribosil-1-pyrophosphate in
red blood cells, J. Lab. Clin. Med. 79: 1021 (1972).

THE HORMONAL REGULATION OF PURINE BIOSYNTHESIS:

BASAL LEVELS OF DIFFERENT HORMONES IN PRIMARY GOUT

F. Aleo[1], S. Comandé[2], R. Marcolongo[1], E. Marinello[1], and U. Valentini[3]

[1]Dept. Biochemistry and Rheumatology, University of Siena and Brescia (Italy)
[2,3]Dept. of Physics and Chemistry, EULO, Brescia, Italy

INTRODUCTION

No complete analysis of the hormonal situation of gouty patients is now available. In the present experiments, the basal levels of follicle-stimulating hormone (FSH), luteinizing hormone (LH), 17-β-estradiol, and testosterone have been determined in normal and gouty men of various ages.

MATERIAL AND METHODS

40 normal subjects (between 46 and 71 years), and 36 gouty men (between 22 and 78 years) were considered. They had received no hypouricemic drug for at least one month before the determination of hormones, and were kept on a free mixed diet (2500 calories/die) one week before. The hormones were assayed by radioimmunological methods (RIA).[1,2,3,4,5,6].

Statistical analysis of results consisted of the following:

(1) A study of the relation between the hormone basal levels and age by the method of least squares, and by the correlation coefficient r.
(2) A "comparison between groups," performed by the U test of Wilcoxon-Mann-Whitney.[7] Only P values lower than 1% were considered statistically significant.

RESULTS AND DISCUSSION

Results concerning normal subjects and gouty patients are reported in Tables 1 and 2, respectively. The statistical correlation (r values) between hormone basal levels and age is reported in Table 3.

Our observations lead to the following conclusions:

(1) No correlation exists between all the hormones taken into consideration and the age of the patients.
(2) The ratio testosterone/17-β-estradiol is significantly higher in gouty patients than in normal subjects: this might indicate a lower transformation testosterone → 17-β-estradiol in gout.
(3) FSH, LH, and 17-β-estradiol are significantly decreased in gouty men.

In our precedent case material,[8] the values of FSH, LH, 17-β-estradiol, and testosterone were also lower in gouty patients, but the difference was not significant.

The discrepancy may be due to the fact that, in the present case material, there is a more uniform age distribution in both groups. The great variability of normal values must also be considered.

Our present observations seem to confirm the fact that gout may be linked to an endocrine disorder. Clinical data (the

Table 1. Basal levels of different hormones in normal subjects of different ages. Results are the mean value ± SD. The number of cases is given in parentheses.

Age	FSH (mU.I./ml)	LH (mU.I./ml)	Testosterone (ng/ml)	17-β-Estradiol (pg/ml)
40–50	12.0±7.2 (3)	9.3±1.5 (3)	5.1±2.0 (3)	43.0±15.5 (3)
50–60	12.5±5.6 (16)	10.4±4.2 (16)	3.9±1.9 (14)	48.0±19.8 (14)
60–70	10.4±5.5 (20)	11.2±6.0 (20)	3.6±1.0 (16)	58.0±24.0 (16)
>70	22.0 (1)	12.0 (1)	———	———

Table 2. Basal levels of different hormones in gouty patients of different ages. Results are the mean value ± SD. The number of cases is given in parentheses.

Age	FSH (mU.I./ml)	LH (mU.I./ml)	Testosterone (ng/ml)	17-β-Estradiol (pg/ml)
≤ 40	8.0±7.4 (16)	8.0±3.3 (16)	5.0±1.8 (13)	24.8±9.6 (16)
40-50	9.4±6.1 (12)	7.4±2.8 (12)	4.1±1.5 (9)	26.7±9.1 (12)
50-60	5.0 (1)	8.0 (1)	3.0 (1)	20.0 (1)
60-70	6.0±4.9 (4)	8.2±3.3 (4)	3.6±1.1 (4)	36.5±4.2 (4)
> 70	5.0±2.0 (3)	6.6±2.3 (3)	4.3±1.6 (3)	34.6±16.8 (3)

Table 3. Statistical correlation of hormone basal values with age. The hormone values (HV) are related to the age by the following formula: $HV = a_0 + a_1 \cdot age$. r is the correlation coefficient, and only values of $r > 0.5$ were considered significant.

Hormone	Gouty patients			Normal subjects		
	a_0	a_1	r	a_0	a_1	r
FSH	11.784	−0.085	−0.185	17.021	−0.088	−0.085
LH	8.731	0.022	−0.104	−41.020	0.909	0.312
Testosterone	6.359	−0.044	−0.391	9.640	−0.096	−0.320
17-β-Estradiol	25.414	0.106	0.082	6.411	0.765	0.190

difference in incidence of gout between the sexes) strongly support
this hypothesis and are more consistent than laboratory data,
which are few and contradictory.

Our results give the first screening of different hormones in
gouty patients: study of a larger number of cases will lead to more
definite conclusions on this promising subject.

REFERENCES

1. Rosa, U. I metodi radiochimici nella pratica clinica. Minerva
        Medica, 61, 59-60, 3255 (1970).
2. Yallow, R.S., and Berson, S.A. Immunoassay of endogenous
        plasma insulin in man. J. Clin. Invest. 39, 1157 (1960).
3. Midgley, A.R., Jr. Radioimmunoassay for human follicle stimula-
        ting hormones. J. Clin. Endocr. and Met. 27, 295 (1967).
4. Midgley, A.R., Jr. Radioimmunoassay: a method for human
        chorionic gonadotropin and human luteinizing hormone,
        Endocr. 79, 10 (1966).
5. Doerr, P. Hapten radioimmunoassay of plasma estradiol. Acta
        Endocrinologica, 72, 330-334 (1973).
6. Forest, M.G., Cathiard, A.M., and Bertrand, J.A. Total and
        unbound testosterone levels in the newborn and in normal
        and hypogonadal children: use of a sensitive radio-
        immunoassay for testosterone, J. Clin. Endocr. and Met.
        36, 1132 (1973).
7. Wilcoxon, F., Mann, H.B., and Witney, D.R. Individual com-
        parison by ranking methods. Biometrics Bull. 1, 80
        (1945).
8. Valentini, U., Riario-Sforza, G., Marcolongo, R., and Marinello,
        E. Hormonal aspects of gouty patients. Purine Metabolism
        in Man III, Plenum Press, New York (1979), p. 65.

REGULATION OF DE NOVO PURINE BIOSYNTHESIS IN CHINESE HAMSTER CELLS

Milton W. Taylor, Suman Olivelle, Kathy Coy,
Roy A. Levine, and Howard Hershey

Department of Biology
Indiana University
Bloomington, Indiana 47402

INTRODUCTION

It has been proposed that in mammalian cells (1), as in the case of E. coli (2), de novo purine biosynthesis is controlled by both feedback inhibition by the end products of the pathway and by repression. In procaryotes, repression appears to involve an interaction between a specific nucleotide (ATP or GTP) and a repressor protein binding to the operator region of the early purine genes (2). We have recently isolated a series of PurR⁻ mutants in E. coli, in which repression of the de novo pathway is abolished (3). However such mutants are still subject to feedback inhibition.

We thus decided to re-examine the possibility of repression of purine biosynthesis in mammalian cells. Martin and Owen (1) proposed a repression/derepression mechanism in rat hepatoma cells in culture, on finding that de novo purine biosynthesis was inhibited in the presence of adenine and that there was no restoration of de novo purine biosynthesis when adenine was removed in the presence of actinomycin D or cycloheximide. This suggested that the de novo pathway was controlled by a repressor molecule and that new mRNA synthesis was required to initiate de novo purine biosynthesis. However, Martin & Owens (1) did not find repressed enzyme levels in the early enzymes of the de novo purine biosynthetic pathway in cells grown in high adenine concentrations. In agreement with Martin and Owens (1), we find that none of the early enzymes of the de novo purine biosynthetic pathway are repressed in Chinese hamster fibroblast grown in high adenine concentration. Thus we have examined the effect of

441

metabolic inhibitors on nucleotide pools since increases in pools
may act, via feedback inhibition, to reduce PRPP synthetase and
possibly amidophosphoribosyltransferase (4, 5). This is supported
by findings of decreased levels of PRPP in cells treated with
inhibitors.

MATERIALS AND METHODS

All experiments were done with V79, a Chinese hamster lung
cell, and CHO, a Chinese hamster ovary cell. De novo purine
biosynthesis was measured by a modification of the method of
Hershfield and Segmiller (6). PRPP synthetase, PRPP levels and
amidophosphoribosyltransferase were measured as described (7).
High performance liquid chromatography of nucleotides was
performed on two µBondapak C_{18} columns in tandem (8).

RESULTS

Effect of adenine on de novo purine biosynthesis: De novo
purine biosynthesis in CHO and V79 cells is very sensitive to
regulation by exogenous adenine. Data presented in Table 1
illustrate that 10 µM adenine is sufficient to inhibit de novo
purine biosynthesis by more than 60% as measured by the
incorporation of [^{14}C] formate in total purines in a 2-h labeling
period. The response of Chinese hamster cells to inhibition by
adenine is approximately 5-fold more sensitive than that of E.
coli. However, growth of Chinese hamster fibroblasts is not
affected until concentratons of 500 µM adenine are in the medium.
Under these conditions, there is a slight inhibitory effect.

Effect of growth in adenine on PRPP synthetase and
amidophosphoribosyltransferase: If adenine inhibited de novo
purine biosynthesis by a repression type mechanism, the specific

Table 1

% Inhibition of de novo purine biosynthesis in Chinese hamster
cells by adenine.

Adenine levels	CHO cells	V79 cells
µM	%	%
1	11	0
5	7	11
10	80	60
50	89	88
100	89	88

activities of the first enzymes of the pathway should be reduced following growth in adenine. However, growing cells in adenine concentrations (50 μM) which strongly inhibit de novo purine synthesis for 10 h or even 7 days had no effect on the in vitro activiies of either PRPP synthetase or amidophosphoribosyl-transferase.

Effect of actinomyin D, cycloheximide, and azacytidine on de novo purine biosynthesis: Table 2 illustrates that all three inhibitors, actinomycin D, cycloheximide, and azacytidine inhibit de novo purine biosynthesis as measured by the incorporation of C^{14}-formate into total purines. Labelling was for 2 hours, and the results are the average of three experiments. Controls had between 25,000 and 30,000 cpm/10^6 cells. Actinomycin D had no effect on in vitro assayed PRPP amidotransferase or PRPP synthetase activities for cell pretreated for 2 hours with the drug.

Acid-soluble pools: Semiconfluent monolayers of V79 and CHO were treated with actinomycin D (5 μg/ml) or cycloheximide (10 μg/ml) for 60 and 120 min. The cells were trypsinized and acid-soluble nucleotides removed by extracting the washed cells with 0.3 M formic acid. Under these conditions there is no degradation of nucleotide triphosphate pools. Using ^{3}H-ATP as a control, there is 90% recovery of the triphosphate from the column. Nucleotide pool levels were measured in an equal number of cells by analysis on a C_{18} μBondapak reversed phase column. In the actinomycin D-treated cells, there was an increase in the guanosine triphosphate pool, with a slightly smaller increase in the ADP pool (Fig. 1 (a), (b)). In the presence of cycloheximide, as with actinomycin D, there was a change in nucleotide pool levels, with the largest increase being in ATP concentration. These results were confirmed by measuring the relative increases

Table 2

Inhibition of total purine biosynthesis by metabolic inhibitors

Treatment	% inhibition	
	V79	CHO
Control	0	0
Actinomycin D	79	82
Cycloheximide	60	67
Azacytidine	53	67

in radioactive nucleotides in prelabeled cells following 2 h
incubation in actinomycin D. As shown in Table 3 there was a 2-4
fold increase or turnover into ATP and GTP pools following
treatment with actinomycin, azacytidine, or cycloheximide (Table
3).

Effect of inhibitors on PRPP levels: In order to confirm
that the inhibition of de novo purine biosynthesis was by a
similar mechanism with all three inhibitors, the levels of PRPP
were measured in cells treated for 1 or 2 h with actinomycin D,
cycloheximide and azacytidine. Data from these experiments are
presented in Table 4. All three inhibitors lower the level of
PRPP in cells to the same range. Thus the decrease in de novo
purine biosynthesis in those cells may be a consequence of lowered
PRPP concentrations due to feedback inhibition resulting from the
increased nucleotide levels.

Table 3

Amount of C^{14} in nucleotides following incubation for 2 hrs in
actinomycin D, cycloheximide and azacytidine.

	Actinomycin D		Azacytidine		Cycloheximide (cpm)	
	Control	Treated	Control	Treated		Control	Treated
GDP	290	332	589	1092	GDP	570	980
GTP	570	1167	1583	4335	GTP	760	2420
ADP	660	802	2114	2374	ADP	2090	3810
ATP	3,310	4,688	11,990	20,822	ATP	7700	20,360

Table 4

Effect of inhibitors on PRPP concentrations

Experiment	N mol PRPP/mg protein	% Decrease
Control	0.113	---
Actinomycin D--60 min	0.082	28
Actinomycin D--120 min	0.048	58
Cycloheximide--60 min	0.058	49
Cycloheximide--120 min	0.067	41
Azacytidine--60 min	0.080	30
Azacytidine--120 min	0.024	79

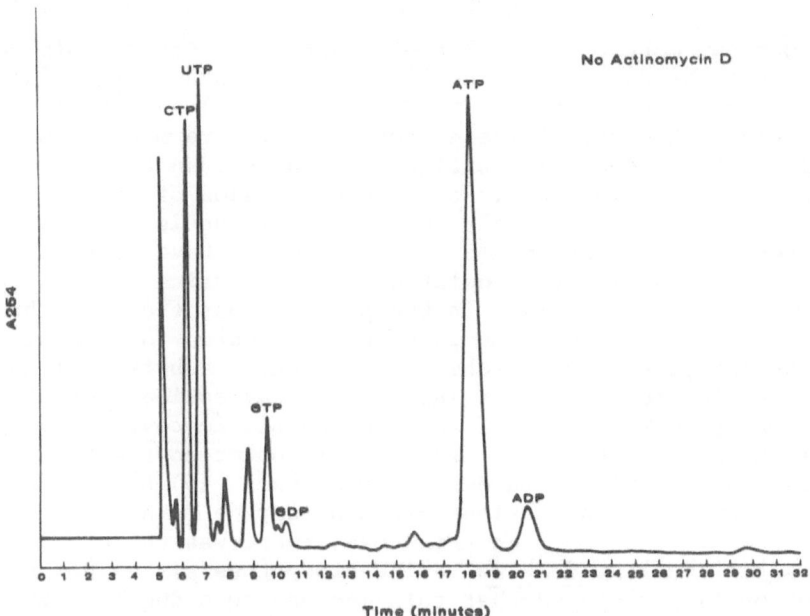

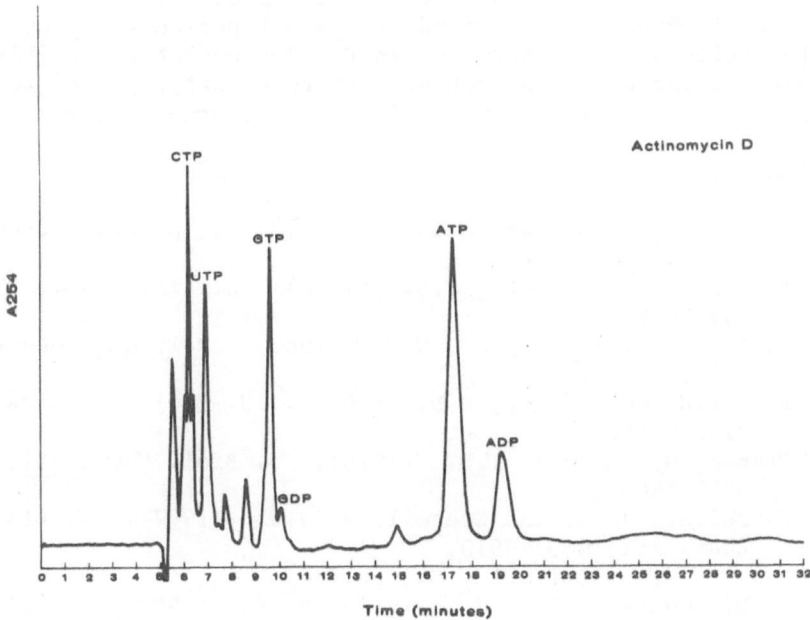

Fig. 1. HPLC of nucleotide pools of CHO cells.

DISCUSSION

Our own work (3) and that of others (2) with E. coli have shown that the de novo purine biosynthetic pathway is regulated by both a repressor molecule (pur R gene product) and by feedback inhibition. However, Chinese hamster cells are much more sensitive to feedback inhibition by adenine than E. coli and, unlike the situation in E. coli, no repression of PRPP amidotransferase or formyglycinamide biosynthesis could be detected. If repression did occur, it would have to be by a mechanism not normally associated with the purine biosynthetic pathways or at a site late in the purine biosynthetic pathway. Moreover, the nucleotide pools of cells treated for 2 h with with actinomycin D or cycloheximide showed a substantial increase in nucleotide levels. This increase in nucleotide concentration is probably sufficient in itself to inhibit de novo purine biosynthesis by feedback inhibition without recourse to a repression mechanism. Snyder and Henderson (10) have also reported an effect of actinomycin D on purine metabolism in Ehrlich ascites cells. In this case, there was no large effect (11% inhibition) on de novo purine biosynthesis. Snyder and Henderson (10) proposed that this decrease was due to a 29% reduction in PRPP levels as a result of increased (1.3-fold increase in ATP and 2.8-fold increase in GTP) nucleotide pools. These observations are consistent with our data in which a 58% decrease in PRPP level is found over a 2-h period in Chinese hamster cells grown in actinomycin D. The extent of inhibition in Chinese hamster cells is much greater than that reported for Ehrlich ascites cells and may reflect a difference between cells.

REFERENCES

1. Martin, D.W., Jr. and Owen N.T. (1972). J. Biol. Chem. 247, 5477-5485.
2. Koduri, R.K. and Gots, J.S. (1980). J. Biol. Chem. 255, 9594-9598.
3. Levine, R.A. and Taylor, M.W. (1981). Mol. Gen. Genet. 181, 313-320.
4. Fox, I.H. and Kelley, W.N. (1971). J. Biol. Chem. 246, 5739-5748.
5. Planet, G. and Fox, I.H. (1976). J. Biol. Chem. 251, 5839-5844.
6. Hershfield, M.S. and Seegmiller, J.E. (1977). J. Biol. Chem. 252, 6002-6010.
7. Taylor, M.W., Olivelle, S., Levine, R.A., Coy, K., Hershey, H., Gupta, K.C. and Zawistowich, L. (1982). J. Biol. Chem. 257, 377-380.

8. Taylor, M.W., Hershey, H.V., Levine, R.A., Coy, K., and Olivelle, S. (1981). J. Chromatogr. 219, 133–139.
9. Taylor, M.W., Gupta, K.C., and Zawistowich, L. (1980). In Purine metabolism in man (Rapado, A.; Watts, R.W.E.; and DeBruyn, C.H.M.M., eds.) Vol. 3B, Plenum Press, N.Y.
10. Synder, F.F. and Henderson, J.F. (1974). Can. J. Biochem. 52, 263–267.

REGULATORY STUDIES ON PURINE BIOSYNTHESIS

IN THE CHICK

Mary M. Welch and Frederick B. Rudolph

Department of Biochemistry
Rice University
Houston, TX 77001 U.S.A.

Uricotelic animals use synthesis of the purine ring system as the pathway for excretion of excess nitrogen. This results in a high level of de novo purine synthesis (1). When birds are subjected to stress situations that lead to higher nitrogen excretion, the rate of purine synthesis is increased (2,3) and changes in levels of adenylosuccinate lyase, xanthine dehydrogenase and amidophosphoribosyltransferase occur (2,4). Livers from chickens treated with β-estradiol can provide a model system for a rapidly growing, normal cell. Treatment with this hormone causes a rapid increase in liver weight and in protein, RNA and DNA content in preparation for the production of egg proteins (5). Weber and co-workers (6-8) have shown that the level of certain purine biosynthetic enzymes is increased in regenerating rat liver, another system that is used as a model for normal, rapid growth. This is also true in certain rat liver hepatomas (6-8). The chick liver can therefore be used as a model for study of regulation of purine biosynthesis under conditions where requirements for nucleotides vary.

MATERIALS AND METHODS

White leghorn roosters of two, three and four weeks of age were used as experimental animals. β-Estradiol was dissolved in propylene glycol at a concentration of 25 mg/ml and 0.2 ml/100 gm body weight was injected intramuscularly into the upper thigh of each chick. Control chicks were matched by weight to the experimental animals and injected with an equivalent volume of propylene glycol (5). The animals were sacrificed twenty-four and 48 hours after treatment.

449

The amount of 5-phospho-α-D-ribosyl pyrophosphate (PRPP) in chicken liver was measured using the procedure of LaLanne and Henderson (9). The soluble purine base levels were determined using freeze-clamped liver that was extracted with perchloric acid, centrifuged and then boiled to hydrolyze the soluble nucleotides to the free bases. The bases were separated and quantitated on a Partisil pressure liquid chromatography column (Whatman).

Livers removed from chickens injected intraperitoneally with radioactive formate 30 minutes prior to sacrifice were homogenized in perchloric acid and boiled for one hour to hydrolyze the RNA and DNA and solubilize the nucleotides formed, converting them to bases.

Adenylosuccinate synthetase was assayed using endogenous adenylosuccinate lyase and adenylate kinase to trap the adenylosuccinate synthesized in the form of adenine di- and triphosphate nucleotides. Adenylosuccinate lyase activity in the chicken liver was measured by spectral analysis using the conditions described by Carter and Cohen (10). The AMP deaminase assay used was modeled after the procedure of Schultz and Lowenstein (11). The activities of amidophosphoribosyltransferase and GMP synthetase were measured by monitoring glutamate formation using DEAE-cellulose disks. The assay for IMP dehydrogenase activity was as described by Anderson and Sartorelli (13). The activity of xanthine dehydrogenase was measured according to Krenitsky et al. (14). GMP reductase activity was not detected.

RESULTS AND DISCUSSION

The activities of seven enzymes described above involved in purine biosynthesis, interconversion and catabolism were measured in liver extracts obtained from normal and β-estradiol treated chickens using assays specially developed for the chicken system. The enzyme levels present in the chicken homogenates do not change after hormonal treatment. This is in direct contrast to the regenerating rat liver system in which amidophosphoribosyltransferase activity doubles and IMP dehydrogenase activity is elevated five-fold on a similar time scale (6-8). The observed activity levels for amidotransferase, adenylosuccinate lyase and xanthine dehydrogenase indicate the high capacity of chick liver for IMP biosynthesis relative to the branch pathways to AMP and GMP.

The overall rate of purine biosynthesis was determined by the incorporation of formate into cellular purines (both acid soluble and those incorporated into RNA and DNA). The rate of de novo purine synthesis (Table I) is increased at twelve hours after hormone treatment and remains at a high level for at least 48 hours

(p > 0.002). At 12 hours more of the newly synthesized IMP is being channelled into the GMP branch of the interconversion pathway,
as shown by the adenine:guanine ratios. By 48 hours, the ratio of
GMP and AMP synthesis has returned almost to normal, even though
the overall rate of synthesis remains high. Thus, it appears that
the rapid growth of the liver induced by the treatment with β-estradiol is supported by an increase in the rate of de novo purine
synthesis. The branches of the purine interconversion pathways
are more equally expressed during the initial 24 hours of this
increase in de novo synthesis than under normal growth conditions.

PRPP, which is a substrate of the first committed step in
purine biosynthesis, has often been implicated in the regulation
of the rate of de novo synthesis (15,16). Amidophosphoribosyltransferase from pigeon liver shows sigmoidal kinetics at PRPP
concentrations below its K_m, so that small changes in cellular
levels of PRPP could greatly influence the rate of purine biosynthesis (16,17). Determination of the level of PRPP in livers of
chickens treated with β-estradiol shows that after 24 hours the
amount of PRPP in the tissues has doubled (Table II).

Table I. Measurement of De Novo Purine Biosynthesis in Chick
         Livers from ^{14}C-Formate after β-Estradiol Administration

Time after β-Estradiol Administration	Total ^{14}C Formate Incorporated (10^3 dpm/g wet weight)	Ratio of Adenine/Guanine
12 hours (6)	77.5 ± 19.5	2.6
24 hours (6)	75.0 ± 26.5	3.7
48 hours (6)	90.0 ± 22.0	4.0
Control (7)	32.0 ± 15.5	4.9

Table II. Levels of PRPP in Livers of β-Estradiol-Treated Chicks

Time After β-Estradiol Injection	nmoles PRPP/mg protein
24 hours (6)	8.3 ± 2.8
48 hours (6)	5.0 ± 2.5
Control (10)	4.2 ± 1.5

The results of this study suggest that the major response of purine metabolism in chick liver to rapid growth is to increase the level of PRPP thereby increasing the flux through the de novo pathway. Changes in regulation at the IMP branch occur as GMP synthesis increases. Neither enzyme levels nor soluble purine pool sizes change. Regulation of chick liver purine metabolism in a rapid growth situation is different than that observed in ureotelic species.

ACKNOWLEDGEMENT

This work was supported by Grant #14030 awarded by the National Cancer Institute and Grant #C-582 awarded by the Robert A. Welch Foundation.

REFERENCES

1. Lipstein, B., Boer, P., and Sperling, D. (1978) Biochim. Biophys. Acta 521, 45–54.
2. Katunuma, N., Katsunuma, T. Towatari, T., and Tomino, I. (1973) Regulatory Mechanisms of Glutamine Metabolism in Enzymes of Glutamine Metabolism (Prusiner, S. and Stadtman, E.R., eds.) Academic Press, Inc., New York and London, 227–258.
3. Burns, R.A. and Buttery (1981) Arch. Biochem. Biophys. 208, 468–476.
4. Brand, L.M. and Lowenstein, J.M. (1978) J. Biol. Chem. 253, 6872–6878.
5. Jost, J., Keller, R., and Kierks-Ventling, C. (1973) J. Biol. Chem. 248, 5262–5266.
6. Jackson, R.C., Morris, H.P., and Weber, G. (1975) Biochem. Biophys. Res. Comm. 66, 526–532.
7. Jackson, R.C., Weber, G., and Morris, H.P. (1973) Nature 256, 331–333.
8. Katunuma, N. and Weber, G. (1974) FEBS Letters, 49, 687–703.
9. LaLanne, M. and Henderson, J.F. (1974) Anal. Bioc. 62, 121–123.
10. Carter, C.E. and Cohen, L.H. (1956) J. Biol. Chem. 222, 17–30.
11. Schultz, V. and Lowenstein, J.M. (1976) J. Biol. Chem. 251, 485–492.
12. Martin, D.W., Jr. (1972) Anal. Bioc. 46, 239–243.
13. Anderson, J.H. and Sartorelli, A.C. (1968) J. Biol. Chem. 243, 4762–4768.
14. Krenitsky, T.A., Tuttle, J.V., Cattar, E.L., Jr., and Wang, P. (1974) Comp. Biochem. Physiol. 49B, 676–703.
15. Henderson, J.F. (1972) Regulation of Purine Biosynthesis, American Chemical Society Monograph 170, Washington, D.C.
16. Wyngaarden, J.B. (1976) Adv. Enz. Reg. 14, 25–42.
17. Rowe, D.B. and Wyngaarden, J.B. (1968) J. Biol. Chem. 243, 6373–6383.

THE CHRONOLOGICALLY SYNCHRONOUS ELEVATION OF PHOSPHORIBOSYL -

PYROPHOSPHATE AND CYCLIC AMP IN REGENERATING RAT LIVER

Mitsuo Itakura and Kamejiro Yamashita

Institute of Clinical Medicine
The University of Tsukuba
Ibaraki, 305 JAPAN

INTRODUCTION

Glucagon increases DNA synthesis in regenerating rat liver[1] or even in nonoperated rat liver[2] and an elevated level of plasma glucagon has been documented after hepatectomy.[3] Other investigations have demonstrated that the concentration of 5-phosphoribosyl 1-pyrophosphate(PP-ribose-P), which is an important regulator of purine biosynthesis de novo, and the rate of purine biosynthesis de novo itself are increased by glucagon administered in mice[4] and in isolated rat hepatocytes.[5,6]

It is hypothesized that one effect of endogenous glucagon in regenerating rat liver is to increase the rate of purine biosynthesis de novo through elevating PP-ribose-P level. In this study, this hypothesis has been tested by assaying plasma concentration of glucagon, tissue concentration of cyclic AMP and PP-ribose-P in liver and the rate of purine biosynthesis de novo chronologically in hepatectomized and sham-operated rats.

MATERIALS

Eight week old male Wistar rats were used.

METHODS

Hepatectomy and Sham-operation

Seventy percent hepatectomy was performed according to Higgins' method.[7] Sham-operation was performed in a similar way but liver lobes were returned to the peritoneal cavity without ligation. Both

453

Table 1. Chronological Course of Plasma Glucagon, Rate of Purine
Biosynthesis de novo, Hepatic Cyclic AMP and Hepatic PP-
ribose-P in Hepatectomized (HTX) and Sham-operated (SHAM)
Rat

(HOUR)	Plasma Glucagon[a] (AV ± SEM) (pmole/ml)		Rate of Purine Biosynthesis de novo[b] (CPM/μmole adenine)		Hepatic Cyclic AMP[c] (AV ± SEM) (μmole/g prot)		Hepatic PP-ribose-P[d] (AV ± SEM) (nmole/g prot)	
Control	130 ± 8 (n=5)		1175 1196 1989	1265 1201	3.97 ± 0.18 (n=27)		25.7 ± 1.2 (n=30)	
	HTX	SHAM	HTX	SHAM	HTX	SHAM	HTX	SHAM
1	332 ±50 (n=6)	213 ±37 (n=4)	2572 2293	2786 2271	7.90 ±1.39 (n=3)	3.67 ±0.20 (n=3)	33.1 ±2.2 (n=3)	31.2 ±7.1 (n=3)
3	413 ±26 (n=6)	213 ±32 (n=4)	–	–	6.97 ±0.82 (n=3)	4.13 ±0.22 (n=3)	37.4 ±7.2 (n=3)	26.9 ±6.1 (n=3)
6	565 ±75 (n=6)	176 ±16 (n=4)	7983 10023	6417 5577	6.05 ±0.62 (n=3)	3.67 ±0.14 (n=3)	51.8 ±5.3 (n=3)	27.8 ±5.9 (n=3)
12	996 ±257 (n=5)	159 ±9 (n=4)	12523 14973	5783 5737	7.87 ±0.76 (n=5)	4.73 ±0.13 (n=3)	79.2 ±6.5 (n=3)	27.7 ±4.0 (n=4)
18	703 ±42 (n=5)	158 ±9 (n=4)	8896 6787	3301 3985	6.67 ±0.69 (n=3)	4.47 ±0.13 (n=3)	50.9 ±3.4 (n=3)	33.1 ±1.5 (n=3)
24	423 ±88 (n=4)	147 ±18 (n=6)	2703 2738	1131 1471	4.80 ±0.48 (n=3)	3.90 ±0.40 (n=3)	43.7 ±5.8 (n=3)	21.6 ±5.8 (n=3)
36	–	–	1625 3158	641 918	–	–	–	–
48	343 ±27 (n=3)	115 ±3 (n=3)	2116 2932	2416 1914	5.53 ±0.29 (n=5)	3.70 ±0.35 (n=3)	29.3 ±6.7 (n=3)	34.5 ±4.4 (n=3)
72	239 ±16 (n=4)	143 ±15 (n=4)	2422 2531	1358 2187	5.09 ±0.12 (n=3)	3.33 ±0.33 (n=3)	44.1 ±8.4 (n=3)	29.8 ±6.1 (n=3)

[a]Differences between HTX and SHAM are significant at all the time
points assayed.
[b]The data described represent the actual CPM/μmole adenine obtained
from each rat at different time points.
[c]Differences between HTX and SHAM are significant at all the time
points assayed except at 24 hours.
[d]Differences between HTX and SHAM are significant at 6, 12 and 18
hours.

surgical operations were performed between 9:00 and 12:00 AM.

Plasma Glucagon Assay

Plasma glucagon was assayed by radioimmunoassay kit.

Rate of Purine Biosynthesis de novo

Thirty minutes after intravenous administration of 5 μCi of ^{14}C glycine, liver tissue was sampled. Adenine nucleotides extracted by acid were degraded to adenine and it was purified by silver nitrate precipitation and Dowex 50 column. The specific activity of adenine thus obtained served as the index of the rate of purine biosynthesis de novo.[8]

Cyclic AMP Assay

Cyclic AMP was assayed by radioimmunoassay kit.

PP-ribose-P Assay

PP-ribose-P was extracted by acid[9] and assayed according to the published method.[10]

RESULTS

Table 1 summarizes the data obtained. Basal plasma glucagon concentration was 130.9 pg/ml, which increased and reached its peak of 996.0 at 12 hours after hepatectomy. Plasma glucagon concentrations in sham-operated rats did not show significant elevation.

Basal rate of purine biosynthesis de novo is 1,365 CPM/μmole adenine (n=5), which increased and reached its peak of 13,748 (n=2) at 12 hours after hepatectomy. In sham-operated rat liver, the rate of purine biosynthesis de novo also increased and it reached its peak of 5,997 (n=2) or 5,760 (n=2) at 6 and 12 hours after sham-operation respectively. When compared either with control or sham-operated rat, the rate of purine biosynthesis de novo in regenerating rat liver is significantly increased at 3, 6, 12 and 18 hours after hepatectomy with the largest difference at 12 hours.

Basal cyclic AMP concentration in liver was 3.97 μmole/gprot., which increased suddenly and reached its first peak of 7.90 at 1 hour after hepatectomy. After decreasing transiently at 6 hours, it again increased and reached its second peak of 7.87 at 12 hours.

Basal concentration of PP-ribose-P in liver was 25.7 nmole/gprot., which increased and reached its peak of 79.2 at 12 hours after hepatectomy.

CONCLUSIONS

1.   The chronological synchrony of plasma glucagon, hepatic cyclic AMP, hepatic PP-ribose-P and the rate of purine biosynthesis de novo with the exception of the first peak of hepatic cyclic AMP suggests that the endogenous glucagon is responsible for the fluctuation of the other three factors in regenerating rat liver except for the first peak of cyclic AMP.

2.   One of the direst role of endogenous glucagon in regenerating rat liver is increasing the rate of purine biosynthesis de novo through elevating PP-ribose-P level, which is a necessary prerequisite of increasing DNA synthesis and liver proliferation.

REFERENCES

1. N. L. R. Bucher and M. N. Swaffield, Regulation of hepatic regeneration in rats by synergistic action of insulin and glucagon, Proc. Nat. Acad. Sci. U.S.A. 72:1157 (1975)
2. J. Short, K. Tsukada, W. A. Rudert and I. Lieberman, Cyclic adenosine 3':5'-monophosphate and induction of deoxyribonucleic acid synthesis in liver, J. Biol. Chem. 250:3602 (1975)
3. H. Leffert, N. M. Alexander, G. Faloona, B. Rubalcava and R. Unger, Specific endocrine and hormonal receptor changes associated with liver regeneration in adult rats, Proc. Nat. Acad. Sci. U.S.A. 72:4033 (1975)
4. M. Lalanne and J. F. Henderson, Effects of hormones and drugs on phosphoribosyl pyrophosphate concentration in mouse liver, Can. J. Biochem. 53:394 (1975)
5. T. Hisata, N. Katsufuji and M. Tatibana, Glucagon or cyclic AMP-stimulated synthesis of 5-phosphoribosyl 1-pyrophosphate in isolated hepatocytes and inhibition by antimicrotubular drugs, Biochem. Biophy. Res. Comm. 81:704 (1978)
6. C. C. Rosiers, M. Lalanne and J. Willenot, Purine synthesis de novo and its regulation in rat hepatocytes, Can. J. Biochem. 58:599 (1980)
7. G. M. Higgins and R. M. Anderson, Restoration of the liver of the white rat following partial surgical removal, Arch. Path. 12:186 (1931)
8. M. Itakura, R. L. Sabina, P. W. Heald and E. W. Holmes, Basis for the control of purine biosynthesis by purine ribonucleotides, J. Clin. Invest. 67:994 (1981)
9. T. Hisata, An accurate method for estimating 5-phosphoribosyl 1-pyrophosphate in animal tissues with the use of acid extraction, Analyt. Biochem. 248:2529 (1973)
10. C. B. Thomas, W. J. Arnold and W. N. Kelley, Human adenine phosphoribosyltransferase: Purification and properties, J. Biol. Chem. 248:2529 (1973)

ETHANOL INDUCED ALTERATIONS OF URIC ACID METABOLISM*

Jason Faller** and Irving H. Fox

The Human Purine Research Center
Departments of Internal Medicine and
Biological Chemistry
University Hospitals
Ann Arbor, Michigan

INTRODUCTION

Early studies by Lieber and MacLachlan presented evidence suggesting that ethanol led to hyperlactic acidemia and decreased the excretion of uric acid.[1,2] Delbarre indicated the possibility of increased urate production[3] and a study by Grunst, et al. suggested that ethanol infusion lead to increased release of uric acid from the liver.[4] We have reexamined whether ethanol-induced hyperuricemia may be related to decreased renal excretion of uric acid, increased urate production by either accelerated degradation of nucleotides or increased de novo synthesis or both mechanisms together.

RESULTS

Five gouty patients were studied on an isocaloric purine free diet, with or without ethanol 1.8 grams/kilogram/day in four daily doses. There was an increase in serum urate concentration to 10.3 mg/dl, alcohol levels were maintained at a modest level of 12.8 mg/dl and there was a doubling of the blood lactate level (Table 1).

---

* Supported by USPHS grant AM19674 and 5 M01 RR00042 from Michigan Heart Association and Warner/Lambert-Parke/Davis.
** Dr. Faller is supported by an Arthritis Foundation Postdoctoral Fellowship.

Table 1.   Urate, Alcohol and Lactate Value in Five
           Patients During Oral Ethanol Studies

Test	No Drug	Ethanol
Serum Urate (mg/dl)	9.0 ± 0.4[a]	10.3 ± 1.0
Blood Alcohol (mg/dl)	---	12.8 ± 0.9
Blood Lactate (mM)	1.3 ± 0.3	3.1 ± 0.7

[a]Mean ± standard error

There was no evidence for a decrease in the urate clearance to creatinine clearance.   In fact, this ratio increased to a mean of 145% of
the baseline value suggesting a mild uricosuric effect (Table 2).
The urinary uric acid to creatinine ratio increased to 151% of the
baseline value, the urinary oxypurine to creatinine ratio increased
to 605% of the baseline value and the urate turnover increased by
170% of the baseline value (Table 3).

These observations provide evidence to suggest that the major
change in the serum urate concentration during ethanol ingestion was
related to an increase in the production of uric acid.   The changes
on the serum may have been counteracted by an increase in the renal
clearance of uric acid.

Table 2.   Uric Acid Clearance To Creatinine Clearance
           During Oral Ethanol Administration

Patient	Baseline (Cur/Ccr x 100)	Ethanol (Cur/Ccr x 100)	Ethanol (% Control)
JA	2.5	5.4	225
NP	3.8	2.7	71
TW	2.4	3.4	142
LG	2.9	3.6	124
WK	4.7	7.6	162
Mean ± SE	3.2 ± 0.4	4.5 ± 0.9	145 ± 25

Cur, urate clearance;   Ccr, creatinine clearance

Table 3. Effect of Oral Ethanol on Uric Acid Metabolism

Patient	Urinary Oxypurines			Urinary Uric Acid			Urate Turnover		
	Baseline (mol/g creat)	Ethanol	% Control	Baseline (mg/g creat)	Ethanol	% Control	Baseline (mg/d)	Ethanol	% Control
JA	43	950	2209	373	650	174	1050	1470	145
NP	155	86	55	238	238	100	—	—	—
TW	80	199	249	170	363	214	1020	1406	138
LG	139	470	338	289	368	127	1174	2460	210
WK	55	97	176	228	320	140	864	1610	186
Mean	94	360	605	260	388	151	1016	1737	170*
± SE	± 22	± 163	± 404	± 34	± 70	± 20	± 63	± 245	± 17

*Significant at p = .05

We next undertook to study the mechanism for increased production of uric acid. Increased production of urate could result from accelerated purine nucleotide degradation or accelerated purine bio-synthesis de novo. We examined the second possibility by prelabeling the adenine nucleotide pool by administering radioactive adenine to the patients and in a few days later infusing ethanol. Ethanol was infused for a period of two hours at 0.25 to 0.35 grams/kilogram/hour as a 15 to 20% solution in isotonic saline. Saline was infused for two hours prior to and following ethanol adminstration. Under these conditions if ethanol increases the synthesis of uric acid or its precursors, a concomitant elevation of urinary radioactivity will indicate an origin from accelerated ATP turnover. On the other hand, an increase of uric acid or its precursors without a change in urine radioactivity would indicate an origin from other sources such as de novo purine synthesis.

There was no change in the serum urate concentration during the infusion and only a modest elevation of the blood lactate levels (Table 4). Blood ethanol levels reached a mean value of 105 mg/dl during the infusion period.

The urine uric acid to creatinine ratio increased to 114 and 123% of the preinfusion baseline during successive hours of ethanol administration, and this was similar to the increase in the urate clearance to 108 and 116 percent respectively of the baseline values. The urine oxypurine to creatinine ratio increased to 341, 409, 405 and 415 percent of baseline during ethanol infusion and the following infusion. Urinary radioactivity increased to 127 and 149 percent of baseline during the ethanol infusion and remained elevated at 141 percent of baseline for two successive hours after the infusion.

Table 4. Urate, Alcohol and Lactate Values in Six Patients During Intravenous Ethanol

	Pre	During	Post
Peak Serum Urate (mg/dl)	8.4 ± 0.2[a]	8.3 ± 0.2	8.5 ± 0.2
Peak Blood Alcohol (mg/dl)	< 0.1	105 ± 88	88 ± 12
Peak Blood Lactate (mM)	1.14 ± 0.1	1.39 ± 0.1	1.42 ± 0.1

[a]Mean ± Standard Error.

These data confirm that at this rate of ethanol administration, uric acid excretion did not decrease. The increased excretion of uric acid precursors suggests, in fact, that there is increased flux through the pathways of purine nucleotide degradation to uric acid. Excretion of labeled degradation products derived from the adenine nucleotide pool is significantly accelerated and suggests accelerated ATP or adenine nucleotide degradation.

The mechanism for accelerated nucleotide degradation during ethanol administration is not clear. It has been proposed that ethanol leads to an accelerated breakdown of preformed purines by mechanisms similar to fructose-induced hyperuricemia. Ethanol may cause a depletion of inorganic phosphate and ATP in liver cells as the result of an accumulation of alpha-glycerolphosphate analagous to the accumulation of fructose-1-phosphate after intravenous fructose administration.[5,6] However, the depletion of intracellular ATP levels seems to be contrary to other observations which suggest that elevated NADH generated during alcohol metabolism may actually cause an increased synthesis of ATP.[5] Careful examination of ethanol metabolism suggests a possible additional pathway for ATP consumption. All ethanol is oxidized to acetate. The acetate then enters intermediary metabolism by initial conversion to an acetyl-CoA, a reaction which requires the conversion of ATP to AMP and inorganic pyrophosphate. Since virtually all ethanol is converted to acetyl-CoA and oxidized to carbon dioxide and water, this cycle represents a mechanism for massive ATP turnover since 2 moles of high energy phosphate are utilized for each mole of ethanol metabolized. Therefore, if only a small proportion of the ATP turned over entered the pathway of purine nucleotide degradation, this could account for the substantial increase in uric acid synthesis.

SUMMARY AND CONCLUSIONS

Our observations have shown that chronic oral ethanol administration was associated with increased serum urate, urine uric acid excretion, urine uric acid clearance and oxypurine excretion. The daily rate of uric acid turnover was significantly increased.

Intravenous ethanol administration was associated with increased uric acid excretion, increased uric acid clearance and significantly increased oxypurine excretion. Excretion of radioactivity derived from intravenously administered adenine increased significantly.

We conclude that hyperuricemia related to ethanol consumption at lower blood ethanol levels (less than 150 mg/dl) results from increased production of uric acid probably secondary to accelerated degradation of adenine nucleotides.

REFERENCES

1. C.S. Lieber, D.P. Jones, M.S. Losowsky and C.S. Davidson, J.
        Clin. Invest. 41:1863 (1962).
2. M.J. MacLachlan, G.P. Rodnan, Am. J. Med. 42:38 (1967).
3. F. Delbarre, C. Auscher, H. Brouilhet and A. deGery, Sem. Hop.
        Paris 43:659-664 (1967).
4. J. Grunst, G. Dietze and M. Wicklmayr, Nutr. Metab. 21 (Suppl. 1)
        138 (1977).
5. R.L. Veech, R. Guyinn and D. Veloso, Biochem. J. 127:387 (1972).
6. J. Grunst, G. Dietze and M. Wicklmayr, Leber. Verh. Dt. Ges. inn
        Med. 79:941-917 (1973).

INFLUENCE OF ETHANOL ON THE PRODUCTION OF ALLANTOIN
BY THE PERFUSED RAT LIVER

G. Van den Berghe, F. Bontemps and H.G. Hers

Laboratoire de Chimie Physiologique, Université de
Louvain and International Institute of Cellular and
Molecular Pathology, B-1200 Brussels, Belgium

INTRODUCTION

Ingestion of alcohol is traditionally considered a predis-
posing or precipitating factor in the gouty attack. Studies
performed in man by Lieber et al. (1962) have shown that the
administration of ethanol provokes an increase in the serum
concentration of uric acid, which was attributed to a decrease in
its renal excretion. More recently, ethanol infusion has been
shown to enhance the production of uric acid by the liver, as
measured from arteriovenous concentration differences during
hepatic catheterisation in human volunteers (Grunst et al., 1973).

The present work was undertaken in order to elucidate the
mechanism of this direct stimulatory effect of ethanol on the
hepatic production of purine catabolites. Isolated preparations
from rat liver were used, in which, due to the presence of uricase, allantoin constitutes the end-product of purine breakdown.

RESULTS

Fig. 1 shows the effect of the addition of ethanol on the
production of uric acid and allantoin by the isolated liver of a
fed rat, perfused with a recirculating Krebs-Ringer bicarbonate
buffer containing 10 mM glucose. Before the addition of ethanol
there was a gradual increase in the concentration of uric acid,
until a steady state was obtained. The production of allantoin
was linear in function of time and reached about 12 nmol/min/g of
tissue. The addition of 20 mM ethanol provoked a brisk elevation
of the concentration of uric acid, which increased to a new steady
state. This was followed by an approx. 2.5-fold enhancement in the
rate of production of allantoin.

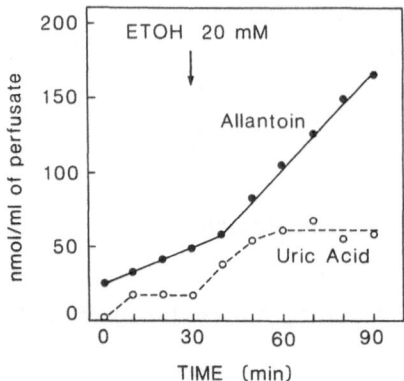

Fig. 1. Influence of the addition of ethanol to the perfusion
medium of an isolated liver from a fed rat. Further
details are given in the text.

Measurements of metabolites in freeze-clamped tissue samples,
taken at various time intervals from perfused livers of fed rats
(Table I), revealed that the concentration of ATP decreased

Table 1. Influence of the addition of ethanol on metabolite
concentrations in perfused livers from fed rats

The metabolite concentrations, expressed as μmol per g of tissue
wet weight, were determined in freeze clamped biopsies that were
taken after the addition of 25 mM ethanol to the perfusion
medium, at the times indicated. Values shown are means ± S.E.M.
of 3 to 4 experiments. α-G-P = α-glycerol-P.

	0 min	5 min	10 min	20 min
ATP	2.92 ± 0.14	3.12 ± 0.07	1.63 ± 0.30	1.38 ± 0.10
ADP	1.02 ± 0.08	1.24 ± 0.14	1.31 ± 0.02	1.18 ± 0.02
AMP	0.22 ± 0.02	0.28 ± 0.05	0.68 ± 0.08	0.59 ± 0.04
GTP	0.35 ± 0.02		0.21 ± 0.04	0.18 ± 0.02
Pi	6.54 ± 0.59	5.85 ± 0.94	5.77 ± 0.30	6.22 ± 0.75
α-G-P	0.94 ± 0.25	2.55 ± 1.09	4.50 ± 1.02	6.39 ± 1.67

rapidly by about 50 % between 5 and 10 min after the addition of
ethanol. Thereafter, a slower decay was recorded. The concentration of ADP was barely modified and that of AMP increased 3-fold.
This resulted in an approx. 25 % decrease of the total adenine
nucleotide pool. The concentration of GTP decreased in parallel
with that of ATP and the concentration of P_i did not change.

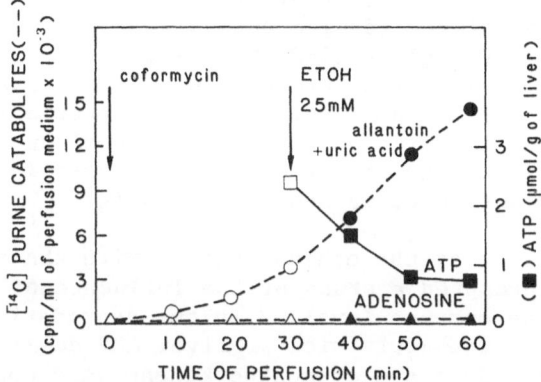

Fig. 2. Influence of 2.5 x 10^{-7} M coformycin on the catabolism
        of the adenine nucleotides induced by ethanol in the
        perfused liver of a fed rat. The purine catabolites in the
        medium were determined by thin-layer chromatography. The
        concentration of ATP was measured in freeze-clamped
        biopsies. Further explanations are given in the text.

        Fig. 2 depicts an experiment in which the hepatic adenine
nucleotides were labelled with [^{14}C]adenine, prior to the isolation of the liver for perfusion. Coformycin was added to the
recirculating medium at the beginning of the perfusion. This
resulted in a tissue concentration of the inhibitor, measured
after 30 min, equal to 2.5 x 10^{-7} M. This concentration was
previously shown to inhibit maximally hepatic adenosine deaminase,
without influencing AMP deaminase (Van den Berghe et al., 1980). A
release of radioactive uric acid and allantoin was measured in the

perfusion medium that was also stimulated approx. 3-fold by the addition of ethanol. No accumulation of adenosine was observed either before or after the addition of ethanol.

From further investigations it appeared that the effect of ethanol on the production of purine catabolites was not observed when the liver was isolated from rats that had been starved for 24 hours. Moreover, the addition of ethanol to the incubation medium of isolated hepatocytes from fed as well as from fasted animals did not modify their rate of production of allantoin.

Additional analysis of the experiments revealed that the first change detected after the administration of ethanol to perfused livers of fed rats, occuring prior to the modification of the concentrations of the adenine nucleotides, was an elevation of the concentration of α-glycerol-P. This increase proceeded progressively, from basal values around 1 μmol/g of tissue, to an average value of about 6 μmol/g after 20 min (table I). In the liver preparations in which no effect of ethanol on the production of purine catabolites was observed, the concentration of α-glycerol-P did not increase above 1.5 μmol/g.

The association of the ethanol effect with the accumulation of α-glycerol-P prompted a study of the influence of this phosphate ester on the enzymes involved in the degradation and the resynthesis of AMP. The activities of liver AMP deaminase, cytoplasmic 5'-nucleotidase and adenosine kinase were not influenced by concentrations of α-glycerol-P up to 10 mM.

DISCUSSION

A decrease in the concentration of liver ATP has been repeatedly documented after chronic administration of ethanol to rats (Walker and Gordon, 1970; Bernstein et al., 1973; Gordon, 1977). Although several authors have not observed an acute effect of ethanol on hepatic ATP (Williamson et al., 1969; Veech et al., 1972) our results are in agreement with those of Soboll et al. (1978) who have shown an approx. 30 % decrease in the concentration of this nucleotide after the addition of ethanol to the perfused liver of fed rats. The finding (Fig. 2) that the effect of ethanol on the production of purine catabolites was still observed in the presence of coformycin at a concentration that inhibits selectively adenosine deaminase (Van den Berghe et al., 1980) indicates that the degradation of the adenine nucleotide pool, induced by ethanol, does not proceed by way of the dephosphorylation of AMP but involves AMP deaminase.

The observation that the concentration of AMP increases several fold after the addition of ethanol (Table 1) suggests that the elevation of the substrate of AMP deaminase constitutes the primary trigger in the ethanol-induced enhancement of purine catabolism. The decrease in GTP, one of the physiological inhibitors of the enzyme (Van den Berghe et al., 1977),may constitute an additional factor, once ATP has started to decline.

Extrapolating from experiments performed by Thurman and Scholz (1977), that have shown that ethanol increases the uptake of oxygen in perfused livers from fed, but not from fasted rats, the following mechanism can be hypothesized to explain the initial loss of ATP induced by the addition of ethanol to the perfused liver of fed rats. These authors have proposed that in the latter preparation, the maintainance of the concentration of ATP depends in part on its extramitochondrial generation by the glycolytic pathway. Ethanol, by increasing the NADH/NAD$^+$ ratio, inhibits glycolysis at the glyceraldehyde-3-P dehydrogenase step. This would decrease the generation of ATP in the cytosol and enhance the supply of ADP to the respiratory chain, manifested by an increased oxygen uptake. Through the action of myokinase, this enhanced supply of ADP may also elevate the concentration of AMP, resulting in an increased activity of AMP deaminase.

The absence of effect of ethanol on purine catabolism in the perfused liver of fasted rats, could be explained by the near-absence of glycolysis in this preparation (Woods and Krebs, 1971), resulting in the disappearance of the interaction with the respiratory chain.

The absence of effect of ethanol on purine catabolism in isolated hepatocytes from fed rats may be related to a lower rate of glycolysis (Seglen, 1974) or to its lesser inhibition by ethanol, as suggested by the limited increases in α-glycerol-P. Experiments are in progress in order to verify this hypothesis.

ACKNOWLEDGEMENTS

This work was supported by the Belgian FRSM. GVDB is Maître de Recherches of the Belgian FNRS and FB is Research Fellow of the International Institute of Cellular and Molecular Pathology. Coformycin was a generous gift of Professor H. Umezawa (Institute of Microbial Chemistry, Tokyo, Japan).

REFERENCES

Bernstein, J., Videla, L. and Israel, Y., 1973, Biochem. J., 134: 515-521.
Gordon, E.R., 1977, Biochem. Pharmacol., 26:1229-1234.
Grunst, J., Dietze, G. and Wicklmayr, M., 1973, Verh. Dt. Ges. inn. Med. 79:914-917.

Lieber, C.S., Jones, D.P., Losowsky, M.S. and Davidson, C.S., 1962, J. Clin. Invest. 41:1863-1870.

Seglen, P.O., 1974, Biochim. Biophys. Acta, 338:317-336.

Soboll, S., Scholz, R. and Heldt, H.W., 1978, Eur. J. Biochem., 87:377-390.

Thurman, R.G. and Scholz, R., 1977, Eur. J. Biochem., 75:13-21.

Van den Berghe, G., Bronfman, M., Vanneste, R. and Hers, H.G., 1977, Biochem. J., 162:601-609.

Van den Berghe, G., Bontemps, F. and Hers, H.G., 1980, Biochem. J., 188:913-920.

Veech, R.L., Guynn, R. and Veloso, D., 1972, Biochem. J., 127:387-397.

Walker, J.E.C. and Gordon, E.R., 1970, Biochem. J., 119:507-514.

Williamson, J.R., Scholz, R., Browning, E.T., Thurman, R.G. and Fukami, M.H., 1969, J. Biol. Chem., 244:5044-5054.

Woods, H.F. and Krebs, H.A., 1971, Biochem. J., 125:129-139.

THE PATHWAY OF AMP CATABOLISM AND ITS CONTROL IN ISOLATED RAT HEPATOCYTES SUBJECTED TO ANOXIA

M.-F. Vincent, G. Van den Berghe and H.G. Hers

Laboratoire de Chimie Physiologique, Université de Louvain and International Institute of Cellular and Molecular Pathology, B-1200 Brussels, Belgium

INTRODUCTION

In anoxic conditions, the hepatic concentrations of high energy phosphates, most notably ATP, decrease to a marked extent, whereas those of AMP and P_i increase several-fold (Deuticke and Gerlach, 1966; Busch et al., 1968; Hems and Brosnan, 1970; Jackson et al., 1976). The exact pathway of the further catabolism of AMP in this situation has been a subject of controversy : whereas Deuticke and Gerlach (1966) have proposed that the initial degradation of AMP occurs by way of AMP deaminase and is followed by the dephosphorylation of IMP by 5'-nucleotidase, Busch et al. (1968) have asserted that the degradation of AMP involves a prior dephosphorylation by the same enzyme, followed by deamination of adenosine by adenosine deaminase. The present study was undertaken with isolated rat hepatocytes in order to try to solve this controversy. As in our previous work, coformycin was used to determine the pathway of degradation of AMP. In the liver, this inosine analog inhibits selectively adenosine deaminase at concentrations around 0.1 μM, whereas at concentrations around 50 μM, it also inhibits AMP deaminase (Van den Berghe et al., 1980). If the initial degradation of AMP occurs by way of 5'-nucleotidase, the low concentration of coformycin would decrease the formation of the terminal products of adenine nucleotide catabolism. If, however, the initial step of the degradation of AMP is catalyzed by AMP deaminase, the decrease in the production of purine catabolites would be observed only at high concentrations of coformycin. Further details concerning the experiments reported and a description of the methodology can be found in Vincent et al. (1982).

RESULTS AND DISCUSSION

Fig. 1 shows the effect of the replacement of the O_2/CO_2 gas phase by N_2/CO_2 on the concentrations of ATP and AMP in

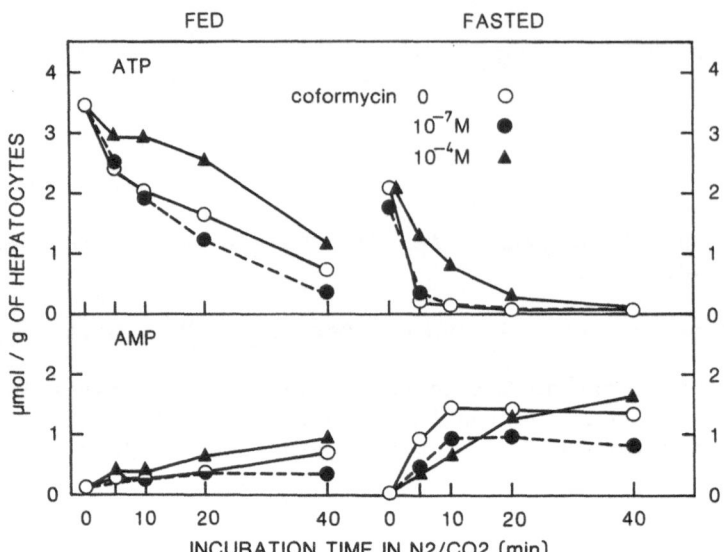

Fig. 1. Influence of anoxia on the concentration of ATP and AMP in
        isolated hepatocytes. The O_2/CO_2 gas phase was repla-
        ced by N_2/CO_2 at time zero in hepatocytes that had
        been preincubated in the absence or in the presence of
        coformycin. Note that in the hepatocytes from fasted rats,
        the high concentration of coformycin was 5.10^{-5}M.

isolated rat hepatocytes of a fed rat, as compared to a fasted
animal. The depletion of ATP proceeded slowly in the fed state
but occured rapidly in the fasted state. This difference may be
explained by the presence of glycogen stores in hepatocytes from
fed animals, that allow the generation of ATP by anaerobic glycolysis. In both nutritional states, the degradation of ATP was
barely influenced by prior incubation with 10^{-7}M coformycin
but significantly delayed by higher concentrations of the
inhibitor.

    The depletion of ATP was accompanied by an increase in the
concentration of AMP, that proceeded more rapidly and reached
higher levels in the hepatocytes from fasted animals. In both
conditions, a loss of the sum of the adenine nucleotides was
recorded, that occured rapidly during the first 5 min of anoxia,
but proceeded much more slowly thereafter. These data indicate
that potent control mechanisms restrict the further degradation of
the hepatic adenine nucleotide pool in anoxic conditions.

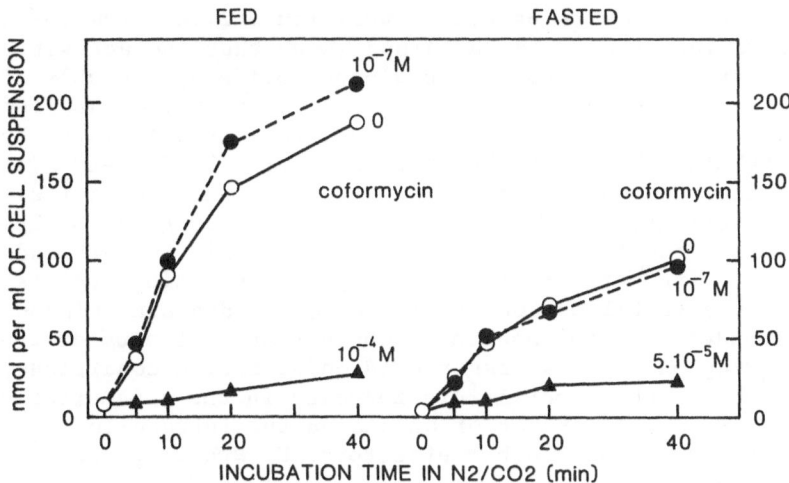

Fig. 2. Influence of anoxia on the concentration of uric acid in
        isolated hepatocyte suspensions. Same experiment as in
        Fig. 1.

Determination of the purine catabolites in the suspensions of
isolated hepatocytes incubated in N_2/CO_2, revealed a striking
accumulation of uric acid (Fig. 2). This compound is normally not
measurable in hepatocyte suspensions incubated in O_2/CO_2, in
which allantoin constitutes the end-product of purine catabolism.
The appearance of uric acid in anoxic conditions is explained by
the oxygen-dependence of urate oxidase (Keilin and Hartree, 1936).
The production of uric acid was not significantly influenced by
10^{-7}M coformycin. It was, however, nearly completely suppressed by the higher concentrations of the inhibitor. These data
provide convincing evidence that the initial degradation of AMP in
anoxic conditions proceeds by way of the sequential action of AMP
deaminase and 5'-nucleotidase, thus confirming the results of
Deuticke and Gerlach (1966).

Further studies were aimed at the elucidation of the mechanisms whereby the hepatic adenine nucleotide pool is preserved in
anoxic conditions, especially in the fasted state. In hepatocytes
from fed rats, indeed, this protection is mainly due to a better
better maintenance of the ATP concentration by anaerobic glycolysis. In hepatocytes from fasted animals, however, it is chiefly
caused by a restriction of the degradation of AMP, as evidenced by
the more marked accumulation of this nucleotide (Fig. 1) and the
lower production of uric acid (Fig. 2) in comparison with the fed
state.

Previous work with partially purified enzymes from rat liver (Van den Berghe et al., 1977a,b) had shown that the activities of AMP deaminase and of cytoplasmic 5'-nucleotidase (the only 5'-nucleotidase that could qualify for a role in purine catabolism) are regulated by ATP, GTP and P_i. AMP deaminase is stimulated by ATP and submitted to a synergistic inhibition by GTP and P_i. At physiological concentrations of AMP (0.1-0.2 mM) and of its effectors, the activity of AMP deaminase is inhibited by 95 %. The cytoplasmic 5'-nucleotidase is stimulated by ATP and GTP and inhibited by P_i. Its activity in physiological conditions (*) is offset by the reutilization of adenosine by adenosine kinase. In order to understand the mechanism whereby the activity of the AMP-degrading enzymes was restricted under anoxic conditions, notwithstanding the several-fold increase in the concentration of their substrate, the effect of anoxia on the intracellular concentrations of their other effectors, P_i and GTP, was also investigated.

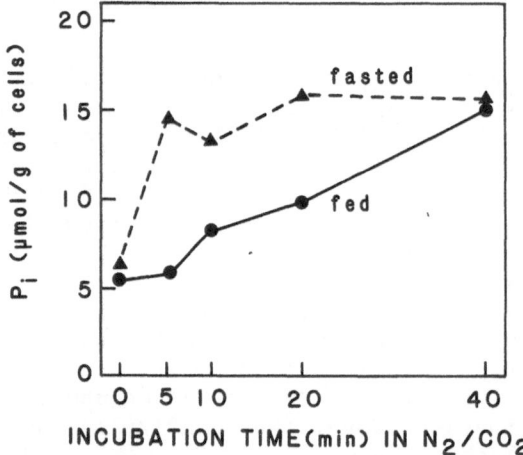

Fig. 3. Influence of anoxia on the intracellular concentration of P_i inside isolated hepatocytes. Each point represents the mean of 2 experiments.

(*) Recent results (see Bontemps et al., this volume) have shown that, contrary to our previous conclusion, the cytoplasmic 5'-nucleotidase operates in hepatocytes in basal conditions and participates in a futile cycle. This activity could be confirmed when a highly purified cytoplasmic 5'-nucleotidase, prepared according to Itoh (1981) was assayed in physiological conditions (see also Fig. 4), instead of the partially purified enzyme we used previously.

As depicted in Fig.3, the concentration of P_i increased
3-fold in hepatocytes subjected to anoxia. This increase proceeded
more rapidly in hepatocytes from starved rats than in those from
fed animals, as a consequence of the faster degradation of ATP in
the former situation. Measurements of the concentration of GTP in
the hepatocytes showed that it decreased in parallel with that of
ATP under both nutritional conditions (results not shown).

The limited degradation of increased concentrations of AMP by
AMP deaminase in anoxic conditions, may thus be explained by the
decrease in its stimulator ATP and the increase in P_i, one of
its physiological inhibitors. The better preservation of AMP in
fasted hepatocytes can be accounted for by the fact that the
variation of both effectors occurs more promptly in the fasting
state.

Since the reutilization of adenosine by adenosine kinase is
likely to be impaired in ATP-depleted cells, inhibition of the
5'-nucleotidase appears essential to explain the lack of significant degradation of AMP by the dephosphorylation-deamination
sequence, evidenced by the lack of effect of 10^{-7} M coformycin
on the production of uric acid (Fig. 2).

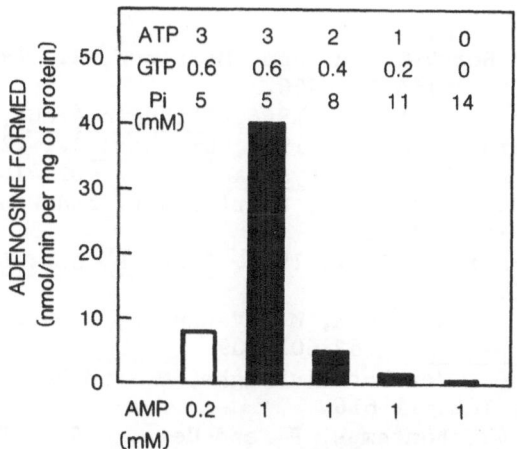

Fig. 4. Influence of conditions prevailing in anoxia on the
activity of purified rat liver cytoplasmic 5'-nucleotid-
ase. The enzyme was prepared according to Itoh (1981). The
open bar depicts the enzymic activity in the conditions
observed in control liver. The filled bars represent the
enzymic activity at 1 mM-AMP in the presence of various
mixtures of the effectors, mimicking the progress of
anoxia.

Fig. 4 shows the influence of the conditions prevailing in isolated hepatocytes subjected to anoxia, as compared to the control situation, on the activity of purified rat liver cytoplasmic 5'-nucleotidase. The enhancement of the enzymic activity, resulting from the increase in the concentration of AMP, was completely offset and even reversed, when the concentration of the stimulators was decreased and that of the inhibitor increased, so as to mimic the progress of anoxia.

It can thus be concluded that the regulation of both AMP-degrading enzymes appears uniquely designed to preserve the adenine nucleotide pool when the liver is subjected to anoxic conditions. The importance of this preservation is evidenced by experiments that show that AMP, but not IMP, may be completely reconverted into ATP upon reoxygenation of the hepatocytes (Vincent et al., 1982).

ACKNOWLEDGEMENTS

This work was supported by the Fonds de la Recherche Scientifique Médicale. M.F.V. is Aspirant and GVDB Maître de Recherches of the Belgian Fonds National de la Recherche Scientifique. Coformycin was a generous gift of Prof. H.Umezawa (Institute of Microbial Chemistry, Tokyo, Japan).

REFERENCES

Busch, E.W., Von Borcke, I.M. and Martinez, B., 1968, Biochim. Biophys. Acta., 166:547-556.
Deuticke, B. and Gerlach, E., 1966, Pflügers Arch., 292:239-254.
Hems, D.A. and Brosnan, J.T., 1970, Biochem. J., 120:105-111.
Itoh, R., 1981, Biochim. Biophys. Acta, 657:402-410.
Jackson, R.C., Boritzki, T.J., Morris, H.P. and Weber, G., 1976, Life Sci., 19:1531-1536.
Keilin, D. and Hartree, E.F., 1936, Proc. R. Soc. London, Ser B, 119:114-140.
Van den Berghe, G., Bronfman, M., Vanneste, R. and Hers, H.G., 1977a, Biochem. J., 162:601-609.
Van den Berghe, G., Van Pottelsberghe, C. and Hers, H.G., 1977b, Biochem.J., 162:611-616.
Van den Berghe, G., Bontemps, F. and Hers, H.G., 1980, Biochem.J., 188:913-920.
Vincent, M.F., Van den Berghe, G. and Hers, H.G., 1982, Biochem. J., 202:117-123.

STIMULATION BY ADENOSINE OF ADENINE NUCLEOTIDE TURNOVER IN
ISOLATED HEPATOCYTES: EVIDENCE FOR A FUTILE CYCLE BETWEEN AMP AND
ADENOSINE

F. Bontemps, G. Van den Berghe and H.G. Hers

Laboratoire de Chimie Physiologique, Université de
Louvain and International Institute of Cellular and
Molecular Pathology, B-1200 Brussels, Belgium

INTRODUCTION

Addition of adenosine to isolated rat hepatocytes, as well as to
other liver preparations, provokes marked increases in the
intracellular concentration of ATP and total adenine nucleotides
(Chagoya de Sanchez et al., 1972; Lund et al., 1975; Wilkening et
al., 1975), that are explained by the utilisation of adenosine by
adenosine kinase. The present work was initiated as a search for
a mechanism whereby the rate of degradation of the adenine
nucleotide pool would adapt to an increased rate of synthesis. It
led to the unexpected ascertainment that, under normal conditions,
there is a continuous formation of adenosine by the hepatocytes.
This production does, however, not contribute to the formation of
allantoin but is part of a futile cycle operating between AMP and
adenosine.

RESULTS AND DISCUSSION

1. Influence of adenosine on the catabolism of the hepatic adenine
   nucleotides

Fig. 1. depicts the sequence of events taking place after
addition of 0.5 mM-adenosine in hepatocyte suspensions that had
been preincubated with [^{14}C]adenine in order to label their
adenine nucleotide pool (Van den Berghe et al., 1980). Concentrations of the adenine nucleotides in the hepatocytes and of adenosine and allantoin in the cell suspension are shown on the left,
whereas the radioactivity in these compounds is displayed on the
right. Cells incubated without adenosine are compared to cells incubated with 0.5 mM-adenosine, in the absence and in the presence

of 50 μM coformycin. This concentration of the inosine analog
inhibits completely adenosine deaminase and also causes a profound
reduction of the activity of AMP deaminase (Van den Berghe et al.,
1980).

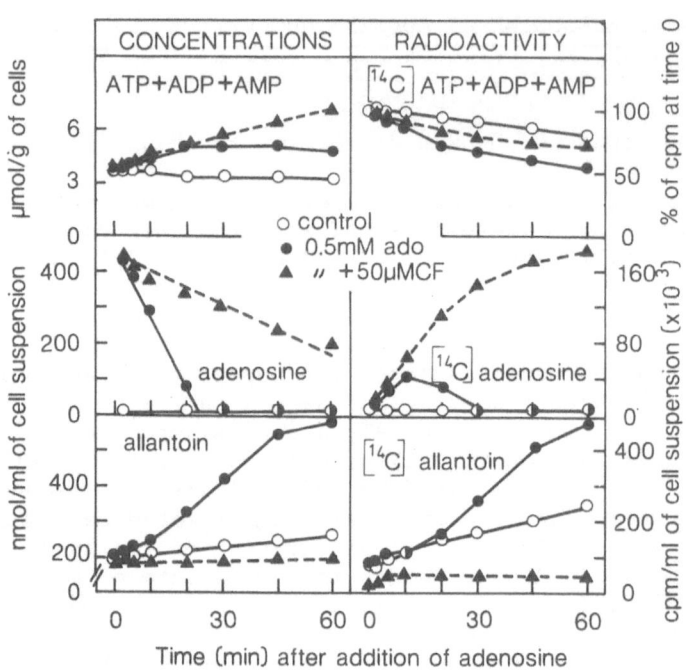

Fig. 1. Influence of adenosine on the metabolism of adenine
        nucleotides in isolated hepatocytes. Adenosine (0.5
        μmol/ml) (●,▲) or NaCl (O) was added at zero time to
        hepatocytes that had been preincubated for 15 min in the
        presence of 1 μM[14C]adenine without (O,●) or with (▲)
        50 μM coformycin (CF). The concentrations of metabolites
        are given in the left pannels and the radioactivity found
        in these metabolites, in the right pannels.

Addition of adenosine alone resulted in an increment of the
total adenine nucleotide content, mainly under the form of ATP,
which proceeded only as long as adenosine was present. It was
accompanied by an approx. 6-fold increase in the production of
allantoin. In the presence of 50 μM coformycin, the concentration
of the adenine nucleotides continued to increase over 60 min. This
is explained by the inhibition of adenosine deaminase, resulting

in the complete recovery of adenosine in the adenine nucleotides.
As a consequence of the inhibition of both adenosine deaminase and
AMP deaminase, the production of allantoin became negligible.

In control cells, the radioactivity in the adenine nucleotides decreased linearly with time, reflecting their rate of degradation. A more rapid loss of this radioactivity was recorded in
the presence of adenosine, indicating a stimulation of the rate of
degradation of the prelabelled adenine nucleotide pool. In the
presence of coformycin, this stimulation was approx. halved. Since
AMP deaminase is maximally inhibited at 50 μM coformycin, this
finding suggested that the degradation of the adenine nucleotides
also proceeded by way of 5'-nucleotidase. This was confirmed by
the determination of radioactivity in adenosine. Whereas no radioactive adenosine was detected in control conditions, the addition of unlabelled adenosine induced a prompt appearance of radioactivity in this nucleoside, which was transient in the absence of
coformycin, but increased steadily in its presence. The formation
of radioactive allantoin in the three experimental conditions was
similar to that of the non-radioactive catabolite.

Determination of the specific activities of the adenine
nucleotides and summation of the radioactivity appearing in all
catabolites (adenosine, inosine, hypoxanthine, xanthine, uric acid
and allantoin) allowed the calculation of the rates of catabolism
of the adenine nucleotides in the three experimental conditions.
In control conditions, the degradation of the adenine nucleotides
reached 12 nmol/min per g of hepatocytes. As shown in previous
work, it involves only AMP deaminase (Van den Berghe et al.,
1980). Upon addition of 0.5 mM adenosine, the rate of catabolism
of the adenine nucleotides was increased to 55 nmol/min per g.
This could be due to a stimulation of AMP deaminase as well as of
5'-nucleotidase, as evidenced by the formation of radioactive
adenosine in this condition. In the presence of 50 μM coformycin,
the rate of degradation of the adenine nucleotides was increased
only approx. to 29 nmol/min per g. Since at this concentration of
coformycin, AMP deaminase is maximally inhibited, the observed
rate reflects the participation of 5'-nucleotidase in the
adenosine-induced catabolism. It could thus be calculated that the
participation of AMP deaminase accounted for 26 nmol/min per g of
hepatocytes, which represents a 2-fold stimulation as compared to
the basal condition.

Evaluation of the stimulatory effect of ATP on the activity
of partially purified AMP deaminase, in the presence of
physiological concentrations of its substrate and inhibitors, GTP
and P_i (Van den Berghe et al., 1977a), revealed that this
activity was increased about 2-fold when ATP was raised from its
concentration in the control situation (approx. 2.5 mM), to the
concentration prevailing after the administration of adenosine

(close to 5 mM) (results not shown). An assessment of the activity of the cytoplasmic 5'-nucleotidase present in rat liver
which is stimulated by ATP and GTP, and inhibited by P_i (Van
den Berghe et al., 1977b) was similarly performed. It also showed
an approx. 2-fold stimulation of the enzymic activity in the conditions resulting from the addition of adenosine, as compared to
the activity in the control conditions (*).

     It can thus be concluded that mechanisms exist at the level
of the AMP degrading enzymes that counteract the enhancement of
the rate of synthesis of the adenine nucleotides induced by
adenosine.

2. Evidence for a futile cycle between AMP and adenosine

     The formation of radioactive adenosine from labelled adenine
nucleotides, revealed by the addition of the unlabelled nucleoside
to the hepatocyte suspension, as well as the reassessment of the
activity of the cytoplasmic 5'-nucleotidase in physiological
conditions, raised the possibility that adenosine may be continuously formed from adenine nucleotides in basal conditions but not
contribute to the production of allantoin because of its reutilization by adenosine kinase. This futile cycle would explain our
previous observation that the production of allantoin by isolated
hepatocytes is not influenced by coformycin at a concentration
that inhibits selectively adenosine deaminase (Van den Berghe et
al., 1980). The arrest of the reutilization of adenosine would
cause an irreversible loss of adenine nucleotides under the form
of this nucleoside, that would be further degraded to allantoin.

     In order to test this hypothesis, experiments were performed
with 5-iodotubercidin, an inhibitor of adenosine kinase (Henderson
et al., 1972). As evidenced by measurements of the incorporation
of [^{14}C]adenosine into the intracellular adenine nucleotides,
5-iodotubercidin provoked a dose-dependent inhibition of adenosine
kinase inside the hepatocytes, that reached 98 % at a concentration of 100-200 μM. It also provoked a dose-dependent, progressive loss of intracellular ATP (results not shown) and accumulation of adenosine and allantoin (Fig. 2).

(*) A previous study, performed with a partially purified enzyme
preparation, had shown that the cytoplasmic 5'-nucleotidase was
inactive on AMP in physiological conditions (Van den Berghe et
al., 1977b). The present finding of an easily measurable activity
towards AMP, although it represented only a few percent of the
activity towards the same concentration of IMP, can be explained
by the use of a highly purified enzyme, prepared according to Itoh
(1981).

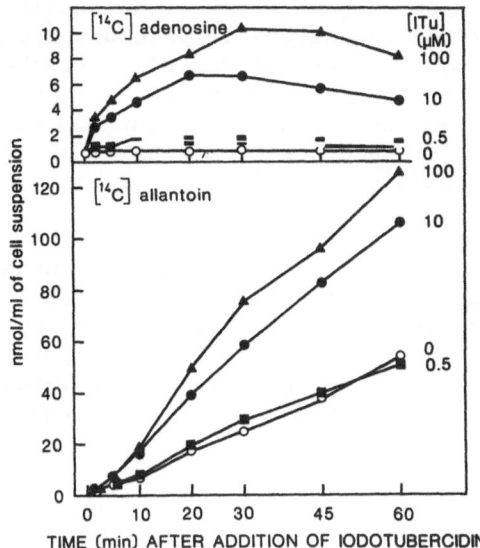

Fig. 2. Influence of iodotubercidin on the concentrations of
adenosine and allantoin in suspensions of isolated hepatocytes. The cells were preincubated for 15 min with 1 μM
[^{14}C]adenine and iodotubercidin (ITu), at the concentrations indicated, was added at zero time. Concentrations
of adenosine and allantoin were calculated from the specific radioactivity of the adenine nucleotides. The rate of
production of allantoin in the 4 experimental conditions
reached 14.7 (O), 15.2 (■), 31 (●) and 37 (▲) nmol/min per
g of hepatocytes.

        These observations provide convincing evidence that adenosine
is continuously formed by the hepatocytes, and has to be rephos-
phorylated in order to maintain the adenine nucleotide pool. The
results depicted in Fig. 2 were only slightly modified when a mixture of AOPCP and β-glycerophosphate was added to the cell suspensions in order to inhibit the membranous 5'-nucleotidase and aspe-
cific phosphatases. This rules out a significant participation of
the former enzyme, as well as of the extracellular catabolism of
adenine nucleotides released by dammaged hepatocytes, in the formation of adenosine. The production of adenosine thus most likely
results from dephosphorylation of AMP inside the cells, by the
cytoplasmic 5'-nucleotidase. From the observation that this formation of adenosine does not contribute to the physiological
production of allantoin, it can be concluded that both the formation and utilization of adenosine proceed at the same rate in
control conditions, and thus constitute a "futile" cycle. From

the increase in the rate of production of allantoin observed at maximally inhibitory concentrations of 5-iodotubercidin (Fig. 2) it can be deducted that the rate of recycling reaches approx. 23 nmol/min per g of cells in the basal state. As evidenced by the increase in the adenine nucleotide pool, the addition of adenosine results in a stronger stimulation of adenosine kinase than of the cytoplasmic 5'-nucleotidase.

Lomax and Henderson (1973) have briefly alluded to the possibility that simultaneous breakdown and resynthesis of AMP may be a normal occurence in Ehrlich ascites tumor cells, and Arch and Newsholme (1978) have discussed the theoretical implications of this substrate cycle in detail. The discovery that the futile cycle between AMP and adenosine operates in isolated hepatocytes in basal conditions should provide further impetus for the study of its physiological significance.

ACKNOWLEDGEMENTS

This work was supported by the Belgian FRSM. G.V.D.B. is Maître de Recherches of the Belgian FNRS and F.B. is Fellow of the International Institute of Cellular and Molecular Pathology. Coformycin and 5-iodotubercidin were generously given by Prof. H. Umezawa (Tokyo) and Dr. L.B. Townsend (Ann Arbor) respectively.

REFERENCES

Arch, J.R.S. and Newsholme, E.A., 1978, Essays in Biochemistry, 14:82-123.
Chagoya de Sanchez, V., Brunner, A. and Pina, E., 1972, Biochem. Biophys. Res. Commun., 46:1441-1445.
Henderson, J.F., Paterson, A.R.P., Caldwell, I.C., Paul, B., Chan, M.C. and Lau, K.F., 1972, Cancer Chemotherapy Rep., Part 2, 3:71-85.
Itoh, R., 1981, Biochim. Biophys. Acta, 657:402-410.
Lund, P., Cornell, N.W. and Krebs, H.A., 1975, Biochem. J., 152: 593-599.
Lomax, C.A. and Henderson, J.F., 1973, Cancer Res., 33: 2825-2829.
Van den Berghe, G., Bronfman, M., Vanneste, R. and Hers, H.G., 1977a, Biochem. J., 162:601-609.
Van den Berghe, G., van Pottelsberghe, C., and Hers, H.G., 1977b, Biochem. J. 162:611-616.
Van den Berghe, G., Bontemps, F. and Hers, H.G., 1980, Biochem. J., 188:913-920.
Wilkening, J., Nowack, J. and Decker, K., 1975, Biochim. Biophys. Acta, 392:299-309.

EFFECTS OF FRUCTOSE ON PURINE NUCLEOTIDE METABOLISM IN ISOLATED RAT HEPATOCYTES

S. Brosh, P. Boer and O. Sperling

Department of Clinical Biochemistry, Beilinson Medical Center, Tel Aviv University School of Medicine Petah-Tikva, Israel

In recent years it became apparent that fructose metabolism in the liver tissue has considerable effects on purine nucleotide metabolism. There is ample evidence that fructose metabolism is associated with acceleration of nucleotide degradation (1-5) and evidence is accumulating (6-8), indicating an acceleration effect of fructose also on purine synthesis.

The present investigation was undertaken in an attempt to further clarify the effects of fructose on purine nucleotide metabolism. Isolated rat hepatocytes were chosen as a model system, since they allow the study of each pathway of purine metabolism separately and without interference from other tissues.

EXPERIMENTAL PROCEDURES

Isolated rat liver hepatocytes were prepared as described by Seglen (9), except that the cells were suspended and incubated in Krebs-Henseleit bicarbonate buffer (K-H buffer), supplemented as specified.

Purine synthesis de novo was gauged by the rate of (^{14}C) formate incorporation into total purines produced (10). 1 mM oxonic acid was added to block allantoin formation (11).

5-phosphoribosyl-1-pyrophosphate (PP-rib-P) was extracted from the cells by heat (85°C for 2 min) and assayed by an enzymatic radiochemical method (12).

Cellular PP-rib-P availability for metabolic reactions, was gauged

by measuring the incorporation of (8-^{14}C) adenine (1μCi) into the
intact cell total nucleotide pool and into allantoin (10).

Degradation of purine nucleotides was gauged by the labeling of
the degradation end-product allantoin from nucleotides prelabeled
with (^{14}C) adenine, or during labeling with the purine precursor
(^{14}C) formate.

ATP content was assayed by a bioluminescent method (LKB-Wallace
Luminometer 1250).

Pi concentration was assayed according to Lowry and Lopez (13).

RESULTS

When cells, prelabeled with (8-^{14}C) adenine, were incubated
for 10 min with glucose, there was an 8% decrease in the radioactivity of the nucleotides. In contrast, in cells incubated with
fructose the decrease was 28%, i.e. 3.5-fold greater. The increase
during that period in the radioactivity of allantoin was in presence
of fructose 7-fold that in presence of glucose.

In accordance with the results presented above, the incorporation of (^{14}C) formate into allantoin in presence of fructose was
approximately 2-fold that in presence of glucose (Table 1).

TABLE 1: Effect of glucose and of fructose on degradation of purine
nucleotides during labeling with (^{14}C) formate.

Supplement (28 mM)	Incorporation of label into allantoin* (% of total labeled purines produced) Incubation time (min)	
	10	30
Glucose	20%	33%
Fructose	47%	67%

*the difference between the incorporation values in presence and
absence of oxonic acid.

The rate of de novo purine synthesis in presence of fructose
was only 55% of that in presence of glucose and preincubation of
the cells with fructose (before the addition of (^{14}C) formate)
resulted in even greater deceleration of purine synthesis, being

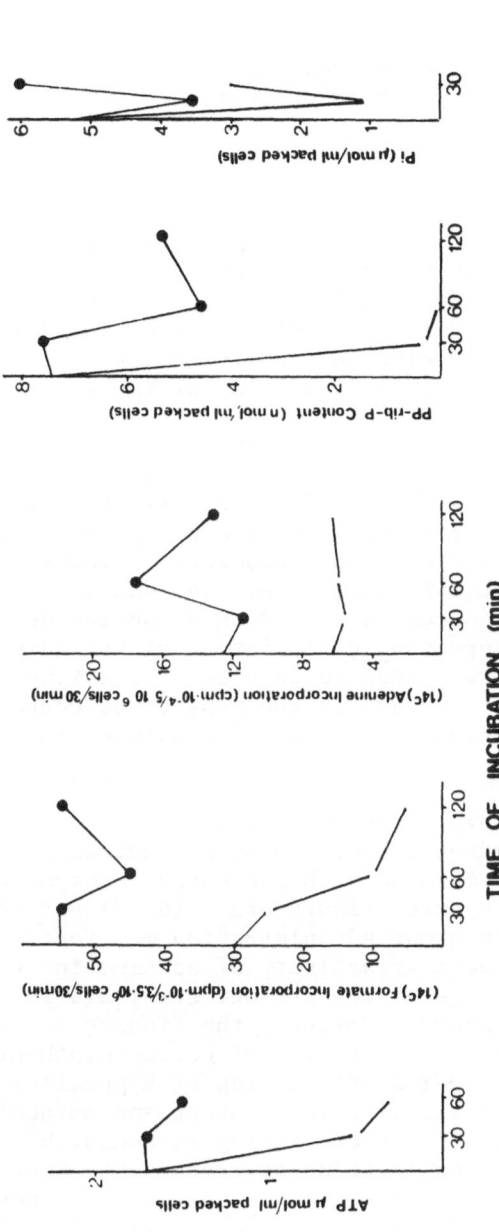

TIME OF INCUBATION (min)

Fig. 1. - Effect of fructose on isolated rat hepatocytes content of ATP, PP-rib-P and Pi, on the rate of de novo purine synthesis ((^{14}C) formate incorporation) and on PP-rib-P availability for metabolic reaction ((^{14}C) adenine incorporation). The time of incubation indicates the time of incubation with 28mM glucose or fructose before the assay of the specified parameter. The measurement of the rate of incorporation of (^{14}C) formate and of (^{14}C) adenine required additional incubation of 30 min, following the incubation period indicated in the figure. Thus, for these two measurements, 0 time of incubation values refer to values obtained during the 30 min assay incubation period alone (without pre-incubation). ● - incubation with glucose; ○ - incubation with fructose.

following 2 h of preincubation only 9% of that in cells preincubated with glucose. In addition, fructose reduced cellular
content of PP-rib-P, of ATP and of Pi (Fig. 1).

Incubation of the hepatocytes, with fructose or glucose, up
to 2 h, did not affect the number of hepatocytes, nor their viability, as determined by the trypan blue exclusion test.

DISCUSSION

The main pathway of purine nucleotide catabolism in the
liver cell was demonstrated to be that from AMP to IMP, inosine,
hypoxanthine and by several additional steps to end-product allantoin (5). This pathway is regulated mainly through the activity
of AMP deaminase, and probably also through the activity of HGPRT.
The activity of these two enzymes is regulated by the availability
of substrates, but in addition, AMP deaminase is subjected to
regulation by several affectors such as activator ATP and inhibitors GTP and Pi (5), whereas HGPRT is regulated by the availability
of cosubstrate PP-rib-P (14). The acceleration effect of fructose
on purine nucleotide degradation was demonstrated by others (4, 5)
to be mediated through activation of AMP deaminase. The results
of the present study support this mechanism. In addition, they
suggest that the fructose-induced acceleration of purine degradation is also mediated by decreased reutilization of hyoxanthine for
IMP synthesis. This effect was deduced in view of the finding of
the marked fructose-induced decrease in the hepatocyte content of
PP-rib-P (Fig. 1), a regulating cosubstrate for salvage IMP synthesis (14).

Acceleration of de novo purine synthesis by fructose was
deduced from results of studies in men, in whom fructose administration was found to be associated with increased incorporation
of labeled precursor glycine into urinary urate (6, 7) and from
studies in mice, in which fructose administration was found to be
associated with increased specific activity of adenine in the
liver tissue (8). The findings of the present study are incompa-
tible with those described above. However, the finding in our
study of the fructose-induced deceleration of purine synthesis is
in full accordance with the additional finding of a parallel
fructose-induced decrease in the cellular content and metabolic
availability of PP-rib-P. (This latter finding is compatible with
the finding of fructose-induced lowering of erythrocyte PP-rib-P
content following administration of fructose to men (2)). Several
mechanisms could underly the fructose-induced lowering of cellular PP-rib-P content. Evidently, the most obvious mechanism is
that through the fructose-induced lowering of cellular concentration of Pi, a potent activator of the PP-rib-P synthetase (15).

An additional possible mechanism is that through inhibition of the above enzyme by fructose metabolites. Such an effect for fructose--1-6-diphosphate was demonstrated by others (16) and confirmed in our laboratory. The same or other fructose metabolites may also inhibit the PP-rib-P amidotransferase, directly affecting the rate of purine synthesis. However, no data is available as yet to support this latter possibility. Another mechanism that should be considered to simultaneously underly the lowering of PP-rib-P level and the deceleration of purine synthesis is that the derangement in purine nucleotides, caused by the fructose-induced depletion of ATP and GTP, could result in altered nucleotide concentration profile, which is more inhibitory to both the PP-rib-P synthetase and the amidotransferase, than the normal profile.

REFERENCES

1.   Perheentupa, J. and Raivio, K. (1967) Lancet 2, 528-531.
2.   Fox, I.H. and Kelley, W.N. (1972) Metabolism 21, 713-721.
3.   Simkin, P.A. (1972) Metabolism 21, 1029-1036.
4.   Smith, C., Rovamo, L. and Raivio, K. (1977) Adv. Exp. Med. Biol. 76A, 535-541.
5.   Van Den Berghe, G., Bronfman, M., Vanneste, R. and Hers H.G. (1977) Biochem. J. 162, 601-609.
6.   Emmerson, B.T. (1974) Ann. Rheum. Dis. 33, 276,-280.
7.   Raivio, K.O., Becker, M.A., Meyer, L.J., Greene, M.L., Muki, G. and Seegmiller, J.E. (1977) Metabolism 24, 861-869.
8.   Itakura, M., Sabina, R.L., Heald, P.W. and Holmes, E.W. (1981) J. Clin. Invest. 67, 994-1002.
9.   Seglen, P.O. (1976) Methods Cell Biol. 13, 29-83.
10. Zoref, E., De Vries, A. and Sperling, O. (1975) J. Clin. Invest. 56, 1093-1099.
11. Johnson, W.Y., Stravic, B. and Chartrand, A. (1969) Proc. Soc. Exp. Med. Biol. 131, 8-12.
12. Fox, I.H. and Kelley, W.N. (1971) J. Biol. Chem. 246, 5739-5748.
13. Lowry, O.H. and Lopez, J.A. (1946) J. Biol. Chem. 162, 421-428.
14. Zoref, E., Sivan, O. and Sperling, O. (1978) Biochim. Biophys. Acta 521, 452-458.
15. Hershko, A., Razin, A. and Mager, J. (1969) Biochim. Biophys. Acta 184, 64-76.
16. Fox, I.H. and Kelley, W.N. (1972) J. Biol. Chem. 247, 2126-2131.

EFFECT OF HORMONES ON ADENINE AND ADENOSINE METABOLISM

IN MAMMARY GLAND IN VITRO

J. Barankiewicz, P. Tysarowski and P. Chomczynski

Department of Biochemistry
Institute of Animal Physiology
Warsaw University of Agriculture
Warsaw, Poland

Little is known regarding the effect of hormones on purine metabolism. It has been shown that injection of some hormones, such as insulin, glucagon, increase PRPP concentration in mouse liver[1]. Also integrative action of insulin in purine metabolism has been reported[2]. According to these authors, hepatic PRPP amidotransferase activity in diabetic rats decreased 50%, and hepatic concentrations of ATP and GTP decreased markedly. Injection of insulin restored enzyme activity and ribonucleotide level increased. Insulin treatment of rat hepatocytes also increased adenine nucleotide pool[3].

On the other hand, it has been reported that ribonucleosides stimulated secretion and biosynthesis of insulin in mouse pancreatic islets[4].

The mammary gland is an ideal tissue for studying hormonal effects on metabolism. It is a tissue in which cellular structure and metabolic activities change markedly between three physiological states: virginity, pregnancy and lactation. It has been clearly demonstrated that a number of physiological and biochemical events leading to the full development of mammary glands are induced by the coordinated action of insulin, prolactin and hydrocortisone[5].

Our recent studies have shown that overall purine metabolism increased considerably in pregnancy and lactation of mouse[6]. The aim of present work was to study how nucleotide synthesis and their metabolism is affected by hormones taking part in activation of mammary gland.

Explants of mammary gland of pregnant mice were cultured with insulin, prolactin and hydrocortisone in different combinations and incubated with radioactive adenine and adenosine. Radioactivity incorporated and measurements of apparent activities of several enzymes of purine metabolism were done by the methods of Henderson et al[7].

Insulin as well as prolactin when used separately enhanced adenine salvage into AMP (Table 1). The effect of insulin was stronger than prolactin. Insulin also resulted in increased AMP conversion to ATP, increased deamination of AMP, and to a smaller extent increased guanine nucleotide synthesis (results not presented). Prolactin also increased markedly adenine salvage into AMP, but more of the synthesized AMP was deaminated than phosphorylated. Hydrocortisone had no significant effects on this system.

Table 1. Adenine metabolism in mammary gland explants of pregnant mice treated with various hormones

Treatment	Apparent Activity (nmoles/mg of tissue)				
	A-PRTase	AMP Kinase	ADP Kinase	AMP Deaminase	IMP Dehydrogenase
Control	329	101	61	189	112
Insulin	749	343	261	228	136
Prolactin	536	107	28	267	198
Hydrocortisone	412	100	48	202	142

Explants (1 mg) of mammary gland were cultured for 24 hrs with insulin, prolactin or hydrocortisone (1 µg of each hormone per ml M-199 medium), were washed twice and incubated in Fisher's medium with (8-^{14}C) adenine (0.1 mM, 40 mCi/mmole) for 2 hrs. After incubation samples were homogenized in cold 0.3 M HClO4 and radioactive purine compounds were extracted and determined[7].

The rate of adenine metabolism was enhanced, when insulin was used with prolactin or hydrocortisone, and it was highest when both of these hormones were added together (Table 2).

Table 2. Adenine metabolism in mammary gland explants of pregnant mice treated with various combination of hormones

Treatment	Apparent Activity (nmoles/mg of tissue)				
	A-PRTase	AMP Kinase	ADP Kinase	AMP Deaminase	IMP Dehydrogenase
Insulin	634	321	264	163	108
Insulin+ Hydro-cortisone	835	335	252	296	174
Insulin+ Prolactin	1219	681	541	378	252
Insulin+ Prolactin +Hydrocortisone	1894	921	870	376	251

Explants of mammary gland were cultured for 24 hrs with insulin alone, and then for the following 24 hrs with insulin, prolactin and hydrocortisone in different combinations. Explants were then incubated for 2 hrs with $(8-{}^{14}C)$ adenine (0.1 mM, 40 mCi/mmole).

When adenosine was used as a substrate, insulin markedly increased AMP, ADP and ATP synthesis (Table 3). Prolactin enhanced these reactions to a much smaller extent and hydrocortisone had no effect. Adenosine deamination and inosine cleavage were not affected by any of the hormones.

The rate of adenosine metabolism was enhanced, when insulin was used with prolactin and hydrocortisone, and it was highest when both of these hormones were added together (Table 4).

Table 3. Adenosine metabolism in mammary gland explants of
pregnant mice treated with various hormones

Treatment	Apparent Activity (nmoles/mg of tissue)				
	Adenosine Kinase	AMP Kinase	ADP Kinase	Adenosine Deaminase	Inosine Phosphorylase
Control	439	221	154	17400	7022
Insulin	1564	1009	850	14400	4380
Prolactin	664	345	266	17569	7360
Hydrocortisone	536	254	165	11002	6880

Explants of mammary gland were cultured for 24 hrs with
insulin, prolactin or hydrocortisone, then incubated with
($8-^{14}$C) adenosine (1 mM, 20 mCi/mmole). For details see Table
1.

This study demonstrates that some hormones active in mammary
gland development: insulin, prolactin, hydrocortisone, also
affect adenine and adenosine metabolism in mammary gland in
vitro. The most pronounced effect of insulin, and to a lesser
extent prolactin, was on the salvage pathway of adenine and
adenosine into AMP and its phosphorylations. The rates of
adenine and adenosine conversions are markedly enhanced when
insulin was used together with prolactin and hydrocortisone. In
contrast, hydrocortisone alone had no significant effect. In
vitro studies have shown that the presence of insulin, glucocorticoid (hydrocortisone) and prolactin is needed for the complex
sequence of cellular events in epithelium leading to the full
development of mammary gland[5]. Elevation of several cellular
processes require increased synthesis of ATP as a source of
energy as well as a substrate for nucleic acid synthesis.
Therefore, increased salvage of adenine and adenosine and following ATP formation by lactogenic hormones may be important in
developing mammary gland.

Table 4. Adenosine metabolism in mammary gland explant of pregnant mice treated with various combination of hormones

	Apparent Activity (nmoles/mg of tissue)				
	Adenosine Kinase	AMP Kinase	ADP Kinase	Adenosine Deaminase	Inosine Phosphorylase
Insulin	1023	9352	8204	15034	14971
Insulin +Hydrocortisone	1641	788	625	18970	18970
Insulin +Prolactin	1988	1528	1261	12804	12600
Insulin +Prolactin +Hydrocortisone	2998	2442	2211	17370	17042

Explants of mammary were cultured for 24 hrs with insulin alone and then for the following 24 hrs with insulin prolactin and hydrocortisone in different combinations. Incubation was continued with (8-^{14}C) adenosine (1 mM, 20 mCi/mmole for 2 hrs. For details see Table 1.

Acknowledgement: This work was supported by the Polish Academy of Sciences within project 09.7. The excellent technical assistance of Mrs. Halina Nasiegniewska is gratefully appreciated.

REFERENCES

1. M. Lalanne and J.F. Henderson, Effect of hormones and drugs on phosphoribosylpyrophosphate concentrations in mouse, liver, Can. J. Biochem. 53:394 (1924).
2. G. Weber, D. Tzeng, M.S. Lui, H. Kizaki and T. Shiotani, Integretise action of insulin on purine and pyrimidine metabolism, Fed. Proc. 39:636 (1980).
3. U.N. Jeejeebhoy, J. Ho, R. Mehra and A. Bruce-Robertson, Hepatotrophic effects of insulin on glucose, glycogen and adenine nucleotides in hepatocytes isolated from fed adult rats, Can. J. Biochem. 58:1009 (1980).

4. A. Andersson, Opposite effects of starvation on oxidation of
       [^{14}C] adenosine and adenosine-induced insulin release
       by isolated mouse pancreatic islets, <u>Biochem. J.</u>
       176:619 (1978).
5. Y.J. Topper and C.S. Freemen, Multiple hormone interactions
       in the developmental biology of the mammary gland,
       <u>Physiol. Rev.</u> 60:1049 (1981).
6. J. Barankiewicz, H. Trembacz and L. Zwierzchowski, Purine
       metabolism in isolated mammary epithelial cells of
       mice, (in preparation).
7. J.F. Henderson, J.H. Fraser and E.E. McCoy, Methods for the
       study of purine metabolism in human cells in vitro,
       <u>Clin. Biochem.</u> 7:338 (1979).

NUCLEOSIDE UTILIZATION BY S AND G_1 CELLS

J. Rigau-Lloveras, E. Olivé-Morros,
M. P. Rivera-Fillat, and M. R. Grau-Oliete

Instituto de Farmacología del C.S.I.C.,
C/. Jorge Girona Salgado s/n, Barcelona (34) España

The present study was undertaken to obtain more information on the uptake and intracellular metabolism of guanosine in murine leukemic lymphoblast (L5178Y).

The study was performed on intact cells in different life cycle phases in order to measure the relative rates of the alternative pathways in relation to a wide range of nucleoside extracellular concentrations, the highest of which inhibited the growth and DNA synthesis.

METHODS

Murine leukemic lymphoblasts (L5178Y) were harvested from BDF_1 hybrid mice (C57BL/6 ♀ x DBA2 ♂) 10 days after i.p. tumor inoculation of 2×10^6 cells.

Cells in S and G_1 life cycle stages were obtained by a volume selection method described elsewhere.[1]

Studies were performed on suspension cultures of selected L5178Y lymphoblasts, incubated "in vitro" (2×10^6 cells/ml) for 1 hour in Fischer medium at 37ºC and in the presence of increasing concentrations of guanosine (1 µM-1 mM) labelled with radioactive ($8-^3H$)guanosine at 5 µCi/ml.

The total amount of radioactivity incorporated into the cells and its distribution in cold acid soluble and insoluble fractions were determined by liquid scintillation counting. The acid soluble metabolites were separated by chromatography on PEI-cellulose thin-layer plates.[2]

RESULTS AND DISCUSSION

The overall uptake of guanosine and the fate of the label in S and G_1 cells at different nucleoside concentrations are shown in Figure 1 and Table I respectively.

Despite the larger amount of precursor taken up by S cells in relation to G_1 cells (Fig.1), the percentage of radioactivity found as unmetabolized guanosine is similar in both phases and increases in proportion to the nucleoside extracellular concentration.

The percentages of precursor found as acid insoluble fraction and nucleotides are high in G_1 and S cells, and they both decrease when the extracellular concentration of guanosine increases. The level of labelling in nucleotides in S cells exceeded that of the insoluble fraction, while in G_1 cells this relationship was inverted.

In all the concentrations used the percentages of labelled bases (guanine plus xanthine) were higher in G_1 cells than in S cells and in both phases only traces of hypoxanthine, xanthosine, adenine and adenosine were found.

The results seem to indicate that purine nucleoside phosphorylase does not restrict the flow of nucleoside through the salvage pathway and that the larger amount of guanosine taken up by

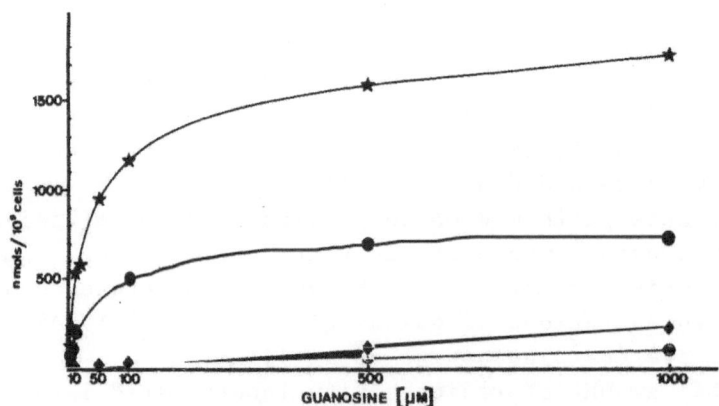

Figure 1: Uptake of guanosine by L5178Y cells in S phase (★) and G_1 phase (●). Unmetabolized guanosine in S cells (◆) and G_1 cells (◎).

TABLE I: Distribution of radioactivity (%).

	S PHASE						G₁ PHASE				
	GUANOSINE EXTRACELLULAR CONCENTRATION μM						GUANOSINE EXTRACELLULAR CONCENTRATION μM				
	1	10	100	500	1000		1	10	100	500	1000
GUANOSINE	1.6	1.7	2.4	7.0	14.7		1.4	1.1	6.1	10.7	15.9
GUANINE	2.8	3.1	9.1	14.3	18.5		3.8	12.0	16.6	10.9	10.3
XANTHINE	0.9	2.2	8.9	8.4	4.4		3.0	6.8	17.1	21.0	20.1
INOSINE	0.1	0.1	0.1	1.1	2.7		0.2	0.5	2.6	4.9	4.6
NUCLEOTIDES	51.9	49.4	44.6	41.6	30.6		38.6	33.5	26.4	15.9	19.5
INSOLUBLE FRACTION	40.8	41.6	33.4	25.8	24.0		51.6	42.0	27.2	26.5	23.0

S cells with respect to G₁ cells could be related to the later metabolic steps.

In Figure 2 we represent the anabolism and catabolism of guanosine in S and G₁ cells against the total amount of precursor taken up by the cells. The anabolism is considered to be equivalent to the total amount of guanosine converted to nucleo-

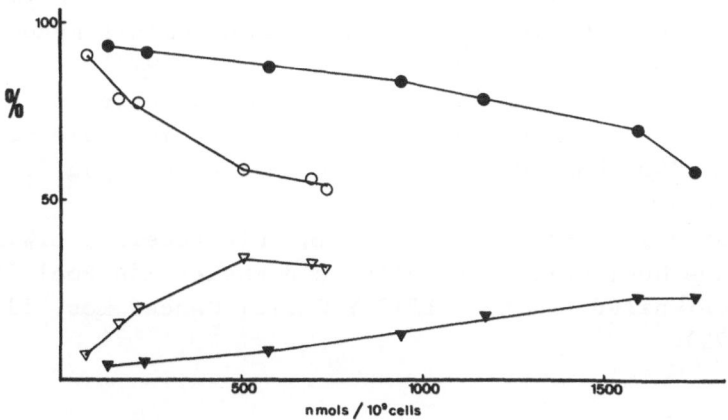

Figure 2: Anabolism and catabolism of guanosine. Abscissa: total amount of precursor taken up by the cells. Ordinate: percentages of anabolism and catabolism. Anabolism in S (●) and G₁ (O) cells. Catabolism in S (▼) and G₁(▽) cells.

tides, which includes the percentage of labelling found in nucleic acids, acid soluble nucleotides and in the nucleosides and
bases that are formed via the catabolism of nucleotides. The catabolism includes the sum of radioactivity found as guanine and
xanthine, since L5178Y cells have no xanthine oxidase activity.[3]

As can be seen, although anabolism predominates in both
cycle phases studied, in G_1 cells the percentage of catabolism
is higher and detected earlier than in S cells.

In S phase the catabolism increases when the total amount
of guanosine taken up by the cells exceeds 600 nmol/10^9 cells,
which coincides with a decrease of thymidine incorporation into
DNA[4]. The greater importance of catabolic route in G_1 cells may
be an expression of the lower guanine phosphoribosyl transferase activity in this stage of the cell cycle.[5]

REFERENCES

1. J. Rigau,M.P.Rivera,M.R.Grau,F.G.Valdecasas. Comportamiento
   de células leucémicas de ratón (L5178Y) en gradientes de
   ficoll. Arch. Farmacol. Toxicol., 4:164 (1978).
2. A. Goday,M.R.Grau,I.Jadraque,M.P.Rivera. Purine Salvage Pathway in leukemic cells. In "Purine Metabolism in Man-III",
   A.Rapado,R.W.E.Watts,C.H.M.M.De Bruyn ed. Plenum Press
   New York and London (1979).
3. E. M.Scholar,P.Calabresi. Identification of the Enzymatic
   Pathways of Nucleotide Metabolism in Human Lymphocytes
   and Leukemia Cells. Cancer Res. 33:94 (1973).
4. A. Goday,M.R.Grau,M.P.Rivera. Efecto de precursores purínicos
   de ácidos nucleicos sobre la proliferación celular. Sim-
   posio "40 años de Farmacología en España", pag.239.Fac.de
   Medicina ed. Barcelona (1981).
5. F. F.Snyder,J.F.Henderson,S.C.Kim,A.R.P.Paterson,L.W.Brox.
   Purine Nucleotide Metabolism and Nucleotide Pool Sizes in
   Synchronized Lymphoma L5178Y Cells. Cancer Res. 33:2425
   (1973).

EVIDENCE FOR AN ADENOSINE RECEPTOR IN HUMAN TISSUES*

Laureen Kurpis, David John** and Irving H. Fox

The Human Purine Research Center
Departments of Internal Medicine and
Biological Chemistry
University Hospitals
Ann Arbor, Michigan

INTRODUCTION

Adenosine is an intermediate of the pathway of purine nucleotide degradation. Many biological properties of adenosine have been identified: It is toxic to mammalian and bacterial cells, and its presence is associated with inhibition of the immune response, coronary vasodilation, delayed neurotransmission, sedation, inhibition or stimulation of hormone secretion and changes in the metabolism of a number of tissues.[1]

Studies of the effect of adenosine and its analogs on adenylate cyclase have defined two distinct adenosine sensitive sites.[2] One extracellular site, the "R" site, may either activate or inhibit adenylate cyclase and this site requires the integrity of the ribose ring for activity. The other site, an intracellular site, is directly associated with the adenylate cyclase itself and has been termed the "P" site. This mediates inhibition of adenylate cyclase and requires integrity of the purine ring for activity.

---

*   Supported by USPHS grant AM19674 and 5 MO1 RR00042 and a grant
    from Michigan Heart Association and Warner/Lambert-Parke/Davis.
** Dr. John is supported by an Arthritis Foundation Postdoctoral
    Fellowship.

Two subclasses of external adenosine receptors have been
defined, Ra and Ri[3,4] or Al and A2 receptors. Both receptors are
GTP dependent and are inhibited by methylxanthines. The Ri or Al
receptor is characterized by having inhibitory properties toward
adenylate cyclase. Adenosine and its analogs have affinity constants
in the range of 10 to 100 nM for Al receptors. In contrast, the Ra
or A2 receptor is a low affinity extracellular adenosine receptor.
Adenosine and adenosine analogs have affinity constants of 2 to 20 μM
for this receptor. High affinity specific binding of adenosine analogs has been observed in guinea pig, bovine and rat brain membranes.[5-9]

We have evaluated the human placental microsomal fraction for
properties related to an adenosine receptor. Our studies provide
evidence to support the existance of a high affinity inhibitory
receptor in this tissue.

METHODS

The binding assay was performed in triplicate in a total volume
of 850 μl containing 400 μg protein of a crude membrane preparation,
20 nM [2-^{3}H]chloroadenosine and incubation buffer consisting of 50 mM
Tris-HCl pH 7.5, 2.5 mM $MgCl_2$ and 2.5 mM $CaCl_2$. The mixture was
incubated at 4°C for 40 minutes. Duplicate sets of incubation mixtures had 10 μM nonisotopic chloroadenosine added. At the end of the
incubation period 4 ml of cold incubation buffer were added to each
incubation medium and which was then rapidly filtered by vacuum on
Whatman GF/C filters. The filters were washed twice with 4 ml of
cold incubation buffer and dried. The filters were then counted in
a liquid scintillation spectrometer system. In these experiments,
specific binding is defined as the cpm of total chloroadenosine
binding minus cpm of fraction with 10 μM chloroadenosine. Specific
binding amounted to 65-80% of the total bound counts.

Adenylate cyclase was assayed by a modification of Salomon, et
al.[10]

RESULTS

Saturation of binding sites occurred with 300 fmol ligand bound/
400 μg protein or 0.75 pmol ligand bound/mg protein. At a ligand
concentration of approximately 30 nM, half the binding sites were
occupied. This value represents an estimate of the dissociation
constant of [^{3}H]chloroadenosine for the binding sites. Analysis of
this data by Scatchard plot indicates a single class of binding sites
with a Kd of 56 nM. This analysis gives an estimate of the concentration of binding sites of 433 pM for the 400 μg protein in the
incubation medium or 1.1 pmol ligand bound/mg protein.

Table 1.   Adenosine Binding to Human Crude
           Membrane Preparations

Tissue	Analogue	Specific Binding (fmol)	Specific Binding (%)
Brain	[^{3}H]2-Chloro-adenosine (20 nM)	126	53
Brain	[^{3}H]-Cyclohexyl-adenosine (10 nM)	56	49
Testis	[^{3}H]2-Chloro-adenosine (20 nM)	186	84
Spleen	[^{3}H]2-Chloro-adenosine (20 nM)	60	62

The inhibition of [^{3}H]chloroadenosine binding by adenosine receptor antagonists was studied.   Isobutylmethylxanthine was a more potent inhibitor than theophylline with a Kd of 23 μM as compared to 154 μM.

Chloroadenosine binding to placental microsomes reached equilibrium at approximately 20 minutes with a $T_{\frac{1}{2}}$ of 4 minutes.   The observed forward rate constant was 0.24 min^{-1} giving a second order rate constant (k_1) of 0.71 x 10^8.   At equilibrium, the addition of 10 μM chloroadenosine resulted in a rapid displacement of [^{3}H]chloroadenosine which was completed by 5 minutes.   The first order rate constant (k_2) for the reversal of [^{3}H]chloroadenosine binding was -1.17 min^{-1}. The kd calculated from the reverse rate constant was 17 nM which was the same order of magnitude as the Kd estimated from equilibrium data (56 nM).

There is evidence to suggest substantial binding of adenosine analogs to other human tissues (Table 1).

Preliminary studies indicate possible inhibition of adenylate cyclase by chloroadenosine (Table 2).   However, more work needs to be accomplished before this can be considered proven.

Table 2.   Effect of Chloroadenosine on Adenylate
           Cyclase (Preliminary Study)

Condition	Adenylate Cyclase (pmol CAMP/15 min)
No addition	4
Isoproterenol 100	7
Isoproterenol 100 +	5.5
Chloroadenosine 500 nM	

SUMMARY AND CONCLUSIONS

Our observations suggest that [^{3}H]chloroadenosine, an adenosine receptor agonist, identifies binding sites in human placenta with characteristics of the high affinity, adenosine receptor. The binding is time dependent, reversible and saturable. The potency series of adenosine receptor methylxanthine antagonists in displacing [^{3}H] chloroadenosine is appropriate.

ACKNOWLEDGEMENTS

The authors wish to thank Holly Gibson for excellent typing of the manuscript.

BIBLIOGRAPHY

1. I.H. Fox and W.N. Kelley, Ann. Rev. Biochem. 47:655 (1978).
2. C. Londos and J. Wolff, Proc. Nat. Acad. Sci., USA, 74:5482 (1977)
3. C. Londos, D.M.F. Cooper, W. Schlegel and M. Rodbell, Proc. Nat.
     Acad. Sci., USA, 75:5362 (1978).
4. C. Londos, D.M.F. Cooper and J. Wolff, Proc. Nat. Acad. Sci., USA,
     77:2551 (1980).
5. R.F. Bruns, J.W. Daly and S.H. Snyder, Proc. Nat. Acad. Sci., USA,
     77:5547 (1980).
6. T. Trost and U. Schwabe, Mol. Pharmacol., 19:228 (1981).
7. L. Kurpis and I.H. Fox, Clin. Res. 29:705A (1981).
8. P.H. Wu, J.W. Phillis, K. Bulls and B. Rinaldi, Can. J. Physiol.
     Pharmacol., 58:576 (1980).
9. U. Schwabe and T. Trost, Arch. Pharmacol., 313:179 (1980).
10. Y. Salomon, C. Londos and M. Rodbell, Anal. Biochem., 58:541,
     (1974).

ADENOSINE RECEPTORS ON HUMAN INFLAMMATORY CELLS

Gianni Marone, Rosaria Petracca and Mario Condorelli

Department of Medicine, University of Naples II School
of Medicine, Via S. Pansini 5, 80131 Naples, Italy

Adenosine, a naturally occurring purine nucleoside, plays an important role in controlling a variety of biological activities. Adenosine is formed within cells from AMP or from the degradation of S-adenosylhomocysteine and can be removed from cells by release into the extracellular fluid, by deamination to inosine, or by reutilization to form AMP (1). Adenosine is present in human blood and in most of the extracellular fluid at levels (3×10^{-7} M to 3×10^{-6} M) which in vitro modulate several biochemical and biological functions (2).

Evidence that purine metabolism is important in the immune response has been obtained from the observation that markedly reduced or absent adenosine deaminase (ADA) activity in man has been casually associated with an autosomal recessive form of severe combined immunodeficiency disease (3). Recently, ADA levels in lymphocytes from patients with untreated chronic lymphatic leukemia have been found to be consistently lower than in lymphocytes from normal subjects (4). Children with ADA deficiency and immunodeficiency have been shown to have increased levels in plasma, urine, lymphocytes and erythrocytes of adenosine, adenine, deoxyadenosine, adenine nucleotides, and deoxyadenine (5, 6). Although the exact biochemical mechanism(s) is unknown, elevated levels of adenosine, and/or deoxyadenosine and their metabolites are thought to be selective inhibitors of both differentiation and effector function of lymphocytes (7, 8). Adenosine was known to inhibit the PHA-induced blastogenesis of human peripheral blood lymphocytes (9) even before the discovery of the first ADA-deficient child. In addition, elevated levels of cyclic AMP (cAMP) were known to be inhibitory for lymphocyte-mediated cytotoxicity (7). Since adenosine was known to cause elevations in cAMP in a variety of cell types (7, 10, 11), the immunodeficiency associated with ADA deficiency was initially postulated to arise through adenosine accumulation and subsequent elevation of cAMP in human lymphocytes. Several investigators demonstrated that cAMP, when directly added to the culture medium, was capable of inhibiting blastogenesis and the addition of inhibitors of ADA potentiated its toxicity (12). However, after careful analysis of the many effects of cAMP, it was concluded that the toxicity of cAMP arose from its slow conversion to adenosine in the tissue culture medium (13).

It was subsequently shown that adenosine itself was able to increase intracellular cAMP and to inhibit lymphocyte-mediated cytolysis (7). More recently, exogenous adenosine has been demonstrated to inhibit several immunological functions such as

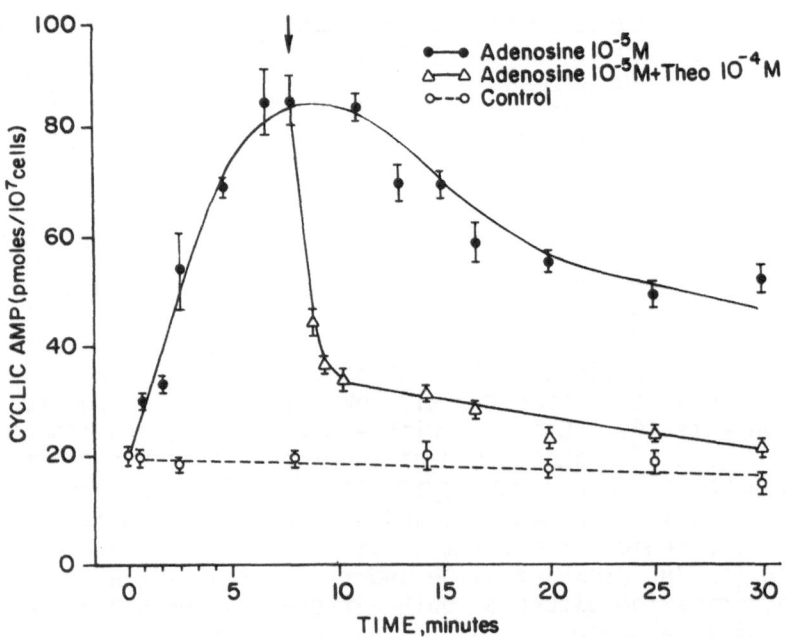

Fig. 1   Cyclic AMP (cAMP) levels in human lymphocytes incubated
         with (●———●) or without (O-----O) adenosine for varying
         time intervals. cAMP levels were also determined in
         lymphocytes incubated with theophylline (Δ-----Δ).

monocyte chemotaxis (14), lymphocyte transformation (9, 12), human basophil and mast cell histamine release (15, 16). Although increased intracellular concentrations of cAMP have clearly been shown to inhibit all the above mentioned activities, the possible role of increased cAMP content in the immunodeficiency seen in ADA-deficient patients is not well established.

In recent studies directed in this field, we have explored the effects of adenosine alone and of adenosine derivatives on the lymphocyte cAMP levels. We have found that exogenous adenosine and 2-chloroadenosine (10^{-7} M - 10^{-4} M) induce a dose-dependent increase in human peripheral blood lymphocyte cAMP levels (10, 17). These experiments were carried out with a 10 min incubation period based on kinetic studies such as those shown in figure 1. The intracellular concentration of cAMP increases within the first minute of exposure to the nucleoside; the peak level is achieved between 7 and 10 min and the level remains elevated for at least 20-30 min, but decreases promptly if low concentrations of theophylline are added. We have in fact, shown that methylxanthines, such as 1-isobutyl-1-methylxanthine (IBMX) and theophylline, in low concentrations specifically block the accumulation of cAMP induced by adenosine in lymphocytes. We have also demonstrated that theophylline is a competitive inhibitor of the effect of adenosine, with an estimated dissociation constant of the theophylline-receptor complex of approximately 6.3×10^{-7} M (10, 18). Our results suggest that adenosine and 2-chloroadenosine increase the intracellular cAMP content of human lymphocytes by interacting with a specific membrane receptor causing activation of adenylate cyclase. This adenosine receptor resembles the A2 receptor for adenosine according to the biochemical and pharmacological characteristics recently described (19). Whether these findings are relevant to the immunosuppressive and modulatory role of adenosine in normal physiological and ADA-deficient patients remains to be seen.

Since the basophil release reaction is one of the central steps in inflammatory reactions, we have also investigated the ability of exogenous adenosine to modulate IgE-mediated histamine release from human basophils. We have demonstrated that adenosine and its poorly metabolized derivative 2-chloroadenosine cause a dose-dependent inhibition of IgE-mediated histamine release from human basophils (15,16). Adenosine and 2-chloroadenosine appear to act on the cell surface of human basophils since 1) dipyridamole, which markedly inhibited ^{3}H-adenosine uptake by human leukocytes, did not affect adenosine-induced inhibition of histamine secretion and 2) theophylline blocks the inhibition of release caused by the nucleoside. We have concluded that endogenous adenosine is probably one of the several autacoids that modulate the release reaction of human basophils.

Intracellular adenosine may also react with L-homocysteine to form S-adenosyl-L-homocysteine which is a potent inhibitor of

S-adenosyl-L-methionine-mediated reactions (1) involving the transfer
of a methyl group to various acceptors. Methylation of membrane
phospholipids seems to play some role in the control of histamine
release. We have recently explored the possible role of intracellular
methylation in the control of histamine release from human basophils.
We have found that the adenosine analogues, 3-deazaadenosine and 5'-
deoxy-5'-S-isobutyladenosine, which are potent inhibitors of transmethylation, inhibit IgE-mediated histamine release when incubated
with L-homocysteine thiolactone (16). In addition, we have found
that exogenous adenosine, in the presence of an ADA inhibitor and L-
homocysteine thiolactone, markedly inhibits IgE-mediated secretion
from human basophils (16). Thus, histamine secretion can be inhibited
by two separate effects of adenosine: interaction with an A2 receptor
linked to adenylate cyclase and inhibition of intracellular transmethylation.

Recently, Lichtenstein and co-workers have studied the effect
of adenosine on IgE-mediated histamine release from mast cells
purified from human lung (20). Adenosine also inhibits IgE-mediated
histamine secretion in a dose-dependent fashion within this system.
These data suggest that human lung mast cells, like peripheral
basophils, probably have an A2 receptor linked to adenylate cyclase.

The characterization of a specific A2 receptor linked to
adenylate cyclase on several human inflammatory cells (including
lymphocytes, basophils and mast cells) suggest that adenosine might
increase intracellular cAMP and thereby block a variety of effector
function (2, 7, 9, 10, 13, 15).

Recent studies have shown the presence of two types of extracellular adenosine receptors on the membrane of several tissues by
which adenosine and its derivatives regulate adenylate cyclase. The
A2 site mediates the excitatory effects on adenylate cyclase:adenosine
and its analogues have a low affinity for this extracellular site,
and the stimulatory effects are blocked by theophylline (10, 19).
The A1 receptor apparently mediates an inhibitory effect of adenosine,
N^6-L-phenylisopropyl-adenosine (L-PIA) and 2-Chloroadenosine on
adenylate cyclase (19). Low concentrations of these nucleosides also
block the cAMP accumulation induced by agonists of adenylate cyclase,
such as isoproterenol and PGE_2 (21). Adenosine and its derivatives
have an high affinity for this site and their effects are also
blocked by xanthines such as caffeine and theophylline (22).

We have recently explored the possible presence of the A1
receptor on human lymphocytes by studying the effect, in peripheral
lymphocytes, of low concentrations of L-PIA alone and in combination
with known agonists of adenylate cyclase, such as isoproterenol,
PGE_1, and histamine. As shown in figure 2, L-PIA, in low concentrations
(10^{-9} M - 10^{-7} M), has a modest effect on the basal level of cAMP in

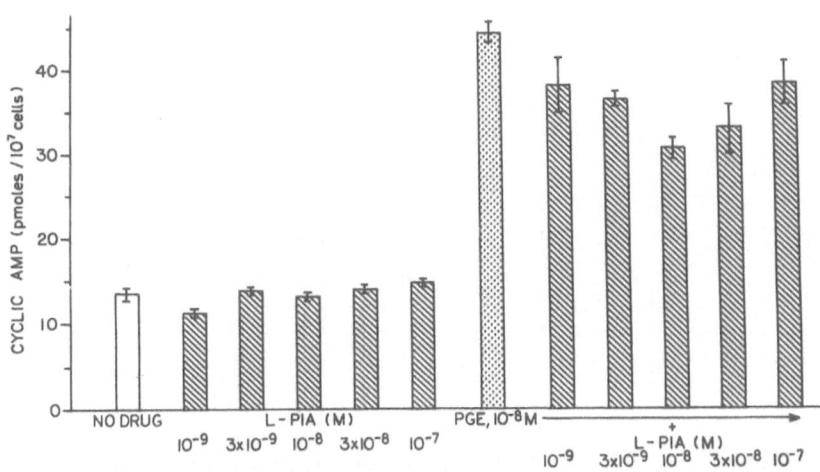

Fig. 2 The effect of L-PIA on the elevation by PGE_1 (10^{-8} M) of the level of cAMP in human lymphocytes.

human peripheral lymphocytes. PGE_1, at suboptimal concentration increases cAMP content approximately four-fold. Preincubation of lymphocytes with low concentration of L-PIA partially blocks the stimulatory effect of PGE_1. It is important to observe that the inhibitory effect of L-PIA on cAMP accumulation induced by agonists of adenylate cyclase in human peripheral lymphocytes is incomplete, but is usually 10 - 25 % of the stimulatory effect of these agonists. It has been previously suggested that adenosine receptors only exist in a minor part of the population ($\simeq$ 10%) of human peripheral lymphocytes and that the majority of these cells ($\simeq$ 80%) were T lymphocytes (24). Our results, indicating a small inhibitory effect of L-PIA on the accumulation of cAMP by agonists of adenylate cyclase, are compatible with this observation. Further studies on the selective distribution of adenosine receptors on subpopulations of human peripheral blood lymphocytes might provide additional information on the biological relevance of adenosine in the control of human inflammatory and immunological reactions.

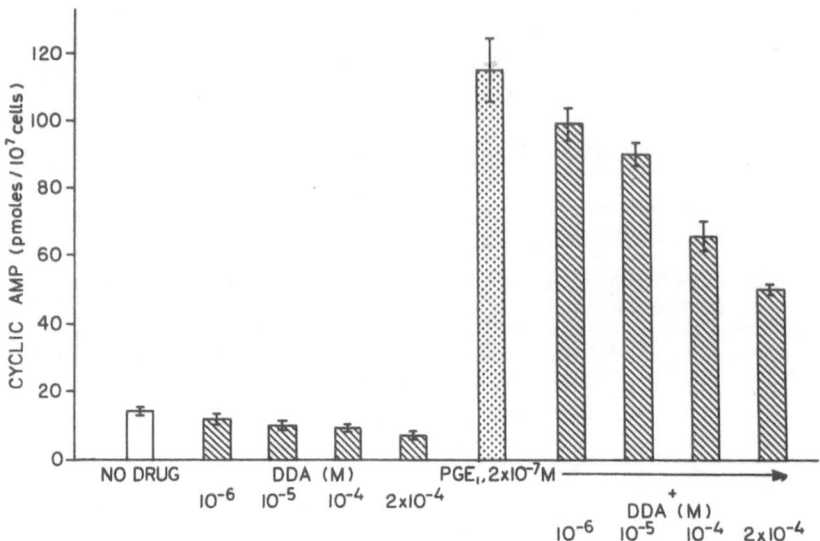

Fig. 3   The effect of DDA on the elevation by PGE₁ (2 x 10^{-7} M) of the level of cAMP in human lymphocytes.

It has also been suggested that adenosine and its derivatives might influence adenylate cyclase by an intracellular regulatory unit (P-site). We have investigated the effect of a P-site effector, 2',5'-dideoxyadenosine (DDA) on human lymphocytes. DDA (3 x 10^{-6} - 2 x 10^{-4} M) decreases the cAMP content of human peripheral lymphocytes, thus confirming its inhibitory effect on adenylate cyclase observed in other tissues (26). We have also studied the effect of DDA on the cAMP accumulation induced by agonists of adenylate cyclase. Figure 3 shows that DDA dose-dependently inhibits the stimulatory effect of PGE₁ on cAMP accumulation in human lymphocytes. We have also observed that the inhibitory effect of DDA is extremely rapid (T½ ≃ 2 min). Furthermore, the inhibitory action of DDA is not blocked by theophylline.

Recently, we have explored the effect of DDA on another secretory system, such as the platelet release reaction and aggregation response. In these studies we found that DDA, by itself, has no effect on both platelet aggregation and PF4 release induced by several stimuli (ADP, collagen, thrombin, and adrenalin). Interestingly enough, DDA was able to block the inhibitory effect of PGE_1 and adenosine itself on platelet aggregation and release. These studies suggest that the P-site, explored by DDA, may well be relevant in the modulation of some inflammatory reactions.

Several observations can be drawn from these studies. We and others have demonstrated that adenosine and its analogues, presumably by the activation of A2 receptors increase intracellular cAMP of human inflammatory cells and modulate a variety of immunological reactions. Additional studies have also emphasized the importance of intracellular adenosine in the control of S-adenosyl-L-methionine-mediated reactions. More recent studies suggest that at least a small percent of peripheral lymphocytes possess A1 receptors which block the effect of agonists of adenylate cyclase. We have also presented some evidence that the P-site effector, DDA, negatively modulates adenylate cyclase and agonists of this enzyme in human peripheral lymphocytes and probably platelets. Although adenosine and its derivatives have already many biological properties in vitro, the in vivo relevance of these observations in pathophysiological conditions is, however, still unclear.

In conclusion, the role of adenosine in the modulation of the immune system is emphasized by both the association of inherited (3) and acquired (4, 2)) immunodeficiencies with alterations of its metabolism, and by its marked biological and biochemical effects. This nucleoside appears to control the immune system in vitro both by interacting extracellularly with A1 and A2 membrane receptors and also intracellularly by inhibiting transmethylation and by acting on a P-site. Although adenosine is not the only endogenous modulator of the immune function, the above cited effects and the observation of patients with ADA deficiency suggest that this nucleoside may well have a significant role in the control of human inflammatory and immune reactions.

ACKNOWLEDGMENTS

This work was supported in part by Grants 79.02392.65 and 80. 00501.04 from the C.N.R. (Rome, Italy).

REFERENCES

1. F. Salvatore, E. Borek, V. Zappia, H.G. Williams-Ashman, and F. Schlenk, "The Biochemistry of Adenosylmethionine", Columbia University Press, New York (1977).
2. G. Marone, M. Plaut, and L.M. Lichtenstein, The role of adenosine in the control of immune function, Ric. Clin. Lab. 10:303 (1980).

3.   R. Hirschhorn and D.W.Jr. Martin, Enzyme defects in immunodefi-
     ciency diseases, Semin. Immunopathol. 1: 299 (1978).
4.   F. Ambrogi, B.Grassi, S. Ronca-Testoni, and G. Ronca, Blood
     lymphocytes in chronic lymphocytic leukaemia and Hodgkin's
     disease: immunological features and enzymes of nucleoside meta-
     bolism, Clin. Exp. Immunol. 28: 80 (1977).
5.   F.C. Schmalstieg, J.A. Nelson, G.C. Mills, T.M. Monahan, A.S.
     Goldman, and R.M. Goldblum, Increased purine nucleotides in a
     adenosine deaminase-deficient lymphocytes, J. Pediat. 91: 48
     (1977).
6.   J.F. Kuttesch, F.C. Schmalstieg, and J.A. Nelson, Analysis of
     adenosine and other adenine compounds in patients with immuno-
     deficiency diseases, J. Liquid Chromatogr. 1: 97 (1978).
7.   G. Wolberg, T.P. Zimmerman, K. Hiemstra, M. Winston, and L.C.
     Chu, Adenosine inhibition of lymphocyte-mediated cytolysis:
     possible role of cyclic adenosine monophosphate, Science 187:
     957 (1975).
8.   J. Uberti, J.J. Lightbody, and R.M. Johnson, The effect of
     nucleosides and deoxycoformycin on adenosine and deoxyadenosine
     inhibition of human lymphocyte activation, J. Immunol. 123:
     189 (1979).
9.   R. Hirschhorn, J. Grossman, and G. Weissmann, Effect of cyclic
     3',5'-adenosine monophosphate and theophylline on lymphocyte
     transformation, Proc. Soc. Exp. Biol. Med. 133: 1361 (1970).
10. G. Marone, M. Plaut, and L.M. Lichtenstein, Characterization of
     a specific adenosine receptor on human lymphocytes, J. Immunol.
     121: 2153 (1978).
11. J.N. Fain and C.C. Malbon, Regulation of adenylate cyclase by
     adenosine, Mol. Cell. Biochem. 25: 143 (1979).
12. D.A. Carson and J.E. Seegmiller, Effect of adenosine deaminase
     inhibition upon human lymphocyte blastogenesis, J. Clin. Invest.
     57: 274 (1976).
13. J.W. Smith, A.L. Steiner, and C.W. Parker, Human lymphocyte
     metabolism. Effects of cyclic and non-cyclic nucleotides on
     stimulation by phytohemagglutinin, J. Clin. Invest. 50: 442
     (1971).
14. M.C. Pike, N.M. Kredich, and R. Snyderman, Requirement of S-
     adenosyl-L-methionine-mediated methylation for human monocyte
     chemotaxis, Proc. Natl. Acad. Sci. 75: 3928 (1978).
15. G. Marone, S. Findlay, and L.M. Lichtenstein, Adenosine receptor
     on human basophils: modulation of histamine release, J. Immunol.
     123: 1473 (1979).
16. G. Marone, R. Snyderman, and L.M. Lichtenstein, Modulation of
     IgE-mediated basophil histamine release by adenosine and by two
     adenosine analogs, in: XI Congress of the European Academy of
     Allergology and Clinical Immunology, 1980.
17. G. Marone, M. Valentine, and L.M. Lichtenstein, Modulation of
     cyclic AMP in human leukocytes by physiological doses of adeno-
     sine, Fed. Proc. 37: 1688 (1978).

18. G. Marone, M. Plaut, S. Findlay, and L.M. Lichtenstein, The
     adenosine receptor on human leukocytes, in: First International
     Colloquium on Receptors, Capri, 1979.
19. J.W. Daly, R.F. Bruns, and S.H. Snyder, Adenosine receptors in
     the central nervous system: relationship to the central actions
     of methylxanthines, Life Sciences 28: 2083 (1981).
20. L.M. Lichtenstein, Mediators and mechanisms of release from
     purified human basophils and mast cells, in: The American
     Academy of Allergy, Montreal, 1982.
21. D. van Calker, M. Muller, and Hamprecht, Adenosine inhibits the
     accumulation of cyclic AMP in cultured brain cells, Nature
     276: 839 (1978).
22. R.F. Bruns, J.W. Daly, and S.H. Snyder, Adenosine receptors in
     brain membranes: binding of N^6-cyclohexyl(^{3}H)adenosine and 1,3-
     diethyl-8-(^{3}H)phenylxanthine, Proc. Natl. Acad. Sci. 77: 5547
     (1980).
23. T. Trost and U. Schwabe, Adenosine receptors in fat cells.
     Identification by (-)-N^6-(^{3}H)phenylisopropyladenosine binding.
     Molec. Pharmacol. 19: 228 (1981).
24. C. Moroz and R.H.Stevens, Suppression of immunoglobulin production
     in normal human B lymphocytes by two T-cell subsets distinguished
     following in vitro treatment with adenosine, Clin. Immunol. Im-
     munopathol. 15: 44 (1980).
25. C. Londos and J. Wolff, Two distinct adenosine-sensitive sites
     on adenylate cyclase, Proc. Natl. Acad. Sci. 74:5482 (1977).
26. R.J. Haslam, M.M.L. Davidson, and J.V. Desjardins, Inhibition of
     adenylate cyclase by adenosine analogues in preparations of
     broken and intact human platelets. Biochem. J. 176: 83 (1978).
27. R. Tung, R. Silber, F. Quagliata, M. Conklyn, J. Gottesman, and
     R. Hirschhorn, Adenosine deaminase activity in chronic lymphocytic
     leukemia. Relationship to B- and T-cell subpopulations. J.Clin.
     Invest. 57: 756 (1976).

BIOCHEMICAL MECHANISMS FOR SEX SPECIFIC DiFFERENCES OF RAT LIVER XANTHINE OXIDASE

Dennis J. Levinson and Douglas E. Decker

Division of Rheumatology
Michael Reese Hospital and Medical Center
Chicago, IL

Xanthine oxidase (EC 1.2.3.2) catalyzes the irreversible oxidation of hypoxanthine and xanthine to uric acid. Recently, we have shown that rat liver xanthine oxidase enzyme activity is, in part, dependent on both age-and sex-specific differences.[1] Immaturity in both sexes, adult females and pubertal male castrates demonstrate a basal or feminine pattern of enzyme activity. Androgen is required in the pubescent period for the full expression of hepatic xanthine oxidase activity in the adult male. The effect of androgen exposure on hepatic enzyme activity, however, remains speculative.

Various mechanisms controlling enzyme activity have been proposed.[2-5] These include interconvertible forms of xanthine oxidase; an NAD^+ utilizing dehydrogenase (D-form) which is readily converted into an oxidase (O-form) by proteolytic treatment, purification, and incubation in anaerobic conditions. The low aerobic rate of oxidation by the D-form is stimulated significantly by NAD^+. Other factors include immunological cross reactive material without catalytic activity in chick liver organ culture, and induced enzyme synthesis in starved rats fed high protein diets. A final mechanism controlling xanthine oxidase activity could be a direct sex steroid-enzyme interaction. The present study was undertaken to investigate further a causal relationship between the sexual dimorphism of rat liver xanthine oxidase and the various mechanisms known to affect enzyme activity. For this purpose, we have developed a sensitive and specific radioimmunoassay (RIA) for quantitation of enzyme protein.

The purification scheme for rat liver xanthine oxidase was modified from the procedure described by Johnson, et al, and is shown in Table 1.[6] Antisera was prepared in adult New Zealand white rabbits immunized with purified xanthine oxidase emulsified in Freund's

Table 1.   Purification of Crude Rat Liver Xanthine Oxidase

Step	Procedure	Specific Activity (units/mg prot/30 min)	Total Enzyme Activity (units/100g prot/30 min)	Percent Recovery
1.	Heat Inactivation 50 Percent $(NH_4)_2$ $SO4$ Fractionation	1100	880,000	100
2.	Acetone Fractionation 33–43 Percent	3376	765,181	87
3.	DEAE Sephacel Ion Exchange Chromatography	29,894	165,424	19
4.	Sephacryl S-300 Gel Filtration	71,052	112,102	13
5.	Disc Gel Electrophoresis	92,300	39,647	5

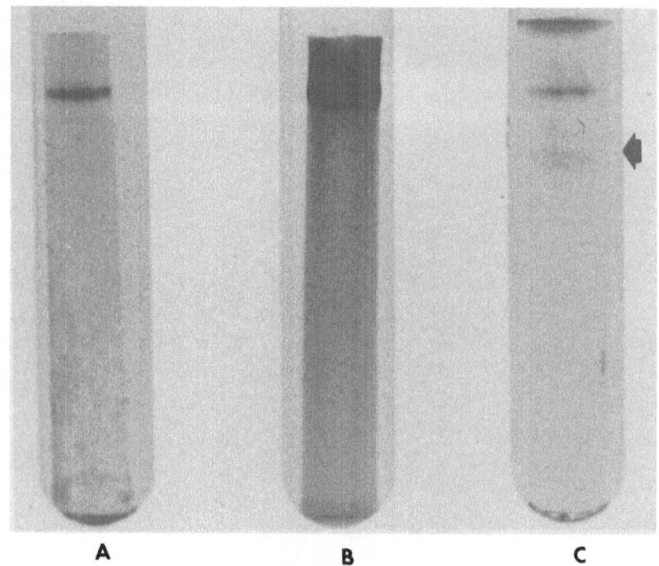

Figure 1. PAGE of purified rat liver xanthine oxidase. (A) Coomassie blue staining. (B) Enzyme specific stain. (C) Coomassie blue stain after Sephacryl S-300 showing protein contamination (arrow).

adjuvant. Polyacrylamide gel electrophoresis (PAGE) was required for the final purification step to remove trace contaminants that could otherwise be visualized by Coomassie blue staining (Fig. 1C). Homogenous xanthine oxidase (Table 1, Step 5) and an immunoprecipitate prepared from crude rat liver homogenate reacted with rabbit antisera were dissociated and subjected to SDS-PAGE. A single protein band (Fig. 2B) coincident with purified xanthine oxidase dimer (Fig. 2A) was seen. Also, a single immunoprecipitin band with enzyme activity was seen when antisera was reacted with crude homogenate in Ouchterlony double immunodiffusion (Fig. 3). In addition to immunizing rabbits, the purified protein was also used for ^{125}I-labelling with Bolten-Hunter reagent and for the standards in the RIA. ^{125}I-labelled xanthine oxidase was shown to have the same immunoreactivity as enzyme in crude liver homogenates as determined by parallelism between a standard curve and serial dilutions of liver homogenate (Fig. 4). The standard curve was linear from 5 ng to 250 ng of enzyme protein. Crude rat liver xanthine oxidase mass, determined by RIA, showed a linear correlation with enzyme specific activity further assuring antibody specificity (r = .843, p < .001).

Employing the RIA, the mature male rat was found to have 20 percent greater hepatic xanthine oxidase protein than mature female (mean ± SE; male, 2.72 ± 0.34 µg/mg protein; female, 2.26 ± 0.27

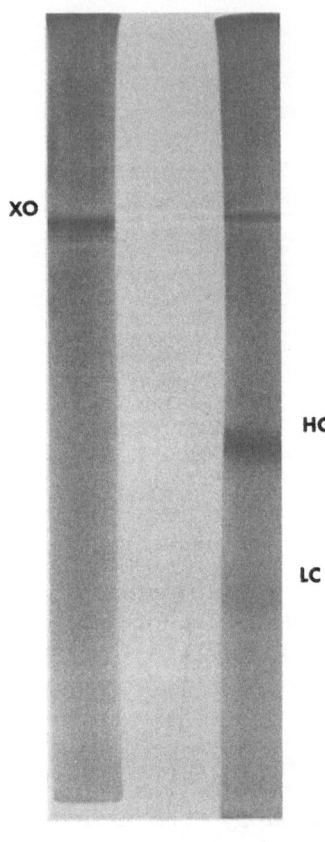

A   B

Figure 2. SDS-PAGE showing a single protein band (XO) of purified
xanthine oxidase coincident with an immunoprecipitate. Forty μg of
purified enzyme (A) and 60 μg of an immunoprecipitate (B) prepared
from antiserum and crude liver homogenate were applied to gels. Gels
were stained with Coomassie blue. Immunoglobulin heavy (HC) and
light (LC) are seen.

μg/mg protein; p < .002) (Fig. 5). Both the linear correlation between enzyme mass and specific activity, and the similar absolute
specific activities for male and female xanthine oxidase based on
microgram of enzyme protein as determined by the RIA (mean ± SE;
male, 50 ± 3 nmole/μg protein n^{-1}; female, 52 ± 2 nmole/μg protein
h^{-1}; p > .05) support the notion that enhanced activity in the male
is due to a greater enzyme concentration. The data do not support
the existence of an immunologically identical, but inactive enzyme
species.

Interconvertible forms of xanthine oxidase were examined by

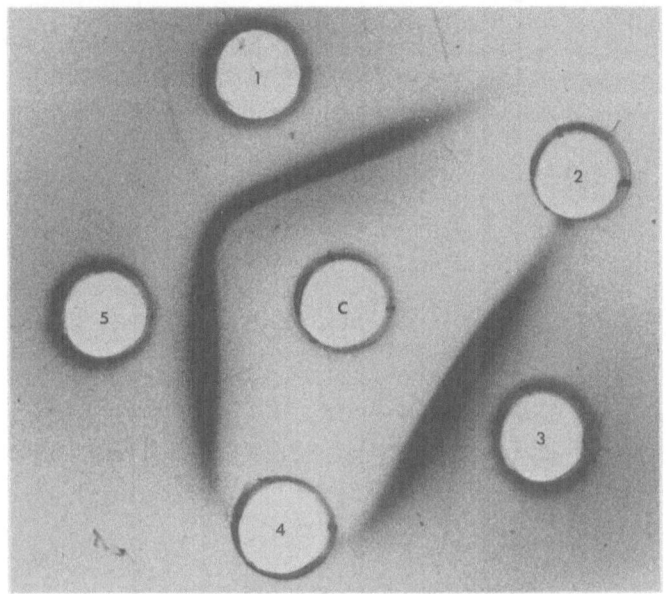

Figure 3. Ouchterlony immunodiffusion. (C) Crude liver homogenate.
(1,3,5) Antiserum from a single rabbit. (2,4) Control Sera. Single
immunoprecipitin band was stained for xanthine oxidase activity.

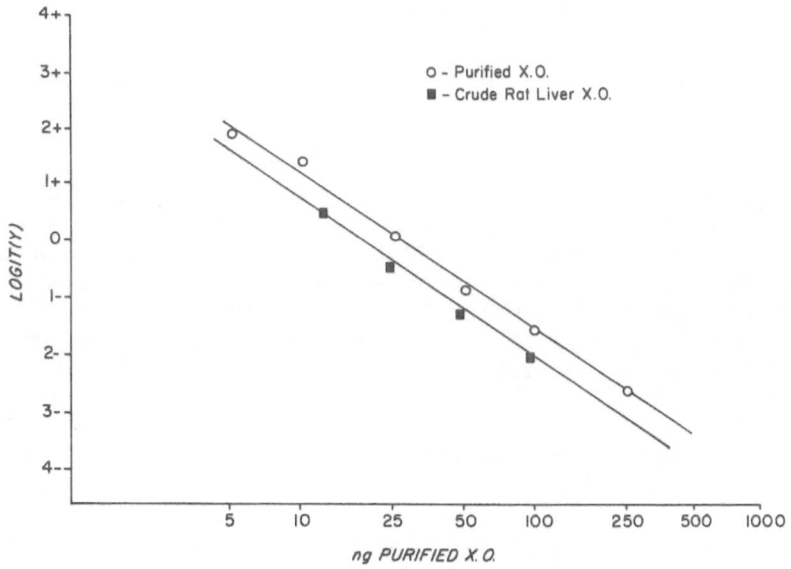

Figure 4. Standard curve showing parallelism between purified xanthine oxidase and serial dilutions of crude liver homogenate.

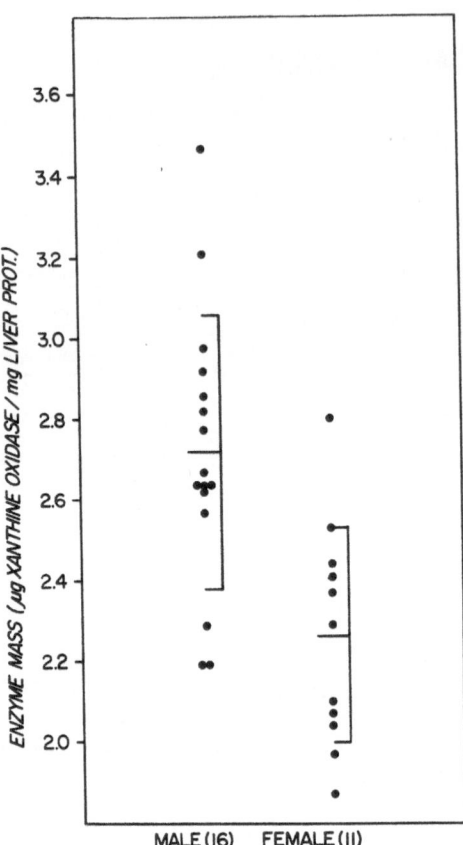

Figure 5.   Sex differences in the level of rat liver xanthine oxidase.
Mean ± SD.   Number of rats in parenthesis.

assaying enzyme activity in the presence and absence of NAD$^+$ and an
NAD$^+$-regenerating system.   A significant linear correlation was found
between the O-and D-forms (r = .896, p < .001).   Mean values for
enzyme activity were greater in males whether the enzyme was assayed
in the O-form (mean ± SE; male, 227 ± 30 μmole/mg protein h^{-1}; female,
179 ± 18 μmole/mg protein h^{-1}; p < .001) or in the D-form (mean ± SE;
male, 531 ± 70 μmole/mg protein h^{-1}; female, 409 ± 39 μmole/mg protein h^{-1}; p < .001).

Five μM testosterone and estradiol-17β were compared for their
effect on the O-and D-forms of the purified protein.   The D-form was
prepared in the presence of 10 mM dithiothrieitol and 0.1 mM EDTA in
0.05 M phosphate buffer, pH 7.8.   Without the addition of NAD$^+$, only
28 percent of total activity could be recovered following reduction
of sulfhydryl groups in the presence of dithiothrieitol.   No consistent effect of either sex hormone on the O-and D-forms could be shown.

In conclusion, increased rat liver xanthine oxidase in the male corresponds to a greater quantitative complement of enzyme protein. Lower activity in the female cannot be exaplained by immunologically cross-reactive material without enzyme activity, a direct sex-steroid enzyme interaction or interconversion between the O-and D-forms. Confirmation of the mean value of 2.51 ± 0.39 µg xanthine oxidase/mg liver ptoein (range, 1.87 to 3.47 µg), and the observed sexual dimorphism in mature rat liver will required additional study. Also, requiring more study is the mechanism of hormonal influence such as the imprinting phenomenon and sex hormone binding proteins in liver cytosol.

REFERENCES

1. D. J. Levinson and D. Chalker, Rat hepatic xanthine oxidase: Age and sex specific differences, Arthritis Rheum. 23:77 (1980).
2. E. Della Corte and F. Stirpe, The regulation of rat liver xanthine oxidase: Involvement of thiol groups in the conversion of the enzyme activity from dehydrogenase (Type D) into oxidase (Type O) and purification of the enzyme, Biochem. J. 126: 739 (1972).
3. J. M. Thompson, J. S. Nickels, and J. R. Fisher, Synthesis and degradation of xanthine dehydrogenase in chick liver, Biochim. Biophys. Acta. 568:157 (1979).
4. P. B. Rowe and J. B. Wyngaarden, The mechanism of dietary alterations in rat hepatic xanthine oxidase levels, J. Biol. Chem. 241:5571 (1966).
5. E. G. McQuarrie and A. T. Venosa, The effect of dietary protein intake on the xanthine oxidase activity in rat liver, Science. 101:493 (1945).
6. J. L. Johnson, W. R. Wand, H. J. Cohen, and K. V. Rajagopalan, Molecular basis of the biological function of molybdenum: Molybdenum-free xanthine oxidase from liver of tungsten-treated rats, J. Biol. Chem. 249:5056 (1974).

IMPRINTING BY THE NEONATAL TESTIS: A MECHANISM FOR SEXUAL

DIMORPHISM OF RAT LIVER XANTHINE OXIDASE

Dennis J. Levinson and Douglas E. Decker

Division of Rheumatology
Michael Reese Hospital and Medical Center
Chicago, IL

We have shown that adult rat liver xanthine oxidase exhibits sexual dimorphism.[1] Males have greater enzyme activity by comparison to females. The increased enzyme activity in the male corresponds to a greater mass of xanthine oxidase protein.[2] Immature animals of both sexes have similar enzyme activity.

Sex steroid and drug metabolizing enzymes in adult rat liver also conform to a pattern of activity consistent with sexual dimorphism.[3-5] The sexual dimorphism of these enzymes has been shown to arise as a result of exposure to testicular androgen in the neonatal period. Since rat liver xanthine oxidase also exhibits sexual dimorphism in adult animals, the involvement of testicular androgen in the neonatal period as a mechanism for the observed sex differences in the adult was studied.

Xanthine oxidase activity was measured by the conversion of ^{14}C xanthine to ^{14}C uric acid as previously described.[1] Neonatal (2 days of age) and pubertal (30 days of age) castrations were performed under light ether anesthesia. Xanthine oxidase activity in adult animals undergoing hypophysectomy (60 days of age) was measured two weeks following surgery. All animals were sacrificed between 65 and 75 days of age. Where indicated, androgen replacement was given for one week prior to sacrifice by daily subcutaneous injections of 100 μg of testosterone.

To determine the influence of neonatal androgen on the expression of xanthine oxidase activity in the adult, experiments using neonatal and pubertal castrates were carried out. When xanthine oxidase activity of control animals was compared to adult males undergoing neonatal castration, significant differences were ob-

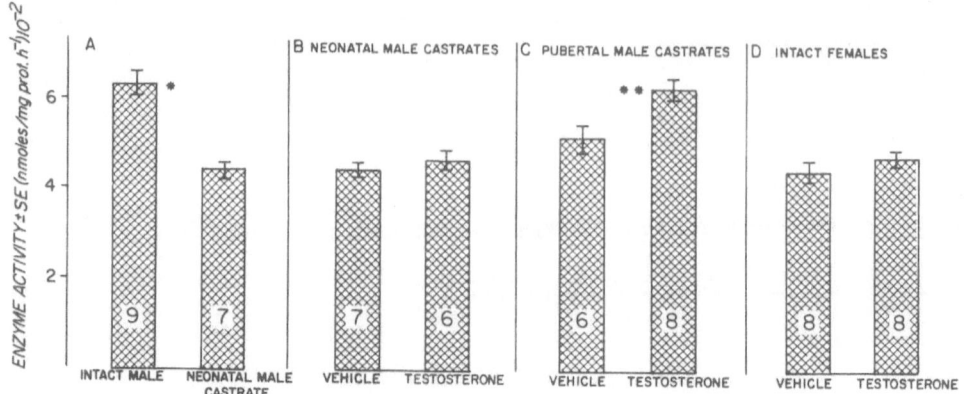

Figure 1.   Response of adult hepatic enzyme activity to neonatal or
pubertal castration and androgen replacement.   (A) the effect of neonatal (2-day) orchiectomy is compared to intact controls.   Testosterone, 100 µg/day for seven days prior to sacrifice was given to neonatal (2-day) (B), and pubertal (30-day) castrates (C).   For comparison, intact females (D) received testosterone, 100 µg/day for 14 days
prior to sacrifice.   Controls received vehicle alone.   Significant
differences are represented by asterisks (* p < .001, ** p < .002).
Number of animals studied is indicated in each bar.   Values are the
mean ± SE.

served (Fig. 1A) (p < .001).   The neonatal castrate had a 28 percent
decrease in xanthine oxidase activity.   This decrease in enzyme activity was not due to adult androgen deficiency alone, since androgen
replacement was without effect in restoring enzyme levels to values
seen in control animals (Fig. 1B).   Furthermore, the enzyme activity
was not significantly different from neonatal castrates in which
androgen replacement had not been given (p > .05).   The lack of neonatal androgen exposure prevented the characteristic sexual dimorphism
normally observed in adult animals.   However, in animals castrated
at 30 days of age and presumably following the imprinting period, a
significant response to androgen therapy was found (mean ± SE; castrates, 518.3 ± 9.27 nmole/mg prot. h^{-1}; castrates plus testosterone,
624.6 ± 20.3 nmole/mg prot. h^{-1}; p < .002) (Fig. 1C).   These results
indicate that by 30 days of age, the ability of androgen to establish
sex specific differences in adult enzyme activity has taken place.
Intact adult females are unresponsive to androgen therapy, perhaps
through the lack of androgen exposure in the neonatal period (Fig.
1D).

     To establish the kinetics of this imprinting process, male
animals were castrated at either 2, 4, 7, 10 or 14 days of age.   All
groups received one week of androgen replacement therapy before sac-

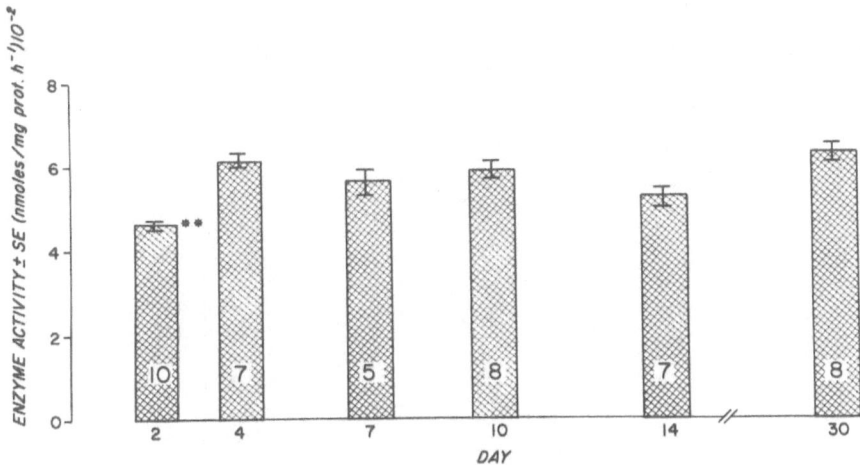

Figure 2.   The kinetics of neonatal imprinting on adult hepatic enzyme
activity.   Castrations were performed on the days indicated.   All
groups received testosterone, 100 μg/day for seven days prior to sacrifice.   A significant difference in results from 2-day castrates and
all other groups are represented by asterisks (** p < .001).   Values
are the mean ± SE.

rifice at 65 to 70 days of age.   Using the Newman-Keuls multiple
range statistic, individual group comparisons with the 2 day castrates
were obtained.   All other castrated groups were significantly different from day 2 castrates (Fig. 2).   From this data, we conclude
that by four days of age the neonatal imprinting of the adult male
pattern of enzyme activity has taken place.

Sex steroid and drug metabolizing enzymes of rat liver also
demonstrate adult sexual dimorphism as a result of neonatal androgen
exposure.   These enzymes require the presence of an intact pituitary
for the maintenance of the sex related differences.   Experiments were
carried out to determine the requirement of an intact pituitary in
the adult for the maintenance of hepatic xanthine oxidase sexual
dimorphism.   Hypophysectomy in male rats resulted in a 29 percent
decrease in enzyme activity by comparison to controls (Fig. 3A)
(p < .001).   The value in hypophysectomized males was not significantly different from either intact or hypophysectomized females
(Fig. 3B) (F-value, 0.59, p > .05).   When hypophysectomized males
received androgen replacement, a 40 percent increase in xanthine
oxidase activity resulted (p < .001).   This data suggest that sex
related differences in xanthine oxidase in adult animals is independent of an intact pituitary so long as androgen replacement is given.
The enzyme activity in adult females is unresponsive to either hypophysectomy or androgen therapy.

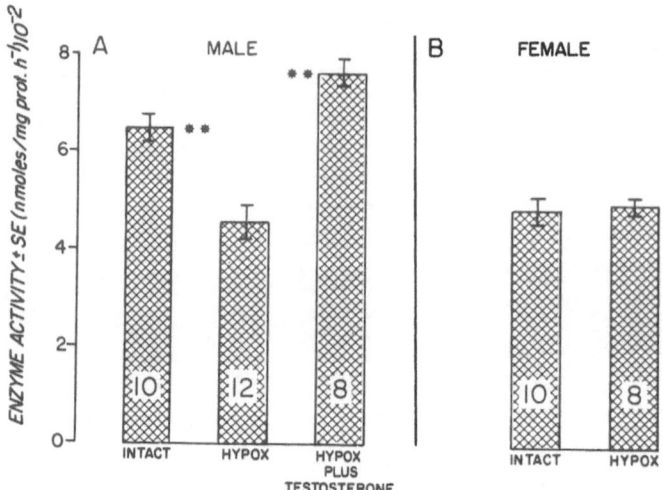

Figure 3. The response of adult hepatic enzyme activity to hypophysectomy (hypox) and androgen replacement. (A) hypophysectomy was
performed at 60 days of age. A group of hypophysectomized animals
received testosterone, 100 μg/day for seven days prior to sacrifice.
Results are compared to both intact and hypophysectomized females (B).
Significant differences from the results of hypophysectomy alone in
males are represented by asterisks (** p < .001). Number of animals
is indicated in each bar. Values are the mean ± SE.

        In conclusion, the sexual dimorphism observed for rat liver
xanthine oxidase activity in the male appears to result from the
effect of androgen exposure in the neonatal period. This imprinting
phenomenon is complete by four days of age and does not require an
intact pituitary in the presence of androgen replacement for full
expression of adult sexual dimorphism. The adult female is unresponsive to androgen therapy and hypophysectomy. In the female,
this unresponsiveness may result because of the lack of androgen exposure in the neonatal period.

REFERENCES

1. D. J. Levinson and D. Chalker, Rat hepatic xanthine oxidase
        activity: Age and sex specific differences, Arthritis Rheum.
        27:77, 1980.
2. D. E. Decker and D. J. Levinson, Quantitation of rat liver xan-
        thine oxidase by radioimmunoassay: A mechanism for sex spe-

cific differences, Arthritis Rheum. 25:326, 1982.
3. C. A. Lamartiniere, C. S. Dieringer, E. Kita, and G. Lucier, Altered sexual differentiation of hepatic uridine diphosphate glucuronyltransferase by neonatal hormone treatment in rats, Biochemistry J. 180:313, 1979.
4. C. Denef and P. DeMoor, Sexual differentiation of steroid metabolizing enzymes in rat liver: Further studies on predetermination by testosterone at birth, Endocrinology. 91:374, 1972.
5. N. P. Illsley and C. A. Lamartiniere, The imprinting of adult hepatic monoamine oxidase levels and androgen responsiveness by neonatal androgens, Endocrinology. 107:551, 1980.

PYRIMIDINE METABOLISM IN FOLATE DEFICIENT LYMPHOBLASTS

Martin B. Van Der Weyden, Ronald J. Hayman, and
Ian S. Rose
Monash University Department of Medicine,
Alfred Hospital, Prahran, Victoria, Australia

Thymidine kinase (EC 2.7.1.2), the initial enzyme of the
pyrimidine salvage pathway, catalyzes the phosphorylation of
thymidine to thymidine monophosphate (1).   Although increases in
thymidine kinase activity parallel corresponding changes in
cellular DNA synthesis (2-4), the exact mechanisms for control of
this activity are ill understood.   Of some relevance, in this
regard, has been the increased thymidine kinase activities occurring in either dormant (5) or proliferating human folate deficient
lymphocytes (6) with the augmented activity under these circumstances postulated to reflect enzyme induction secondary to a
presumed reduced thymidylate pool (5,6).   Recently we have
developed an in vitro system for the production of folate deficiency
in human lymphoblasts by mitogen stimulation of peripheral blood
lymphocytes in folate restricted RPMI 1640 medium (7).   The changes
of thymidine kinase activities in relationship to controlled
fluctuations of the cellular thymidine triphosphate pool in this
system are reported here.

The activities of thymidine kinase and dihydrofolate reductase
during the cell cycle of human peripheral blood lymphocytes
stimulated with phytohemagglutinin in either folate restricted or
supplemented medium are shown in Fig. 1.   Dihydrofolate reductase
activity, not detectable in unstimulated cells, reached a maximum
in cells 72 hr in culture and was independent of cellular folate
status.   The temporal increase in thymidine kinase activity
coincided with that for dihydrofolate reductase but was dependent
on cellular folate status being, in lymphoblasts at 72 hr in
culture, 1.8 fold higher in folate deficient compared to replete
cells.   The cellular TTP levels under these conditions were
16 ± 3 and 24 ± 3 pmoles/10^6 cells for folate deficient and replete

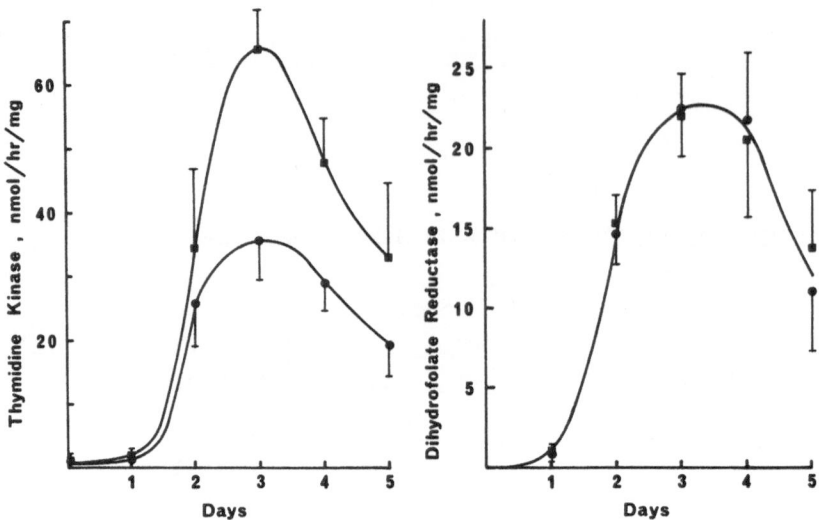

Fig. 1.   Thymidine kinase and dihydrofolate reductase activities
in PHA stimulated lymphocytes cultured in RPMI 1640 folate restricted
(■———■) or folate replete (●———●) medium (7).   Each value
represents mean ± 1 SE for eight determinations at day 3 and four
determinations at the other indicated times.   Thymidine kinase was
assayed as described previously with the final $[^{14}C]$ thymidine
concentration modified to 25 μM (6).   Dihydrofolate reductase was
assayed as described previously using $[U-^{3}H]$ dihydrofolate as the
radiolabelled substrate (9).   The folate content of lymphoblasts in
either folate depleted or supplemented medium is 70 ± 40 and
4600 ± 400 pg/10^{6} cells respectively (7).

cells respectively (Table 1).   The addition ab initio of thymidine
or inosine to folate replete cultures had no significant effect at
72 hr on either thymidine kinase activities or TTP levels (Table 1).
In folate restricted cultures, the addition of thymidine but not
inosine, resulted at 72 hrs in levels of thymidine kinase or TTP
which were not significantly different to those observed in folate
replete cells (Table 1).   Neither thymidine nor inosine had a
significant effect on cell numbers or viability.

    The mechanisms for the diverse cellular aberrations observed
in human folate or cobalamin deficiency remain controversial.   Of
prime importance is thought to be interference in one or another
step of DNA synthesis consequent to impaired de novo thymidylate
synthetase activity which in turn reflects decreased cellular
5,10-methylenetetrahydrofolate levels (7,8,13).   At variance with
these data have been the normal TTP levels in either marrow cells
(13) or mitogen stimulated lymphocytes (14) of individuals with
in vivo acquired folate or cobalamin deficiency.   Under the

Table 1. Effect of Thymidine on Thymidine Kinase and TTP
Levels of PHA Stimulated Lymphocytes in Folate
Restricted or Replete RPMI 1640 Medium.

Culture conditions	Thymidine kinase activity (nmoles/hr/mg protein)		TTP level (pmol/10^6 cells)	
Folate restricted				
(folate 1 ng/ml)[a]	65 ± 9 (n=8)[b]		16 ± 3 (n=10)[b]	
+ 50 µM TdR[a]	32 ± 9 (n=4)	p< 0.05[c]	32	(n= 1)
+ 100 µM TdR	32 ± 6 (n=6)	p< 0.05		
+ 100 µM IR[a]	56 ± 14(n=4)			
Folate replete				
(folate 1 µg/ml)	35 ± 5 (n=8)	p< 0.05	24 ± 3 (n=10)	
+ 50 µM TdR	35 (n=1)			
+ 100 µM TdR	34 ± 7 (n=3)		28	(n= 1)
+ 100 µM IR	36 ± 9 (n=3)			

[a] TdR, thymidine; IR, inosine; [b] mean ±1 SE; [c] student t test,
comparison with folate deficient cells. Number in parenthesis
equals number of determinations. Thymidine or inosine were added
to the cultures in aliquots (50 µl of RMPI 1640 medium). Cells
were harvested and deoxynucleotides were extracted in 60% methanol
at -20°C overnight. Thymidine triphosphate levels were assayed
in cell extracts by the method of Lindberg and Skoog (10).
Radioactivity was determined in a liquid scintillation counter and
an isotopic dilution correction technique was used to obtain final
TTP concentrations (11). Protein was determined by the method of
Lowry et al (12).

controlled metabolic conditions of the model system developed in
our laboratory for in vitro acquired folate deficiency, TTP levels
in deficient lymphoblasts are consistently reduced and accompanied
by elevated thymidine kinase levels. The reduced TTP levels
appear not to reflect decreased cellular proliferation as cell
numbers or viability are not impaired and [^{3}H]thymidine incorporation into DNA is increased (7). In addition the increased
thymidine kinase activity in these cells appears to be specific
as the activity of another enzyme, dihydrofolate reductase, was
independent of cellular folate status. The increased levels of
thymidine kinase in folate deficient lymphoblasts, demonstrated in
this study, are consistent with changes for this enzyme in either
resting (5) or mitogen stimulated peripheral blood lymphocytes (6)
of individuals with in vivo acquired folate deficiency.

The basis for these changes is unclear but is thought to
reflect enzyme induction consequent to cellular thymidylate lack

(5,6,7).   The inverse relationship between enzyme activity and TTP
levels disclosed in this study, and the reduction of elevated
activities pari passu with increased TTP levels in folate deficient
lymphoblasts cultured with thymidine are all compatible with this
hypothesis.   The exact molecular mechanism involved in this type
of enzyme regulation is currently being explored.

ACKNOWLEDGEMENTS

This study was supported in part by the Anti Cancer Council of
Victoria and the National Health and Medical Research Council of
Australia.

REFERENCES

1.   R. Okazaki and A. Kornberg.   Deoxythymidine kinase of
      Escherichia coli.   I. Purification and some properties of the
      enzyme.   J. Biol. Chem. 239: 269, 1964.
2.   R. Adler and B.R. McAuslan.   Expression of thymidine kinase
      variants is a function of the replicative state of cells.
      Cell 2: 113, 1974.
3.   B. Munch-Petersen and G. Tyrsted.   Induction of thymidine
      kinases in phytohaemagglutinin-stimulated lymphocytes.
      Biochim. Biophys. Acta. 478: 364, 1977.
4.   S. Kit.   Thymidine kinase, DNA synthesis and cancer.   Mol.
      & Cell Biochem. 11: 161, 1976.
5.   P.H. Ellims, R.J. Hayman and M.B. Van Der Weyden.   Expression
      of fetal thymidine kinase in human cobalamin or folate
      deficient lymphoblasts.   Biochem. Biophys. Res. Commun.   89:
      103, 1979.
6.   J.W.L. Hooton and A.V. Hoffbrand.   Thymidine kinase in
      megaloblastic anaemia.   Brit. J. Haematol. 33: 527, 1976.
7.   R.J. Hayman and M.B. Van Der Weyden.   Phytohemagglutinin
      stimulated normal human peripheral blood lymphocytes in folate
      depleted medium: an in vitro model for megaloblastic hemopoiesis.
      Blood 55: 803, 1980.
8.   M.H.N. Tattersall, A. Lavoie, K. Ganeshaguru, E. Tripp and
      A.V. Hoffbrand.   Deoxyribonucleoside triphosphates in human
      cells: changes in disease and following exposure to drugs.
      Europ. J. Clin. Invest. 5: 191, 1975.
9.   R.J. Hayman, R. McGready and M.B. Van Der Weyden.   A rapid
      radiometric assay for dihydrofolate reductase.   Anal. Biochem.
      87: 400, 1978.
10.   U. Lindberg and L. Skoog.   A method for the determination of
      dATP and dTTP in picomole amounts.   Anal. Biochem. 34: 152, 1970.
11.   B. Munch-Petersen, G. Tyrsted and B. Dupont.   The deoxy-
      ribonucleoside-5'-triphosphate (dATP and dTTP pool) in phyto-
      hemagglutinin stimulated and non stimulated human lymphocytes.
      Exp. Cell Res. 79: 249, 1973.

12. O.H. Lowry, N.J. Rosebrough, A.L. Farr and R.J. Randall.
     Protein measurement with the folin phenol reagent. J. Biol.
     Chem. 193: 265, 1951.
13. A.V. Hoffbrand, K. Ganeshaguru, A. Lavoie, M.H.N. Tattersall
     and E. Tripp. Thymidylate concentration in megaloblastic
     anaemia. Nature 248: 602, 1974.
14. K. Ganeshaguru and A.V. Hoffbrand. The effect of deoxy-
     uridine, vitamin B$_{12}$, folate and alcohol on the uptake of
     thymidine and on the deoxynucleoside triphosphate concentrat-
     ions in normal and megaloblastic cells. Brit. J. Haematol.
     40: 29, 1978.
15. P. Ekker. Activities of thymidine kinase and deoxyribo-
     nucleotide phosphatase during growth of cells in tissue
     culture. J. Biol. Chem. 241: 659, 1966.

PYRIMIDINE METABOLISM IN RAT BRAIN CORTEX AND LIVER

G.J. Peters and J.H. Veerkamp

Department of Biochemistry, University of Nijmegen
P.O. Box 9101, 6500 HB Nijmegen
The Netherlands

INTRODUCTION

Pyrimidine nucleotide synthesis proceeds via a salvage pathway and a de novo pathway. In rat liver all enzymes involved in UMP synthesis from bicarbonate have a considerable activity (1-4), but in rat brain not all enzymes of the OA-pathway (orotic acid) have been demonstrated, although a significant incorporation of [^{14}C]bicarbonate into OA was found (5). A considerable activity of uridine kinase is present in brain (6,7). OA and uracil can not pass the blood-brain barrier (8), but uridine can be taken up (9). In this study we compare the de novo and salvage pathways by measuring the incorporation of aspartate into OA and assaying the activities of DHOdehydrogenase (dihydroorotic acid dehydrogenase), OPRT (orotic acid phosphoribosyltransferase), ODC (orotidylate decarboxylase), uridine kinase and uridine phosphorylase.

METHODS AND MATERIALS

Origin of most materials is described in Tax et al. (10). Young adult Wistar rats were killed by cervical dislocation. Brain cortex and liver for the incorporation assay were immediately removed. 10% Homogenates were prepared in ice-cold buffer (250 mM sucrose, 50 mM Tris-HCl, pH 7.4) and centrifuged at 600g. The reaction mixture (105 μl) contained 75 μl supernatant, 0.5 mM NAD$^+$, 0.5 mM MgCl$_2$, 1 mM azauridine, 0.5% glucose, 2.5 mM carbamyl phosphate and 0.9 mM L-[^{14}C]aspartate (5.7 mCi/mmol). After 60 min incubation at 37^{0}C the reaction mixture was heated at 100^{0}C. Substrates and products were separated by high-voltage paper-electrophoresis (HVPE) in acetic acid/formic acid/water (17/17/1000, by vol; pH 2.5 with 2 M NaOH); and by cellulose TLC with ether/formic acid/water (7/2/1, by vol).

531

Livers used for the assays of enzyme activities were perfused via the vena porta with ice-cold saline before removal out of the animal. Enzyme activities were determined in 100 000g supernatants as described for lymphocytes (10); uridine kinase and phosphorylase were assayed at 5 mM uridine, 5 mM ATP and 5 mM MgCl$_2$. Mitochondria were isolated as described (11,12) and subfractionated using a Lubrol W solution (1.9%, w/v). For assay of DHOdehydrogenase in various cell fractions 50 mM Tris-HCl (pH 7.4) buffer containing 1 mM [6-^{14}C]-DHO was used.

RESULTS AND DISCUSSION

Two separation procedures were used to identify all products that could be formed from [4-^{14}C]aspartate. With the HVPE the end-product, OA, could be separated from the substrate and all other products, including citric acid cycle intermediates. Malate and fumarate had the same mobility as carbamyl aspartate and DHO, respectively, but could be separated by TLC. The identity of OA was confirmed by conversion of eluted radioactivity with partially purified yeast OPRT and ODC to OMP and UMP. With brain cortex the rate of OA synthesis from aspartate was 52 ± 12 and with liver 179 ± 35 nmol/h per g wet tissue (means ± SD of 7 and 4 experiments, respectively); expressed per mg protein these values were 0.81 ± 0.21 and 1.12 ± 0.46, respectively. With both tissues about 10% of the label was found in citric acid cycle intermediates, and with cortex and liver about 1% and 10% of radioactivity was recovered as $^{14}CO_2$.

Subcellular fractionation of rat liver demonstrated that DHOdehydrogenase was located in the mitochondrion (Fig. 1). Subfractionation of the mitochondrial fraction showed that DHOdehydrogenase is

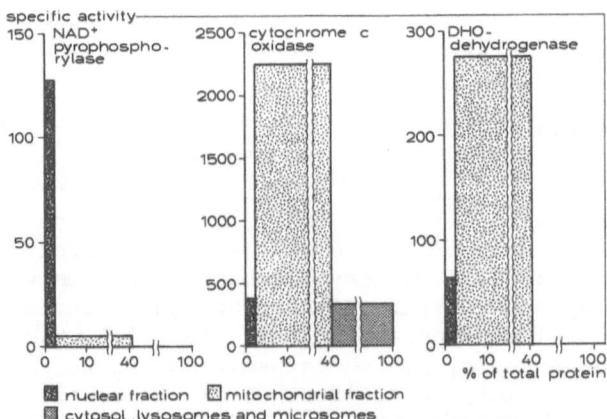

Fig. 1. Distribution pattern of enzymes in subcellular fractions of rat liver. Specific activities are given in nmol/mg protein per min for cyt oxidase and per h for the other enzymes.

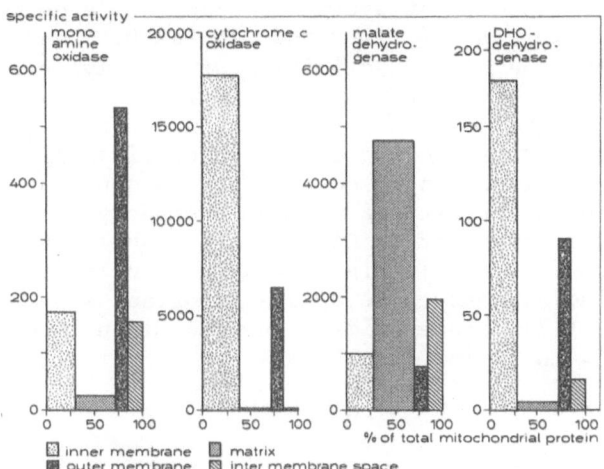

Fig. 2. Distribution pattern of enzymes in submitochondrial fractions of rat liver. Specific activities are expressed in nmol/mg protein per h for DHOdehydrogenase and monoamine oxidase and per min for the other enzymes.

mainly located in the inner membrane (Fig. 2) as previously shown for murine and rat liver (13,14) and rat brain (15). Activity of the enzyme in whole homogenates was 6.7 ± 2.2 µmol/h per g wet liver (mean ± SD of 6 rats, about 38 nmol/h per mg protein).

Differential centrifugation showed that the activities of OPRT, ODC, uridine kinase and phosphorylase are nearly completely located in the cytosol. The activities of OPRT and ODC are comparable in liver and cortex (Table 1). Cortex OPRT showed biphasic kinetics with Km values of 273 and 30 µM OA. The activity of uridine kinase is much higher than that of uridine phosphorylase in cortex in contrast to in rat liver when both enzymes are investigated in the same system without additional Pi. Addition of 40 mM Pi increased only the phosphorylase activity 50–100%.

Table 1. Activities of enzymes involved in pyrimidine metabolism

Enzyme	Cortex	Liver
OPRT	0.66 ± 0.15	0.84 ± 0.20
ODC	1.63 ± 0.56	2.09 ± 0.95
Uridine kinase	3.08 ± 0.36	0.79 ± 0.16
Uridine phosphorylase	1.38 ± 0.31	3.40 ± 1.23

Values (in µmol/h per g wet tissue) are means ± SD for 4-7 rats. Protein contents in 100 000g supernatants were about 30 and 40 mg/g for brain cortex and liver, respectively.

The overall capacity of pyrimidine nucleotide de novo synthesis appears to be higher in rat liver than in rat brain. This can also be concluded from the higher activities of carbamyl phosphate synthetase II and aspartate transcarbamylase in liver (2,4). The liver primarily depends on the de novo pathway for nucleotide synthesis. With liver slices pyrimidine nucleotides are predominantly derived from OA; uridine is mainly catabolyzed to uracil and β-alanine (16) in agreement with high activity of uridine phosphorylase. With brain slices uridine was superior to CO_2 or OA in labelling RNA (8). This concords with the relatively high activity of uridine kinase. In vivo, however, cytidine appears to be a more important substrate for nucleotide synthesis (17), since uridine in predominantly catabolyzed by various tissues, including liver.

ACKNOWLEDGEMENTS

We thank A. Oosterhof, W. Ghijsen, G.E.J.M. van der Laar and W.J.M. Tax and miss J. Peters for their contributions to this work.

REFERENCES

1. Hager SE & Jones ME (1967) J. Biol. Chem. 242, 5674-5680
2. Yip MCM & Knox WE (1970) J. Biol. Chem. 245, 2199-2204
3. Pausch J, Wilkening J, Nowack J & Decker K (1975) Eur. J. Biochem. 53, 349-356
4. Aoki T, Morris HP & Weber G (1982) J. Biol. Chem. 257, 432-438
5. Tremblay GL, Jimenez U & Crandall DE (1976) J. Neurochem. 26, 57-64
6. Appel SH & Silberberg DH (1968) J. Neurochem. 15, 1437-1443
7. Krenitsky TA, Miller RL & Fyfe JA (1974) Biochem. Pharmac. 23, 170-172
8. Hogans AF, Guroff G & Udenfriend S (1971) J. Neurochem. 18, 1688-1710
9. Nakagawa S & Guroff G (1973) J. Neurochem. 20, 1143-1149
10. Tax WJM, Peters GJ & Veerkamp JH (1979) Int. J. Biochem. 10, 7-10
11. Schnaitman C & Greenawalt JW (1968) J. Cell Biol. 38, 158-175
12. Chan TL, Greenawalt TL & Pedersen PL (1970) J. Cell Biol. 45, 291-304
13. Kensler TW, Cooney DA, Jayaram HN, Schaeffer C & Choie DD (1981) Anal. Biochem. 117, 315-319
14. Chen J-J & Jones ME (1976) Arch. Biochem. Biophys. 176, 82-90
15. Calcagnotto AM, Villela GG & Ribiero CP (1980) IRCS Med. Sci.-Biochem. 8, 904-905
16. Harley EH & Losman MJ (1981) Int. J. Biochem. 13, 247-249
17. Moyer JD, Oliver JT & Handschumacher RE (1981) Cancer Res. 41, 3010-3017

EVIDENCE FOR A DISTINCT DEOXYPYRIMIDINE 5'-NUCLEOTIDASE IN HUMAN TISSUES

D.A. Hopkinson, D.M. Swallow, V.S. Turner and I. Aziz

MRC Human Biochemical Genetics Unit,
Galton Laboratory, UCL,
London, England

INTRODUCTION

A pyrimidine-specific 5'nucleotidase was first identified in human erythrocytes by Valentine and his colleagues during the investigation of patients with a particular form of non-spherocytic haemolytic anaemia associated with basophilic stippling of the erythrocytes (1). This nucleotidase was shown to catalyse the hydrolysis of pyrimidine 5'ribomonophosphates (2) such as uridine 5'monophosphate (UMP) and cytidine 5'monophosphate (CMP) but was found to be relatively ineffective with purine ribonucleotides such as adenosine 5'monophosphate (AMP). Subsequent examination of partially purified material from human erythrocytes (3) confirmed the identity and properties of this nucleotidase. The genetically determined enzyme deficiency is inherited as an autosomal recessive character and the enzyme is presumed to be determined by a single autosomal structural gene locus.

We have explored the possibility of mapping this locus to a human chromosome using the technique of somatic cell hybridization and in the course of this work obtained unexpected evidence for the occurrence of two distinct pyrimidine nucleotidases in rodent and human tissues. Both enzymes catalyse the hydrolysis of UMP and are referred to here as uridine monophosphate hydrolases with the abbreviation UMPH, in keeping with the previous designation in the human gene mapping literature (4). One enzyme appears to correspond with the erythrocyte pyrimidine-specific nucleotidase described above and is designated UMPH-1. The other has a broader specificity and shows optimum activity with deoxyribonucleotides and is designated UMPH-2.

MATERIALS AND METHODS

Tissue homogenates were prepared from laboratory stocks of mice, rats and Chinese hamsters and from human post mortem material using a Silverson mixer emulsifier. Cell lines of rodent and human origin were cultured by standard methods and homogenates were prepared by brief sonication of cell pellets. Haemolysates were prepared by freeze-thawing saline washed erythrocytes.

Horizontal starch gel electrophoresis was carried out at pH 7.4 using TEMM buffer (0.1 M Tris, 0.01 M Edta, 0.01 M $MgCl_2$ and 0.1 M Maleic anhydride in the bridge and tenfold dilution for the gel). The gels also contained mercaptoethanol at a final concentration of 10 mM. Nucleotidase isozymes were identified by a two stage staining system which depends on the detection of inorganic phosphate liberated from the nucleotide substrate (5). Phosphatase activity of the nucleotidase isozymes against β-naphthyl phosphate (βNP) and 4-methylumbelliferyl phosphate (4MUP) was demonstrated by standard methods (5).

Nucleotidase activity was measured in human erythrocyte lysates which had been dialysed overnight against Tris/HCl buffer, pH 8.0 containing 20 mM mercaptoethanol and then diluted 1:5 in the same buffer. The assay mixture contained 0.5 ml haemolysate, 0.2 ml 0.1 M $MgCl_2$ and 0.2 ml 0.04 M 5'ribonucleotide substrate in 0.05 M Tris/HCl buffer pH 8.0 and was incubated at $37^{\circ}C$ for 1-2 hrs before adding 0.5 ml 20% trichloroacetic acid. Phosphate determination was carried out on the clear supernatant using the Chen modification (6) of the Fiske-SubbaRow procedure.

RESULTS

Electrophoresis of human and rodent tissue homogenates

Electrophoretic analysis of human erythrocytes and most other human tissues reveals a single major UMPH isozyme (not illustrated). This is demonstrable with a very broad range of substrates (Table 1) including pyrimidine nucleotides (UMP and CMP), deoxypyrimidine nucleotides (dUMP, dTMP and dCMP), certain deoxypurine nucleotides (dIMP, dAMP, dGMP) and the synthetic substrates 4MUP and βNP. Highest levels of activity are observed in skeletal muscle and in this tissue the major isozyme is associated with a minor more anodal secondary isozyme. The finding of a single major UMPH isozyme in human tissues is in agreement with previous reports on the electrophoretic properties of human erythrocyte UMPH (3,4,7,8,9).

More complex UMPH isozyme patterns are observed in rodents. The patterns vary according to the tissue examined, the substrate used to demonstrate the isozymes and the species tested. In the

Table 1.   Relative activities of rodent and human nucleotidase
           isozymes with various substrates.

	Rodent		Human
	UMPH-1	UMPH-2	UMPH
Pyrimidine nucleotides:			
UMP	++	+	++
CMP	++	<u>+</u>	++
Deoxypyrimidine nucleotides:			
dCMP	++	+	+
dUMP	+	+++	++
dTMP	+	+++	++
Purine nucleotides:			
GMP	−	−	−
AMP	−	−	−
IMP	−	+	<u>+</u>
Deoxypurine nucleotides:			
dGMP	−	+	+
dAMP	−	<u>+</u>	+
dIMP	−	++	+
Synthetic substrates:			
4-methylumbelliferyl phosphate	−	++	+
β-naphthyl phosphate	−	++	+

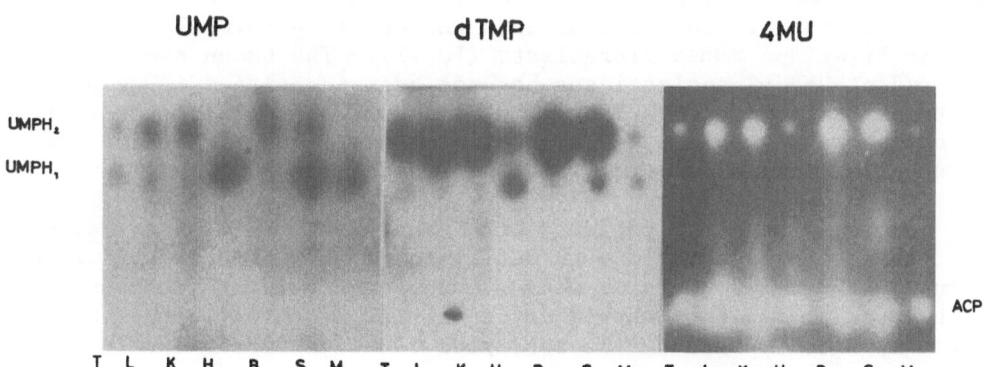

Fig. 1.   Nucleotidase isozymes detected with various substrates
          after starch gel electrophoresis of homogenates from
          mouse tissues: testis (T), liver (L), kidney (K), heart
          (H), brain (B), spleen (S) and skeletal muscle (M).

mouse for example most tissues show two major isozymes, designated
UMPH-1 and UMPH-2 (Fig. 1). The least anodal mouse isozyme (UMPH-1)
was found to have a narrow substrate specificity. It shows highest
levels of activity with the pyrimidine nucleotides UMP and CMP and
the deoxy analogue of the latter (dCMP) but only moderate activity
with the deoxypyrimidines dUMP and dTMP and no activity with the
other substrates tested (Table 1). The more anodal mouse isozyme
(UMPH-2) has a much broader substrate specificity. It hydrolyses
the synthetic substrates 4MUP and βNP and several pyrimidine and
some purine nucleotides. The highest levels of mouse UMPH-2 activity were found with the deoxypyrimidine nucleotides dUMP and dTMP
and with the deoxypurine nucleotide dIMP.

Two major UMPH isozymes are also detected in Chinese hamster
and rat tissues (not illustrated here). As in the mouse they differ
in their substrate specificity and show consistent tissue differences in expression. However in these rodents the more anodal isozyme is designated UMPH-1 since it exhibits a narrow specificity and
prefers the pyrimidine substrates UMP, CMP and dCMP; the less anodal
isozyme is designated UMPH-2 in hamster and rat since it displays
activity with a broad range of substrates and preference for the
deoxy nucleotides dUMP, dTMP and dIMP.

The simplest interpretation of the rodent nucleotidase isozyme
patterns is that there are two forms of UMPH determined by separate
structural gene loci. Rodent UMPH-1 is specific for pyrimidine
nucleotides and is probably homologous with the pyrimidine specific
nucleotidase first identified in human erythrocytes by Paglia and
Valentine (2). Rodent UMPH-2 hydrolyses pyrimidine nucleotides,
some purine nucleotides and the synthetic phosphate esters 4MUP and
βNP; it is most active with the deoxyribonucleotides dUMP, dTMP and
dIMP. It probably corresponds with the enzyme previously described
in rat liver and mouse fibroblasts (10,11). The human homologue of
the rodent UMPH-2 nucleotidase has not been identified. One possibility is that human UMPH-2 overlaps the human UMPH-1 isozyme in its
electrophoretic mobility. The broad substrate specificity of the
human UMPH isozyme supports this interpretation.

The occurrence of patients with genetically determined haemolytic anaemia associated with deficiency of erythrocyte pyrimidine
5'-nucleotidase provides us with a means of testing this hypothesis.
Such cases should show marked deficiency of erythrocyte nucleotidase
activity with UMP and CMP as substrate but significant levels of
activity with dUMP and dTMP as substrates. The second part of this
paper reports the study of such a case.

Human erythrocyte 5'-nucleotidase deficiency

The patient was a 13 year old Japanese boy of first cousin

parents. He was moderately anaemic (Hb 10.8 G/100 ml) but not
jaundiced (serum bilirubin 4.3 mg/100 ml), with a reticulocytosis
of 23% and his red cells showed marked basophilic stippling.

On electrophoresis his erythrocytes showed a complete deficiency of nucleotidase activity with UMP, CMP and dCMP as substrates
(Fig. 2). However, significant levels of activity were demonstrable
with dUMP and dTMP as substrates (Fig. 2). The isozyme detected in
the patient appeared to be just slightly slower in anodal electrophoretic mobility than in the control specimens. The deoxyribo-
nucleotidase isozyme was also detectable, with normal levels of activity, with 4MUP as substrate.

The assay data are summarized in Table 2. The patient's erythrocytes show marked deficiency of pyrimidine nucleotidase activity
with CMP and UMP but normal levels of activity with dUMP and dTMP as
substrates.

DISCUSSION

The results obtained by electrophoresis and quantitative assay
are complementary and support the hypothesis that there are two
types of pyrimidine nucleotidase in human erythrocytes. The patient
shows deficient activity with UMP, CMP and dCMP, the preferred substrates of rodent UMPH-1 and this agrees with the conclusion that
rodent UMPH-1 and the human pyrimidine-specific nucleotidase identified by Valentine et al. (1) and purified by Torrance et al. (3)
are homologous.

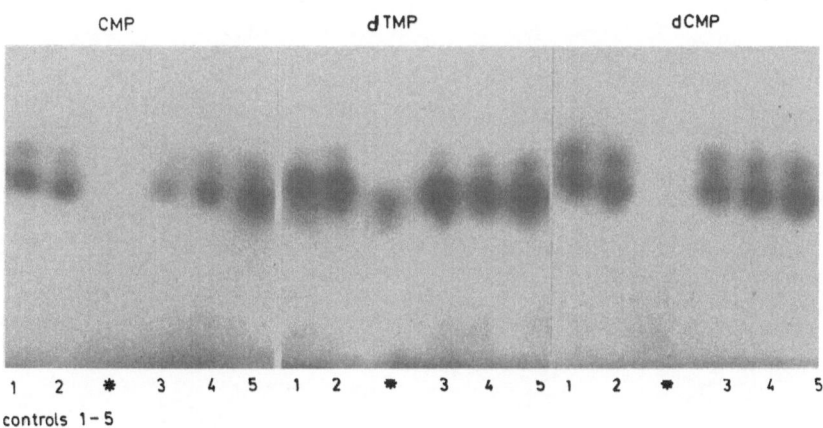

Fig. 2. Nucleotidase isozymes in erythrocytes from the patient
         with pyrimidine 5'-nucleotidase deficiency (*) and five
         healthy controls (1-5) detected using cytidine monophos-
         phate (CMP), thymidine monophosphate (dTMP), deoxycytidine
         monophosphate (dCMP) as substrates.

4. Anderson, J.A., Teng, Y.-S. & Giblett, E.R. Stains for six
    enzymes potentially applicable to chromosomal assignment by
    cell hybridization. Rotterdam Conf. (1974): 2nd International
    Workshop on Human Gene Mapping, Birth Defects: Original
    Article Series 11, 295-9 (1975).
5. Harris, H. & Hopkinson, D.A. Handbook of Enzyme Electrophoresis
    in Human Genetics. North Holland, Amsterdam (1976).
6. Chen Jr., P.S., Toribara, T.Y. & Warner, H. Microdetermination
    of phosphorus. Analyt.Chem. 28: 1756-60 (1956).
7. Rosa, R., Valentine, C. & Rosa, J. Electrophoretic character-
    isation of pyrimidine 5'-nucleotidase of human erythrocytes
    and its distinction from acid phosphatase. Clin.Chim.Acta
    79: 115-8 (1977).
8. Rosa, R., Rochant, H., Dreyfus, H., Valentine, C. & Rosa, J.
    Electrophoretic and kinetic studies of human erythrocytes
    deficient in pyrimidine 5'-nucleotidase. Hum.Genet. 38:
    209-15 (1977).
9. Fujii, H., Nakashima, K., Miwa, S. & Nomura, K. Electrophoretic
    and kinetic studies of a mutant red cell pyrimidine 5'-
    nucleotidase. Clin.Chim.Acta 95: 88-92 (1979).
10. Fritzon, P. & Smith, I. A new nucleotidase of rat liver with
    activity towards 3' and 5'-nucleotides. Biochim.Biophys.
    Acta 235: 128-41 (1971).
11. Magnusson, G. Deoxyribonucleotide phosphatase from rat liver
    and cultured fibroblasts. Eur.J.Biochem. 20: 225-30 (1971).

ABBREVIATIONS

AMP, CMP, GMP, IMP, TMP : 5'-monophosphates of adenosine,
cytidine, guanosine, inosine and thymidine and the prefix d indicates
the 2'deoxyribonucleotide. 4MUP and βNP : monophosphate esters of
4-methylumbelliferone and β-naphthol.

ACKNOWLEDGEMENT

We are grateful to Professor S. Miwa (Tokyo) for the blood
samples from the patient with pyrimidine 5'-nucleotidase deficiency
and healthy Japanese subjects.

THE EFFECT OF 5-FLUOROOROTIC ACID ON THE EARLY LABELLING OF NUCLEO-
TIDES AND RNA IN WHOLE LIVER AND OF RNA IN SUBNUCLEAR FRACTIONS
IN RATS GIVEN ^{3}H-OROTIC ACID

Göran Eriksson and Unne Stenram

Department of Pathology
University Hospital
S-22185 Lund, Sweden

ABSTRACT

Male white rats were given either orotic acid or 5-fluorooro-
tic acid 60 min, and ^{3}H-orotic acid 3 1/2 min before sacrifice.
Liver nucleotides were analyzed by isotachophoresis. The (F)UTP
pool increased to the same extent in both groups and showed the
same specific labelling. The RNA synthesis, as measured by nmol of
UTP entering RNA-UMP/g wet liver tissue was significantly lower in
fluoroorotic acid treated rats.In the nucleolar fraction there was
an increase in the RNA/DNA ratio, a decrease in the specific RNA
labelling and an essentially unaltered specific labelling of RNA/
μg/DNA. The effect on the specific labelling of RNA/μg DNA was the
same in the nucleolar fraction and the nucleolar-free pellet, suggesting that the transcription of RNA in these two fractions proceeded in a similar way. There was a slight increase in the specific RNA labelling of the supernatant subnuclear fraction, containing low-molecular weight RNA. Incorporation into UDP-hexoses
and UDP-N-acetylhexoseamine was unchanged.

INTRODUCTION

We found that high doses of 5-fluorouracil enlarged the nuc-
leoli in rat liver cells and altered their ultrastructure. They
also depressed the RNA labelling of the liver after administration
of a labelled precursor[1]. At low doses there was a delay or block
in the maturation of pre-ribosomal RNA and no ribosomes or labelled ribosomal RNA appeared in the cytoplasm[2,3]. These observations
have been greatly extended by several groups of workers and also
include the effects of 5-fluoroorotic acid and similar fluoro compounds. The block in the nucleolar RNA metabolism has generally

been localized to the post-transcriptional processing steps[4,5].
There seems to be no impairment of either the labelling of heterogeneous (messenger) RNA[6,7] or low-molecular weight RNA[7,8]. Transcription of nucleolar 45S RNA is essentially unaltered[5], but may
be depressed[4]. These studies have mainly been performed with labelled precursors, but the flow of labelling from, for example, orotic
acid via UTP into both RNA, UDP-N-acetylhexoseamine, UDP-hexose
and UDP-glucuronic acid has not been analyzed and, therefore, the
rate of RNA synthesis has been difficult to estimate. Isotachophoresis has made it possible to analyze these nucleotides in the
liver[9]. Using ^{3}H-orotic acid we describe in this paper the effects
of 5-fluoroorotic acid on the labelling of nucleotides and RNA in
whole liver and of RNA in subnuclear fractions.

EXPERIMENT

    Rats were given either 1500 nmol of orotic acid or 5-fluoro-
orotic acid per 100 g body weight 60 min and 0.50 mCi ^{3}H-orotic
acid (spec act 20 Ci/nmol) per 100 g body weight 3 1/2 min before
sacrifice.

    Nucleotides and RNA were extracted from total liver. RNA and
DNA was analyzed in subnuclear fractions.

    The nucleotides were run in isotachophoresis according to
Eriksson[9].

RESULTS

    The RNA synthesis, as measured by nmol of UTP entering RNA-
UMP/g wet liver tissue, was significantly lower in fluoroorotic
acid treated rats. In the nucleolar fraction there was an increase
in the RNA/DNA ratio, a decrease in the specific RNA labelling and
an essentially unaltered specific labelling of RNA/µg DNA. The
effect on the specific labelling of RNA/µg DNA was the same in the
nucleolar fraction and the nucleolar-free pellet, suggesting that
the transcription of RNA in these two fractions proceeded in a
similar way. There was a slight increase in the specific RNA labelling of the supernatant subnuclear fraction, containing low-molecular weight RNA. Incorporation into UDP-hexoses and UDP-N-acetylhexoseamine was unchanged.

    A detailed report is published in the Proceedings of the 3rd
International Symposium on Isotachophoresis, Goslar, G.F.R., June
1-4, 1982.

REFERENCES

1. U. Stenram, Cytological, radioautographic and ultrastructural studies on the effect of 5-fluorouracil on rat liver, Z. Zellforsch. mikroskop. Anatom. 71:207-216 (1966).
2. R. Willén, Polyacrylamide-agarose electrophoretic pattern of the RNA labelling in liver cytoplasm and total liver of 5-fluorouracil treated rats, Hoppe-Seyler's Z. Physiol. Chem. 351:1141-1150 (1970).
3. U. Stenram and R. Willén, The effect of 5-fluorouracil on ultrastructure and RNA labelling in the liver of rats following partial hepatectomy, Chem.-Biol. Interactions 2:79-88 (1970).
4. D. S. Wilkinson, T. D. Tlsty and R. J. Hanas, The inhibition of ribosomal RNA synthesis and maturation in Novikoff Hepatoma cells by 5-fluorouridine, Cancer Res. 35:3014-3020 (1975).
5. A. A. Hadjiolov and K. V. Hadjiolova, The effect of 5-fluoropyrimidines on the processing of ribonucleic acids in liver, FEBS Symposium 57:77-84 (1979).
6. D. S. Wilkinson, A. Cihak and H. C. Pitot, Inhibition of ribosomal ribonucleic acid maturation in rat liver by 5-fluoroorotic acid resulting in the selective labeling of cytoplasmic messenger ribonucleic acid, J. Biol. Chem. 216:6418-6427 (1971).
7. K. V. Hadjiolova, E. V. Golovinsky and A. A. Hadjiolov, The site of action of 5-fluoroorotic acid on the maturation of mouse liver ribonucleic acids, Biochim. Biophys. Acta 319: 373-382 (1973).
8. R. I. Glazer and K. D. Hartman, The effect of 5-fluorouracil on the synthesis and methylation of low molecular weight nuclear RNA in L1210 cells, Molec. Pharmacol. 17:245-249 (1980).
9. G. Eriksson, A method for measuring nucleotides in the liver of rat with isotachophoresis, Analyt. Biochem. 109:239-246 (1980).

THE EFFECT OF DIPYRIDAMOLE ON PLASMA ADENOSINE LEVELS AND SKIN MICRO-CIRCULATION IN MAN

A. Sollevi, J. Östergren, P. Hjemdahl, B.B. Fredholm, and B. Fagrell

Dept. of Pharmacology, Karolinska Institutet and Dept. of Medicine, Danderyds Hospital, Stockholm, Sweden

Dipyridamole (Dip) is a vasodilator, which probably acts by elevating endogenous adenosine. We have studied the influence of Dip on plasma levels of adenosine and on finger skin circulation. Thirteen healthy males were given Dip (0.4 mg/kg i.v.); 5 of them received the drug and saline double-blind. Heart rate, blood pressure, digital pulse amplitude, skin temperature and capillary blood cell velocity (CBV, measured by videophotometric capillaroscopy) were recorded continuously. Reactive hyperemia in the capillaries following arterial occlusion for 15 and 60 sec was measured. The plasma concentrations of noradrenaline, adrenaline, Dip and adenosine were determined by HPLC methods.

RESULTS

Plasma Dip was 2.52 ± 0.18 µM 15 min and 0.81 ± 0.08 µM 75 min after injection. Basal plasma adenosine was 0.15 ± 0.03 µM. Dip doubled this level from 15 to 60 min after injection. Dip transiently increased heart rate by 13 ± 2 beats/min ($p < 0.01$) and decreased diastolic blood pressure by 6 ± 2 mmHg ($p < 0.05$). CBV, digital pulse amplitude, skin temperature and catecholamines were unaffected. In the double-blind group 3 of 5 subjects showed an increase in post occlusion capillary hyperemia after Dip but not after saline ($p < 0.05$). The other parameters were uninfluenced by placebo.

CONCLUSION

Clinical doses of Dip double plasma adenosine, reduce diastolic blood pressure and increase post occlusion skin CBV. The lack of effect on basal skin CBV suggests that factors other than adenosine are major determinants of skin capillary flow under basal conditions.

Index